Physics of Low Dimensional Systems

Physics of Low Dimensional Systems

Edited by

J. L. Morán-López
Instituto Potosino de Investigación Científica y Tecnológica
San Luis Potosí, Mexico

Kluwer Academic / Plenum Publishers
New York, Boston, Dordrecht, London, Moscow

Library of Congress Cataloging-in-Publication Data

Physics of low dimensional systems/edited by J.L. Morán-López.
p. cm.
Includes bibliographical references and index.
ISBN 0-306-46566-3
1. Low-dimensional semiconductors—Congresses. I. Morán-López, J. L. (José L.), 1950–

QC611.8 .L68 2001
530.4′12—dc21

2001020366

Proceedings of an International Workshop on Physics of Low Dimensional Systems, held January 16–20, 2000, in Oaxaca, Mexico

ISBN 0-306-46566-3

233 Spring Street, New York, New York 10013

http://www.wkap.nl

10 9 8 7 6 5 4 3 2 1

A C.I.P. record for this book is available from the Library of Congress

Printed in the United States of America

Professor Hans Christoph Siegmann

Preface

Oaxaca, Mexico, was the place chosen by a large international group of scientists to meet and discuss on the recent advances on the understanding of the physical properties of low dimensional systems; one of the most active fields of research in condensed matter in the last years. The International Symposium on the Physics of Low Dimensions took place in January 16-20, 2000. The group of scientists converging into the historical city of Oaxaca, in the state of the same name, had come from Argentina, Chile, Venezuela, several places in Mexico, Canada, U.S.A., England, France, Italy, Germany, Russia, and Switzerland. The presentations at the workshop provided state-of-art reviews of many of the most important problems, currently under study. Equally important to all the participants in the workshop was the fact that we had come to honor a friend, Hans Christoph Siegmann, on his sixty-fifth birthday.

This *Festschrift* recognizes the intellectual leadership of Professor Siegmann in the field and as a sincere homage to his qualities as an exceptional friend, college and mentor. Those who have had the privilege to work closely with Hans Christoph have been deeply impressed by his remarkable analytic mind as well as by his out of range kindness and generosity.

Hans Christoph has contributed to the understanding of the difficult and very important problem of the magnetic properties of finite systems: surfaces, thin films, heterostructures. His ideas have crystallized in the construction of many experimental apparatus dispersed around the world, to measure the elusive magnetic properties of solids. One can recognize that the recent development of some of the electronic devices rests on his pioneering studies. His group at the Swiss Institute of Technology in Zurich is known as one of the world leaders in this area.

This volume opens with a contribution of our honoree on the production of spin-polarized electrons and its potential application to new magnetic devices. He advances some ideas to be used in the future generation of magnetic devices. The next chapters are dedicated to the study of the physical properties of nanoparticles: superconductivity, magnetism, energetics, and stability. Hans Christoph has also contributed to the understanding of the physical properties of suspended particles (aerosols). Some contributions on the study of the combustion of hydrocarbons and its effect on human health are discussed. An application to volcanic activity is also included.

The physical properties of one dimensional systems, like quantum wires, adsorbate systems, single electron transistors are treated in a few chapters. Increasing the dimensionality of the systems, two dimensional systems like quasicrystals, ultrathin films, and interfaces, are considered in the next chapters. Studies on the transport of spin-polarized electrons and tunneling magnetoresistance are also discussed.

Finally a set of studies, theoretical and experimental, on the magnetic properties of extended solids closes the *Festschrift*. The very last chapter contains a summary of the meeting.

I wish to thank all the participants for their enthusiasm and support in the realization of the Workshop. Special thanks go to the other members of the organizing committee: Santos F. Alvarado, Alejandro Díaz-Ortiz, Maria Luisa Marquina, and Dan Pierce. I am particularly grateful to Juan Martin Montejano-Carrizales for his invaluable assistance in the preparation of this volume. Finally I acknowledge the financial support of Consejo Nacional de Ciencia y Technología, Consejo Potosino de Ciencia y Tecnología, Sociedad Mexicana de Física, Universidad Autónoma de San Luis Potosí, and Universidad Nacional Autónoma de México.

August 2000

J. L. Morán-López
San Luis Potosí, S.L.P., Mexico

Contents

Physics of Low Dimensional Systems

Spin-Polarized Electrons and Magnetism 2000

H. C. Siegmann

Solid State Physics Laboratory
Swiss Federal Institute of Technology
8093 Zürich
SWITZERLAND

Abstract

The spin polarization P of the electron states of a ferromagnet is obtained from photoemission experiments provided one corrects for the spinfiltering in the transport of the photoexcited electrons to the surface. It turns out that there exist two types of ferromagnetic metal: one with positive P at the Fermi energy E_F (Fe, Gd, $La_{0.7}Sr_{0.3}MnO_3 \cdots$) and the other with negative $P(E_F)$ (Ni, Co, $Fe_3O_4 \cdots$) This together with progress in understanding the role of the metal-oxide interface in determining the spin polarization of the current through magnetic tunnel junctions makes possible the injection of polarized electron currents of either sign from ferromagnetic emitters into quantum wells, insulators, or other ferromagnetic metals leading to a wealth of new insight as well as applications such as realized in GMR-(Giant magnetoresistance) and TMR-(Tunneling magnetic resistance) devices. Generally, in such spin electronics, the electric currents are manipulated through the spin state of the electrons. Basic to the understanding of spin electronics as well as electron emission from solids is the understanding of the spin attenuation in transport of the electrons. We have performed two different experiments to elucidate the underlying spin dependent electron-electron scattering. In the first experiment, the electrons are excited with a femtosecond laser pulse and their spin dependent relaxation is observed via a subsequent laser pulse inducing photoemission. In the second experiment, spin-polarized electrons from a GaAs-type of electron source are injected into a ferromagnetic film and their absorption and spin rotation is observed.

I. Introduction

In their famous book on the theory of metals and alloys, Mott and Jones[1] explain the electrical conductivity in transition metals by electron-electron scattering from the

Physics of Low Dimensional Systems
Edited by J. L. Morán-López, Kluwer Academic/Plenum Publishers, New York 2001

s-band to the d-band. The probability of this scattering is proportional to the number of unoccupied states or holes in the d-band. With a ferromagnetic metal such as Ni only half the electrons contribute namely those with magnetic moment antiparallel to the magnetization (the states of the minority spin electrons) because all places in the d-band with spin parallel to the magnetization (the states of the majority spin electrons) are occupied. This leads to a decrease of the resistivity on lowering the temperature below the Curie point T_c in agreement with the observations. Today, the electron-electron scattering in transition metals can be studied in spin-polarized electron spectroscopies, in pump/probe type of experiments using femtosecond laser pulses, and in transmission of spin-polarized electron beams through extremely thin ferromagnets. It is the purpose of this paper to discuss these recently developed techniques and the results obtained so far. The absorption and spin motion of electrons travelling through solids and specifically through ferromagnets is crucial to the development of spin electronics constituting a new and very promising application of magnetism in which the electric currents are manipulated through the spin state of the electron.

In 1968, Fert and Campbell[2] developed Mott's two current model in ferromagnets further, in particular they found that the transitions from one spin channel to the other are comparatively rare. Generally, spin flips can occur in scattering of electrons on spin waves as well as in e-e scattering. It has been found that the cross section for generation or absorption of spin waves by electrons is much smaller compared to the electron-phonon cross section.[3] Hence it is clear that these spin flip processes may be neglected at temperatures below the Curie point. However, in e/e-scattering, the absence of spin flips is not obvious. To illustrate this, we take the example of Co rather than Ni. In Co, the majority d-band is fully occupied just like in Ni, but the minority band exhibits 1.7 holes/atom. Consequently, the scattering of electrons is still fully spin selective but stronger than in Ni with only 0.5 holes/atom. The Co-atoms will be in a mixture of $3d^8$- and $3d^9$-configurations. For the $3d^9$-configuration, we can predict the events following the absorption of a minority spin electron as follows:

$$e\downarrow + \mathrm{Co}3d^9\uparrow \;\rightarrow\; \mathrm{Co}3d^{10} + U \begin{array}{lll} \rightarrow & (\mathrm{a}) & e\downarrow + \mathrm{Co}3d^9\uparrow, \\ \rightarrow & (\mathrm{b}) & e\uparrow + \mathrm{Co}3d^9\downarrow + \Delta E. \end{array} \qquad (1)$$

First, the absorption of the minority spin electron leads to an electrically charged Co-site with no magnetic moment. The Coulomb energy U is partly screened and will be of the order of 1 eV; U must come from the kinetic energy of the incident electron. The resulting metastable configuration can subsequently decay in two ways: a) by reemitting a minority electron which brings us back to the initial state except for possible excitation of multiplets in the Co$3d^9$-atom left behind; b) by emitting a majority $e\uparrow$. In this case, the spin of the Co$3d^9$-atom is inverted which increases the energy by the exchange splitting ΔE. Additionally, excitation of multiplets might also occur. The branching ratio a:b should be 1, but b) is weaker because there is the energy barrier of ΔE. The scattering described by Eq. (1b) must be blocked as it would destroy the spontaneous magnetization, and also would be in contradiction to the two current model as b) involves a transition from a down spin to an up spin electron. The results obtained 20 years after the work of Fert and Campbell on magnetoresistance of magnetic multilayers have splendidly verified the two current model. In Fe for instance, the electrical resistivity in one spin channel is $\sim$ 5 times larger compared to the other at room temperature.[4] This phenomenon is now known as giant magnetoresistance (GMR) and has lead to significantly improved reading heads in magnetic recording. It is important to keep in mind at this point that from resistivity alone one can not determine which spin state has the shorter relaxation time. One can however predict that the resistivity must be smallest when the two ferromagnetic films, through which the electrons travel, are magnetized in parallel. If the two ferromagnetic metals are separated by a thin oxide layer instead of a non-magnetic metal, the electrons can tunnel between the two ferromagnets. If a bias voltage V_B is applied across the oxide, the electrons will tunnel into the unoccupied levels above the Fermi-energy E_F. In analogy to the GMR-experiment, one expects that the tunneling magneto-resistance (TMR) is lowest in the limit of $V_B = 0$ if two identical ferromagnets are magnetized in parallel. This is called normal TMR. Yet, depending on the oxide, and on the ferromagnetic metal, one has also observed inverse TMR.[5] This paradoxical result finds its explanation by the observation that tunneling barriers depend on the parentage of the electron states and can also be spin dependent.

By applying $V_B \neq 0$ to a magnetic tunnel junction, one is able to perform electron spectroscopy of the unoccupied electron states. Additionally, one is able to supply enough energy to switch on the scattering Eq. (1) leading to inverted magnetic moments.

One experiment that can determine the spin direction of the tunneling electrons absolutely, not only relative to the second metal, was introduced by Tedrow and Meservey.[6] Here, the electrons tunnel in a magnetic field from the ferromagnetic metal through an Al_2O_3 oxide barrier into superconducting Al. However, this experiment yields a paradoxical result too. The electrons tunneling from Ni and Co were mainly majority spin electrons while one knows that the electrons at E_F are predominantly of the minority type in both these ferromagnetic metals.

II. Electron-Electron Scattering in Electron Spectroscopy

The band theory of ferromagnetism predicts that there exist two types of ferromagnetic metal: one in which the spin state of the electrons at E_F is predominantly of the majority type and the other in which it is of the minority type. Fe belongs to the first, and Ni and Co to the second class. It has been realized already in the fifties that this must lead to a large spin-polarization with a characteristic dependence on binding energy of the electrons emitted in either field- or photoemission.[7] However, experiments performed in a number of laboratories[8–10] seemed to show that the spin-polarization of electrons field- or photoemitted from Fe, Co, and Ni is zero. We reached the conclusion at the time that this result might be due to a predominance of the unpolarized electrons from the *s*-*p*-bands in the emission.[11] However, this proved to be true only for field emission into vacuum. In the classical models describing electron emission, one expects that tunneling spectroscopy and threshold photoelectron spectroscopy deliver identical results. This turns out that it does not always apply. Hence additional phenomena, namely electron-electron scattering and, in the case of the tunneling, the dependence of the tunneling barrier on the parentage and spin state of the electron must be taken into account to understand the experimental results. In the following, we discuss what one can learn about *e*-*e* scattering by means of electron spectroscopy.

II.1. Photoemission of Electrons Near Photoelectric Threshold

The simplest case of spin-polarized photoemission is threshold photoemission in which the photon energy $h\nu$ is close to the photoelectric work function ϕ. The photons are absorbed by the electrons in the valence bands generating electron-hole pairs. The electrons in the energy window $h\nu - \phi$ may escape into vacuum and can be collected into an electron beam. The spin-polarization is then measured in Mott scattering after acceleration to suitable energies. Energy and momentum analysis of the photoelectrons is automatically present as only those electrons are able to escape over the surface barrier potentials that travel perpendicular to the surface. Furthermore, one can generate large electron currents with the available sources of UV-light such as pulsed lasers, excimer lamps and high pressure arcs. On the theoretical side it is important to realize that the hole state left behind after excitation of the photoelectron does not need to be renormalized as it is already close to E_F; therefore in the limit of $h\nu \rightarrow \phi$ multiplets or other excitations can not occur.

The energy window in which one expects negative spin polarization depends on the magnitude of the exchange splitting ΔE. In the calculations available at the time when the first experiments on Ni were done,[12–14] $\Delta E = 0.5$ eV. The experimental results are clearly not in agreement with this splitting, but a nice fit of the data is obtained with $\Delta E = 0.33$ eV irrespective of the crystal plane from which the electrons were emitted.[14] This splitting has been verified later in a number of independent experiments,[15] yet it still needs to be corrected for the effects of *e*-*e* scattering that are weaker, but still present in Ni as well.

However, very early it has been realized that threshold photoemission from Co cannot at all be explained in this simple model.[16] In this case, the energy window of the minority spins should be large, of the order of 1 eV and more, but what one observes is nothing but majority spins right up to $h\nu \rightarrow \phi$ where the emission ceases. It turns out that photoemission of electrons from Co can only be understood if one takes into account the effects of electron-electron scattering in transport of the electrons to the surface. Equation (1) shows that this scattering removes the minority spins. We will show that the majority spin produced in branch b) of the scattering has insufficient energy to escape through the energy window $(h\nu - \phi)$ into vacuum. This then constitutes a spin filter in which the minority spins are selectively absorbed by the holes in the $3d$-shell.

The absorption of electrons travelling through a solid depends on the thickness t of the film through which the electrons penetrate. It is given by

$$I = I_0 e^{-\sigma t}, \tag{2}$$

where I_0 is the intensity of the electron beam entering the solid and I the intensity leaving it without suffering any scattering. σ is the absorption cross section which in turn is related to the total scattering cross section q and the atom density N by $\sigma = Nq$. In a ferromagnet, we have two different absorption cross sections σ^+ and σ^- for majority and minority spin electrons, respectively. With $I^+ = I_0 e^{-\sigma^+ t}$ and $I^- = I_0 e^{-\sigma^- t}$, one obtains the spin polarization generated in transport of the electrons through a film of thickness t:

$$A(t) = \frac{I^+ - I^-}{I^+ + I^-} = \frac{e^{\Delta\sigma t} - 1}{e^{\Delta\sigma t} + 1}. \tag{3}$$

It is seen that the spin polarization $A(t)$ generated in transport depends only on the difference $\Delta\sigma = \sigma^+ - \sigma^-$ of the absorption cross sections. This difference is exactly the difference in the occupancy of the $3d$-shell, hence it must be proportional to the number of Bohr magnetons n_B:

$$\Delta\sigma = \text{const} \cdot n_B. \tag{4}$$

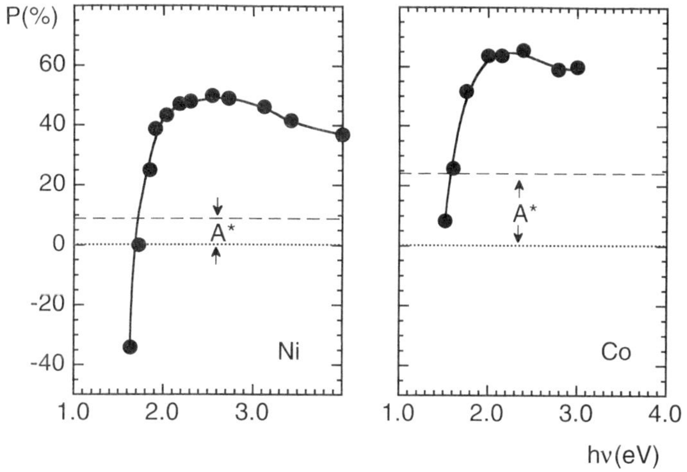

Figure 1. Spin polarization of electrons *vs.* width of the window ($h\nu - 1.6$ eV) from the cesiated Ni(111) and Co(000 ⊥) surfaces. The data are adapted from Ref. 19. The photoelectric threshold is $\phi = 1.6$ eV. The integrated transport polarization A^* is calculated from Eq. (5) to be 9% for Ni and 24% for Co.

The constant in Eq. (4) is the absorption cross section σ_d of the d-shell, const $= \sigma_d$. It is reasonable to expect that σ_d does not vary substantially across Ni, Co and Fe. This expectation is met by the experimental results.[17] From the observed σ_d one obtains q$\cong 10^{-16}$ cm^2 which is consistent with the assumption that the $3d$-shell is absorbing the electrons.[18]

It will be shown below that the experiment has also verified yet another simple consequence of Eq. (3) which is $\lim A(t) = 1$ for $t \to \infty$.

To understand the spin polarization of the photoelectrons, one must take into account that the electron-hole pairs are excited at various distances from the surface and that the probability of escape depends on t according to Eq. (2). This yields for the integral polarization A^* acquired in transport to the surface:

$$A^* = \frac{\int_0^\infty A(t)I(t)\,dt}{\int_0^\infty I(t)\,dt} = \frac{\sigma^+ - \sigma^-}{\sigma^+ + \sigma_-} = \frac{1}{2}\cdot\frac{\Delta\sigma}{\sigma}. \tag{5}$$

To obtain the spin polarization P of the photoemitted electrons, one can add A^* to the polarization P_0 in the band, as long as $A^* < 1$:

$$P = P_0 + A^*. \tag{6}$$

It turns out that $A^* = 0.09$ for Ni and $A^* = 0.24$ for Co.[17]

Figure 1 shows how Eq. (6) explains why one can observe negative polarization in Ni but not in Co. The data are from single crystalline surfaces in a more recent experiment to verify the older historical results.[19] It is also evident that P happens to be almost identical for energy windows of 0.4 and 0.8 eV from threshold in the case of Ni as observed in the first experiment on Ni.[12] It is interesting to also look at the results with ferro-, ferri-, and antiferromagnetic insulators that have been investigated in detail as well. In this case, one expects that the electron-electron scattering is blocked due to the existence of the forbidden energy zone. And indeed, it has been found that all the threshold photoemission results could be understood in detail from the spin-polarized band structure comprising the highly polarized $4f$- or $3d$-states, the magnetic and nonmagnetic impurity states, and the unpolarized valence bands.[20] Even fine details like the difference in the ionic energy at A- and B-sites in the ferrites were obtained in agreement with other work. Convincing proof for the absence of significant e-e scattering is the fact that negative polarization is found with threshold photoelectrons from magnetite. These electrons are excited from Fe$3d^6$ and therefore must be of the minority type. This result also tells us that magnon creation by spin down electrons is too weak to be observed in photoemission.

II.2. Tunneling Spectroscopy

Alvarado and coworkers[21] have introduced a new form of tunneling spectroscopy in which one can vary the tunneling barrier without introducing exchange coupling and in which one can also determine the direction of the spin of the tunneling electrons just like in Ref. 6. In this experiment, a ferromagnetic metal such as Ni is formed into a tip. The magnetization is either parallel or antiparallel to the tip axis. The ferromagnetic tip is approached to a p-doped single crystalline surface of GaAs. The electrons tunnel from the ferromagnetic metal into the conduction bands of GaAs where they eventually recombine with the holes introduced by p-doping. The photons produced in this fluorescence are right or left circularly polarized depending on the spin of the electrons. By measuring the circular polarization of the fluorescence light one determines the spin polarization of the tunneling electrons. The potential barrier between the metallic tip and the GaAs surface is variable by adjusting the distance of the tip. The electron states can tunnel in an energy window from states at and below E_F. The width of this window is chosen by applying a voltage between the tip and the GaAs. Then, the average polarization of the electrons in the chosen energy window is what is observed. This experiment is therefore very similar to threshold photoemission of electrons in which the energy window is determined by choosing the photon energy. And indeed, one finds the same dependence of the spin polarization on the width of the chosen energy window as demonstrated for the case of Ni.[21] Characteristic for the contribution of $3d$-electrons is the negative spin polarization observed when the energy window is sufficiently narrow to let pass only electrons from a close neighborhood of E_F. Furthermore, Alvarado[22] was able to show that the tunneling barrier is $\sim$ 1eV higher for the $3d$-electrons compared to the electrons from the $4s$, p-bands. Generally, the relative contribution of unpolarized s, p-electrons and d-electrons depends on the distance of the tip to the surface. This explains the very low polarization observed with field emitted electrons[23] and proves that a number of previous experiments were in fact faulty[24] or misinterpreted.[25]

This now brings us back to the early paradoxical result obtained by Tedrow and Meservey in tunneling from Ni and Co into superconducting Al.[26] Recent theoretical[27] and experimental[5] work has shown that the spin direction of the tunneling electrons depends crucially on the thickness and the chemical nature of the tunneling barrier. Therefore, in tunneling spectroscopy it appears to be essential to have control over the tunneling barrier, the chemical bonding at the interface to the metal, and the quantum well states present in nonmagnetic interlayers.[28] Otherwise, tunneling spectroscopy is extremely promising especially in the present context as it makes it possible to switch on e-e scattering by applying a bias voltage to overcome the energy barriers U and ΔE in Eq. (1). The onset of the scattering might be detected for instance by a reduction of the magnetization while the current flows, or it might even be possible to switch the magnetization in small samples. Such a phenomenon has already been observed.[29]

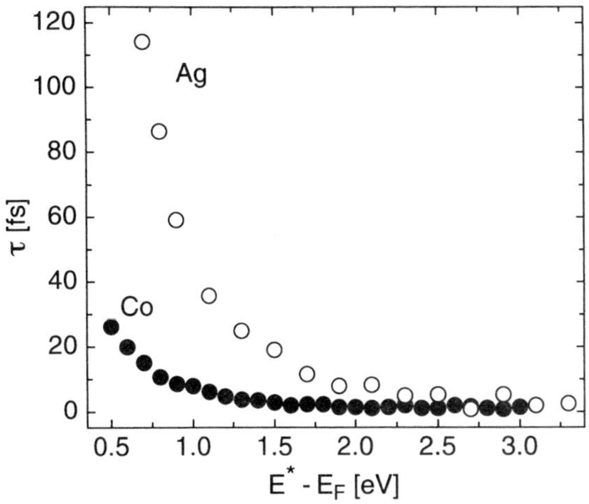

Figure 2. The spin integrated inelastic lifetime of cesiated Co(001) with a work function $\phi = 3.5$ eV and Ag(111) with a work function of 4.1 eV *vs.* the energy above E_F.

II.3. Two-Photon-Photoemission

The lifetime of hot electrons can directly be determined in pump-probe experiments employing femtosecond laser pulses. The absoprtion of the first photon brings an electron to the energy level E^* above the Fermi energy E_F. If a second photon is absorbed while the electron is still at E^*, it can escape into vacuum with a kinetic energy $E_{\rm kin}$. By setting the energy analyzer to transmit electrons at a specific $E_{\rm kin}$, E^* can be varied. If now the delay between first and second photon pulse is increased, one sees a decrease in the electron intensity reflecting the finite life time in the excited state E^*. Figure 2 shows results with the noble metal Ag in which the *d*-band is full, and with Co. One sees that the life time of hot electrons is much smaller in Co than in Ag. It is also seen that the spin averaged life time increases as one approaches E_F in both metals. This proves that *e-e* scattering is the dominant process determining the life time of hot electrons in this experiment. The *e-e* scattering model also postulates that the life time must depend on the spin state in a ferromagnetic metal because there are more holes for minority spin states compared to majority spin states. This is indeed observed by measuring the spin state of the electrons emitted in 2 PPE photoemission.[30] Figure 3 shows the result for the cesiated Co(001) surface. We see that the life time of the majority spins $\tau\uparrow$ is about 2 times longer compared to the one of the minority spins $\tau\downarrow$ at $E^* = 1$ eV. As E^* is lowered, $\tau\uparrow/\tau\downarrow$ decreases. This agrees once more with the notion that *e-e* scattering is dominant in this experiment. For energies closer to E_F, the *e-e* scattering is increasingly blocked due to the energy barrier U in Eq. (1) apparently resulting in a reduction of $\tau\uparrow/\tau\downarrow$. However, it must be emphasized that the 2 PPE-process has still several weaknesses in the interpretation. The number of electrons at E^* that can escape into vacuum is reduced not only by *e-e* scattering, but also by diffusion of the electrons into the bulk. On the other hand side, this number is increased by electrons relaxing from higher states into E^*. The holes in the *d*-band left behind after excitation of the photoelectron will produce Auger electrons complicating further the detailed interpretation. Yet, the 2 PPE experiment is clearly consistent with the *e-e* scattering model, confirming the order of magnitude of the spin dependent lifetimes known from other work.

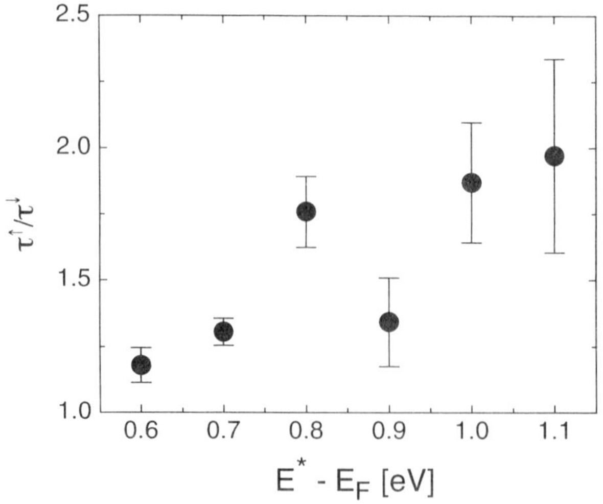

Figure 3. The ratio $\tau\uparrow/\tau\downarrow$ of the lifetime of majority and minority spins respectively *vs.* the intermediate state energy above E_F for the Co-sample of Fig. 2.

III. Absorption and Spin Motion of Electrons in a Ferromagnet

The most direct information on *e-e* scattering in a ferromagnet is obtained by studying the transport of electrons through a ferromagnetic sample. To do this experiment, a film must be prepared through which at least an observable portion of a spin-polarized electron beam can penetrate. The *e-e* scattering produces mean free paths of the order of a few lattice spacings in transition metals. Therefore, the film has to be extremely thin to observe the emerging electron beam. In fact, the experiment cannot be done with a free standing film of a transition metal. Rather, one has to make use of the fact that scattering into the *d*-states is absent in the noble metals, and use for instance Au as a substrate for the ferromagnetic sample. An Au-film of 20 nm thickness for instance will attenuate a low energy eletron beam by only $1/10^6$ making it still possible to measure energy, intensity, and spin polarization of the electron beam after transmission. One can now insert a thin film of a ferromagnetic transition metal such as Co into the Au-film. While the spin averaged absorption of electrons due to the additional Co is not detectable, one can switch the magnetization of the Co-film with respect to the spin polarization of the electron beam and look for spin induced changes of the transmission. The spin dependent part of the attenuation is rather easy to detect even on the huge background of spin independent attenuation.[31]

Figure 4 is a schematic sketch of the experiment. A spin modulated electron beam is prepared using a GaAs-type photocathode. The spin polarization vector $\vec{P}_0$ of the incident electron beam can have any direction in space. It is also possible to produce an unpolarized electron beam with $\vec{P}_0 = 0$ by using unpolarized light at the photocathode. The electron beam impinges along the surface normal onto a trilayer consisting of a supporting Au film of 20 nm thickness, followed by a ferromagnetic film of varying thickness of a few nm, and a capping Au-layer to prevent corrosion of the ferromagnetic film. After the trilayer, a retarding field is applied that rejects all electrons that have lost energy in passage through the trilayer. In this way, and within the limits of the finite energy resolution of the retarding field of ± 0.3 eV, one counts only those electrons that have passed through the trilayer without energy loss.

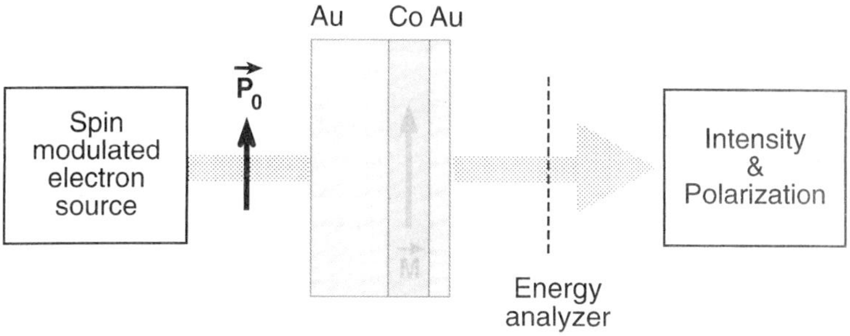

Figure 4. The principle of the experiment measuring the spin dependence of the absorption of electrons in ferromagnetic Co.

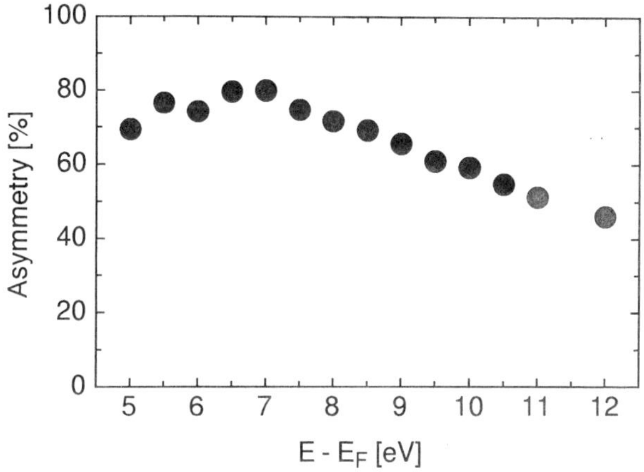

Figure 5. The spin asymmetry A (Eq. 3), in % *vs.* the thickness of the Co-layer at the electron energy of 5 eV above the Fermi-energy of Au.

The spin-orbit interaction in the Au-atom is large, yet no spin asymmetry can be generated by it as the transmitted electron beam is not deflected. Hence all the spin dependent absorption must be due to the ferromagnetic layer. The spin dependent absorption $A(t)$ is given by Eq. (3), where t is now the thickness of the ferromagnetic layer.

Figure 5 shows $A(t)$ in % with polycrystalline *hcp* Co-films at the lowest possible electron energy. We see that very large spin asymmetries are observed, and that, except for the thinner films, the thickness dependence follows nicely the predictions of Eq. (3). Figure 5 is valid for an electron energy of ~ 5 eV with respect to E_F in Au. The asymmetry $A(t)$ decreases with increasing electron energy as shown in Fig. 6 for a Co-film of 4 nm thickness. It is not possible to do the measurement in the energy range of 15–100 eV because the attenuation of electrons in the Au-layer is too large in this energy range. At energies larger than 100 eV the spin dependence of the attenuation can again be observed, but it is at or below the 1%-level. All this is in good agreement with the *e-e* scattering model as the matrix elements for transition into the $3d$-shell must decrease with increasing mismatch of the energy. The experiment yields $\Delta\sigma$ in Eq. (4) as it depends on electron energy. At low energies the total scattering cross

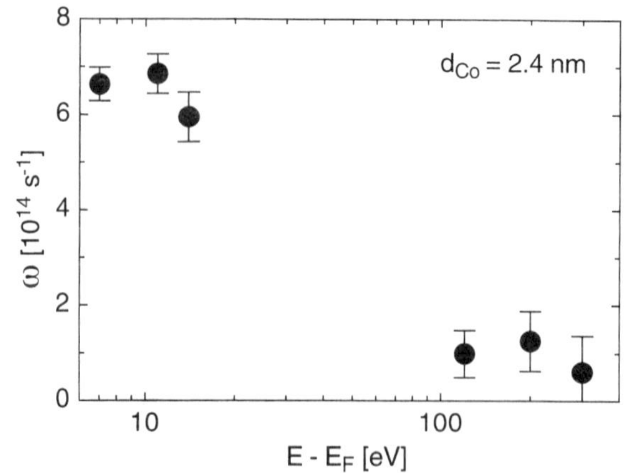

Figure 6. The spin asymmetry A, Eq. (3) in % *vs.* the energy of the electrons above E_F for a Co-film of 4 nm thickness.

section for one hole in the $3d$-shell of Co is $q = 0.25 \times 10^{-16}$ cm^2. To obtain this cross section, the assumption is made that all the spin dependent attenuation is due to inelastic scattering into the $3d$-states. In principle, q is the sum of all spin dependent scattering, elastic and inelastic. In elastic electron scattering, the scattering amplitude is the superposition of the direct scattering amplitude and the exchange scattering amplitude. There is a rigorous experimental test proving that there is no contribution of elastic exchange scattering, and with this test one concludes that elastic electron scattering may be neglected. There are other qualitative arguments given in Ref. 18 leading to the same conclusion.

The test on the contribution of the elastic quantum mechanical exchange of electrons consists of two parts:

1. An unpolarized electron beam passes through the ferromagnet. This produces a spin polarization P' which consists of 2 parts

$$P' = P + \Delta P \tag{6}$$

P is the transport polarization generated by spin selective scattering into the holes of the d-band, while ΔP is the additional polarization generated by elastic spin exchange collisions.

2. An electron beam of initial polarization P_0 passes through the ferromagnet. The intensity asymmetry $A' = (I^+ - I^-)/(I^+ + I^-)$ observed by inverting the relative direction of P_0 from parallel to the magnetization (I^+) to antiparallel to it (I^-) is given by

$$A' = A - \Delta A, \tag{6}$$

where A is the asymmetry due to inelastic spin selective absorption and ΔA is the reduction of this asymmetry because some electrons have flipped their spin in elastic spin exchange collisions and thus avoided spin dependent absorption. We find within the experimental uncertainty of less than 1% that $P' = A'$. Therefore, $\Delta P = \Delta A = 0$ meaning that spin flips without energy loss are not present. From this result we can further conclude that "Stoner excitations" as described by Eq. (1) branch b) do not contribute to the spin polarization generated in

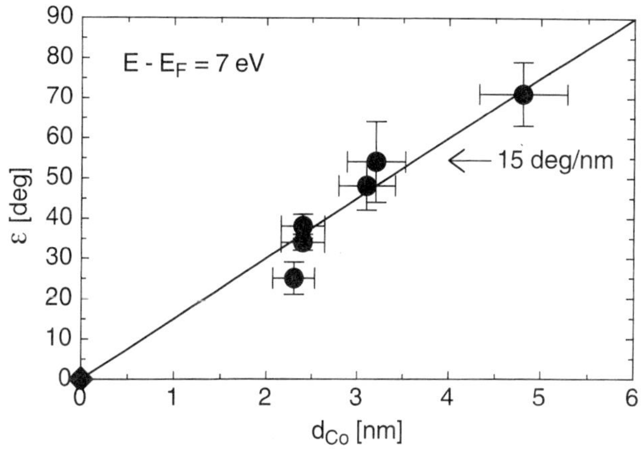

Figure 7. The angle ε of precession of the spin polarization at an electron energy of 7 eV above E_F *vs.* the thickness of the Co-film.

transport of electrons through a ferromagnet with the present energy resolution. This is easily understood as the emerging majority spin electron has lost at least the exchange energy ΔE which is of the order of more than 1 eV in the case of Co.

In yet another version of the very powerful transmission experiment, one chooses the initial polarization P_0 of the electron beam perpendicular to the magnetization of the ferromagnet. One observes a very fast precession of the electron spin about the direction of the magnetization. The precession frequency is given by $\omega = \Delta E/h$, where ΔE is the exchange splitting. The precession angle ε is given by the time t' the electron spends in the ferromagnet, $\varepsilon = \omega t'$, hence it must increase linearly with the thickness t of the film unless there are resonances. Figure 7 shows on the example of Co-films that $\varepsilon = \omega t$ is actually observed. This proves that the precession is not an effect of the surface but tests bulk properties. The observed large angles correspond to magnetic fields in the Megagauss range. Therefore, the precession is not produced by dipolar fields. Figure 8 shows the precession frequency as a function of electron energy. We see that the precession persists up to electron energies of over 200 eV. The precession is due to the phase difference that develops between majority and minority spin wave functions and has to be distinguished from the spin selective absorption. More on this experiment is described in Ref. 31.

The observation of the spin precession is to our knowledge the first experimental proof that the exchange interaction produces a torque $\vec{T}$ on the electron spin, corresponding in magnitude to the exchange splitting, $|T| = \frac{1}{2}\Delta E$, but being in direction perpendicular to the magnetization of the ferromagnet and to the spin of the electron. Hence altogether the polarization vector $\vec{P_0}$ of an electron beam passing through a ferromagnet is given by a rotation of $\vec{P_0}$ into the direction of the magnetization $\vec{M}$, and by a precession of $\vec{P_0}$ around the direction of $\vec{M} \cdot \vec{P_0}$ then moves on a spiral into the direction of $\vec{M}$.

The material properties of the ferromagnet are described by the following filter matrix

$$F = \begin{pmatrix} \sqrt{1 + A(t)}e^{-\frac{i}{2}\omega t} & 0 \\ 0 & \sqrt{1 - A(t)}e^{+\frac{i}{2}\omega t} \end{pmatrix}.$$

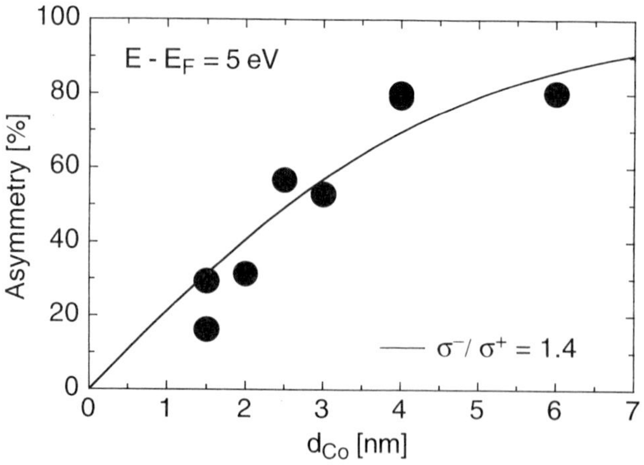

Figure 8. The precession frequency ω calculated from the observed precession angle ε *vs.* the energy of the electrons. In the calculation it is assumed that the effective mass is equal to the free electron mass.

The wave function Ψ' of the electron after spending the time t' in the ferromagnet of thickness t is given by

$$\Psi' = F\,\Psi,$$

where Ψ is the wavefunction at $t' = 0$. For instance, if the electron is incident with the spin perpendicular to the magnetization,

$$\Psi = \frac{1}{\sqrt{2}}\left[\begin{pmatrix}1\\0\end{pmatrix} + \begin{pmatrix}0\\1\end{pmatrix}\right].$$

After passage through the ferromagnet, the wave function has become

$$\Psi' = \frac{1}{\sqrt{2}}\left[\sqrt{1+A}\begin{pmatrix}1\\0\end{pmatrix}e^{-\frac{i}{2}\omega t} + \sqrt{1-A}\begin{pmatrix}0\\1\end{pmatrix}e^{+\frac{i}{2}\omega t}\right].$$

That is, an amplitude and phase difference has developed between majority and minority spin wave functions corresponding to absorptive and phase velocity induced circular dichroism respectively. In real space, this yields the motion of P_0 on a spiral into M. The same matrix F yet with much smaller material constants ω and A describes magneto-optic phenomena. This arises because there is a complete formal analogy between the description of light and spin-polarized electrons in the non-relativistic limit if one uses the Stokes-vector formalism. The material constants are 1–2 orders of magnitude smaller for light because the photon couples to the magnetization only *via* the spin-orbit interaction.

The recognition that the exchange interaction is of an axial nature is important when considering the transfer of electrons across multilayers, and whenever there is a sudden change of the magnetization direction. It has practical applications, for instance, when polarized electrons are injected into a ferromagnetic sample. By virtue of Newtons action = reaction, the magnetization of the sample will also precess, but in the opposite sense compared to the injected electrons.

IV. Summary and Outlook

The basis of present day electronics are p/n junctions made by doping semiconductors. But we can now also think of spin-up/spin-down junctions in which the electric currents are manipulated by the direction of the magnetization rather than by applying a voltage to a gate. This opens up new applications for magnetism. The foundation of this is the progress made in several areas of solid state physics:

1. Knowledge of the spin-polarized electronic structure in ferromagnetic metals,
2. Mechanism and correct interpretation of electron emission processes, *e.g.*, in tunneling of electrons between solids, and in threshold photoemission of electrons,
3. Spin dependent scattering and motion of the spin polarization vector in transport of electrons through semiconductors and ferromagnets.

Additionally, one area of great impact is the imaging of the magnetization with unprecedented temporal and spatial resolution. The highest spatial resolution of magnetization patterns in thin films or at surfaces is achieved at present with spin resolved secondary electron microscopy.[32,33] The contrast in this technique is so great because the spin polarization of the low energy cascade is enhanced due to the spin selective *e-e* scattering discussed here. The most recent imaging technique uses the photoelectron-microscope (PEEM) in connection with a synchrotron light source.[34] It also relies on transitions of core electrons into the empty d-states. PEEM allows one to image the magnetization with element specificity. For the first time, domain walls in antiferromagnets can also be seen.[35] Finally, using the magneto-optic Kerr-effect in combination with femto-second laser pulses, it is now possible to observe the directional changes of the magnetization vector with a time resolution of below 10^{-12} sec. From ferromagnetic resonance one knows that the magnetization vector performs a precessional motion about a magnetic field B at the frequency $\omega \cong (e/m)B$. In thin films the damping of this precession occurs by emission of spin waves at frequencies of $\sim$ 100 GHz. The precessional motion can now be imaged with spatial resolution of a few μm. This is essential to understand the modes of magnetization reversal. All taken together promises great progress in the basic understanding and in the application of magnetism in the near future.

References

1. N. F. Mott and H. Iones, *The theory of the properties of metals and alloys* (Dover Publ., New York 1958).
2. A. Fert and I. A. Campbell, *Phys. Rev. Lett.* **21**, 1190 (1968).
3. M. Plihal, D. L. Mills, and J. Kirschner, *Phys. Rev. Lett.* **82**, 2597 (1999).
4. R. Coehoorn, *Europhys. News* **24**, 43 (1993), and references cited.
5. J. M. De Teresa, A. Barthelemy, A. Fert, J. P. Contour, R. Lyonnet, F. Montaigne, P. Seneor, and A. Vaures, *Phys. Rev.Lett.* **82**, 4288(1999); *Science* **286** 507 (1999), and references cited.
6. P. M. Tedrow and R. Meservey, *Phys.Rev. Lett.* **26**, 192 (1971).
7. E. S. Dayhoff, *J. Appl. Phys.* **30**, 234S (1959).
8. H. A. Fowler and L. Marton, *Bull. Amer. Phys. Soc.* **4**, 235 (1959).
9. R. L. Long Jr., V. W. Hughes, and J. S. Greenberg, *Phys. Rev. A* **138**, 1630 (1965).
10. A. B. Baganov and D. B. Diatroptov, *Sov. Phys. JETP* **27**, 1733 (1968).

11. N. Müller, H. C. Siegmann, and G. Obermair, *Phys. Lett.* **24A**, 733 (1967).
12. U. Bänninger, G. Busch, M. Campagna, and H. C. Siegmann, *Phys. Rev. Lett.* **25**, 585 (1970).
13. W. Eib and S. F. Alvarado, *Phys. Rev. Lett.* **37**, 444 (1976).
14. E. Kisker, W. Goudat, M. Campagna, E. Kuhlmann, H. Hopster, and I. D. Moore, *Phys. Rev. Lett.* **43**, 966 (1979).
15. P. Aebi, T. J. Kreutz, J. Osterwalder, R. Fasel, P. Schwaller, and L. Schlapbach, *Phys. Rev. Lett.* **76**, 1150 (1996), and references cited.
16. G. Busch, M. Campagna, D. T. Pierce, and H. C. Siegmann, *Phys. Rev. Lett.* **28**, 611 (1972).
17. H. C. Siegmann, *Surface Sci.* **307-309**), 1076-1086 (1994).
18. D. Oberli, R. Burgermeister, S. Riesen, W. Weber, and H. C. Siegmann, *Phys. Rev. Lett.* **81**, 4228 (1998).
19. J. C. Gröbli, A. Kündig, F. Meier, and H. C. Siegmann, *Physica B* **204**, 359 (1995).
20. M. Campagna, D. T. Pierce, K. Sattler, and H. C. Siegmann, *J. de Physique* **34**, C6 87 (1973); S. F. Alvarado, W. Eib, F. Meier, D. T. Pierce, K. Sattler, and H. C. Siegmann, *Phys. Rev. Lett.* **34**, 319 (1975); S. F. Alvarado, W. Eib, H. C. Siegmann, and J. P. Remeika, *Phys. Rev. Lett.* **35**, 860 (1975).
21. S. F. Alvarado and P. Renaud, *Phys. Rev. Lett.* **68**, 1387 (1992).
22. S. F. Alvarado, *Phys. Rev. Lett.* **75**, 513 (1995).
23. M. Landolt and M. Campagna, *Phys. Rev. Lett.* **38**, 663 (1977).
24. W. Gleich, G. Regenfus, and R. Sizman, *Phys. Rev. Lett.* **22**, 1066 (1971).
25. R. Wiesendanger, H. -J. Güntherodt, G. Gntherodt, R. J. Gambino, and R. Ruf, *Phys. Rev. Lett.* **65**, 247(1990).
26. R. Meservey and P. M. Tedrow, *Physics Reports* **238**, 173 (1994).
27. J. Mathon and A. Umerski, *Phys. Rev. B* **60**, 1117 (1999).
28. Jagadeesh S. Moodera, Janusz Nowak, Lisa R. Kinder, and Paul M. Tedrow, *Phys. Rev. Lett.* **83**, 3029 (1999).
29. E. B. Myers, D. C. Ralph, J. A. Katine, R. N. Louie, and R. A. Buhrmann, *Science* **285**, 867 (1999).
30. M. Aeschlimann, M. Bauer, S. Pawlik, W. Weber, R. Burgermeister, D. Oberli, and H. C. Siegmann, *Phys. Rev. Lett.* **79**, 5158 (1997).
31. W. Weber, D. Oberli, S. Riesen, and H. C. Siegmann, *New J. of Phys.* **1**, 9.1–9.6 (1999).
32. D. T. Pierce, J. Unguris, R. J. Celotta, and M. D. Stiles, these Proceedings, p. 301.
33. F. Marty, C. Stamm, U. Maier, U. Ramsperger, and A. Vaterlaus, these Proceedings, p. 335.
34. G. Schönhense, these Proceedings, p. 309.
35. Boris Sinkovic, private communication.

Magnetism and Superconductivity in Ultra-Small Particles

B. Mühlschlegel

Institut für Theoretische Physik
Universität zu Köln
D–50923 Köln
GERMANY

Abstract

When electronic correlations in magnetism and pairing in superconductivity are described in the simplest fashion by the Hubbard model and the BCS model one finds remarkable properties of small systems which are absent in the bulk. We discuss here (for a few selected examples) the present status of the theory of cluster magnetism and of superconducting grains.

I. Introduction

If we look back at the many phenomena treated in condensed matter physics during, let's say, the last four decades of the last century, certainly both superconductivity and itinerant magnetism in ordered and disordered *bulk* systems played dominant roles as major research fields. It became clear that three things are most important for these phenomena: pairing of electrons, strong correlations between electrons, and spatial dimensionality. In retrospect we might regard it as a very beautiful feature that pairing can be handled so conveniently by the BCS model, and that strong correlations among electrons can be modeled within the very simplified frame of the Hubbard model for magnetism (in two dimensions perhaps even for high T_C).

The question of how these phenomena of the bulk are affected when we abandon the thermodynamic limit and turn to systems of variable finite size was originally one of a more academic nature. However, it received gradually more attention with the well known advances in experimental studies of cluster beams and of small particles in the nano-size regime. It is the purpose of this contribution to discuss (for a few selected examples) the present status of small-size effects in magnetic and superconducting systems from a theoretical view point. Here, I shall confine myself to the simplest

Physics of Low Dimensional Systems
Edited by J. L. Morán-López, Kluwer Academic/Plenum Publishers, New York 2001

possible scheme given by the Hubbard model and the BCS model. In Sec. 2 we consider cluster magnetism and its dependence on geometry. In Sec. 3 we turn to size effects in superconductivity. Further remarks and discussions are in the final Sec. 4.

II. Cluster Magnetism

The Hubbard model for correlated itinerant electrons on a lattice is given by the Hamiltonian[1]

$$H = -t \sum_{\langle ij \rangle \sigma} c_{i\sigma}^{\dagger} c_{j\sigma} + U \sum_{i} n_{i\uparrow} n_{i\downarrow} \,. \tag{1}$$

The first term is the kinetic energy caused by electron hopping of strength $t > 0$ between nearest-neighbor sites i and j. The second term is an approximation of the general electron-electron interaction by an on-site Coulomb interaction $U > 0$. This term renders H a nontrivial many-body problem. There are well known limiting cases. If one has one electron per site, H describes an ordinary half–filled metallic band for vanishing U, whereas it leads to antiferromagnetic Heisenberg spins with coupling $J = 4t^2/U$ in the opposite limit of vanishing t. Thus, a transition from a metal to a magnetic insulator is contained in Eq. (1) by changing the dimensionless parameter U/t.

For a finite number N of sites, H can also serve as a model of correlated itinerant electrons in a cluster. The case $N = 2$ is a nice textbook problem[2] which for two electrons illustrates the difference between mean field, Heitler-London and the here exact Hund-Mulliken treatment of the hydrogen molecule. For $N > 2$ we should consider *all* geometrical structures with N sites which are defined by equal nearest-neighbor distances. The lowest energy of H then selects the stable structure.

Let us illustrate this for the simplest case $N = 3$. The two possible structures are the triangle $\overset{3}{\underset{1 \quad 2}{\triangle}}$ with three bonds, and the linear chain $\underset{1\;\;2\;\;3}{\bullet\!-\!\bullet\!-\!\bullet}$ with only two bonds which means that the term $-tc_{1\sigma}^{\dagger}c_{3\sigma}$ is absent in the kinetic energy. Figure 1 shows the clusters with the lowest energy of (1) as a function of the electron number ν for varying correlation strength U/t. In a nutshell we recognize already here the interplay of electronic correlation, magnetic behavior and cluster structure. With 4 electrons on the cluster, the increasing Coulomb repulsion U drives the system from an nonmagnetic to a (ferro)magnetic state with maximal spin $S = 1$, accompanied by a structural transition. For 3 electrons (half filling) increasing U will localize the electrons more and more at the sites of the triangle leaving only fluctuations of the local spins.

It is clear what has to be done in order to obtain the N–site Hubbard cluster for a given number of electrons ν and a given interaction U/t. One has first to find *all* allowed structures. Second, for each allowed structure C the ground–state energy E_C of H in (1) must be calculated. Third, the minimal E of these E_C must be chosen to determine the desired stable structure. As in Fig. 1 we can use a ν-U/t diagram and draw in it the respective cluster. This program has been performed in full generality and precision up to $N = 8$ in Ref. 3. I refer to this paper for all details of the rich problem and show only the outcome for $N = 7$ in Figs. 2 and 3. We might recognize here among other things that structural changes at higher U are often connected with strong changes in the magnetic behavior. For half filling $\nu = N$ one has minimal total spin and strong antiferromagnetic correlations. In contrast, for $\nu = N + 1$ the cluster is ferromagnetic for larger U and eventually gets its maximal spin $S = (N - 1)/2 = 3$.

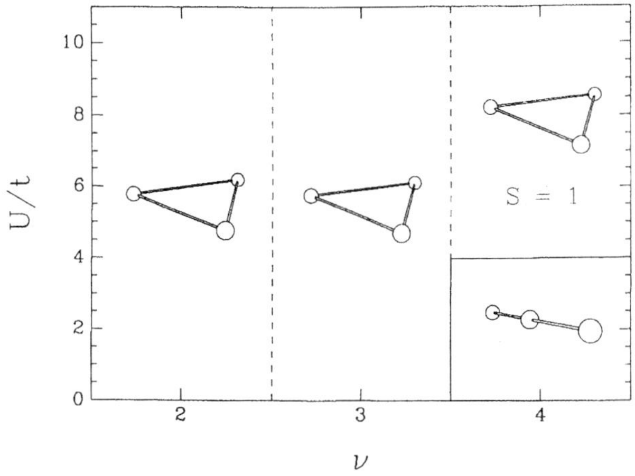

Figure 1. Hubbard clusters having $N = 3$ atomic sites. The ground–state structures corresponding to a Coulomb repulsion U, hopping integral t and number of electrons ν are illustrated. The ground–state spin S is minimal (*i.e.* $S = 0$ or $S = 1/2$) unless indicated. Broken lines separate regions having the same structure but different S. For $U/t \geq 10$ no changes in structure or S are found.[3]

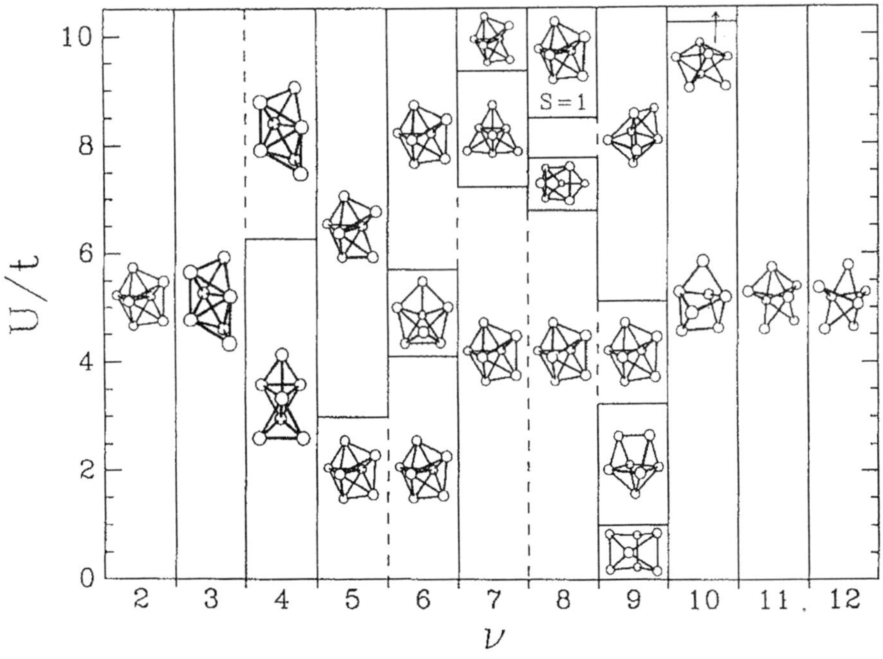

Figure 2. Hubbard clusters having $N = 7$ atomic sites. S is minimal unless indicated. Results for $U/t \geq 10.5$ are given in Fig. 3.[3]

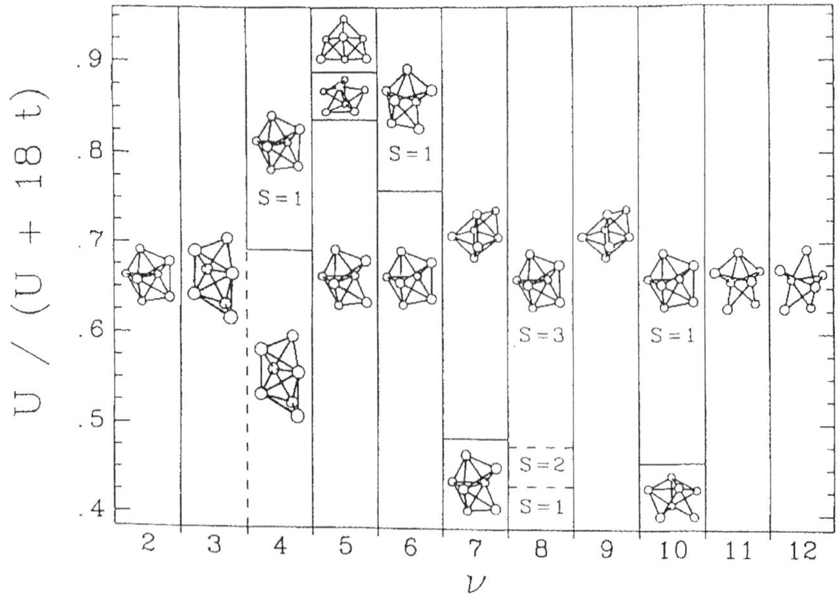

Figure 3. Hubbard clusters having $N = 7$ atomic sites and large values of U/t. As in Fig. 2 the ground–state spin S is indicated only if $S \geq 1$.[3]

This is what can be said about Hubbard clusters. In the concluding section we will come back to them in connection with the present situation for transition-metal clusters.

III. Superconductivity

As we have seen, it is in principle possible to study the modification of magnetism of, let's say, transition metals by decreasing the system size N down to small clusters and even down to a single atom. At first glance this makes no sense for superconductivity where we have characteristic lengths and the criterion of off-diagonal long-range order. To clarify this we should begin with very large N in order to understand size dependent deviations from bulk superconducting behavior.

Consider first N free electrons in a small particle of bulk–like structure with volume $V \sim R^3$. Then $N \sim (k_F R)^3 \gg 1$ where k_F is the Fermi wave vector. For non-degenerate states the average energy-level spacing δ around the Fermi level ϵ_F is given by

$$\delta = \frac{1}{N(0)V} \sim \frac{\epsilon_F}{N} \sim \frac{\hbar v_F}{R} \cdot \frac{1}{(k_F R)^2} \sim \frac{\hbar v_F}{R} N^{-2/3}, \tag{2}$$

with $N(0)$ being the density of states per unit volume at the Fermi energy.

Conventional superconductors with a bulk gap Δ_0 have small $T_C \sim \Delta_0$ and a large pair coherence length $\zeta \sim \hbar v_F/\Delta_0$. In units of the effective particle radius R, the coherence length becomes with Eq. (2)

$$\frac{\zeta}{R} \sim \frac{\delta}{\Delta_0} N^{2/3}. \tag{3}$$

Of course, in the bulk limit both sides of this simple relation will vanish since $\delta \sim N^{-1}$. For an appropriate finite value of the dimensionless size parameter δ/Δ_0, however, it follows that the particle radius is very small compared with the coherence length due to $N^{2/3} \gg 1$ (zero-dimensional superconductor). Spatial variations of the order parameter (if it exists at all) will then be suppressed, and it is very plausible to use a reduced BCS Hamiltonian with discrete single-electron states $|ns\rangle$ and energies ϵ_n as a microscopic model:

$$H = \sum_{ns=\pm} (\epsilon_n - s\mu_B H)\, c^\dagger_{ns} c_{ns} - g\delta \sum_{mn}{}' \, c^\dagger_{m+} c^\dagger_{m-} c_{n-} c_{n+} \,. \tag{4}$$

Here $|n+\rangle$ is time reversed to $|n-\rangle$. The pairing interaction between the states n, m is characterized by a dimensionless coupling constant g, and the prime on the sum indicates a cutoff of states outside an energy shell $\pm\omega_D$ about the Fermi energy. A Zeeman term is included in the single-electron energy, neglecting spin-orbit coupling. Note that Eq. (4) contains the usual BCS model for the bulk by putting $|n+\rangle = |\vec{k}\uparrow\rangle$, $|n-\rangle = |-\vec{k}\downarrow\rangle$.

In the early days, the model (4) was used to study quantum size effects for thermodynamic properties of ensembles of both normal (g = 0) and superconducting small particles. The results are discussed in several review articles.[4–6] More recently, especially after tunneling spectroscopy experiments on *single* grains became possible,[7] the interest in these problems has been greatly revived. For one-electron tunneling into a small particle it matters whether N is even or odd (parity effect). Moreover, for very small grains one should work in the canonical ensemble as was first pointed out by Kubo. This is possible for normal particles[6] but difficult for superconducting grains in the whole temperature regime.

An interesting quantity is the spin susceptibility χ which for normal grains deviates from the Pauli value $\chi_P = 2\mu_B^2/\delta$ and shows marked parity (even/odd) effects.[6] Very recently a calculation of χ for superconducting grains within a parity-projected scheme was performed by di Lorenzo *et al.*[8] With

$$Z = \mathrm{Tr}\, e^{-\beta(H-\mu N)} = \sum_{N \mathrm{even}} Z(N) e^{\beta\mu N} + \sum_{N \mathrm{odd}} Z(N) e^{\beta\mu N} = Z_{\mathrm{even}} + Z_{\mathrm{odd}}, \tag{5}$$

the susceptibility becomes

$$\chi_{\mathrm{even/odd}} = T \frac{\partial^2 \log Z_{\mathrm{even/odd}}}{\partial H^2} \,. \tag{6}$$

The actual calculation employs a functional integral representation of the partition function for which we refer to the original paper. The temperature behavior of the susceptibility for N being odd is most interesting and is shown in Fig. 4. One observes (so far only theoretically) a reentrance phenomenon: With decreasing temperature the system will undergo a "transition" from normal to superconducting, which is broadly smeared out due to large thermal fluctuations of the order parameter. When the temperature is lowered down further, quantum fluctuations will break the pairing influence and drive the system normal again with a Curie behavior for $T \to 0$.

At this point it is appropriate to make a remark about $T = 0$. We should remember that pairing was introduced in nuclear physics by Racah way back in the forties of the last century.[9] Similarities between nuclei and small-particle superconductors have seldom been exploited in the past for the simple reason that quantum size effects in thermodynamic properties of small metal particles are relevant mainly at finite temperatures. This has, however, changed most recently. There is a number of theoretical

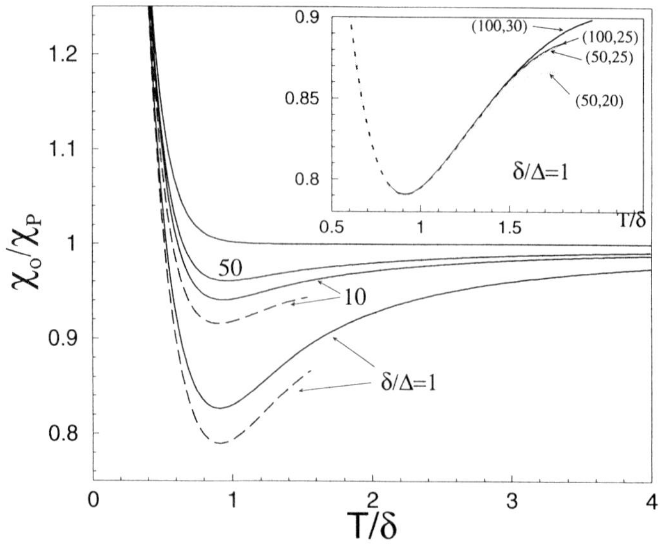

Figure 4. Reentrant spin susceptibility for N odd and $\delta/\Delta_0 \geq 1$. From Ref. 8.

contributions concerning the mathematical structure of the ground state of the Hamiltonian (4) and the effect of pair correlations. For a comparison of an exact solution with various other methods we refer here to the work of von Delft and Braun.[10]

IV. Further Remarks

The model calculations of small Hubbard clusters described in Sec. 2 show very clearly the strong interplay between electronic correlations, geometrical structure and magnetism. These calculations are exact but, unfortunately, it appears to be impossible to extend them to larger systems with the same generality and precision due to formidable calculation problems. An extension to larger N of order 100 using additional approximations would be highly desirable in view of experiments on cluster beams.[11] These experiments show how magnetism develops as the size of iron, nickel and cobalt clusters is increased up to several hundreds of atoms.

The idealized model of Hubbard clusters allows us in principle to take N as a *physical variable* on which magnetism will depend. This feature is more or less lost if we turn to more realistic treatments of magnetic materials which deal with different bond lengths, bond-angle effects, real d-band structure, etc. Some features of a given structure can be handled by a mean-field approach like the spin-density functional method. A discussion of such different aspects can be found in a recent review paper by Pastor and Bennemann on magnetic properties of transition metal clusters.[12]

In Sec. 3 we have mentioned the usefulness of a BCS-type model for both grains and nuclei. One should however keep in mind that pairing in both cases is of a very different origin. The many-nucleon problem with attractive nuclear forces is very well approximated by the nuclear shell model, and the pairing forces cover the remaining interactions. In a superconductor, in contrast, the pairing among electrons arises from phonon exchange. This is well established in the bulk, but when we go to very small grains Eq. (4) is solely a nice model. With other words, the coupling constant g and the cutoff ω_D, which will certainly be affected by size reduction (and also by the surroundings of the grain), are taken as adjustable parameters since we have no theory of their size dependence.

Let us add another side remark. Our purpose in Sec. 2 and 3 was to discuss the very different phenomena of magnetism and superconductivity by introducing appropriate interactions. In Sec. 3 also the $g = 0$ case of normal particles (with still bulk-like structure) was mentioned. What happens now when we also decrease U in Sec. 2 and arrive at a "normal" atom cluster? We obtain of course the same electronic system for different size: N big and N small. If the ions are smeared to a jellium sphere of radius $R \sim N^{1/3}$, one has an *electronic* shell model of an atom cluster. This shell model has turned out to be quite successful for electrons in clusters of *simple* metal atoms.[13]

At the end it should be emphasized that the grains and clusters considered here are free objects. This applies for clusters in a beam or clusters floating around, but in general the surroundings should be taken into consideration, especially for small sizes. In addition, there are a number of examples where granular or powdered small-particle structures and cluster matter exhibit interesting properties not present in the homogeneous state. This is certainly exciting also from a theoretical view point, but beyond the scope of the present short contribution.

References

1. J. Hubbard, *Proc. R. Soc. London Ser. A* **276**, 238 (1963); **281**, 401 (1964).
2. N. W. Ashcroft and N. D. Mermin, *Solid State Physics* (Holt, Rinehart, and Winston, New York, 1976).
3. G. M. Pastor, R. Hirsch, and B. Mühlschlegel, *Phys. Rev. B* **53**, 10382 (1996).
4. W. P. Halperin, *Rev. Mod. Phys.* **58**, 533 (1986).
5. S. Kobayashi, *Phase Trans.* **24-26**, 463 (1990).
6. B. Mühlschlegel, in "Percolation, localization and superconductivity", edited by A. M. Goldman and S. A. Wolf, *Nato Adv. Study Inst. Ser. B* **109**, (Plenum Press, New York 1984).
7. D. C. Ralph, C. T. Black, and M. Tinkham, *Phys. Rev. Lett.* **76**, 688 (1996); **78**, 4087 (1997).
8. A. di Lorenzo, Rosario Fazio, F. W. J. Hekking, G. Falci, A. Mastellone, and G. Giaquinta, *Phys. Rev. Lett.* **84**, 550 (2000); A. Maier and W. Zwerger, private communication.
9. A. L. Fetter and J. D. Walecka, *Quantum Theory of Many-Particle Systems* (McGraw Hill, New York, 1971); P. Ring and P. Schuck, *The Nuclear Many-Body Problem* (Springer, Berlin, Heidelberg, New York, 1980).
10. J. von Delft and F. Braun, *cond-mat*/9911058 4 Nov 1999.
11. I. M. L. Billas, A. Chatelain, and W. A. de Heer, *Science* **265**, 1682 (1994).
12. G. M. Pastor and K. H. Bennemann, in *Metal Clusters*, edited by W. Ekardt, (Wiley, New York, 1999), p. 211.
13. W. A. de Heer, *Rev. Mod. Phys.* **65**, 611 (1993).

Thermal Properties of Magnetic Clusters

J. L. Ricardo-Chávez, F. López-Urías, and G. M. Pastor

Laboratoire de Physique Quantique
UMR 5626 du CNRS
Université Paul Sabatier
118 route de Narbonne
F-31062 Toulouse
FRANCE

Abstract

The magnetic properties of clusters are investigated by applying exact diagonalization methods to the Hubbard and Anderson Hamiltonians. The average magnetic moment per atom $\langle S \rangle$ and the spin correlation functions $\langle \mathbf{S}_i \cdot \mathbf{S}_j \rangle$ of ferromagnetic clusters are determined as a function of temperature T in the weakly, intermediate, and strongly correlated regimes. The effect of electronic spin excitations and structural changes on the cluster magnetization are quantified. Properties of a single magnetic impurity in finite non-magnetic clusters are studied. The temperature dependences of the specific heat C_p and the magnetic susceptibility χ_f of the impurity are interpreted as the finite-size equivalent of the Kondo effect. The Kondo temperature of the cluster is identified and its behavior as a function of size and conduction-band filling is discussed.

I. Introduction

Small clusters show specific magnetic properties that distinguish them from solids, surfaces, and thin films. In past years, cluster magnetism has been the subject of a remarkable research activity.[1,2] Stern-Gerlach deflection experiments in ferromagnetic transition-metal (TM) clusters such as Fe_N, Co_N and Ni_N reveal that at low temperatures the average magnetic moments per atom $\bar{\mu}_N$ are enhanced with respect to the corresponding bulk magnetizations, showing some oscillations as a function of cluster size N. In addition, the temperature dependence of $\bar{\mu}_N$ derived from experiment is far from simple and depends qualitatively on the considered TM.[1] Mean-field ground-state calculations have been quite successful in reproducing, and sometimes even in predicting, the experimental observations at low temperatures.[2] However, very little is still known about the theory at finite temperatures. The problem remains a

Physics of Low Dimensional Systems
Edited by J. L. Morán-López, Kluwer Academic/Plenum Publishers, New York 2001

considerable challenge, particularly due to the strong interdependence of electron correlations, magnetism, and cluster structure. On the one side, as in solids, an accurate treatment of electron correlations is required in order to determine the electronic excitations, leading to spin fluctuations and to changes in the total spin. On the other side, one should also take into account the possibility of temperature-induced changes or fluctuations of the cluster geometry, on which the magnetic properties of itinerant electrons are known to depend strongly. In Sec. II we review some recent investigations of the finite-temperature magnetic properties of clusters by considering electronic and structural degrees of freedom on the same footing.

While numerous studies have been focused on itinerant-electron TM magnetism,[1–3] very little is known at present about clusters involving rare-earth atoms, where the magnetic moments have a strongly localized character.[4–6] In solids, the rare-earth compounds show remarkable many-body phenomena that are intrinsically related to the localized f electrons and to their interactions with the conduction band.[7–9] It would be therefore very interesting, from the point of view of cluster physics, to understand how these interactions are affected by the finite size of the system and how the resulting magnetic properties depend on the structure and composition of the cluster.

A single rare-earth impurity in a metallic cluster is probably the simplest physical situation for studying valence fluctuations and magnetic screening in finite systems. In solids the Kondo effect is characterized by the formation of a singlet ground-state between the localized magnetic moment of the impurity and the conduction-electron spins.[8,9] At low temperatures the local moment is masked by the conduction electrons, which build a spin-polarization cloud around the impurity. At temperatures higher than the Kondo temperature T_K the singlet state is broken and the local moment may be observed directly in a magnetic susceptibility measurement. $k_B T_K$ is of the order of the singlet-triplet gap ΔE which measures the binding energy of the Kondo singlet. Physically, one expects to observe cluster-specific properties when the radius of the cluster is comparable or smaller than the spatial extension ξ_K of the screening cloud (Kondo length). Moreover, interesting size and structural effects should be observed, since ΔE depends crucially on the hybridizations between the impurity and the size-sensitive delocalized valence states at the Fermi energy. In Sec. III we discuss a few representative results on magnetic impurities in small clusters, that were obtained using the Anderson impurity model[10] and exact diagonalization methods. A more detailed account of our investigations may be found elsewhere.[11]

II. Ferromagnetic Clusters

In order to achieve rigorous conclusions on the temperature dependence of the magnetic properties of ferromagnetic clusters we consider the Hubbard Hamiltonian, which in the usual notation is given by

$$H = -t \sum_{\substack{\langle i \neq j \rangle \\ \sigma}} \hat{c}^{\dagger}_{i\sigma} \hat{c}_{j\sigma} + U \sum_{i} \hat{n}_{i\downarrow} \hat{n}_{i\uparrow} \,. \tag{1}$$

The first term is the kinetic energy operator, that describes electronic hopping between nearest neighbors (NN). The hopping integrals are taken to be negative ($t > 0$). This corresponds to s-like local orbitals, and yields for compact structures a higher density of states for positive energies as observed in the TM d-series.[12] The second term takes into account the intra-atomic Coulomb repulsion which is the dominant contribution from the electron-electron interaction ($U \geq 0$). The magnetic properties of clusters have already been investigated in several papers using this model.[13–16] These studies include an analytic solution of the complete spectrum for a four-site tetrahedral cluster,[13] numerical studies at $T = 0$ for a few assumed structures,[14,15] and exhaustive geometry optimizations at $T = 0$ for $N \leq 8$.[16] Finite temperature properties have been determined by Callaway, Chen, and Tang for $N = 6$ atoms by considering a few representative structures.[15] In all these works[14–16] the electron correlation problem was solved exactly. However, to our knowledge, no true geometry optimization at $T = 0$ or sampling at finite T has been performed so far.[17]

The Hubbard model for small clusters is solved numerically by expanding its eigenfunctions $|\Psi_l\rangle$ in a complete set of basis states $|\Phi_m\rangle$ that have definite occupation numbers $n_{i\sigma}^m$ at all orbitals $i\sigma$, *i.e.*, $\hat{n}_{i\sigma}|\Phi_m\rangle = n_{i\sigma}^m|\Phi_m\rangle$ with $n_{i\sigma}^m = 0$ or 1. The values of $n_{i\sigma}^m$ satisfy the usual conservation of the number of electrons $\nu = \nu_\uparrow + \nu_\downarrow$ and of the z component of the total spin $S_z = (\nu_\uparrow - \nu_\downarrow)/2$, where $\nu_\sigma = \sum_i n_{i\sigma}^m$. The expansion coefficients α_{lm} of $|\Psi_l\rangle = \sum_m \alpha_{lm}|\Phi_m\rangle$ are determined by standard numerical diagonalization procedures.[16] In this way both ground-state and excited-state properties are obtained exactly in the framework of the model.[3]

The thermal properties are derived from the canonical partition function Z over electronic and structural degrees of freedom. Z is given by

$$Z = \sum_g \sum_l e^{-\beta \epsilon_l(g)}, \tag{2}$$

where $\epsilon_l(g)$ is the lth eigenenergy corresponding to a cluster structure or geometry g. T stands for the temperature of the cluster source that defines the macroscopic thermal bath with which the small clusters are in equilibrium before expansion. Thermal average refers then to the ensemble of clusters in the beam. Moreover, keeping ν and N fixed (canonical ensemble) corresponds to the experimental situation in charge- and size-selected beams.[1] In order to sample all relevant cluster geometries we note that in the Hubbard model only a single s-like orbital per site and NN hoppings are taken into account. In this case the hopping integrals t_{ij} between sites i and j take only two possible values, namely, $t_{ij} = t$ if $R_{ij} = R_0$ and $t_{ij} = 0$ if $R_{ij} > R_0$, where R_{ij} refers to the inter-atomic distance and R_0 to the NN distance. Thus, only the topological aspect of the structure is relevant to the electronic properties. Taking into account NN hoppings with fixed bond lengths results in a discretization of the configurational space. The sampling of cluster geometries can then be performed within the graph space. Notice, however, that the number of site configurations increases extremely rapidly with N.[16] The physical properties are obtained from Z and its derivatives. In particular the average cluster spin $\langle S\rangle$ is obtained from $\langle S\rangle^2 + \langle S\rangle = \langle \hat{S}^2\rangle$, where

$$\langle \hat{S}^2\rangle = \frac{3}{Z} \sum_g \sum_l \langle \hat{S}_z^2\rangle e^{-\beta \epsilon_l(g)}. \tag{3}$$

In Fig. 1 the temperature dependence of the local moments $\langle \hat{S}_i^2\rangle$ and of the NN spin-correlation functions $\gamma_{ij} = \langle \mathbf{S}_i \cdot \mathbf{S}_j\rangle$ are shown for clusters having $N = 6$ atoms and $\nu = 7$ electrons. This band filling ($\nu = N + 1$) is particularly interesting since it is known to result in ferromagnetism for large U/t. The value of $U/t = 64$ is representative of the limit of strong electron correlation for which the ground state is ferromagnetic (FM) with saturated moment [$S = (N - 1)/2$ for $\nu = N + 1$]. The considered cluster geometry, shown in the inset, is the most stable one at $T = 0$. Here we assume for simplicity that the structure is independent of T. The effect of structural fluctuations shall be discussed in the following. One observes that $\gamma_{ij} > 0$

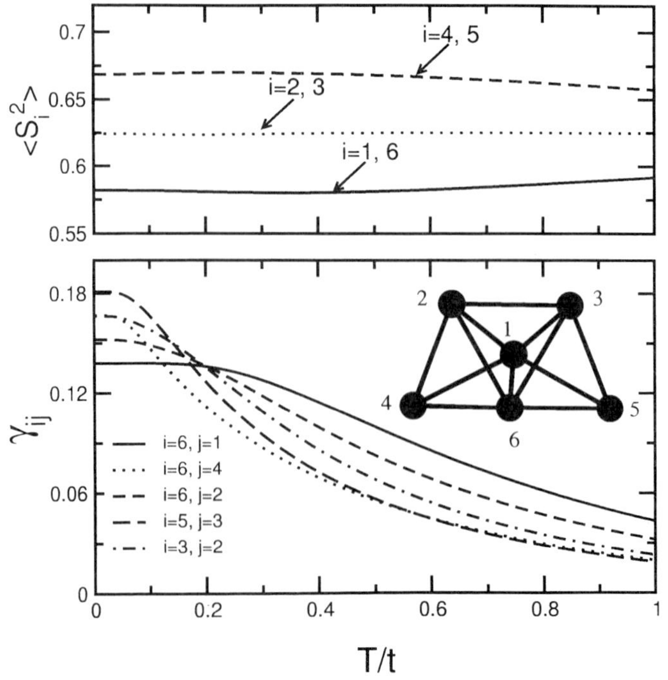

Figure 1. Temperature dependence of the local magnetic moments $\langle S_i^2 \rangle$ and spin correlation functions $\gamma_{ij} = \langle \mathbf{S_i} \cdot \mathbf{S_j} \rangle$ between different NN sites i and j in a cluster having $N = 6$ atoms, $\nu = 7$ electrons, and $U/t = 64$. The results correspond to the most stable structure at $T = 0$ as illustrated in the inset.

as corresponds to FM order. γ_{ij} decreases with increasing T and vanishes in the high-temperature limit. The crossover from the low-temperature ferromagnetically ordered state to the high-temperature disordered state is significantly broadened by the finite size of the cluster. However, a characteristic temperature scale $T_C(N)$ may be identified above which the FM short-range correlations are strongly reduced by thermal spin fluctuations. In contrast, the local moments $\langle \hat{S}_i^2 \rangle$ remain essentially unaffected well above T_C. Taking the inflection point in the $\gamma_{ij}(T)$ curves as a measure of $T_C(N)$ we obtain $T_C/t \approx 0.1$–0.2. This corresponds to about 1/100 of the bulk band width W_b or to 500–600 K if parameters appropriate for TM's are used ($W_b = 16t$ for the *fcc* structure and $W_b = 5$–6 eV for the *d*-band in bulk TM's). Around this temperature the total average magnetization $\langle S \rangle$ shows an inflection point and is significantly reduced with respect to the low-temperature value $S = 5/2$ (see Fig. 2 for $U/t = 64$). Furthermore, other physical properties show qualitative changes for $T \approx T_C = 0.15t$. For example, the magnetic susceptibility $\chi(T)$ has a Curie-Weiss type behavior of the form $\chi \propto 1/(T - T_C)^{-1}$ for $T > T_C$ and the specific heat $C_p(T)$ presents a peak due to magnetic excitations at about this temperature. These effects can be interpreted as precursors of a magnetic phase transition in the infinite solid.[18] $\langle \hat{S}_i^2 \rangle$ and $\gamma_{ij}(T)$ present a very rich environment dependence. The local moments decrease with increasing local coordination number z_i, as already observed in mean-field calculations.[2] For some bonds [*e.g.*, $(i,j) = (5,3)$] the spin correlations are largest at low T but then decrease quite rapidly with increasing T. This corresponds to bonds with the smallest local coordination numbers. On other bonds with larger z_i [*e.g.*, $(i,j) = (6,1)$]

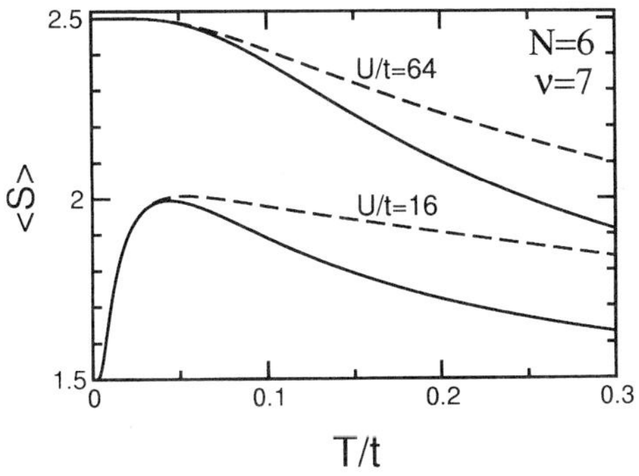

Figure 2. Temperature dependence of the average total spin $\langle S\rangle$ of clusters having $N = 6$ atoms and $\nu = 7$ electrons ($\langle S\rangle^2 + \langle S\rangle = \langle\hat{S}^2\rangle$). The results are obtained by taking into account either the electronic excitations alone (dashed lines), or both electronic and structural excitations (solid lines).

the short-range magnetic order resists better the effects of temperature fluctuations, despite the fact that the ground-state spin correlations are somewhat weaker.

Figure 2 shows the temperature dependence of the average total magnetic moment $\langle S\rangle$ for different values of U/t ($\langle S\rangle^2 + \langle S\rangle = \langle\hat{S}^2\rangle$). For large U/t (saturated FM ground state), $\langle S\rangle$ decreases monotonically with increasing T tending to a remnant value $\langle S\rangle_\infty > 0$ at high temperatures ($T > t$). $\langle S\rangle_\infty$ corresponds to an equally probable occupation of all electronic states within the first Hubbard band, *i.e.*, for a minimal number of double occupations ($T \ll U$). In the present case $\langle S\rangle_\infty = 1.5$. The classical analog of $\langle S\rangle_\infty$ is the average of N random spins which does not vanish in a finite system ($\langle\mu_\infty\rangle = \mu_0/\sqrt{N}$). The difference $\langle S\rangle - \langle S\rangle_\infty$ is a measure of the importance of short-range magnetic order.[19] For smaller values of U/t corresponding to a non saturated FM ground-state ($S = 3/2$ for $U/t = 16$) a completely different, non monotonous temperature dependence of $\langle S\rangle$ is found. For example, for $U/t = 16$ one observes first an increase of $\langle S\rangle$ with increasing T, for $T < 0.05t$, followed by a decrease toward the high temperature limit $\langle S\rangle_\infty$. The increase of $\langle S\rangle$ at low T results from populating low-lying excited states which have higher S than the ground state [$E(S = 5/2) - E(S = 3/2) = 0.02t$ for $U/t = 16$]. If the temperature is further increased, $\langle S\rangle$ decreases since the FM correlations are destroyed in a similar way as for $U/t = 64$ (see Fig. 1). It is interesting to observe that a weak increase of the average magnetization per atom with increasing T has been experimentally observed in large Co_N clusters ($N \simeq 50$–600 and $T \simeq 400$).[1] The temperature dependence of $\langle S\rangle$ in non saturated FM clusters is very sensitive to the value of U/t. If U/t is increased beyond $U/t = 16$, the $S = 5/2$ states are stabilized with respect to the quartet, the maximum in $\langle S\rangle(T)$ shifts to lower temperatures, and eventually the sextet becomes the ground state. In contrast, at smaller U/t (*e.g.*, $U/t = 12$) the excitation energy of high-spin states is too large and low-lying excitations with minimal S are found. Consequently, $\langle S\rangle$ decreases rapidly at low T showing eventually a minimum at intermediate T. The strong sensitivity of the spin excitation spectra on U/t reflects the importance of electron correlations in the finite temperature behavior.

The solid curves in Fig. 2 are obtained by taking into account temperature-induced changes of structure [see Eqs. (2) and (3)]. Structural fluctuations play no role at low temperatures where the ground-state structure dominates ($T/t < 0.05$). However, at higher T they contribute to a more rapid decrease of $\langle S\rangle$. In order to analyze this effect we have computed the temperature dependence of the average coordination number $\langle z\rangle$. With increasing T, $\langle z\rangle$ decreases since breaking NN bonds becomes increasingly probable. The reduction of $\langle z\rangle$ is more important for $U/t = 16$ than for $U/t = 64$, since ferromagnetism tends to stabilize the more compact structures.[16] Consequently, the effect of structural fluctuations on $\langle S\rangle$ is somewhat stronger for moderate U/t, where the ground-state magnetic moments are not saturated. The reduction of $\langle S\rangle$ as a result of structural fluctuations may be qualitatively understood as follows. In open structures the bandwidth for bonding (negative energy) states is smaller. Therefore, the eigenstates with maximal spin—which energy is dominated by the minority electrons [$\nu = N+1$, $S_{\max} = (N-1)/2$]—are comparatively less stable than lower spin states. In other words, the spin-flip energy $\Delta E = E(S_{\max}-1) - E(S_{\max})$ tends to decrease with decreasing coordination number. Thus, $\langle S\rangle(T)$ is reduced as more open structures are populated. Notice, however, that more complex structural effects are expected for other ν/N and U/t. For example, one may find structures having a higher energy *and* a higher ground-state spin S than the optimal structure at $T = 0$.[16] In such cases taking into account structural fluctuations would tend to increase $\langle S\rangle(T)$.[18]

Summarizing, we have shown that thermal induced structural fluctuations often play an important role in the temperature dependence of cluster magnetic properties. In addition, at low temperatures the clusters may show an increase of the average magnetic moment $\langle S\rangle$ with increasing T. More systematic studies will follow in order to extend and further develop the results obtained in this work.

III. Magnetic Impurities in Clusters

In order to study valence fluctuations and magnetic screening in finite systems we consider a N-atom cluster that contains $N-1$ simple-metal atoms and one magnetic impurity. The electronic model Hamiltonian is given by

$$H = H_s + H_f + H_{sf}. \tag{4}$$

The first term,

$$H_s = -t_{ss} \sum_{\langle i,j\rangle,\sigma} c^\dagger_{i\sigma} c_{j\sigma}, \tag{5}$$

describes the delocalized valence electrons of the simple metal and of the impurity atom by using a single-band tight-binding model. As usual, $c^\dagger_{i\sigma}$ ($c_{i\sigma}$) refers to the creation (annihilation) operator of an electron with spin σ at the s orbital of atom i, and t_{ss} to the nearest-neighbor (NN) s-electron hopping integral. Notice that we disregard possible differences between the s levels ε_s in different atoms ($\varepsilon_s = 0$). The second term,

$$H_f = \varepsilon_f \sum_\sigma n_{f\sigma} + U_{ff}\, n_{f\uparrow} n_{f\downarrow}, \tag{6}$$

concerns the degrees of freedom of the impurity.[10] Here $n_{f\sigma} = f^\dagger_\sigma f_\sigma$ is the electron number operator, ε_f the energy, and U_{ff} the Coulomb-repulsion integral of the localized f-like level. Orbital degeneracies are neglected for simplicity. Finally, the third term,

$$H_{sf} = V_{sf} \sum_\sigma \left(f^\dagger_\sigma c_{0\sigma} + c^\dagger_{0\sigma} f_\sigma \right), \tag{7}$$

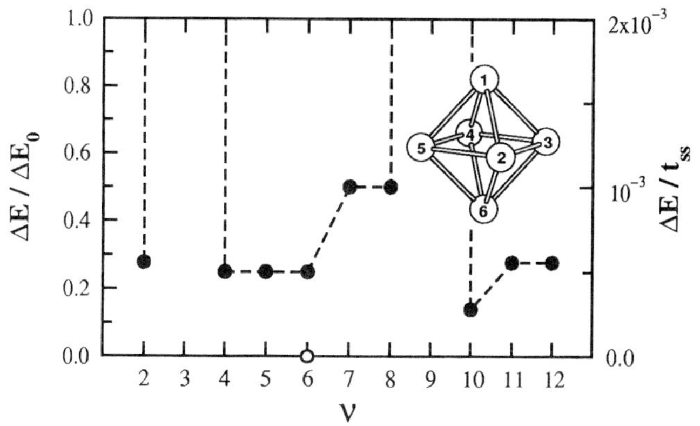

Figure 3. Single-triplet gap ΔE of a 6-atom octahedral cluster with a magnetic impurity in an apical site (*e.g.*, $i = 1$) as a function of the total number of electrons ν. For $\nu = 3$, $\Delta E = 4t_{ss} = 2 \times 10^3 \Delta E_0$, and for $\nu = 9$, $\Delta E = 2t_{ss} = 10^3 \Delta E_0$ ($\Delta E_0 = 2V_{sf}^2/|\varepsilon_f|$, see text).

takes into account the coupling between the f level and the delocalized electrons by means of an intra-atomic sf hybridization at the impurity atom $i = 0$. H may be rewritten in terms of the single-particle eigenstates $a_k^\dagger = \sum_i \alpha_{ik} c_i^\dagger$ of H_s as

$$H = \sum_{k\sigma} \varepsilon_k n_{k\sigma} + \sum_{\sigma} \varepsilon_f n_{f\sigma} + U_{ff} n_{f\uparrow} n_{f\downarrow} + \sum_{k\sigma} V_{kf} \left(f_\sigma^\dagger a_{k\sigma} + a_{k\sigma}^\dagger f_\sigma \right). \quad (8)$$

The s-electron eigenenergies are denoted by ε_k and the kf hybridizations by $V_{kf} = \alpha_{0k} V_{sf}$. Notice that ε_k and V_{kf} are very sensitive to the size and structure of the cluster. The low-temperature magnetic behavior depends crucially on them, and on the band filling or number of electrons ν.

As in the case of the Hubbard Hamiltonian, the Anderson model is solved numerically by expanding its eigenfunctions $|\Psi_l\rangle$ in a complete set of basis states $|\Phi_m\rangle$ which have definite occupation numbers $n_{i\sigma}^m$ at all orbitals $i\sigma$. Ground-state and excited-state properties are calculated exactly by using standard diagonalization procedures.[16] In particular, the spin gap is given by

$$\Delta E = E(S = 1, \tfrac{3}{2}) - E(S = 0, \tfrac{1}{2}), \quad (9)$$

where $E(S)$ stands for the lowest eigenenergy of spin S. Relevant finite-temperature properties are the magnetic impurity susceptibility

$$\chi_f = \frac{\langle S_{zf}^2 \rangle - \langle S_{zf} \rangle^2}{k_B T}, \quad (10)$$

and the specific heat

$$C_p = \frac{\langle E^2 \rangle - \langle E \rangle^2}{k_B T^2}, \quad (11)$$

in the canonical ensemble.[18]

For the calculations we consider parameters representative of the Kondo limit: $\varepsilon_f / t_{ss} = -10, V_{sf}/t_{ss} = -0.1$ and $U_{ff} \to +\infty$. In Fig. 3 results are given for the band-filling dependence of the spin gap ΔE in a 6-atom octahedral cluster with the magnetic impurity at the apical site $i = 1$ (see the inset of Fig. 3). ΔE presents remarkable oscillations as a function of ν, with maxima and minima at specific band

Table I

Single-particle energies ε_k, local component α_{0k} of state k at the impurity site, and kf hybridizations V_{kf} for an octahedral 6-atom cluster ($V_{sf}/t_{ss} = 0.1$).

k	ε_k/t_{ss}	α_{0k}^2	$(V_{kf}/t_{ss})^2$
1	−4.0	1/6	0.00167
2	0.0	1/2	0.00500
3	0.0	0	0.00000
4	0.0	0	0.00000
5	2.0	1/3	0.00333
6	2.0	0	0.00000

fillings. For most ν the ground state is a singlet, and ΔE is of the order of the singlet-triplet gap ΔE_0 in the two level system ($\Delta E_0 = 2V_{sf}^2/|\varepsilon_f|$).[9] In one case ($\nu = 6$, indicated by the open circle) we obtain $\Delta E = 0$. Here, the lowest non-vanishing spin gap is reported. In a few cases ($\nu = 3$ and $\nu = 9$ in the present cluster) ΔE is of the order of the s-electron hopping integral t_{ss}, which indicates that the excitation is related to the formation of an electron-hole pair in the discrete s-electron spectrum ε_k $[(\varepsilon_k - \varepsilon_{k'}) \sim t_{ss}]$. This contrasts with the much smaller values of ΔE found otherwise, and which result from the excitation of spin degrees of freedom without involving charge excitations.[9]

The band filling dependence of ΔE may be qualitatively understood in terms of the cluster specific single-particle spectrum ε_k, and kf hybridizations V_{kf} shown in Table I. First of all, note that in the considered parameter regime (Kondo limit) the number of electrons in the f level is always very close to 1 ($\nu - 1$ electrons in delocalized s-states). For $\nu = 2$ the low-lying states of the cluster are very accurately described by considering, besides the f level, only the lowest s-electron state denoted by $k = 1$ in Table I. This is a consequence of the large gaps in the conduction-electron spectrum ($\varepsilon_2 - \varepsilon_1 = 4t_{ss}$). The low-lying eigenstates of H are accurately given by the singlet and triplet states that can be formed between the spins of the impurity and of a delocalized s-electron in the state $k = 1$.[8] This is the simplest form of the Kondo screening.[9] In fact ΔE is very close to the two-level result $\Delta E' = 2V_{1f}^2/(\varepsilon_1 - \varepsilon_f)$ $[|\Delta E - \Delta E'| \simeq 10^{-4}\Delta E_0]$. For $\nu = 3$ (*e.g.*, $\nu_\uparrow = 2$ and $\nu_\downarrow = 1$) the situation is quite different. Now the $k = 1$ state is full and therefore the kf hybridization may only promote the f-electrons to higher empty states ($k = 2$, for example). There are no low-lying spin excitations. Consequently, the spin at the f level remains essentially frozen ($S = 1/2$), and the first $S = 3/2$ state corresponds to the creation of an electron-hole pair in the s spectrum $[\Delta E \sim (\varepsilon_2 - \varepsilon_1) = 4t_{ss}]$. For $\nu = 4$–8 the 3-fold degenerate levels $k = 2$–4 are partially occupied. Up and down impurity spin configurations are mixed by a process which is second order in V_{kf}, and we recover a Kondo-like ordering of levels with ΔE of the order of the two-level spin gap ΔE_0. A situation similar to the $\nu = 3$ case is observed for $\nu = 9$ when the three-fold degenerate level is completely filled. The degeneracy of the single-particle spectrum and the fact that two of the s levels do not hybridize with the localized orbital ($V_{kf} = 0$ for $k = 3$ and 4, see Table I) lead to the same ΔE for $\nu = 4$ and 5, and for $\nu = 7$ and 8. This is also responsible for the presence of a ground-state spin degeneracy ($\Delta E = 0$) for $\nu = 6$. The step in

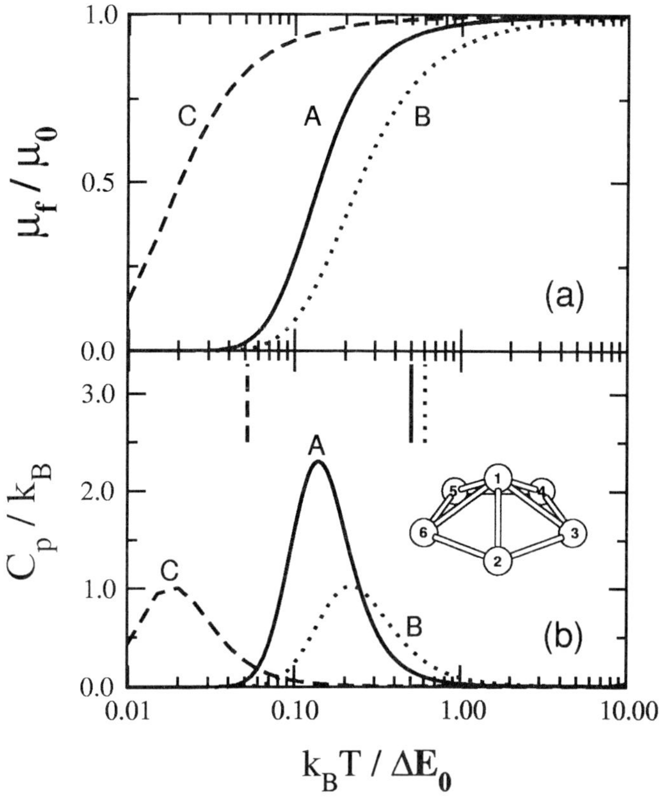

Figure 4. Temperature dependent properties of a single-impurity cluster having $N = 6$ atoms and $\nu = 8$ electrons: (a) effective impurity moment $\mu_f^2 = 3k_BT\chi_f$ in units of the ionic value μ_0, and (b) specific heat C_p. Results for different cluster geometries are shown: (A) octahedral structure with the impurity in an apical site, (B) pentagonal pyramid with apical impurity, and (C) pentagonal pyramid with the impurity atom in the pentagonal ring. The vertical lines indicate the corresponding singlet-triplet energies ΔE ($\Delta E_0 = 2V_{sf}^2/|\varepsilon_f|$, see Fig. 3).

ΔE from $\nu = 4$–6 to $\nu = 7$–8 is a consequence of the larger occupation of the s levels $k = 3$ and 4 as ν increases. The same effect is observed between $\nu = 10$ and 11.[18]

In Fig. 4 results are given for the effective impurity moment μ_f^2, as derived from the magnetic susceptibility χ_f, and for the specific heat C_p as a function of T. These are representative of the finite-temperature behavior that follows from the previously discussed low-energy Kondo-like spin excitations. $\chi_f(T)$ presents a maximum at a temperature of the order of $T_K \simeq \Delta E/k_B$ and a Curie-like decrease, $\chi_f \propto 1/T$, for $T > \Delta E/k_B$. This corresponds to the increase and saturation of the effective moment μ_f as the Kondo singlet is broken by thermal fluctuations (see Fig. 4). In addition, the specific heat $C_p(T)$ shows a maximum for $T \simeq \Delta E/2k_B$ that is also associated to the singlet-triplet excitation. We observe that the positions of the peaks in $C_p(T)$ and $\chi(T)$ scale with ΔE and that they are very sensitive to the cluster structure and to the location of the impurity within the cluster (note the logarithmic temperature scale in Fig. 4). For higher temperatures an exponential increase of C_p is obtained as electron-hole conduction-band excitations are populated.

More systematic studies should follow in order to extend the results presented in this paper. For example, preliminary calculations of ΔE on larger clusters show remarkable dependences as a function of cluster size and structure. Furthermore, the intermediate-valence regime with higher T_K's and the role of orbital degeneracies are also of considerable interest.

Acknowledgments

Two of the authors (F.L.U. and J.L.R.C.) acknowledge support from CONACyT (México). Computer resources were provided by grants from IDRIS (CNRS) and CALMIP (Toulouse, France).

References

1. I. M. L. Billas, A. Châtelain, and W. A. de Heer, *Science* **265**, 1682 (1994); S. E. Apsel, J. W. Emert, J. Deng, and L. A. Bloomfield, *Phys. Rev. Lett.* **76**, 1441 (1996).
2. For a recent review see, *e.g.*, G. M. Pastor and K. H. Bennemann, in *Metal Clusters*, edited by W. Ekardt (Wiley & Sons, New York, 1999), p. 211.
3. F. López-Urías and G. M. Pastor, *Phys. Rev. B* **59**, 5223 (1999).
4. D. C. Douglass, J. P. Bucher, and L. A. Bloomfield, *Phys. Rev. Lett.* **68**, 1774 (1992); D. P. Pappas, A. P. Popov, A. N. Anisimov, B. Reddy, and S. N. Khanna, *Phys. Rev. Lett.* **76**, 4332 (1996); D. Gerion, A. Hirt, and A. Châtelain, *Phys. Rev. Lett.* **83**, 532 (1999).
5. M. Lübcke, B. Sonntag, W. Niemann, and P. Rabe, *Phys. Rev. B* **34**, 5184 (1986); R.-J. Tarento and P. Joyes, *Phys. Rev. B* **41**, 4547 (1990); C. Bréchignac, Ph. Cahuzac, F. Carlier, and J. Ph. Roux, *Z. Phys. D* **28**, 67 (1993).
6. C.-S. Neumann and P. Fulde, *Z. Phys. B* **74**, 277 (1989); M. Koga, W. Liu, M. Dolg, and P. Fulde, *Phys. Rev. B* **57**, 10648 (1998).
7. *Valence Instabilities*, edited by P. Wachter and H. Boppart, (North-Holland, Amsterdam, 1982), and references therein.
8. C. M. Varma and Y. Yafet, *Phys. Rev. B* **13**, 2950 (1976).
9. P. Fulde, *Electron correlations in molecules and solids* (Springer, Berlin, 1993).
10. P. W. Anderson, *Phys. Rev.* **124**, 41 (1961).
11. J. L. Ricardo-Chávez and G. M. Pastor, to be published.
12. The electron-hole transformation $h^{\dagger}_{i\sigma} = c_{i\bar{\sigma}}$ leaves the Hamiltonian formally unchanged, except for an additive constant and a change of sign in the hopping integrals.
13. L. M. Falicov and R. H. Victora, *Phys. Rev. B* **30**, 1695 (1984).
14. Y. Ishii and S. Sugano, *J. Phys. Soc. Jpn.* **53**, 3895 (1984).
15. J. Callaway, D. P. Chen, and R. Tang, *Z. Phys D* **3**, 91 (1986); *Phys. Rev. B* **35**, 3705 (1987).
16. G. M. Pastor, R. Hirsch, and B. Mühlschlegel, *Phys. Rev. Lett.* **72**, 3879 (1994); *Phys. Rev. B* **53**, 10382 (1996).
17. F. López-Urías and G. M. Pastor, to be published.
18. Further details will be published elsewhere.
19. G. M. Pastor and J. Dorantes-Dávila, *Phys. Rev. B* **52**, 13799 (1995).

X-Ray Studies on Co Clusters

Giuseppe Faraci,[1] Agata R. Pennisi,[1] Antonella Balerna,[2] Hugo Pattyn,[3] Gerhard Koops,[3] and Guilin Zhang[4]

[1]*Dipartimento di Fisica - Universitá di Catania*
Istituto Nazionale di Fisica della Materia
Corso Italia 57, 95129 Catania
ITALY

[2]*Laboratori Nazionali di Frascati*
INFN - 00044 Frascati
ITALY

[3]*Instituut voor Kern- en Stralingsfysica*
Physics Department
KU Leuven, Celestijnenlaan 200D, B-3001
BELGIUM

[4]*Shanghai Institute of Nuclear Research*
Chinese Academy of Sciences Shanghai 201800
POPULAR REPUBLIC CHINA

Abstract

Co clusters, obtained in an Ag matrix by Molecular Beam Epitaxy and/or Ion implantation, were observed by Extended X-ray Absorption Fine Structure spectroscopy. In the as-grown samples, we were able to confine small aggregates of size going from 2 up to 20 atoms/cluster. After thermal treatment we observed a size increase of the clusters up to 80 and 140 atoms/cluster. In the present study each configuration was characterized by distinguishing Co-Co and Co-Ag coupling; then this method permitted the investigation of the interior and of the interface contribution. We detected a structural transition from *fcc* to *hcp* when the size of the cluster is larger than about 100 atoms. High stability and preferential agglomeration in dimer ensembles were observed at concentrations in the range 0.10–0.70 at. %. The Co dimers as well as the larger Co clusters were found somewhat contracted (of few hundredths of an Angstrom) with respect to the standard Co-Co bulk distance.

Physics of Low Dimensional Systems
Edited by J. L. Morán-López, Kluwer Academic/Plenum Publishers, New York 2001

I. Introduction

The confinement of a magnetic material in nanostructures and nanoparticles represents today a frontier field in advanced research.[1] In particular, Co clusters exhibit very interesting properties for several fundamental and applied aspects.[1–14] We can mention, *e.g.*, not only the magnetic anisotropies shown by monodispersed ultrafine Co agglomerates, attributed to surface contributions,[2] but also the increase of the magnetic moment per atom both with the applied magnetic field and with the cluster size, or the saturation ferromagnetism theoretically obtained for Ni_{13}, Co_{13} and Pd_{13}, although Rh_{13}, and Ru_{13} seem only partially magnetic.[3] An important role is played by the matrix in which the particles are embedded or deposited on, because of the cluster surface interaction and of the host polarization.[4–6] For granular and ultrathin Co layers grown by molecular beam epitaxy, superparamagnetic behaviour is observed[7] with hysteresis effects during the transition from granular to thick layers depending on the different growth modes in gold or copper. In silver, low concentration of Co introduced by ion implantation shows an initial stage of Co in substitutional and dimer form with subsequent segregation and precipitation.[8] Furthermore, photoelectron spectroscopy studies on Co^- clusters indicate a size dependent transition in the electronic structure at 7 atoms/cluster,[9] assuming a defined geometrical structure. Indeed, the geometry of these clusters is of dramatic importance; although bulk Co at RT has an *hcp* crystal structure, at about 700 K a phase transition to *fcc* occurs; also Co clusters seem to exhibit *fcc* structure, but other possible forms such as truncated octahedra,[10] icosahedra,[11] or chainlike agglomerates[12] have been reported. The formation and evolution of Co dimers is still under debate since it can be strongly dependent on the preparation details such as the implantation dose, the annealing temperature and/or the matrix configuration.[13] As a matter of fact an intense Co-Co interaction should produce Co_2 molecules[14] with a mutual distance of 0.2385 nm much lower than the Co-Co nn distance in the bulk (0.2507 nm).

On this subject many details have still to be clarified and in particular we indicate those related to:

(i) the size and geometrical configuration of the clusters;[11]

(ii) the interaction with the matrix, *i.e.*, the influence of the Debye temperature Θ_D of the matrix on the dynamical behaviour of the clusters and their growth under thermal treatment.

Therefore, the aim of this study is the experimental characterization of Co clusters by Extended X-ray Absorption Fine Structure (EXAFS) spectroscopy, and X-ray Absorption Near Edge Structure (XANES) spectroscopy. These techniques allow the investigation of the geometrical configuration and of the vibrational behaviour of some coordination shells around a Co absorber, permitting to distinguish whether Co faces another cobalt or an interface silver atom; actually, the nearest-neighbour distance, the coordination number, the Debye-Waller factors, and the chemical shifts can be obtained. A complementary investigation of the same samples by Mössbauer spectroscopy will be the subject of a forthcoming paper.

In order to obtain as much as possible a very low size dispersion, several methods were employed including ion implantation and molecular beam epitaxy. Thermal treatments were also performed for producing larger clusters from the initial dimer ensemble. As a host matrix of the Co clusters we chose silver for several reasons:

(i) Co is scarsely soluble in silver and tends to occupy substitutional positions in the Ag lattice;

(ii) At room temperature or lower the Co mobility in Ag is quite low;

(iii) Silver is a rather soft matrix ($\Theta_D^{\mathrm{Ag}} = 225$ K at low temperature) whereas Co is a hard metal at least in its bulk form ($\Theta_D^{\mathrm{Co}} = 440$ K at low temperature); then the question arises whether and how the vibrational behaviour of Co in a cluster configuration can be influenced by the host matrix;

(iv) Many of the relevant above mentioned properties have been attributed to the cluster surface and in the present investigation the interface Co-Ag can be distinguished from the Co-Co interior contributions owing to their different mutual distance and vibrational amplitude.

II. Experiment

In order to introduce clusters in an Ag matrix, we followed three experimental methods:

(i) Ag *e*-gun evaporation in a vacuum chamber at a base pressure of 2×10^{-8} mbar on a Si substrate, previously thermally oxidized over a thickness of 80 nm, concurrently with the ion implantation of Co in the growing Ag layer, with the respective rates adjusted so as to obtain a Co concentration of 1.9 at. %, more or less homogenously distributed over a thickness of about 400 nm (sample A).

(ii) RT implantation at 50 keV of Co in Ag layers, grown in a molecular beam epitaxy (MBE) apparatus on a Si substrate, previously thermally oxidized over a thickness of 80 nm. We repeated 15 times the growth of a Ag layer and the subsequent implantation in order to increase the expected signal to noise ratio. With an optimised Ag layer thickness, we thus could prepare two samples with a rather homogeneous Co depth profile over about 800 nm, one with 0.7 at. % (sample B), the other with 0.1 at. % (sample C).

(iii) MBE co-evaporation of Ag and Co on a Si substrate, previously thermally oxidized over a thickness of 80 nm. The rates were chosen so as to obtain a Co concentration of 6.0 at. % (sample D).

The sample with the lowest concentration, sample C, has a very high Co in Ag dilution, roughly one order of magnitude less than what would give rise to 1 Co atom per 5 Ag shells, if the Co were homogeneously introduced. On the basis of a previous study[15] we selected these concentrations of B and C in order to maximise the population of the primary stage of Co cluster nucleation, namely the Co-dimer.

The experiment was performed at the GILDA beamline of the European Synchrotron Radiation Facility (ESRF), Grenoble, France; the radiation was monochromated and horizontally focused by a double crystal Si(311) monochromator; two Pd mirrors were used to reject the higher harmonics, and focused the beam in the vertical direction; the beam spot on the sample was 1×2 mm^2; the energy resolution was in the order of 10^{-5}. The as prepared samples were investigated by EXAFS and XANES spectroscopy; afterwards, some samples were annealed at 350 C for 30 min in a H_2-Ar atmosphere, and measured again. X-ray spectra were collected around the Co K edge (7709 eV), in the range 7500–8700 eV, and in fluorescence mode using 6 ultrapure Ge detectors cooled at 77 K. Some samples were investigated at different temperatures, namely 77, 148, and 218 K.

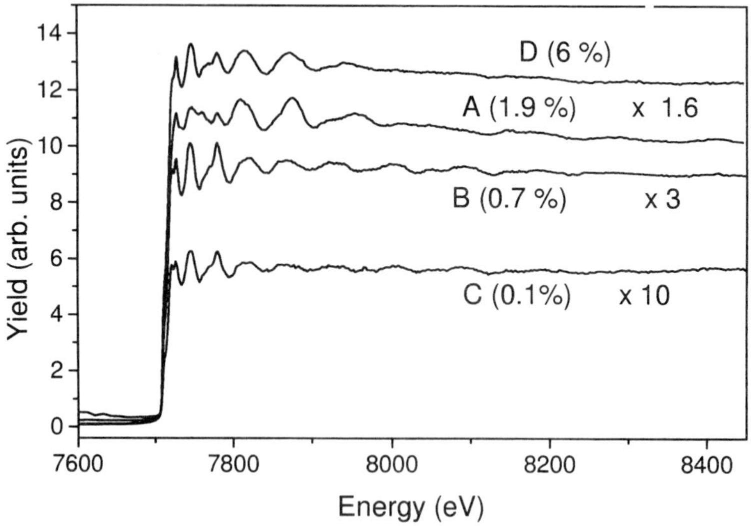

Figure 1. Raw fluorescence spectra for sample A, B, C, and D. Clearly visible the EXAFS oscillations after the threshold.

III. Results

The normalized fluorescence spectra I_{fl}/I_0 were analyzed according to a standard procedure:[16–18] the normalized EXAFS oscillations were extracted removing the background by means of a cubic spline, fitting the continuous component of the spectrum in the k range 2.5–13 Å^{-1}. The EXAFS spectrum $\chi(k)$ is defined as a function of the wavevector $k = \sqrt{2m(E - E_0)}/\hbar$:

$$\chi(k) = \frac{\mu - \mu_0}{\mu_0 d},$$

where $\mu_0 d = J(1 - 8k^2a^2/3)$ with $a = \hbar/[2mE_0]^{1/2}$, E_0 is the threshold energy and J the jump of the spectrum at the edge.

In fact, the theoretical description of an EXAFS spectrum is given by:

$$\chi(k) = \frac{\sum_j S_0^2 N_j F_j(k, R_j)}{kR_j^2} \exp\left[-2\sigma_j^2 k^2 - \frac{2R_j}{\lambda(k)}\right] \sin[2kR_j + \delta_j(k)], \quad (1)$$

where S_0^2 is the many body overlap term, $\lambda(k)$ the electron mean free path; the sum is over the j coordination shells put at a distance R_j from the absorbing atom; $F_j(k, R_j)$ is the atomic backscattering amplitude and δ_j the phase shift of the jth coordination shell, σ_j^2 the corresponding Debye-Waller (DW) factor.

III.1. EXAFS Spectra: As-Prepared Samples

First of all in Fig. 1 we report the raw X-ray spectra, as detected at 77 K, for the as-prepared samples A (1.9 at. %, implanted), B (0.7 at. %, implanted), C (0.1 at. %, implanted), and D (6 at. %, co-evaporated). Clearly visible the EXAFS oscillations after the Co threshold. Here, the jump J of each spectrum is proportional to the local concentration of Co absorbers seen by the X-ray beam; therefore, this jump acts as a

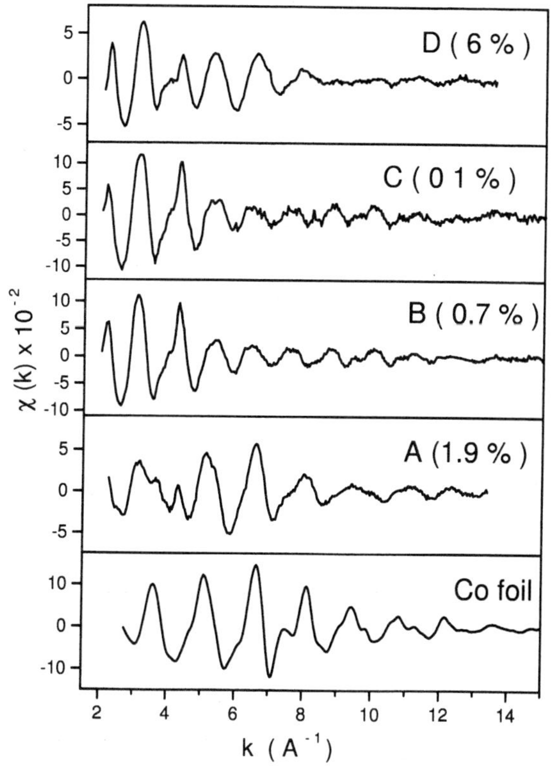

Figure 2. EXAFS oscillations *vs.* k, extracted after the threshold by subtracting the continuous contribution in the raw spectra. For comparison, the EXAFS spectrum of the Co foil is also reported.

test of the total impurity concentration introduced into the silver matrix; the values $J_A = 6850$, $J_B = 2960$, $J_C = 540$, and $J_D = 13\,100$, taking into account the sample thickness, are in agreement with the already mentioned percentages.

After normalization and subtraction of the background we can extract the EXAFS oscillations which are shown in Fig. 2 together with the reference EXAFS of a Co foil. These curves weighted by k^2 have been Fourier transformed: their peaks correspond to the first, second and successive coordination shells due to the local average configuration of each sample. For comparison the Fourier Transform (FT) of a Co metal foil is reported in Fig. 3 together with the FT's of the four samples. Taking into account the phase shift, the first peak of the FT for the reference Co foil corresponds to 12 Co nearest neighbours at 2.507 Å (the two subshells of the *hcp* lattice are not distinguishable, being only 0.01 Å apart), and the second one to 6 atoms at 3.54 Å. The FT plots of the Co/Ag samples show very interesting features: the first shell peak is now splitted in two components denoting that around a Co absorber there are n Co and 12-n Ag atoms, respectively at closer or larger distances. This is an evident symptom of clustering, since on the average a Co atom "sees" around itself partially atoms of the same species and partially atoms of the matrix; the relative weights depend on the size of the cluster and thus on the surface to volume ratio. In particular the FTs of sample B and C look very similar with a very reduced number of Co atoms in the first shell; in sample D the two peaks are of comparable intensity and in sample A

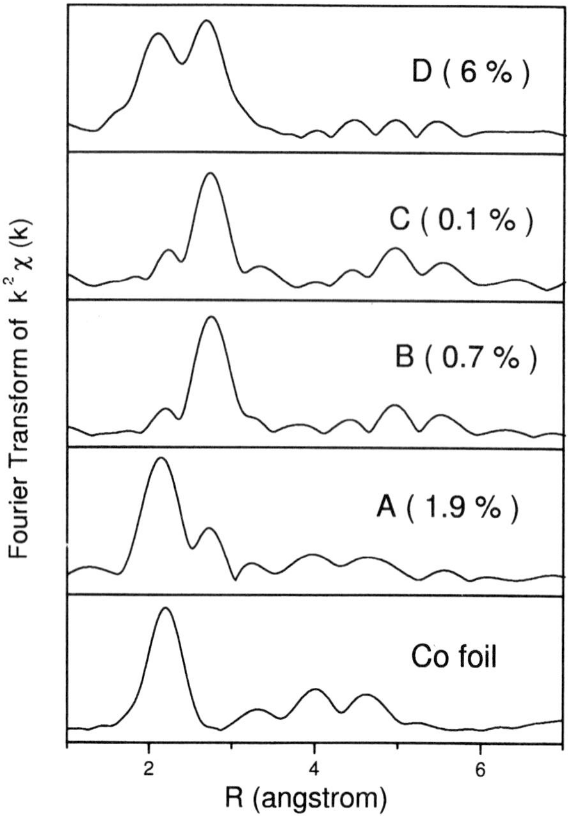

Figure 3. Fourier Transform of the EXAFS spectrum $\chi(k)$ weighted by k^2. In the lowest figure, the FT of the Co metal foil is also shown.

the feature ascribed to Co looks higher than the other one. A fitting procedure on the EXAFS signal, using a very accurate method,[19] includes curved wave effects, multiple scattering paths and inelastic losses. This fit gives the best values for the unknown parameters N, R, σ^2. We collected in Table I the coordination number N, the n-n distance, and the Debye-Waller factor σ^2 for the first shell around the Co absorber. As visible in the table, the first shell around the Co absorber is formed by about 1 Co atom plus 11 Ag atoms for sample B and C: this implies an ensemble of dimers, or perhaps trimers, tetramers, etc. Postponing the discussion about this point, we want to stress the similarity of the B's and C's behaviour although their concentrations differ of a factor 7; we believe that this result is possible only for a very stable configuration, *i.e.*, when the clustering is determined by a strong mutual interaction driving the agglomeration in the most stable (dimer or oligomer) ensemble. Note that in both samples the Co-Co nn distance is slightly contracted with respect to bulk Co, whereas the remaining part of the first shell is represented by Ag atoms which reside at a distance $R_{Co-Ag} = 2.82$ Å.

For sample A and D again we obtain a contracted Co-Co distance and a coordination reduced to 6.9 ± 0.2 and 4.5 ± 0.2, respectively. As we will show afterwards, these values correspond to Co clusters respectively of 20 atoms/cluster and 8 atoms/cluster; an interesting comment on the preparation method should be now pointed out: the

Table I

Fitting parameters of the first shell around a Co absorber obtained by the FEFFIT code; we can distinguish the Co-Co and the Co-Ag contribution. The uncertainty of the coordination number N is $\pm$ 0.2; of the nearest-neighbour distance R is ± 0.02 Å; of the relative squared vibrational amplitude is $\pm 5 \times 10^{-4}$ Å^2.

sample		Co – Co			Co – Ag	
	N	R(Å)	$\sigma^2(\times 10^{-4}$Å$^2)$	N	R(Å)	$\sigma^2(\times 10^{-4}$Å$^2)$
A(1.9%)	6.9	2.47	46	5.1	2.80	70
B(0.7%)	1.2	2.48	...	10.8	2.82	62
C(0.1%)	0.9	2.45	...	11.1	2.82	63
D(6.0%)	4.5	2.45	50	7.5	2.79	74

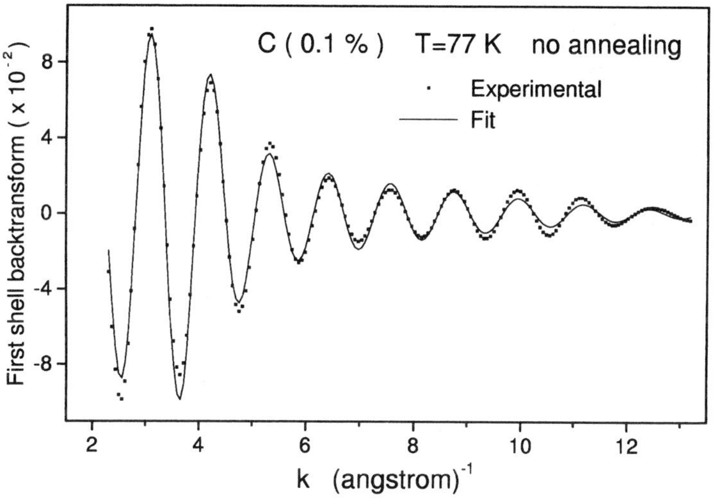

Figure 4. Fourier back transform of the first peak, including Co-Co and Co-Ag contributions, for sample C.

implantation at lower concentration causes larger clustering than MBE evaporation. This effect was expected because of the large energy exchanged during the ion bombardment.

The results shown in Table I can be corroborated by the analysis of the first shell signal obtained by a Fourier back-transform limited to the first shell and compared to the corresponding theoretical simulation. In Fig. 4 we show a typical plot and the parameters extracted for sample C are in close agreement with those of Table I. A similar test of self-consistency was extended to all the cases presented in this manuscript.

III.2. EXAFS Spectra: Annealed Samples

In order to study the stability of each cluster ensemble and the growth mechanism[19] under thermal treatment we annealed the samples A and D at 350 C for 30 min. In

Table II

Fitting parameters of the first shell around a Co absorber for samples A and D after the annealing at 350°C for 30 min. The uncertainty of the coordination number N is ± 0.2; of the nearest-neighbour distance R is ± 0.02 Å; of the relative squared vibrational amplitude is $\pm 5 \times 10^{-4}$ Å^2 for Co-Co, ten times larger for Co-Ag.

sample	Co–Co			Co–Ag		
	N	R(Å)	$\sigma^2(\times 10^{-4}$Å$^2)$	N	R(Å)	$\sigma^2(\times 10^{-4}$Å$^2)$
A_{anneal}	8.3	2.48	44	3.7	2.80	200
D_{anneal}	9.2	2.48	49	2.8	2.80	200

Table III

Fitting parameters of the second and third shell around a Co absorber; we can distinguish the Co-Co and the Co-Ag contribution. The uncertainty of the coordination number N is ± 0.4; of the nearest-neighbour distance R is ± 0.05 Å.

sample	Second shell				Third shell			
	Co–Co		Co–Ag		Co–Co		Co–Ag	
	N	R(Å)	N	R(Å)	N	R(Å)	N	R(Å)
A	1.8	3.62	4.2	4.14	6.0	4.56	18.0	4.97
A_{anneal}	3.0	3.56	3.0	4.13	11.0	4.40	13.0	4.98
B	1.2	3.51	4.8	3.99	2.0	4.65	22.0	4.99
C	1.2	3.64	4.8	4.1	1.4	4.63	22.6	4.98
D	0.9	3.58	5.1	4.02	1.7	4.79	22.3	4.92
D_{anneal}	3.8	3.56	2.2	4.14	14.0	4.38	10.0	4.91

complete analogy with the previous section, we observed the correspondent EXAFS spectra which were analyzed as reported in Table II. The size of the Co clusters in both samples increases: indeed, the first Co-Co coordination reaches a value 8.3 and 9.2, respectively, with an increase of the distance, whereas the Co-Ag shell due to interface contributions at the border of the clusters decreases in coordination with a strong increase of the DW factor. These results can be interpreted as due to the decrease of the surface to volume ratio caused by the presence of larger agglomerates.

III.3. Higher Shell Peaks

For each sample, discussed above, the FT of the EXAFS spectrum showed remarkable peaks corresponding to second, third and fourth coordination shells. Also these features were analyzed; although the uncertainty of the fitted parameters is somewhat larger owing to the reduced amplitude of these higher shells, we report in Table III the second and third shell values of the coordination number and distance which appear quite reliable. Here again each feature is splitted in two components (homo-atomic

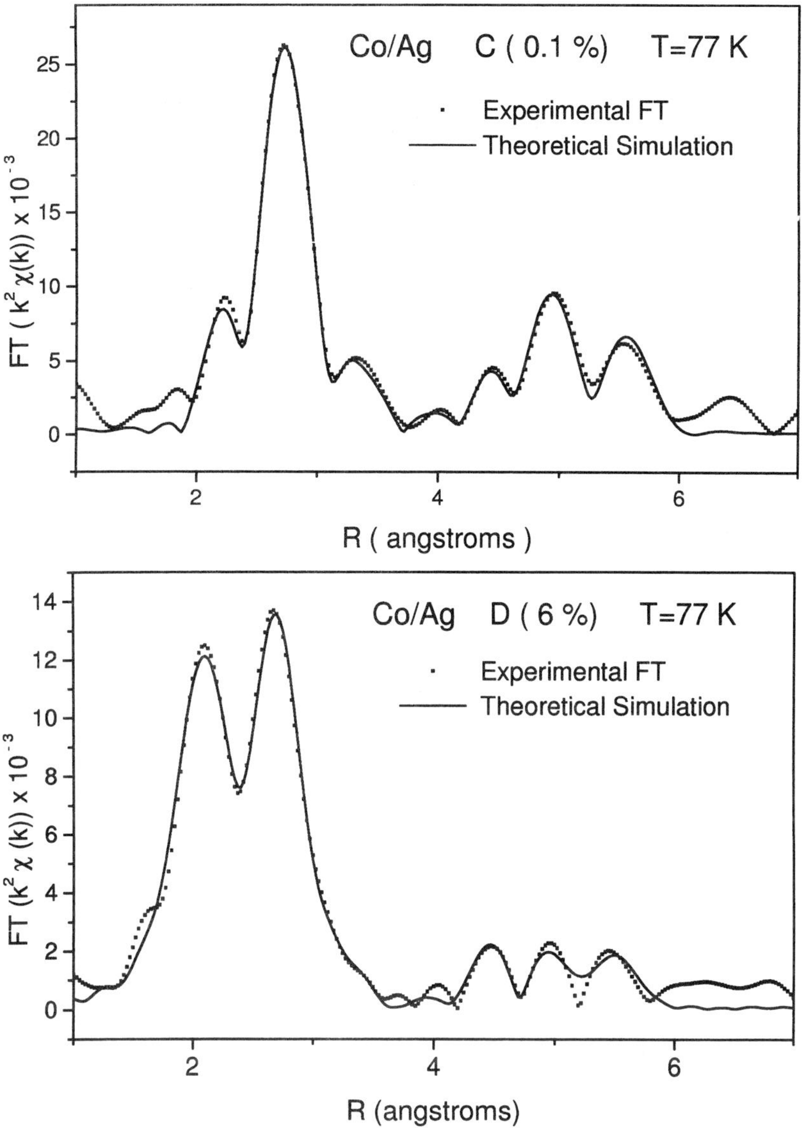

Figure 5. Typical FT spectra with the simulated curves using the fitting procedure and the FEFF code of Ref. 19.

and hetero-atomic). Some typical FT spectra are displayed in Fig. 5 with the theoretical simulation[19] performed using the parameters given in the tables. As visible in the figures, there is a very good agreement between the experimental and the calculated curves. The information given in Table III is of enormous importance to evaluate the configuration of the oligomers: in fact, as discussed later, in order to obtain the size of the clusters, we have to compare the experimental results with the average configuration given by a set of possible Co aggregates, calculating, for each Co atom acting as absorber, how many atoms of the same or of different species belong to the coordination shell under examination.

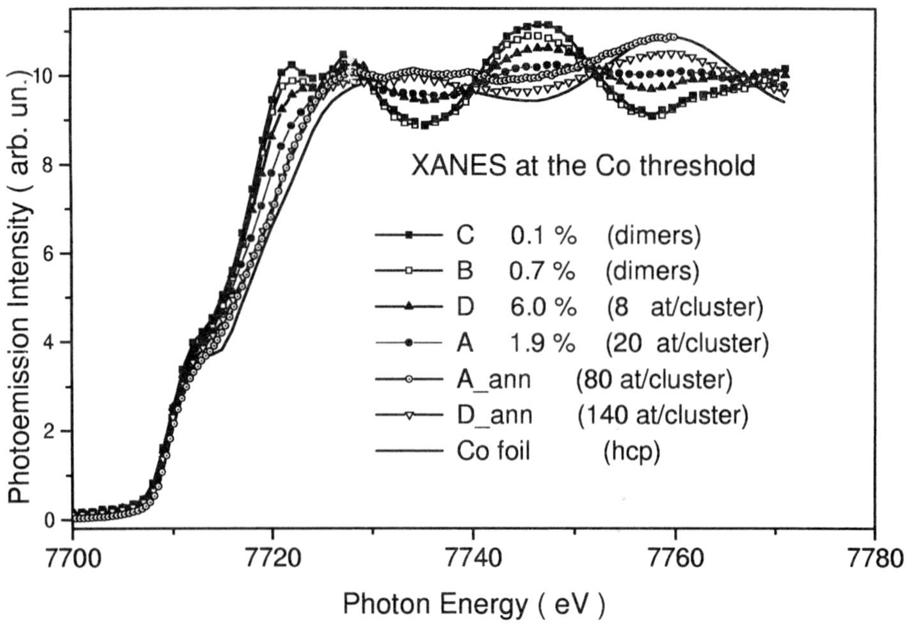

Figure 6. Near edge structure of the spectra showing modifications in the threshold shape and the features shifted with respect to the reference Co foil.

III.4. XANES Spectra

Some complementary information can be obtained by XANES spectroscopy which shows the features of the spectrum close to the threshold. As it is well known, their energy and shape are a clear fingerprint of the chemical environment. In fact, the binding energy can show, with respect to a model, a chemical shift usually ascribed to the topology of the valence band and to modifications of the interatomic mutual bonds. In Fig. 6 we report the XANES features of our samples with respect to the Co metal foil. This last is an *hcp* crystal and its XANES shape is very similar to the largest Co cluster observed after annealing in D_{anneal} sample.

The spectra of the very small clusters as found in sample B, C and D strongly contrast with those of *hcp* Co and of sample D_{anneal} since they are much nearer to a XANES spectrum of a *fcc* Ag environment, suggesting that the major nearest shell contribution is similar to that in an Ag matrix. This is in accordance with our interpretation of the EXAFS spectra of samples B and C, where the Co is observed to be clustered in (an association of) dimers, having a neighbourhood which does not strongly deviate from that in an Ag matrix. The spectra of sample A and A_{anneal} are intermediate between those two groups, which may indicate the prominence in these clusters of *hcp/fcc* stacking faults and/or a combination of *hcp* clusters with a more diluted Co in Ag precursor stage. The latter interpretation finds support in a recent Mössbauer study of similarly prepared samples[20] where at temperatures up to 350°C, Co is found to have precipitated in pure Co form and in the form of a mixed phase. These considerations therefore, make us believe that a transition *fcc* $\rightarrow$ *hcp* was certainly undergone by sample D during annealing: we indicate in particular the phase opposition between the curves D and D_{anneal} in Fig. 6.

IV. Discussion

The main results of the previous presentation are worth of a more detailed discussion, in the light of the numerous papers applying the EXAFS technique to clusters.[21–25] First, we mention the large range going from 0.1 at. % of sample C up to 0.7 at. % of sample B where the same preparation procedure leads to a very similar configuration of Co clustering. The coordination, found for the first shell very close to 1 in both cases, bears to the conclusion that in this range a preferential arrangement of small Co clusters (dimers or chain of dimers) exists, deriving from stable configurations. This interpretation fully supports an earlier Mössbauer investigation of the primary stages of Co precipitation in Ag,[15,26] where the hyperfine parameters of the major component observed in this concentration region pointed to the formation of Co-dimers as initial precipitates. For discriminating the effective cluster ensemble and distinguishing, if possible, between a dimer and other possible small clusters, as a triangular trimer or a tetrahedral tetramer, we have to take into account also the second and third coordination shell, and evaluate the possible models which can give the correct average coordinations in agreement with the experimental ones. In other terms, from the experimental data we should infer the possible number of Co atoms (2,3,4...) in near-substitutional configuration in the Ag lattice clustering together in such a way as to give the average coordinations as those experimentally obtained. It is immediately clear that, if we would have a pure dimer ensemble, the first, second, and third cobalt (Co-Co) coordination should be 1, 0, 0. respectively. This is not the case. A more realistic arrangement is represented, *e.g.*, by three couples of dimers parallelly disposed along three square faces of a *fcc* lattice (coordinates in a conventional cube of edge $a = 2$: 000, 101; 020, 121; 040, 141;). In this last case the average coordinations are 1, 1.33, 1.33 respectively. Instead, if we consider only the first two adjacent dimers we would get 1, 1, 1, and a modified disposition of the third dimer at 211, 312 would give 1.67, 0.66, 2. Very likely these could be the Co configurations in samples B and C. Another multiple dimer configuration is displayed in Fig. 7 with the dimers disposed along the square diagonal in a kind of rotating sticks. This corresponds to average coordinations 1, 0.83, 1.67, very close to the experimental ones.

These considerations suggest possible mixed configurations to be compared with molecular dynamics simulations giving the most stable cohesive energy. Similar simulations can be performed for larger agglomerates. Taking into account both the "internal" and the "surface" contributions, we calculated possible average configurations for each sample. We limit ourselves to give the final results: as already mentioned, in sample A we evaluated the presence of clusters of 20 atoms/cluster, in D 8 atoms/cluster, in A_{anneal} 80 atoms/cluster, in D_{anneal} 136 atoms/cluster.

It is interesting to point out the smaller cluster size in the evaporated D sample and a higher size in the implanted A sample: we ascribe this effect to the large energy exchange during the ion bombardment, that causes a partial coalescence. Conversely, the smaller initial agglomerates in D, more concentrated and very likely more mobile, produce bigger clusters during the annealing than in A_{anneal} where the partial aggregation and the larger distance between the clusters limit the final size.

The reported DW factors permit to evaluate the corresponding Debye temperature Θ_D for the different situations presently described. First of all we checked the consistency of the Debye temperature of the Co metal with that calculated using the experimental DW factor. The agreement is excellent: we obtain $\Theta_D = 418$ K at $T = 77$ K (with a calculated $\sigma^2 = 27 \times 10^{-4}$ Å^2); for the samples A, and A_{anneal} we get $\Theta_D = 285$ K whereas for D and D_{anneal} $\Theta_D = 260$ K. These values bring us to

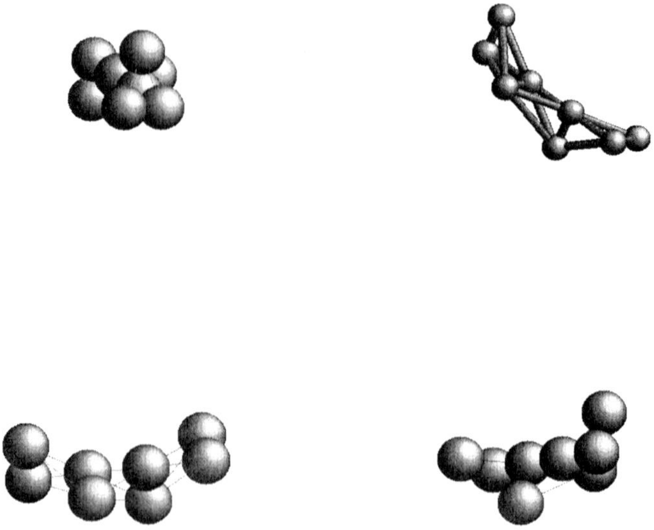

Figure 7. A possible configuration of 4 Co dimers disposed on parallel planes along the square diagonal of a *fcc* lattice, in a kind of rotating sticks orthogonal to each other. For the sake of clarity, this ensemble is shown from different perspectives; however, the Ag atoms in the lattice are not displayed. In this example, the average coordinations are close to the experimental ones.

three observations. First, we observe a marked reduction, compared to the Co bulk value. This implies a less rigid environment, clearly due to the soft Ag host lattice, which influence is felt through the large fraction of Co-Ag interface bonds, quite looser than the Co-Co bonds. This is to be expected since a 140 atoms Co cluster still has a surface to inner atom ratio of about 1. This is further supported by the Co-Ag interface DW factor for the samples A and D, as prepared and annealed, which yields a Θ_D of 200 K, rather close to that of bulk Ag. Second, the Co-Co vibrational amplitude appears to be independent on the cluster size in the range that we studied. This seems to contradict a recent investigation[20] where it was concluded that Θ_D(Co) was markedly dependent on cluster size. This study however spanned a size range that was considerably larger than the one presently studied. Also we may argue that a proper comparison between the as prepared A and D samples and the annealed state is hampered by their rather extreme condition after preparation: sample A contains, and still does after 350 C annealing, a large fraction of its Co in a diluted precursor stage. For sample D we found an initial mean cluster size of only 8 atoms, suggesting that the Co atoms may be arranged in a particular configuration, possibly leading to decreased mean vibrational amplitude. Third, the somewhat larger Θ_D of A_{anneal} compared to D_{anneal}, may be indicative of the tighter embedment of the Co clusters in the Ag matrix after ion implantation (sample A) than in the case of co-evaporation (sample D), where the Co clusters are believed to be quasi-free. The latter state would be comparable to that observed after co-sputtering preparation.[27]

Θ_D derived from σ^2(Co-Ag) of samples B and C, where Co faces 11 Ag atoms in the first shell, is somewhat higher, 220 K, than for the other cases. We believe that this enhancement is caused by the kind of "spring-loaded" configuration of the individual dimers, where the Co-Co neighbours on quasi-substitutional sites, while relaxing towards each other but not achieving the Co-Co distance as in the clusters, are pulling on their Co-Ag bonds, thereby stretching the latters and increasing the interaction force.

Acknowledgments

The authors gratefully acknowledge the excellent collaboration of the ESRF staff during the experiments; in particular we appreciated the contribution of Dr. Stefano Colonna of the GILDA beamline. We acknowledge also the important contribution of Dr. Johan Dekoster and Dr. Stefan Degroote in the sample preparation, as well as the funding, provided by projects BIL98/07 and IUAP contract 4/10. Financial support was provided also from the Centro Siciliano di Fisica Nucleare e Struttura della Materia.

References

1. J. B. Kortright, D. D. Awschalom, J. Stöhr, S. D. Bader, Y. U. Idzerda, S. S. P. Parkin, Ivan K. Schuller, and H. C. Siegmann, *J. Magn. Magn. Mater.* **207**, 7 (1999).
2. M. Respaud, J. M. Broto, H. Rakoto, A. R. Fert, L. Thomas, B. Barbara, M. Verelst, E. Snoeck, P. Lecante, A. Mosset, J. Osuna, T. Ould Ely, C. Amiens, and B. Chaudret, *Phys. Rev. B* **57**, 2925 (1998); S. N. Khanna and S. Linderoth, *Phys. Rev. Lett.* **67**, 742 (1991).
3. B. Piveteau, M.-C. Desjonquères, A. M. Oles, and D. Spanjaard, *Phys. Rev. B* **53**, 9251 (1996).
4. H. A. Dürr, S. S. Dhesi, E. Dudzik, D. Knabben, G. van der Laan, J. B. Goedkoop, and F. U. Hillebrecht, *Phys. Rev. B* **59**, R701 (1999).
5. J. Guevara, A. M. Lois, and M. Weissmann; *Phys. Rev. Lett.* **81**, 5306 (1998).
6. X. Chuanyun, Y. Jinlong, D. Kaiming, and W. Kelin, *Phys. Rev. B* **55**, 3677 (1997).
7. J. Xu, M. A. Howson, B. J. Hickey, D. Greig, E. Kolb, P. Veillet, and N. Wiser, *Phys. Rev. B* **55**, 416 (1997).
8. G. L. Zhang, J. Verheyden, W. Deweerd, G. E. J. Koops, and H. Pattyn, *Phys. Rev. B* **58**, 3026 (1998).
9. H. Yoshida, A. Terasaki, K. Kobayashi, M. Tsukada and T. Kondow, *J. Chem. Phys.* **102**, 5960 (1995).
10. B. C. Guo, K. P. Kerns, and A. W. Castelman Jr., *J. Chem. Phys.* **96**, 8177 (1992).
11. Zhi-qiang Li and Bing-lin Gu, *Phys. Rev. B* **47**, 13611 (1993); V. Dupuis, J. Tuaillon, B. Prevel, A. Perez, P. Melinon, G. Guiraud, F. Parent, L. B. Steren, R. Morel, A. Barthelemy, A. Fert, S. Mangin, L. Thomas, W. Wernsdorfer, and B. Barbara, *J. Magn. Magn. Mater.* **165**, 42 (1997); J. R. Regnard, J. Juanhuix, C. Brizard, B. Dieny, and B Mevel, *Solid State Commun.* **97**, 419 (1996).
12. B. Schleicher, S. Künzel, and H. Burtscher, *J. Appl. Phys.* **78**, 4416 (1995).
13. G. Langouche, M. de Potter, and D. Schroyen, *Phys. Rev. Lett.* **53**, 1364 (1984).
14. I. Shim and K. A. Gingerich, *J. Chem. Phys.* **78**, 5693 (1983).
15. J. Verheyden, G. Zhang, J. Dekoster, A. Vantomme, W. Deweerd, K. Milants, T. Barancira, and H. Pattyn, *J. Phys. D, Applied Physics* **29**, 1316 (1996).

16. P. A. Lee, P. H. Citrin, P. Eisenberger, and B. M. Kincaid, *Rev. Mod. Phys.* **53**, 769 (1981).
17. E. A. Stern, B. A. Bunker, and S. M. Heald, *Phys. Rev. B* **21**, 5521 (1980).
18. G. Faraci, S. La Rosa, A. R. Pennisi, S. Mobilio, and I. Pollini, *Phys. Rev. B* **45**, 9357 (1992); A. Pinto, A. R. Pennisi, G. Faraci, S. Mobilio, and F. Boscherini, *Phys. Rev. B* **51**, 5315 (1995); P. Decoster, B. Swinnen, K. Milants, M. Rots, S. La Rosa, A. R. Pennisi, and G. Faraci, *Phys. Rev. B* **50**, 9752 (1994).
19. J. Mustre de Leon, J. J. Rehr, S. I. Zabinsky, and R. C. Alberts, *Phys. Rev. B* **44**, 4146 (1991); J. J. Rehr, R. C. Alberts, and S. I. Zabinsky, *Phys. Rev. Lett.* **69**, 3397 (1992); J. J. Rehr, S. I. Zabinsky, A. Ankudinov, and R. C. Alberts, *Physica B* **23**, 208, (1995); E. A. Stern, M. Newville, B. Ravel, Y. Yacoby, *Physica B* **23**, 117 (1995).
20. M. Hou, M. El Azzaoui, H. Pattyn, J. Verheyden, G. Koops, and G. Zhang, to be published.
21. J. Zhao and P. A. Montano, *Phys. Rev. B* **40**, 3401 (1989).
22. P. A. Montano, G. K. Shenoy, E. E. Alp, W. Schulze, and J. Urban, *Phys. Rev. Lett.* **56**, 2076 (1986).
23. A. Balerna *et al. Phys. Rev. B* **31**, 5058 (1985).
24. M. A. Marcus, M. P. Andrews, J. Zegenhagen, A. S. Bommannavar and P. Montano, *Phys. Rev. B* **42**, 3312 (1990).
25. P. A. Montano, W. Schulze, B. Tesche, G. K. Shenoy, and T. I. Morrison, *Phys. Rev. B* **30**, 672 (1984).
26. R. N. Nogueira and H. M. Petrilli, *Phys. Rev. B* **60**, 4120 (1999).
27. J. R. Childress, C. L. Chien, M. Y. Zhou, and Ping Sheng, *Phys. Rev. B* **44**, 11689 (1991).

Calculation of Spin-Fluctuation Energies in Fe_N Clusters

J. Dorantes-Dávila,[1] G. M. Pastor,[2,3] and K. H. Bennemann[3]

[1] *Instituto de Física, "Manuel Sandoval Vallarta"*
Universidad Autónoma de San Luis Potosí
78000 San Luis Potosí
MEXICO

[2] *Laboratoire de Physique Quantique*
UMR 5626 du CNRS
Université Paul Sabatier
118 route de Narbonne
F-31062 Toulouse
FRANCE

[3] *Institut für Theoretische Physik*
Freie Universität Berlin
14195 Berlin
GERMANY

Abstract

A functional-integral theory of itinerant d-electron magnetism is applied to small transition-metal clusters. The spin-fluctuation energies $\Delta F_i(\xi)$ at different atoms i in Fe_N ($N \leq 15$) are calculated using a real-space recursive expansion of the local Green's function. The size, structural, and local-environment dependences of $\Delta F_i(\xi)$ are determined. The interplay between fluctuations of the module and orientation of the local exchange fields ξ_i is investigated, and the applicability of phenomenological spin models is discussed.

I. Introduction

The magnetic properties of transition-metal (TM) clusters at finite temperatures are of considerable interest in materials research. In view of possible applications in recording

Physics of Low Dimensional Systems
Edited by J. L. Morán-López, Kluwer Academic/Plenum Publishers, New York 2001

media, it is not enough to identify the clusters showing large average magnetizations per atom at $T = 0$, but it is also crucial to get information on the cluster magnetization as a function of T and on the size and structural dependence of the "Curie" temperature $T_C(N)$ above which ferromagnetic order is destroyed by thermal fluctuations. Recently, the temperature dependence of the magnetization of Fe_N, Co_N and Ni_N clusters has been derived from Stern-Gerlach deflection measurements.[1,2,3] Particularly in the case of Fe_N, the experimental behavior seems quite anomalous. One observes strong deviations from the predictions of the Heisenberg model and rather large high-temperature values of the magnetization at $T > T_C$. From a theoretical point of view, mean-field ground-state calculations have been quite successful in predicting the experimental results for the average magnetic moments per atom at low temperatures.[4] In contrast, very little is still known about the theory of TM cluster magnetism at finite temperatures.

The temperature dependence of the magnetic properties of free clusters is a subject of fundamental importance since the critical behavior of solids and finite systems are intrinsically different.[5] Strong deviations from bulk-like behavior are expected to occur when the correlation length $\lambda(T) \sim (T - T_C)^{-\gamma}$ becomes of the order of the cluster radius R. This is the case at temperatures $T \leq T^*$, where $(T^* - T_C) \sim R^{-1/\gamma}$. The divergence at T_C in the specific heat $C_p(T)$ and in the magnetic susceptibility $\chi(T)$ disappear, since the long wave-length magnetic fluctuations are suppressed by the finite size of the cluster (*i.e.*, $k \geq k_{\min} \sim 1/R$). Instead, these properties present a peak at $T \simeq T_C$ with a size dependent width. Moreover, it is in general difficult to define a unique critical temperature $T_C(N)$, since the position of the peak in $C_p(T)$ and $\chi(T)$ need not be the same.[5] Besides the temperatures $T^*(N)$ and $T_C(N)$, one is interested in the size dependence of the temperature $T_{\mathrm{SR}}(N)$ above which thermal fluctuations destroy the short-range correlations between the local magnetic moments, for example, between nearest neighbor (NN) μ_i. Let us recall that a significant degree of short-range magnetic order (SRMO) is observed in the bulk and near the surfaces of Fe, Co, and Ni. This holds even for $T > T_C$, *i.e.*, after the average magnetization $M(T)$ vanishes $[T_{\mathrm{SR}}(b) > T_C(b)]$. For small clusters having a radius R smaller than the range of SRMO, it is no longer possible to increase the entropy without destroying the energetically favorable local magnetic correlations. In addition, the degree of SRMO is likely to depend on the details of the electronic structure and on cluster size.

The trends in the size dependence of $T_{\mathrm{SR}}(N)$ and $T_C(N)$ seem difficult to infer *a priori*. On the one side, taking into account the enhancement of the local magnetic moments μ_i and of the d-level exchange splittings $\Delta\varepsilon_X(i) = \varepsilon_{i\downarrow} - \varepsilon_{i\uparrow}$, one could expect that $T_{\mathrm{SR}}(N)$ and $T_C(N)$ should be larger in small clusters than in the bulk. However, on the other side, it should be energetically easier to disorder the local magnetic moments in a cluster by flipping or canting μ_i, since the local coordination numbers are smaller. If the later effect dominates, $T_C(N)$ should decrease with decreasing N. Moreover, recent model calculations[6,7] indicate that changes or fluctuations in the cluster structure at $T > 0$ may affect significantly the temperature dependence of the magnetization, in particular for systems like Fe_N and Rh_N which show a remarkable structural dependence of the magnetic properties already at $T = 0$. In order to derive reliable conclusions concerning the size dependence of $T_C(N)$ and $T_{\mathrm{SR}}(N)$, the electronic theory must take into account both the fluctuations of the magnetic moments and the itinerant character of the d-electron states. Simple spin models, for example based on the Heisenberg or Ising model, are not expected to be very predictive, at least until they incorporate the electronic effects responsible for the size dependence of the magnetic moments and their interactions. In fact, surface studies of itinerant magnetism have already shown that the effective exchange interactions J_{ij} between NN moments μ_i and μ_j depend strongly on the local coordination numbers of sites i and j.[8] A similar behavior is also found in clusters, as it will be discussed in the following.

II. Theory

The magnetic properties of TM clusters at finite temperatures are studied by considering a realistic d-band Hamiltonian given by

$$\hat{H} = \hat{H}_0 + \hat{H}_I \ . \tag{1}$$

The first term

$$\hat{H}_0 = \sum_{i,\alpha,\sigma} \varepsilon_i^0 \, \hat{n}_{i\alpha\sigma} + \sum_{\substack{i\neq j \\ \alpha,\beta,\sigma}} t_{ij}^{\alpha\beta} \, \hat{c}_{i\alpha\sigma}^{\dagger} \hat{c}_{j\beta\sigma} \ , \tag{2}$$

describes the single-particle electronic structure of the valence d electrons in the tight-binding approximation.[9] As usual, $\hat{c}_{i\alpha\sigma}^{\dagger}$ ($\hat{c}_{i\alpha\sigma}$) refers to the creation (annihilation) operator of an electron with spin σ at the orbital α of atom i ($\alpha \equiv xy$, yz, zx, x^2-y^2, and $3z^2-r^2$), and $\hat{n}_{i\alpha\sigma} = \hat{c}_{i\alpha\sigma}^{\dagger}\hat{c}_{i\alpha\sigma}$ is the corresponding number operator. ε_i^0 stands for the bare d-level energy of the isolated atom and $t_{ij}^{\alpha\beta}$ for the hopping integrals between atoms i and j. The second term

$$\hat{H}_I = \frac{1}{2} \sum_{\substack{i,\alpha,\beta \\ \sigma,\sigma'}}{}' \, U_{\sigma\sigma'} \, \hat{n}_{i\alpha\sigma}\hat{n}_{i\beta\sigma'} \ , \tag{3}$$

approximates the interactions among electrons by an intra-atomic Hubbard-like model. The Coulomb repulsions between electrons of spin σ and σ' are denoted by $U_{\sigma\sigma'}$. Notice that Eq. (3) does not respect spin-rotational symmetry, since exchange terms of the form $H_{xy} = -\sum_{i,\alpha<\beta} J_{\alpha\beta}(S_{i\alpha}^- S_{i\beta}^+ + S_{i\alpha}^+ S_{i\beta}^-)$ have been dropped.[10] Nevertheless, this is not expected to be a serious limitation in the present work because we are interested in studying the spin fluctuations on top of broken-symmetry ferromagnetic ground states.

The finite temperature magnetic properties of clusters are determined by applying the functional-integral formalism developed by Hubbard and Hasegawa for TM solids.[11–14] The many-body interaction H_I is rewritten as

$$H_I = \frac{1}{2} \sum_i \left[\left(U - \frac{J}{2} \right) \hat{N}_i^2 - 2J \, \hat{S}_{iz}^2 - (U - J) \, \hat{N}_i \right] , \tag{4}$$

where $\hat{N}_i = \sum_{\alpha\sigma} \hat{n}_{i\alpha\sigma}$ is the number of electron operator at atom i, and $\hat{S}_{iz} = (1/2)\sum_\alpha (\hat{n}_{i\alpha\uparrow} - \hat{n}_{i\alpha\downarrow})$ is the z component of the spin operator. $U = (U_{\uparrow\downarrow} + U_{\uparrow\uparrow})/2$ and $J = U_{\uparrow\downarrow} - U_{\uparrow\uparrow}$ stand for the average direct and exchange Coulomb integrals which are taken to be orbital independent. For the calculation of the partition function Z, the quadratic terms in Eq. (4) are linearized by means of a two-field Hubbard-Stratonovich transformation within the static approximation. A charge field η_i and an exchange field ξ_i are thus introduced at each cluster site i. These represent the local finite-temperature fluctuations of the d-electron energy levels and exchange splittings, respectively. Using the notation $\vec{\xi} = (\xi_1, \ldots, \xi_N)$ and $\vec{\eta} = (\eta_1, \ldots, \eta_N)$, Z is given by

$$Z \propto \int d\vec{\eta} \, d\vec{\xi} \ \tilde{Z}(\vec{\xi}, \vec{\eta}) \ , \tag{5}$$

where

$$\tilde{Z}(\vec{\xi},\vec{\eta}) = \exp\{-\beta F(\vec{\xi},\vec{\eta})\} \tag{6}$$

$$= \exp\left\{-\frac{\beta}{2}\sum_i \left(U\eta_i^2 + \frac{J}{2}\xi_i^2\right)\right\} \mathrm{Tr}\left[\exp\{-\beta(\hat{H}_{eff} - \mu\hat{N})\}\right] . \tag{7}$$

$\hat{H}_{\text{eff}}$ describes the dynamics of the d electrons as if they were independent particles moving in a random alloy with energy levels $\tilde{\varepsilon}_{i\sigma}$ given by

$$\tilde{\varepsilon}_{i\sigma} = \varepsilon_i^0 + Ui\eta_i - \sigma\frac{J}{2}\xi_i . \tag{8}$$

The thermodynamic properties of the system are obtained as a statistical average over all possible distributions of the energy levels $\tilde{\varepsilon}_{i\sigma}$ throughout the cluster.

For $T \to 0$ the field configuration $(\vec{\xi}^0, \vec{\eta}^0)$ dominates, which corresponds to the saddle point in the free energy $F(\vec{\xi},\vec{\eta})$. This is determined from the self-consistent equations[15]

$$\left.\frac{\partial F}{\partial \xi_i}\right|_0 = \frac{J}{2}(\xi_i^0 - \langle \hat{S}_i\rangle) = 0 , \tag{9}$$

$$\left.\frac{\partial F}{\partial \eta_i}\right|_0 = -iU(i\eta_i^0 - \langle \hat{N}_i\rangle) = 0 , \tag{10}$$

where $\langle\ldots\rangle$ indicates ground-state average. Replacing Eqs. (9) and (10) in Eq. (8) yields the known mean-field approximation to the energy levels $\varepsilon_{i\sigma}$.[15] Consequently, the present approach is the natural finite-temperature extension of the self-consistent tight-binding theory developed for the ground state in Ref. 15.

Since we are mainly interested in the magnetic properties and since $J \ll U$, we neglect the thermal fluctuations of the charge fields η_i by setting them equal to their exchange-field-dependent saddle-point values $i\overline{\eta}_i = \langle \hat{N}_i\rangle$. This amounts to a self-consistent determination of the charge distribution for each exchange field configuration $\vec{\xi}$. In this way

$$Z \propto \int d\vec{\xi}\, \tilde{Z}(\vec{\xi}) , \tag{11}$$

where

$$\tilde{Z}(\vec{\xi}) = \exp\{-\beta F(\vec{\xi})\}, \tag{12}$$

$$= \exp\left\{-\frac{\beta}{2}\sum_i \left(U\overline{\eta}_i^2 + \frac{J}{2}\xi_i^2\right)\right\} \mathrm{Tr}\left[\exp\{-\beta(\hat{H}_{\text{eff}} - \mu\hat{N})\}\right] , \tag{13}$$

depends only on the most relevant exchange variables ξ_i that describe the fluctuations of the spin degrees of freedom.

The integrand $\exp\{-\beta F(\vec{\xi})\}$ is interpreted as proportional to the probability $P(\vec{\xi})$ for a given exchange-field configuration $\vec{\xi}$. The thermodynamic properties are obtained by averaging over all possible $\vec{\xi}$ with $\exp\{-\beta F(\vec{\xi})\}$ as weighting factor. For example, the local magnetization at atom i is given by the average of $2\langle S_{iz}\rangle = \sum_\alpha \langle \hat{n}_{i\alpha\uparrow} - \hat{n}_{i\alpha\downarrow}\rangle$ which depends on $\vec{\xi}$ and fluctuates at $T > 0$:

$$\mu_i(T) = \frac{1}{Z}\int d\vec{\xi}\, \exp\left\{-\frac{\beta}{2}\sum_i \left(U\overline{\eta}_i^2 + \frac{J}{2}\xi_i^2\right)\right\} \mathrm{Tr}\left[2\hat{S}_i^z \exp\{-\beta(\hat{H}_{\text{eff}} - \mu\hat{N})\}\right]. \tag{14}$$

In terms of the probability

$$P_i(\xi) = \frac{1}{Z}\exp\{-\beta F_i(\xi)\} = \frac{1}{Z}\int \prod_{l\neq i} d\xi_l \exp\{-\beta F(\xi_1,\ldots,\xi_{i-1},\xi,\xi_{i+1},\ldots,\xi_N)\}, \quad (15)$$

of having the value ξ for the exchange field at atom i, one may write

$$\mu_i(T) = \int \xi \, P_i(\xi)\, d\xi \,. \quad (16)$$

Thus, the temperature dependent local magnetization is equal to the average of the local exchange field. Equation (16) justifies the intuitive association between the fluctuations of the local moment $2\langle \hat{S}_{iz}\rangle$ at atom i and those of the exchange field ξ_i. The cluster magnetization per atom is obtained as the average of the local magnetizations $\mu_i(T)$.[14]

A first insight on the magnetic behavior of Fe_N and Ni_N clusters at $T > 0$ can be obtained by considering the low-temperature limit of $F_i(\xi)$. The local free-energy difference

$$\Delta F_i(\xi) = F_i(\xi) - F_i(\xi_i^0)\,, \quad (17)$$

with $\xi_i^0 = \mu_i^0$ being the local magnetic moment of atom i at $T = 0$, represents the energy involved in an exchange-field fluctuation above the Hartree-Fock ground state. $\Delta F_i(\xi)$ has often two minima at $\xi^+ = \xi_i^0$ and $\xi^- \simeq -\xi_i^0$. At finite temperatures the ferromagnetic order is reduced by spin fluctuations involving transitions between these two minima, as well as by Gaussian-like fluctuations around them. For small Fe or Ni clusters directional fluctuations of $\vec{\mu}_i$ are expected to dominate, since the reduction of the local coordination numbers should enhance the local-moment character of the spin excitations. As the cluster size increases the fluctuations of the amplitude of μ_i or ξ_i also become important, indicating a crossover to a more itinerant behavior. The free energy $\Delta F_i(\xi)$ depends strongly on the atomic site i, on the cluster structure, and on the interatomic distances. Moreover, the finite temperature fluctuations of the local moments $\langle \hat{S}_{iz}\rangle$ at different i are not independent from each other. Magnetic correlations are actually present even above the temperature T_C at which the average magnetization vanishes. They are usually referred to as short-range magnetic order (SRMO).[8,16–18] In the framework of the previous functional-integral theory, SRMO manifests itself as correlations between the exchange fields at neighboring sites. For example, in the ferromagnetic case, $\langle \xi_i \xi_j\rangle > \langle \xi_i\rangle\langle \xi_j\rangle$ for NN atoms i and j.

III. Results

In the following, we discuss representative results derived from the spin-fluctuation theory presented in the previous section which takes into account the fluctuations of the magnetic moments and the itinerant character of the d-electron states. The parameters used for the calculations on Fe_N are the same as in Ref. 15, namely, bulk d-band width $W = 6.0$ eV, direct Coulomb integral $U = 6.0$ eV, and exchange integral $J = 0.70$ eV. The effects of relaxation of the cluster bond-length are investigated by considering a few values of the NN distance. In order to investigate the temperature induced spin-fluctuations in small Fe and Ni clusters we determine the low-temperature limit of the local free-energy $F_i(\xi)$ (Eq. 17). For these clusters one obtains $\Delta F_i(\xi) = F_i(\xi) - F_i(\xi{=}\mu_i^0) > 0$ for all $\xi \neq \mu_i^0$, which indicates, as expected,

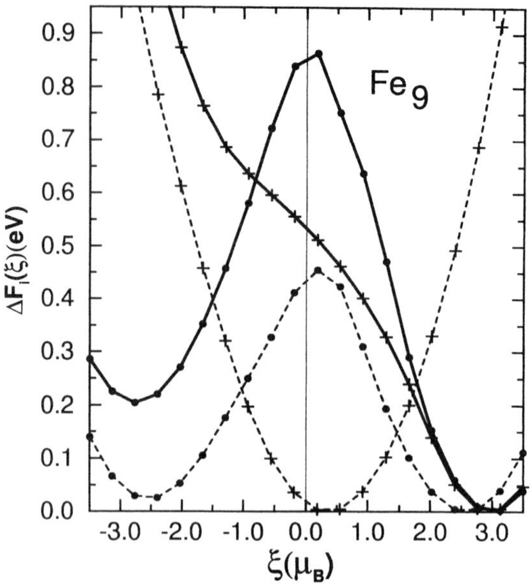

Figure 1. Local spin-fluctuation energy $\Delta F_i(\xi) = F_i(\xi) - F_i(T=0)$ as a function the exchange field ξ at different atomic sites i of Fe_9 with *bcc*-like structure. Plus signs refer to the central atom and dots to any of the eight surface atoms (NN's of the central one). Results are given for $d/d_B = 1.0$ (full curves) and $d/d_B = 0.92$ (dashed curves), where d refers to the NN distance (d_B = bulk NN distance).

that the ferromagnetic order is stable at low temperatures. Comparing the $\Delta F_i(\xi)$ for different sizes and for different atoms i within the cluster provides useful information on the stability of the local magnetizations and its environment dependence.

In Fig. 1 results are given for $\Delta F_i(\xi)$ at the central atom and at one of the surface atoms of an Fe_9 cluster with assumed *bcc*-like structure. For the surface atoms, which have the largest local magnetic moments μ_i^0 at $T = 0$, $F_i(\xi)$ shows two minima located at the molecular fields $\xi^+ = \mu_i^0$ and $\xi^- \simeq -\mu_i^0$. This double minimum structure indicates that the dominant magnetic excitations are flips of the surface magnetic moments keeping their amplitude approximately constant. In contrast, for the central atom, which has a much smaller μ_i^0, one observes a single minimum in $F_i(\xi)$. Thus, the fluctuations of the amplitude of the local moments dominate in this case. At the surface of the cluster only small fluctuations of ξ are possible with an excitation energy $\Delta F_i(\xi)$ smaller than the energy $\Delta F_i(\xi^-) = F_i(\xi^-) - F_i(\xi^+)$ required to flip a local magnetic moment. The probability $P_i(\xi) \propto \exp\{-\beta \Delta F_i(\xi)\}$ has two sharp maxima at $\xi \simeq \xi^+$ and $\xi \simeq \xi^-$ with $P_i(\xi^-) \gg P_i(\xi = 0)$. The fact that very small clusters and particularly cluster atoms having small local coordination numbers z_i show such a Heisenberg- or Ising-like behavior is not surprising. In fact, the kinetic-energy loss caused by flipping a local magnetic moment ($\xi \simeq \mu_i \to \xi \simeq -\mu_i$) is smaller when the local coordination number z_i is reduced ($E_K \propto \sqrt{z_i}$). At the same time, the exchange energy $\Delta E_X = (J/4) \sum_i \mu_i^2$, being a local property, is much less affected by the change of sign of ξ. Even Ni, that in the solid state has a single minimum in $F_i(\xi)$ and is therefore dominated by amplitude fluctuations of ξ,[13] tends to show two minima in $F_i(\xi)$ for sufficiently small N. The results suggest the possibility of a transition from Heisenberg-like to itinerant-like behavior with increasing cluster size.[19] This effect can

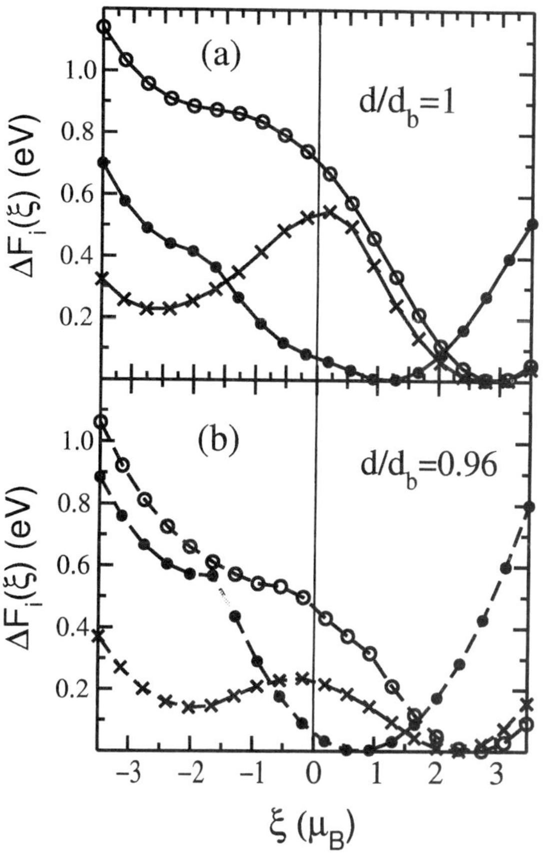

Figure 2. Local spin-fluctuation energy $\Delta F_i(\xi) = F_i(\xi) - F_i(T = 0)$ as a function the exchange field ξ at different atomic sites i of Fe_{15} with *bcc*-like structure. Dots refer to the central atom, crosses to the first NN shell, and open circles to the outermost shell. Results are given for (a) $d/d_B = 1.0$ in full curves and (b) $d/d_B = 0.96$ in dashed curves, where d refers to the NN distance (d_B = bulk NN distance).

be physically interpreted as a consequence of the competition between the Coulomb interaction energy, that is relatively more important in small clusters, and the kinetic d-band energy, that is more important in the bulk.

$\Delta F_i(\xi)$ depends strongly on the local environment of the different atoms within the cluster, as clearly illustrated by the results for Fe_{15} shown in Fig. 2. In addition to the central atom and its first NN's, which show similar behaviors as in Fe_9, we observe that the ferromagnetic order is particularly stable at the outermost shell (open circles in Fig. 2). The larger $\Delta F(-\xi_i^0)$ is favored by the larger local moments μ_i^0 found at these atoms, which compensate the reduction of local coordination number z_i at the cluster surface. However, notice that this trend is not always followed. For example, the atoms at the second shell (crosses in Fig. 2) show a much smaller spin-flip energy $\Delta F_i(-\xi^0)$, despite having similar μ_i^0 and similar z_i as the atoms in the outermost shell. In Ni_N, the spin fluctuations energies are in general smaller at the cluster surface, *i.e.*, as the local coordination number is smaller. This is consistent with the experimentally observed decrease of the cluster Curie temperature with decreasing N.[2]

It is also interesting to determine how $\Delta F_i(\xi)$ depends on the interatomic distances in order to infer the effects of structural distortions on the cluster magnetization curves. If the free-energy is optimized by changing the NN distances and by keeping the cluster symmetry unchanged (uniform relaxation), a bond-length contraction is usually obtained ($d < d_B$).[15] In Fig. 1 results are given for Fe_9 using $d/d_B = 0.92$.[15] In this case one finds no qualitative, but strong quantitative changes in $F_i(\xi)$ as a function of ξ. First, one observes shifts of the positions of the minima at ξ^+ that reflect the reduction of the local magnetic moments μ_i at $T = 0$. Moreover, there is a remarkable reduction (about a factor 10) of the free energy $\Delta F_i(\xi^-) = F_i(\xi^-) - F_i(\xi^+)$ required to flip a local magnetic moment, or to change the exchange field from $\xi^+ = \mu_i$ to $\xi^- \simeq -\mu_i$ at the surface atoms. A similar large reduction of the Curie temperature $T_C(N)$ is expected to occur upon relaxation, since in first approximation $T_C(N) \propto \Delta F_i$. These results clearly show, once more, the strong sensitivity of the magnetic properties of $3d$-TM clusters to the local environment of the atoms. In fact, $\Delta F_i(\xi)$ is much more sensitive to size and structure than the magnetic moments at $T = 0$. The effect is particularly notable in very small Fe_N and Ni_N, where the local magnetic moments are nearly saturated and therefore depend weakly on cluster geometry. Recent calculations including correlations effects exactly within the single-band Hubbard model[6,7] have also revealed the importance of structural changes and structural fluctuations for the temperature dependence of the magnetic properties of clusters. Finally, notice that $F_i(\xi)$ depends strongly on the position of the atom i within the cluster, as already observed in the unrelaxed case. Comparing results for different i one finds that $\Delta F_i(\xi^-)$ does not scale simply with the local coordination number z_i (see Figs. 1 and 2). This implies that the effective exchange couplings between local magnetic moments cannot be transfered straightforwardly from one local environment to another. Electronic structure effect due to the itinerant character of the d-electrons are therefore very important. Similar behaviors have been found near the surfaces of macroscopic TM's.[8]

IV. Discussion

The stability of cluster magnetism at finite temperatures should be characterized by a size dependent "Curie" temperature $T_C(N)$. However, the theoretical understanding of the size dependence of $T_C(N)$ in TM clusters still remains an open problem. As a first approximation one might estimate $T_C(N)$ as being proportional to the energy required to flip local magnetic moments $\langle \hat{S}_{iz} \rangle$ or an exchange fields ξ_i. Thus, $T_C(N) \propto \langle \Delta F_i \rangle$, where $\langle \ldots \rangle$ indicates a cluster average. However, note that this is a crude approximation. Equation (17) for $\Delta F_i(\xi)$ ignores that close to T_C all the exchange fields at different atomic sites fluctuate in some correlated fashion. The low-temperature limit, where all but one field ξ_i are kept equal to the $T = 0$ result, certainly overestimates the thermal average of $\Delta F_i(\xi)$ at $T > 0$ and the values of T_C derived from it. Fluctuations of the ensemble of exchange fields modify the magnetic environment at which individual spin fluctuations occur and should be therefore taken into account. In this framework the Curie temperature is more appropriately defined as the temperature at which it costs no free energy to flip spins.[11] Research in this direction is currently in progress.

Acknowledgments

This work has been financed in part by CONACyT (Mexico) Grant 32085E. G.M.P. acknowledges the support provided by a fellowship of the Alexander von Humboldt Foundation.

References

1. I. M. L. Billas, J. A. Becker, A. Châtelain, and W. A. de Heer, *Phys. Rev. Lett.* **71**, 4067 (1993).
2. I. M. L. Billas, A. Châtelain, and W. A. de Heer, *Science* **265**, 1682 (1994).
3. S. E. Apsel, J. W. Emert, J. Deng, and L. A. Bloomfield, *Phys. Rev. Lett.* **76**, 1441 (1996).
4. See, for instance, G. M. Pastor and K. H. Bennemann in *Metal Clusters*, edited by W. Ekardt (Wiley, New York, 1999) p. 211.
5. P. G. Watson, *Phase Transitions and Critical Phenomena*, edited by C. Domb and M. S. Green, (Academic, London, 1972), Vol. 2, p. 101; M. N. Barber, *Phase Transitions and Critical Phenomena*, edited by C. Domb and J. L. Lebowitz, (Academic, London, 1983), Vol. 8, p. 145.
6. G. M. Pastor, R. Hirsch, and B. Mühlschlegel, *Phys. Rev. Lett.* **72**, 3879 (1994); *Phys. Rev. B* **53**, 10382 (1996); F. López-Urías and G. M. Pastor, *Phys. Rev. B* **59**, 5223 (1998).
7. F. López-Urías and G. M. Pastor, to be published.
8. J. Dorantes-Dávila, G. M. Pastor, and K. H. Bennemann, *Solid State Commun.* **59**, 159 (1986); **60**, 465 (1986).
9. For the sake of clarity, a "hat" (̂) is used to distinguish operators from numbers.
10. B. Mühlschlegel, *Z. Phys.* **208**, 94 (1968).
11. J. Hubbard, *Phys. Rev. B* **19**, 2626 (1979); **20**, 4584 (1979); H. Hasegawa, *J. Phys. Soc. Jpn.* **49**, 178 (1980); **49**, 963 (1980).
12. *Electron Correlation and Magnetism in Narrow-Band Systems*, edited by T. Moriya, Springer Series in Solid State Sciences **29**, (Springer, Heidelberg, 1981).
13. Y. Kakehashi, *J. Phys. Soc. Jpn.* **50**, 2251 (1981).
14. G. M. Pastor, PhD. Thesis, Freie Universität Berlin (1989), unpublished.
15. G. M. Pastor, J. Dorantes-Dávila, and K. H. Bennemann, *Physica B* **149**, 22 (1988); *Phys. Rev. B* **40**, 7642 (1989) 7642.
16. V. Korenman and R. E. Prange, *Phys. Rev. Lett.* **53**, 186 (1984).
17. E. M. Haines, R. Clauberg, and R. Feder, *Phys. Rev. Lett.* **54**, 932 (1985).
18. G. M. Pastor and J. Dorantes-Dávila, *Phys. Rev. B* **52**, 13799 (1995).
19. J. Dorantes-Dávila, G. M. Pastor, and K. H. Bennemann, to be published.

Electronic Relaxation in Metallic Nanoparticles

Martin Fierz

Laboratory for Solid State Physics
Swiss Federal Institute of Technology (ETH)
CH-8093 Zürich
SWITZERLAND

Abstract

We present a new method to measure electronic relaxation in gas-suspended nanoparticles with femtosecond accuracy. We do not observe a big enhancement in the relaxation time in metal nanoparticles down to 10 nm diameter.

I. Introduction

The relaxation time of excited electrons is an interesting fundamental property of a material. In metals, this time is of the order of a few femtoseconds. With the advent of reliable commercial Ti:sapphire ultrafast laser systems, it has become possible to study ultrafast phenomena like the relaxation of electrons in solids directly in the time domain with pump-probe experiments. Many different methods are regularly used, such as transient differential absorption spectroscopy,[1] transient reflectivity measurements[2], two-photon photoemission, time-resolved second harmonic generation,[3] and even the STM[4] has already been used to measure femtosecond relaxation of electrons. It has been suggested that hot electrons live longer in nanoparticles,[5] possibly due to quantum size effects.[6] We set up an experiment to measure the lifetime of hot electrons in nanoparticles in gas suspension. Up to now experiments have been performed on particles embedded in a matrix, measuring for instance the transient change in transmittance of the sample after irradiation with a pump pulse[1] or time-resolved second harmonic generation.[3] Another experimental method is to study particles deposited on a surface, for example silver nanoparticles deposited on HOPG.[7] In all of these experiments, unknown particle–matrix or particle-surface interactions can mask the intrinsic electronic lifetime in the nanoparticle. Our approach uses time-resolved two photon photoemission to determine the lifetime of hot electrons in essentially free nanoparticles. Hot electrons can transfer their energy to molecules

Physics of Low Dimensional Systems
Edited by J. L. Morán-López, Kluwer Academic/Plenum Publishers, New York 2001

Table I

Experimental and theoretical values of k in fsec^2 for the noble metals.

Material	Gold	Silver	Copper
k (Theory)	12.7	12.7	24
k (Experiment)	50	14	60

adsorbed on the surface of the material and can thereby mediate catalytic reactions. Since hot electrons have high energies, they can induce reactions with high activation energies, which cannot be induced by phonons for example. A model experiment which shows this is described in Ref. 8.

II. Theory

Hot electrons can relax through a variety of channels: electron-electron scattering, phonon creation, plasmons, etc. However, for excitations much larger than kT, scattering with cold electrons in the vicinity of the Fermi level is the dominant process. Therefore we only have to discuss this process. Fermi Liquid Theory[9] deals with a single excited electron interacting with a Fermi gas. It predicts a lifetime which is strongly energy dependent:

$$\tau = \frac{k}{\Delta E^2}, \tag{1}$$

where we are assuming that the excess energy $\Delta E = E - E_f$ is much larger than kT. This result is intuitively comprehensible: because of energy conservation, an electron can only scatter with electrons with energies higher than $E_F - \Delta E$, and the phase space for final states increases as the excess energy increases. The inverse-square dependence only holds in 3 dimensions though. From this argument it is clear that the density of states in the vicinity of the Fermi energy has a big influence on the lifetime, it determines the factor k in Eq. (1). In materials with a high density of states the lifetime of hot electrons should be much lower than in materials with a low density of states. The predicted values of k are shown in Table I for the materials we investigated, taken from Ref. 10.

III. Experimental

III.1. Technique

We use time-resolved two photon photoemission in this experiment. The principle of this technique is shown in Fig. 1. A pump pulse with a photon energy $h\nu < \Phi$ creates electron-hole pairs in the particle. After a variable time delay Δt the particle is irradiated with the probe pulse, which in our case has the same photon energy. Electrons with energies $E > \Phi - h\nu$ can be photoemitted by the probe pulse. By measuring the decay of the photoemission as the time delay is increased one can deduce the lifetime of the electrons. The decay times are of the same order of magnitude as

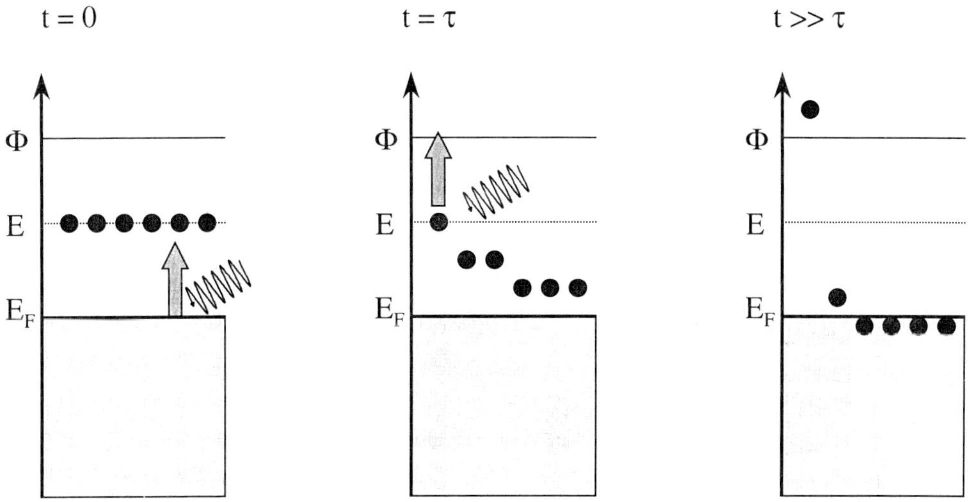

Figure 1. The principle of time-resolved two-photon photoemission: A pump pulse excites electrons above the Fermi level. These states are probed after a variable time delay.

the laser pulse length. This means that we have to take the pulse length into account. For the photoemission rate as a function of the delay time Δt one gets:

$$PE(\Delta t) = I_{\text{pump}}(t)g(t)I_{\text{probe}}(t - \Delta T), \tag{2}$$

where I is the laser intensity and $g(t)$ is the decay function of the hot electrons. In principle one can calculate $g(t)$ by deconvolving the measured photoemission intensity $PE(t)$ with the laser pulse intensity autocorrelation. But since the laser pulse intensity autocorrelation is usually broader than $g(t)$, this procedure is numerically not feasible. Therefore we use a reconvolution method: Assuming a simple exponential decay with time constant τ one gets:

$$PE(\Delta t) = I_{\text{auto}}(\Delta t)\exp(-t/\tau), \tag{3}$$

where I_{auto} is the autocorrelation of the laser pulse intensity. We measure I_{auto} and fit a sech-squared laser pulse to it. Then, we convolve this fitted laser pulse intensity autocorrelation with $\exp(-t/\tau)$, using τ as fit parameter to get a least-squares fit to the measured curve. We stress that this method relies on the fact that the electrons really have the assumed decay characteristic. Recent theoretical developments take into account that this is not necessarily the case: electrons from higher levels can scatter into lower levels and refill them, resulting in a longer apparent decay time of the lower energy level. Furthermore, electron-hole recombination can produce Auger electrons at high energies, again apparently enhancing the lifetime of the pumped states. For a detailed discussion of this subject see Ref. 11. Time-resolved two-photon photoemission has been used to measure electron relaxation at metal surfaces under UHV conditions.[10] In these experiments it is possible to measure the energy of the photoemitted electron. Since our nanoparticles are suspended in a carrier gas at atmospheric pressure, we have no possibility to measure their energy. Therefore, our measured lifetime is an average over the experimentally accessible energy range:

$$\langle \tau \rangle = \frac{\int \tau(E)n(E)dE}{\int n(E)dE}, \tag{4}$$

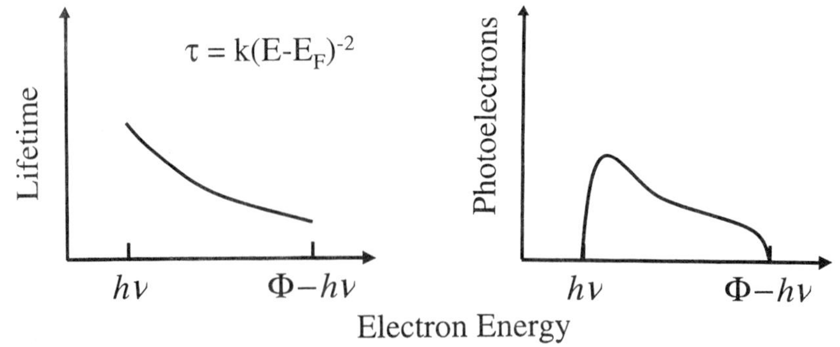

Figure 2. Relaxation time and photoelectron spectrum as a function of energy.

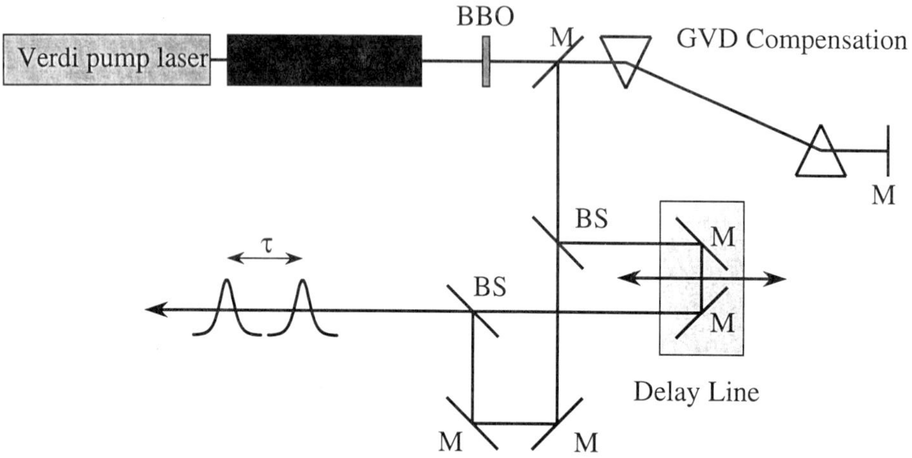

Figure 3. Ultrafast pump-probe laser system setup.

where $n(E)$ is the number of photoelectrons emitted with energy E (see Fig. 2). The photoelectron spectrum $n(E)$ has been measured at surfaces but need not be the same in nanoparticles: if the particles are smaller than the mean free path of the hot electrons, cascade formation is suppressed. Therefore we also expect the photoelectron spectrum to change with changing particle size.

III.2. Setup

Our experiment uses aerosol techniques to handle the particles and an ultrafast pump-probe laser setup.

III.3. The Laser System

Our optical system is shown in Fig. 3. The ultrafast Ti:sapphire laser is pumped by a 10 W solid-state Nd:vanadate laser. It is intra-cavity frequency-doubled to produce 532 nm light. The (infra-)red output power of the Ti:sapphire laser is about 1.1 Watt. The repetition rate is 75 MHz. The wavelength is tunable from 750 to 850 nm. The pulse length, measured in an autocorrelator, is about 50 femtoseconds. The red beam is frequency-doubled in a 0.2 mm thick BBO crystal. We get up to 200 mW average power in the blue beam. We have a group velocity dispersion control in form of a dou-

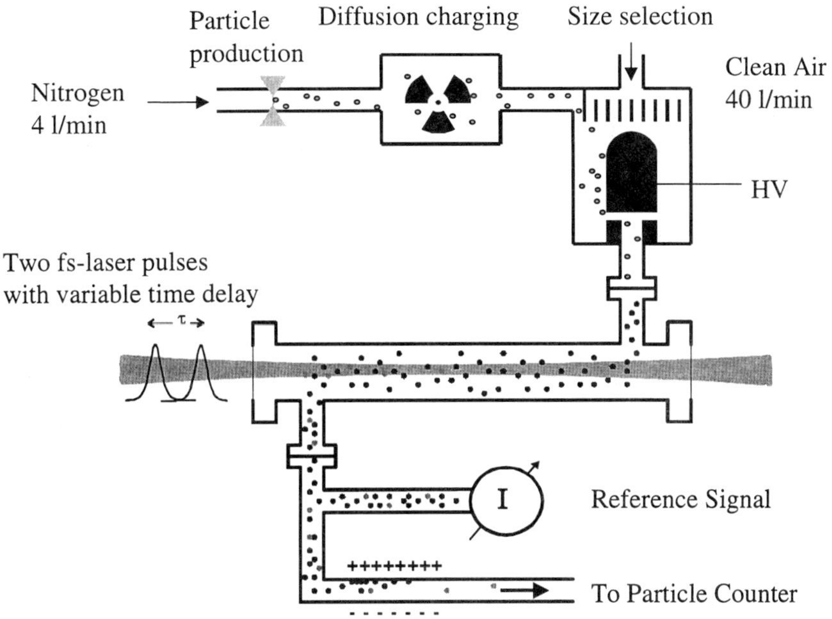

Figure 4. Particle production, size selection and detection system.

ble prism setup to precompensate for the dispersion in the optics. The pulse length of the blue pulses is slightly shorter than that of the red pulses. Typically it is between 40 and 50 fs. The blue beam is split and reassembled in a Michelson-type interferometer. The pump pulse and the probe pulse have equal intensity. Using a mirror setup, the polarization in the fixed-length arm can be rotated by 90 degrees if desired. We can remove coherent effects by using cross-polarized beams. The other arm has a variable length, adjustable in steps of 200 nm. This corresponds a time resolution of 0.66 femtoseconds. In the blue wavelength range there are no commercial autocorrelators available, so we constructed a small vacuum cell with a metal target to measure two-photon photoemission as a function of the delay between the pump and probe pulse. If the lifetime of the hot electrons in the metal target is sufficiently low, this curve can be used to measure the laser pulse length. We tested various materials and found the shortest lifetimes in a Pt-Ir alloy, slightly shorter than in graphite which is known to have a very short lifetime. The pulse length measurement is crucial for the absolute accuracy of this experiment and is also its limiting error.

III.4. The Particles

Our particle production, handling and detection system is shown in Fig. 4. Metal nanoparticles are produced in a spark discharge[12] between two electrodes of the material we wish to investigate. The heat of the discharge evaporates a small part of the electrode. The metal vapor condenses rapidly to small particles. These particles agglomerate to form fractal–like particles. Since we want to measure a size dependence, we heat the agglomerates in a furnace. The particles melt and become nearly spherical and crystalline (see Fig. 5). They do have some defects left though. The result of this process is a gas with suspended metal nanoparticles in it. We can select the particle size with a differential mobility analyzer.[13] The resulting particle stream consists of

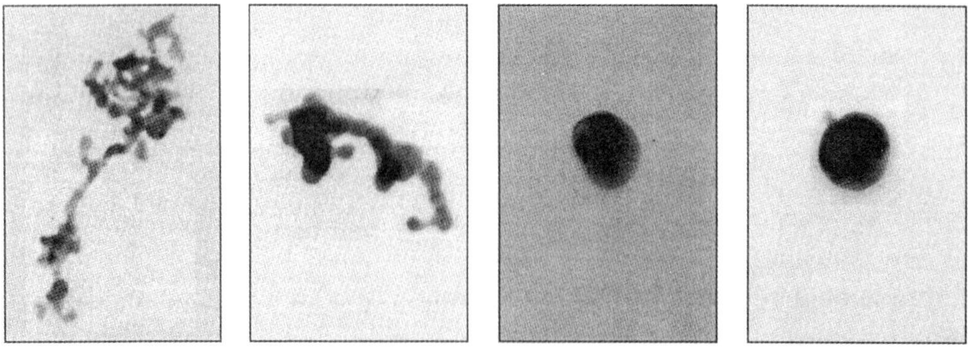

Figure 5. TEM micrographs of 70 nm gold particles, furnace temperature 100, 400, 800, 1300°C, from left to right.

Table II

Relaxation time as a function of the particle diameter for gold nanoparticles.

Particle diameter[nm]	15	20	30	50	90
Relaxation time[fs]	24 ± 5	23 ± 3	19 ± 2	19 ± 2	16 ± 2

particles of equal size, each of them having one negative elementary charge. The particles pass through a photoemission tube where they are irradiated with the pump and probe pulse. If one electron is emitted from the particle, it becomes electrically neutral. Then the particles pass through an electrofilter. This is simply a cylindrical capacitor with a high voltage applied. Here, charged particles feel the electric field and hit the wall of the capacitor and are thereby removed from the gas stream. The remaining particles are counted in a condensation nucleus counter. This commercial device can count single particles. This means that we count one particle in the condensation nucleus counter for every electron which is emitted in the photoemission tube, making the experiment very sensitive.

IV. Results

IV.1. Gold

The first system we investigated was gold. As our nanoparticles are exposed to air it is not quite clear how the surface of the particles looks. With gold particles one can hope that despite having the particles in gas at atmospheric pressure the surface will remain relatively clean, in contrast to silver and copper. We were able to measure the relaxation time in the size range of 15 to 90 nm. Figure 6 shows the photoemission as function of the pulse delay for particles 50 nm in diameter. Table II summarizes the relaxation times as function of the particle size.

To compare these results with bulk measurements, one has to know the work function of the particles. Figure 7 shows a Fowler plot for 70 nm gold particles. We

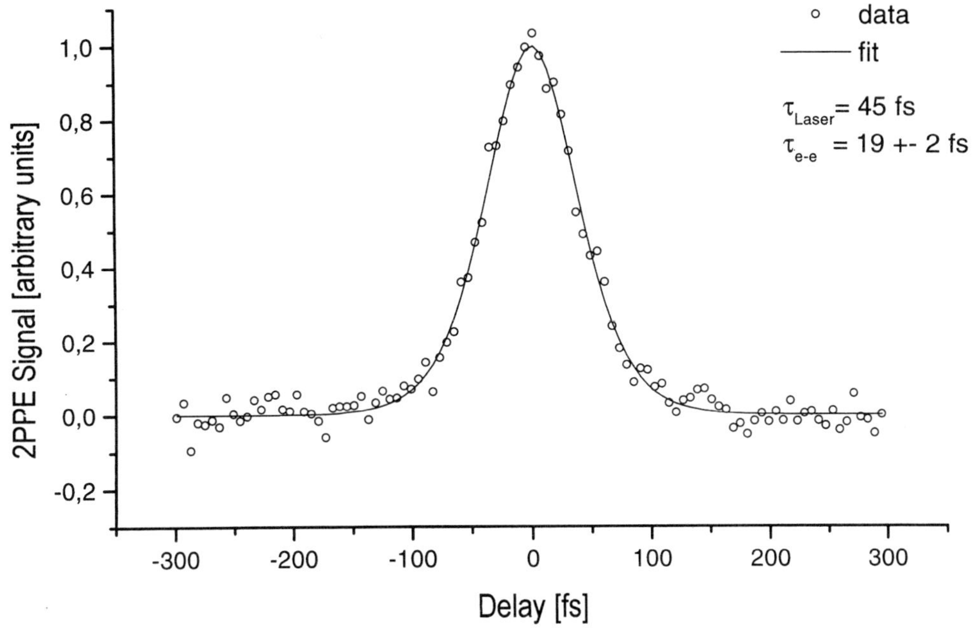

Figure 6. Experimental data for gold particles with 50 nm diameter.

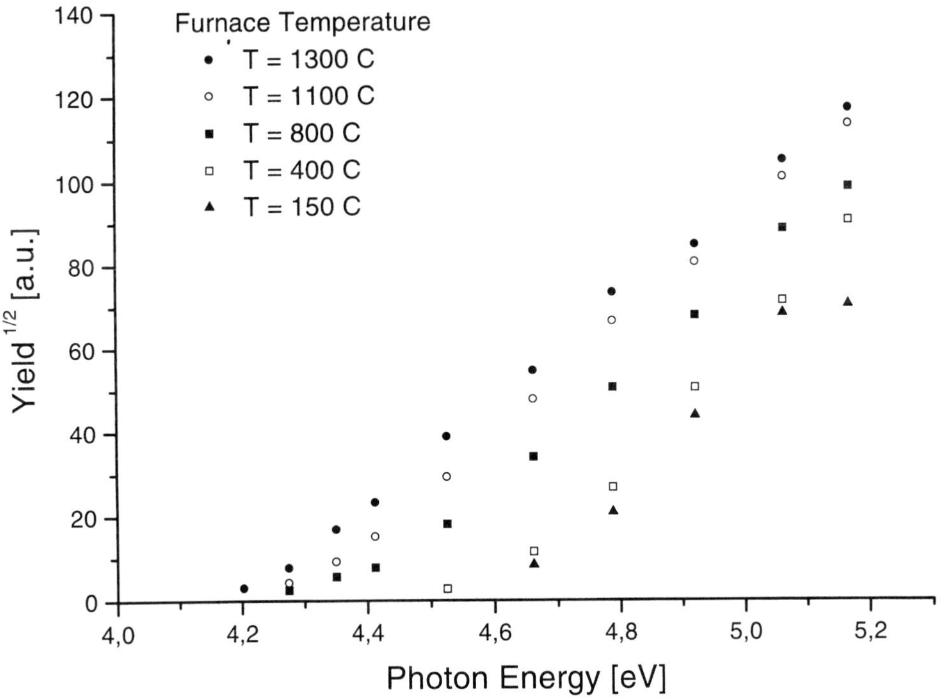

Figure 7. Fowler plot to determine the work function for 70 nm gold particles heated to 1300°C.

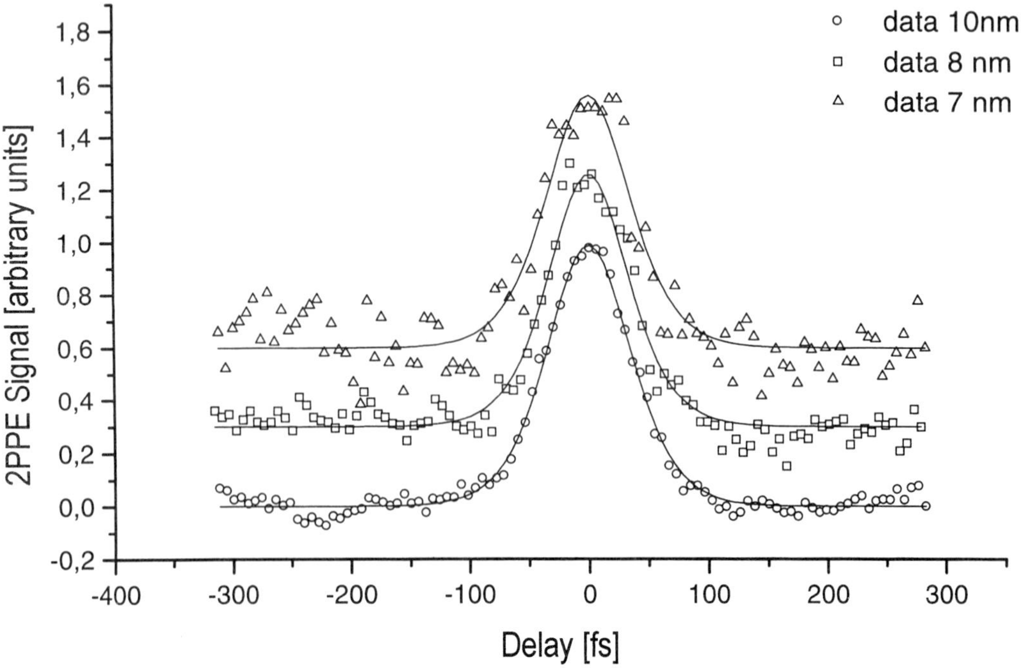

Figure 8. Experimental data for silver particles with 7, 8 and 10 nm diameter.

find that the work function depends on the furnace temperature and is 4.2 eV for a furnace temperature of 1300° C. Using data from bulk measurements[10] and assuming a flat photoelectron spectrum, we calculate the average relaxation time $\langle\tau\rangle$ to be 14.3 fs. A more realistic photoelectron spectrum as in Fig. 2 would give a larger relaxation time because there are more electrons with low energy.

IV.2. Silver

Silver nanoparticles have a huge photoelectric yield.[5] Therefore the signal to noise ratio is improved compared to gold. We were able to measure silver particles down to 7 nm, corresponding to 10 000 silver atoms. The measured curves for 7, 8 and 10 nm are shown in Fig. 8.

IV.3. Copper

The last system we investigated was copper. Copper is known to have a high lifetime of electrons in the bulk. Here we find a large dependece of τ of the furnace temperature. However, transmission-electron microscopy shows that the particles are not nice, round and crystalline as for silver and gold, but seem to be contaminated. An experimental curve is shown in Fig. 9. Obviously the fit is not as good as in the case of silver and gold; an indication that our assumption that $g(t)$ is a monoexponential decay is not correct.

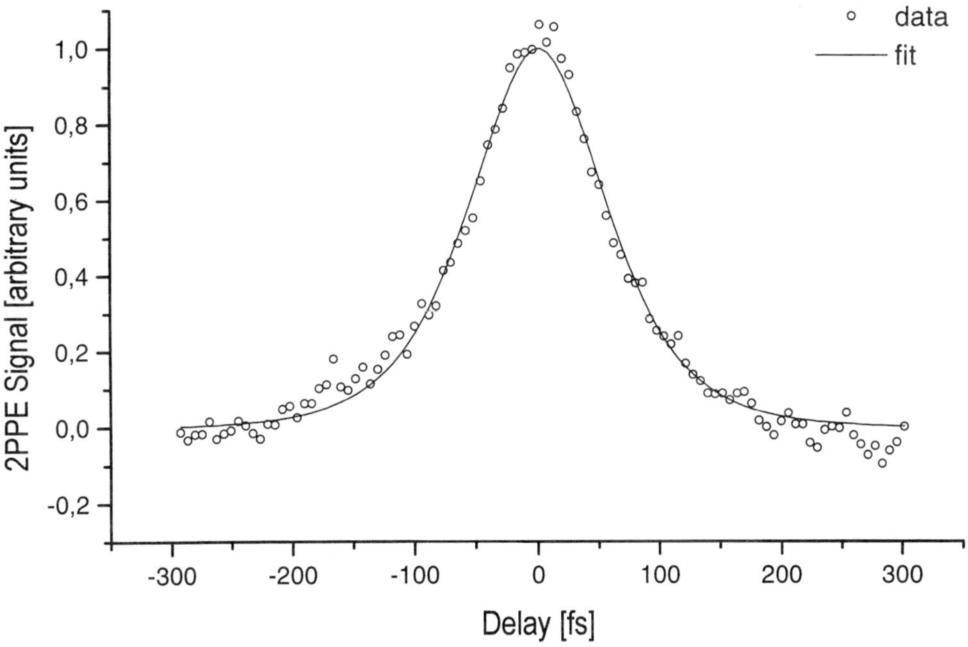

Figure 9. Experimental data for copper nanoparticles with 40 nm diameter.

V. Conclusions

In gold particles we observe a small size dependence: The relaxation time gets larger for smaller particles. However the effect is just slightly larger than the accuracy of the experiment. Two other effects probably influence the observed relaxation time: 1) as the particle gets larger, transport effects may get important. Excited electrons can move away from the exitation region resulting in an apparently smaller relaxation time and 2) as the particle gets larger, more cascade electrons are created. This results in an apparently larger relaxation time. The values we obtain agree well with an averaged value of the relaxation times measured on bulk surfaces. The work function of the gold particles is surprisingly low compared to the literature value of 5.1 eV.[14] Probably even the gold particle surface is contaminated. We hope that although we have a contaminated surface, we still measure the intrinsic electronic properties of gold clusters. Our results show that as far as electron-electron scattering is concerned, these particles still have to be considered as being macroscopic; there is no big size effect. Still, the theoretical question remains: where does one expect the increase in lifetime to set in? In quantum dots one usually applies the following criterion for quantum size effects to occur: as soon as the energy level spacing between two electronic levels gets larger than kT, the discreteness of the electron energy spectrum becomes apparent and one expects something to happen. In this case it means that the particles should contain roughly $N = E_F/kT = 5$ eV$/(1/40)$ eV $= 200$ atoms. Our smallest particles, made of silver, still contain 10 000 atoms. A cluster of 200 atoms corresponds to a particle size of 2–3 nm. It is possible to measure such small particles in principle, however there are still many experimental difficulties to overcome.

References

1. S. Link, C. Burda, Z. L. Wang, and M. A. El-Sayed, *J. Chem. Phys.* **111**, (3), (1999).
2. A. Guerra III, W. E. Bron, and C. Suárez, *Appl Phys. B* **68**, 405 (1999).
3. B. Lamprecht, J. R. Krenn, A. Leitner, and F. R. Aussenegg, *Appl. Phys. B* **68**, 419 (1999).
4. L. Bürgi, O. Jeandupeux, H. Brune, and K. Kern, *Phys. Rev. Lett.* **82**, 4516 (1999).
5. U. Müller, Dissertation 8544, ETH Zürich, 1988.
6. J. A. A. J. Perenboom, P. Wyder, and F. Meier, *Physics Reports,* **78**, 2, (1981).
7. K. Ertel, U. Kohl, J. Lehmann, M. Merschdorf, W. Pfeiffer, A. Thon, S. Voll, and G. Gerber, *Appl. Phys. B* **68**, 439 (1999).
8. M. Bonn, S. Funk, Ch. Hess, D. N. Denzler, C. Stampfl, M. Scheffler, M. Wolf, and G. Ertl*Science,* **285**, 1042, (1999).
9. *The Theory of Quantum Liquids*, edited by D. Pines and P. Nozières, (Benjamin, New York, 1966).
10. M. Aeschlimann, Habilitation Thesis, ETH Zürich 1996.
11. R. Knorren and K. H. Bennemann, *Appl. Phys. B* **68**, 401, (1999).
12. S. Schwyn, E. Garwin, and A. Schmidt-Ott, *J. Aerosol. Sci.* **19**, 639 (1988).
13. E. O. Knutson and K. T. Whitby, *J. Aerosol Sci.* **6**, 443 (1975).
14. *Handbook of Chemistry and Physics*, 72nd Edition, (CRC Press, Boston).

An Energetical Study of Transition-Metal Nanoclusters within the Embedded Atom Method

L. García González and J. M. Montejano-Carrizales[1]

Instituto de Física, "Manuel Sandoval Vallarta"
Universidad Autónoma de San Luis Potosí
78000 San Luis Potosí, S.L.P.
MEXICO
e-mail: [1]jmmc@dec1.ifisica.uaslp.mx

Abstract

The Foiles' version of the Embedded Atom Method (EAM) is used to study the structural stability of Ni, Cu, Ag, and Pd metal clusters. Various structures and sizes are studied. They are based in polyhedra as: tetrahedron, hexahedron, triangular antiprism, trigonal prism, decahedron, and dodecadeltahedron. To construct the various clusters for each one of the polyhedron, top sites over the faces of the polyhedron are used. The faces in the polyhedra can be triangular or squares faces. Face-centered-cubic clusters are also studied. Comparison between the cohesive energy of the various clusters with the same number of atoms but different structures is carried out to find the most stable cluster shape.

I. Introduction

The knowledge of the geometrical shape of metal clusters is very important to understand most of their properties. Various theoretical and experimental studies have been carried out to determine the structural stability of clusters. The dependence of the shape of the structure on size is another intriguing issue.

Theoretical studies of Ni_N clusters, within the Embedded Atom Method (EAM), have been performed to analyze the evolution of the structural stability of large Cu, Ni, Pd and Ag cluster with size up to 5083 atoms;[1] and enhanced stability is found after the completion of certain caps (*umbrellas*) in the size range from 13 to 147 atoms.[2] Other studies reveal that the cluster geometrical shape influence the nonmetal-metal

Physics of Low Dimensional Systems
Edited by J. L. Morán-López, Kluwer Academic/Plenum Publishers, New York 2001

transition in Ni for sizes of the order of 50 atoms,[3] and for Co clusters up to 40 atoms for icosahedral clusters.[4] The relation between the geometrical structure and magnetism of Ni clusters using a simple theoretical model was reported in Ref. 5.

The main goal in this work is to study the energetic structural stability for clusters of a few atoms, within the EAM, and compare with the results for icosahedral and cubo-octahedral clusters obtained within the same method.[1,2] Other structures based on polyhedra that do not belong to the icosahedral like family of clusters might represent also possible structures.

II. The Model

The Embedded Atom Method (EAM) produces different effective interactions between two atoms when those are in different environments in the cluster. Thus the scheme goes beyond the simple pair potential approximations. The EAM, which is related to the effective medium theory, replaces the N-atom problem of the cohesion energy of a metallic system by the sum of the interaction energies of each of the N atoms with a host modeling the effect of the remaining $N-1$ atoms on the one in question. The physical quantity responsible for the interactions is the electron density because those methods are based on the density functional theory.[6]

Daw and Baskes[7] proposed to look at every atom in a metallic system as if it was an impurity in a host formed by the rest of the atoms. The binding energy of the whole system is

$$E_B = \sum_i \left[F_i(\bar{\rho}_i) + \frac{1}{2} \sum_{i \neq j} \phi_{ij}(R_{ij}) \right], \tag{1}$$

where F_i is the energy required to embed atom i into the background electron density $\bar{\rho}_i$ at site i, and $\phi_{ij} R_{ij}$ is the core-core pair interaction between atoms i and j separated by a distance R_{ij}. The host electron density $\bar{\rho}_i$ is approximated by a superposition of the spherically averaged electron densities of all atoms surrounding atom i,

$$\bar{\rho}_i = \sum_{j(\neq i)} \rho_j(R_{ij}), \tag{2}$$

where $\rho_j(R_{ij})$ is the electron density of atom j at the position of the nucleus of atom i. Daw and Baskes[7] obtained the functions F_i and ϕ_{ij} empirically from the physical properties of the solid, although these can be also obtained from first principles.[8] In the version of the EAM by Foiles *et al.*[9] the following function to describe the core-core pair interaction is used

$$\phi_{ij}(R) = \frac{Z_0^2(1+\beta R^\nu)^2 e^{-2\alpha R}}{R}, \tag{3}$$

where Z_0 is the number of outer electrons of the atom, and α, β, and ν are adjustable parameters.

The atomic densities are obtained from the atomic Hartree-Fock calculations of Clementi and Roetti[10] and MacLean and MacLean[11] by writing

$$\rho(R) = n_s \rho_s(R) - (n - n_s)\rho_d(R), \tag{4}$$

where n is the total number of outer (s plus d) electrons, n_s is a measure of the number of outer s-electrons, and ρ_s and ρ_d are the partial densities associated with the s- and d-wave functions, respectively. If the atomic densities $\rho(R)$ and the pair

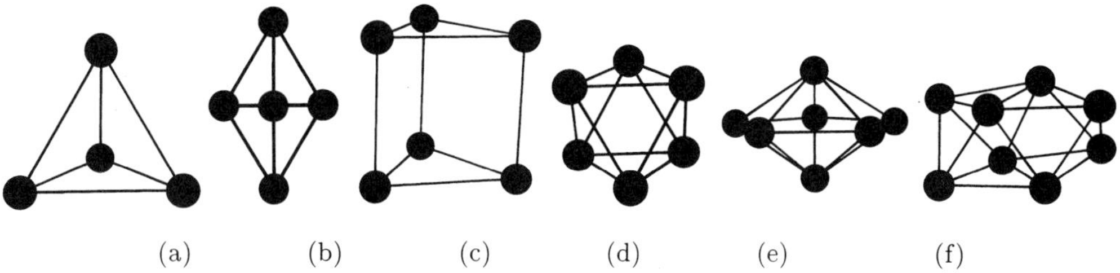

Figure 1. Basic polyhedra used in this work. (a) Tetrahedron, (b) hexahedron, (c) trigonal prism (triangular prism) (d) triangular antiprism (octahedron), (e) decahedron, and (f) dodecadeltahedron.

interactions $\phi(R)$ are both known, the embedding energy can be uniquely defined by requiring that the total energy given by Eq. (1) matches the equation of state of Rose *et al.*[12] for a broad set of values of the lattice parameter. In this way, the function $F_i(\rho)$ is obtained in numerical form.

Foiles *et al.*[9] determined α, β, ν, and n, for the *fcc* metals Cu, Ag, Ni, and Pd, by optimizing the agreement between predicted and experimental data for the elastic constants and vacancy formation energies of the pure metals and for the heats of solution of the associated binary alloys, when information on the latter was available. This procedure ensures perfect agreement with equilibrium bulk lattice constants, cohesive energies, and bulk moduli. The values of the parameters as well as the embedding functions obtained by these authors were used in the cluster calculations of this work.

III. Geometrical Structures

For the calculation of the cohesive energy per atom, within the EAM, various polyhedra were used in this work as basic structures: tetrahedron (Tetra), hexahedron (Hexa), trigonal prism (triangular prism) (Trigo), triangular antiprism (Trian), decahedron (Deca), and dodecadeltahedron (Dode). They are shown in Fig. 1. It is worthwhile to note that the triangular antiprism corresponds to an octahedron. The geometrical characteristics of those polyhedra are enumerated in Table I. The use of the Table is as follows: the polyhedron (first column) is constituted by N_s sites (second column) in a such way that N_f faces are generated (third column); these faces can be triangular and/or square. The following 10 columns give the coordination numbers and the number of sites with those, for the polyhedron. The number of bonds N_B in the cluster is given in the last column.

To increase the number of sites in a given polyhedron, top sites over the faces are added; A top site is one over the center of the face and equidistant to the sites forming the face, these top sites are at first neighbor distance, d_{nn}, and have three or four first neighbors depending of the form of the face. In this way, the covered polyhedron (polyhedron$_c$) is obtained. These polyhedra and their geometrical characteristics are listed in the same Table I with the subindex c. It is worthwhile to mention that it is not always possible to put a top site over each face, in some cases two top sites belonging to two different triangular faces are at a distance smaller or larger than the first neighbor distance. There are two possibilities in these cases: *i*) when the distance is small enough, less than $0.2d_{nn}$ (Figs. 2a and 2b), to get together the two sites as a single site with four first neighbors belonging to the two triangular faces, Fig. 2(c);

Table I

Geometrical characteristics of the polyhedra used in this work. N_s is the number of sites and N_f the number of faces in the polyhedron. The subindex ($_c$) added to the name of the polyhedra represents the covered polyhedra.

Structure	N_s	N_f	Coordination Numbers										N_b
			3	4	5	6	7	8	9	10	11	12	
Tetra	4	4	4										6
Tetra$_c$	8	6	4			4							18
Hexa	5	6	2	3									9
Hexa$_c$	11	12		6		2		3					30
Trigo	6	5	6										9
Trigo$_c$	11	9	2	3		6							27
Trian	6	8		6									12
Trian$_c$	14	12	8					6					36
Octa	6	8		6									12
Octa$_c$	14	24	8					6					36
Deca	7	10		5		2							16
Deca$_c$	19	25				12		5				2	68
Dode	8	12		4	4								18
Dode$_c$	20	22	4		8			4		4			62

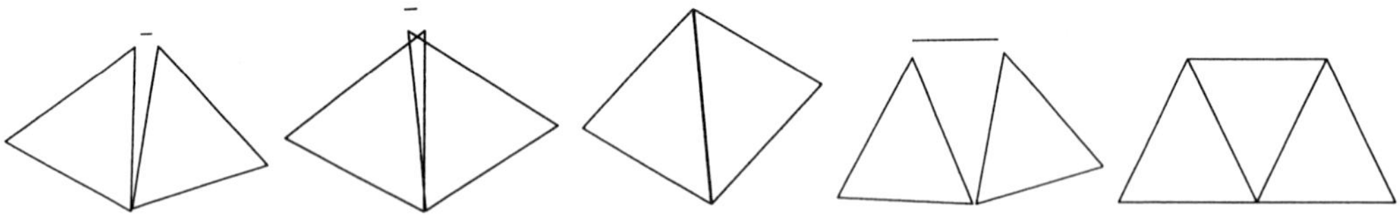

(a) (b) (c) (d) (e)

Figure 2. Arrangements for two top sites separated by a distance different to d_{nn}. (a) and (b) Distance much smaller than d_{nn}, (c) the two sites as a single site. (d) Distance slightly smaller or larger than d_{nn}, and (e) these sites conected between them.

and *ii*) when the distance is not too small, $0.9d_{nn} \leq d \leq 1.1\, d_{nn}$ (Fig. 2d) and it is possible to get two first neighbor sites with coordination four, three with the face's sites and one with the other site, Fig. 2(e). The most of the times the price to pay for increase the size of the cluster is to have first neighbor distances shorter and higher than d_{nn}, the first neighbor distance in the basic polyhedra.

For a given number of sites in a cluster, $N = N_s + N_t$, where N_t is the number of top sites, there are two cases: a) N_s and N_t are kept constant, *i.e.* various arrays for the same polyhedron with different arrangements of the top sites, and b) N_s and N_t are not constants, different structures based in different polyhedra. For the first case there are $N_t = N - N_s$ top sites, $N_t < N_f$, and various ways to put them over the faces of the polyhedron. In Fig. 3 we show the basic polyedra plus one top site. The tetrahedron plus one top site coincides with the hexahedron (see Fig. 3a); The

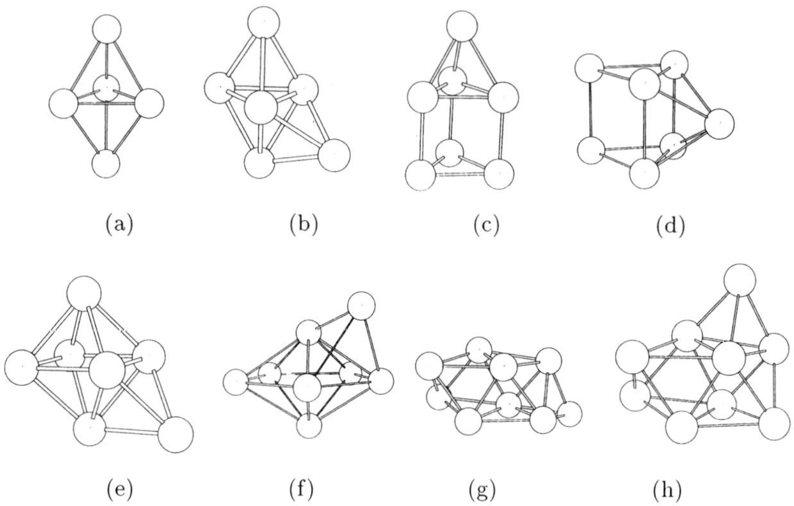

Figure 3. Different arrays for all the basic polyhedra plus one top site.

tetrahedron, the hexahedron, the triangular antiprism and the decahedron show only one array (Figs. 3a, 3b, 3e, and 3f), while the trigonal prism and the dodecadeltahedron show two (Figs. 3c, 3d, 3g, and 3h). That is because these last have two types of faces and the others have only one. The clusters presented in Fig. 3 are not the whole of the arrangements for the polyhedra and one top site, but they are the representative for each one of the basic polyhedra of the equivalent arrays. Two or more arrays are equivalent if they have the same coordination numbers. In Fig. 4 we present the various arrangements of basic polyhedra plus two top sites, and in Fig. 5 all the inequivalent arrays for the hexahedron plus three top sites. Figs. 4(a)–4(c) and Fig. 5 are examples of case a) for $N_s = 5$ and $N_t = 2$ and $N_t = 3$, respectively. For the case b) there are various examples: for $N = 6$, Figs. 1(c), 1(d), and 3(b); for $N = 7$, Figs. 1(e), 3(c)–3(e), 4(a)–4(c); and for $N = 8$, Figs. 1(f), 3(f), 4(d)–4(i).

As it was mentionated above the triangular antiprism corresponds to an octahedron, and this last can be considered as a part of a *fcc* lattice. The increasing of the number of sites in a cluster for these two structures coincides up to 14 sites. The position of top sites in the triangular antiprism are not coincident with sites of a *fcc* lattice for sizes higher than 14 sites. In Table I we can notice that the rows for Trian and Octa are the same, and for Trianc and Octac differ in the number of faces. Then the triangular antiprism and the octahedron, as a piece of a *fcc* lattice, are different for $N > 14$.

IV. Results and Discussion

In order to have a reference to compare the calculated cohesive energy for the whole structures considered here, the data for Ni clusters with 13 to 147 atoms were obtained from Ref. 2. These data were calculated within the EAM for two icosahedral structures, and a relaxation was performed by the steepest-descent strategy. In the cases of Cu, Pd, and Ag, the data for 13, 55 and 147 atoms were obtained from Ref. 1, which were also calculated within the EAM for cubo-octahedral and icosahedral structures.

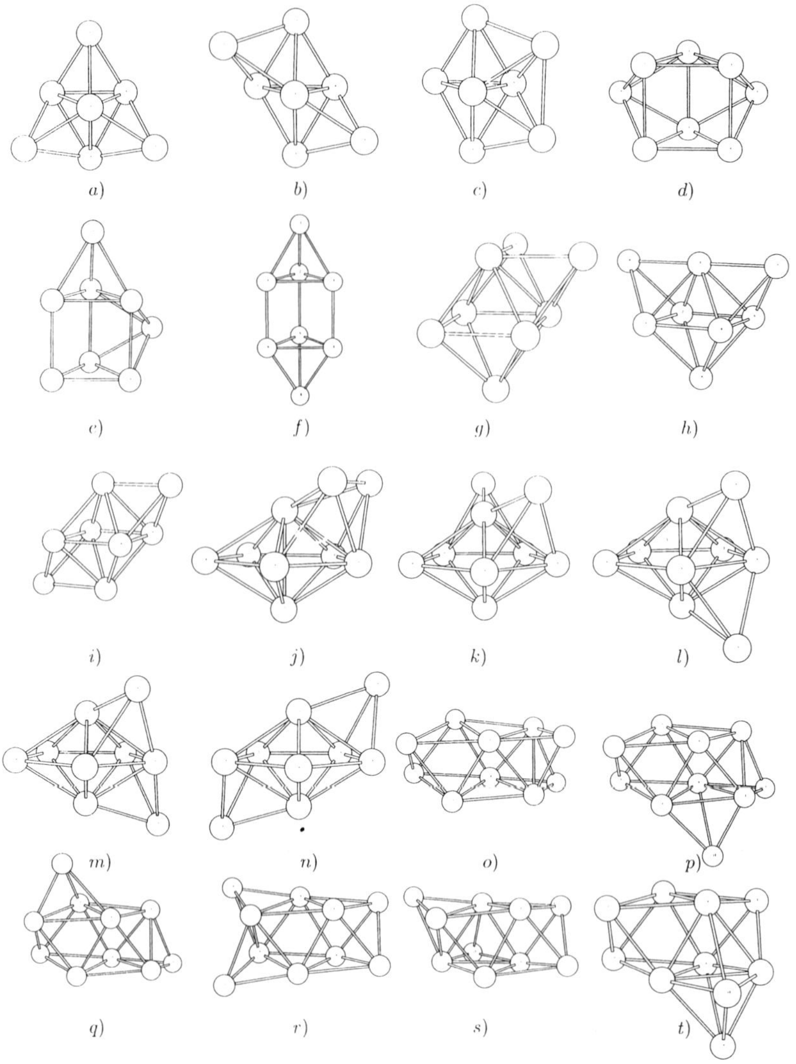

Figure 4. The various arrangements of the basic polyhedra plus two top sites.

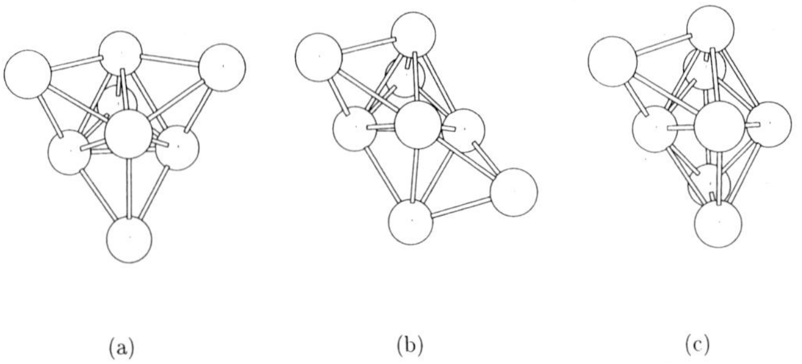

Figure 5. All the inequivalent arrays for the hexahedron plus three top sites.

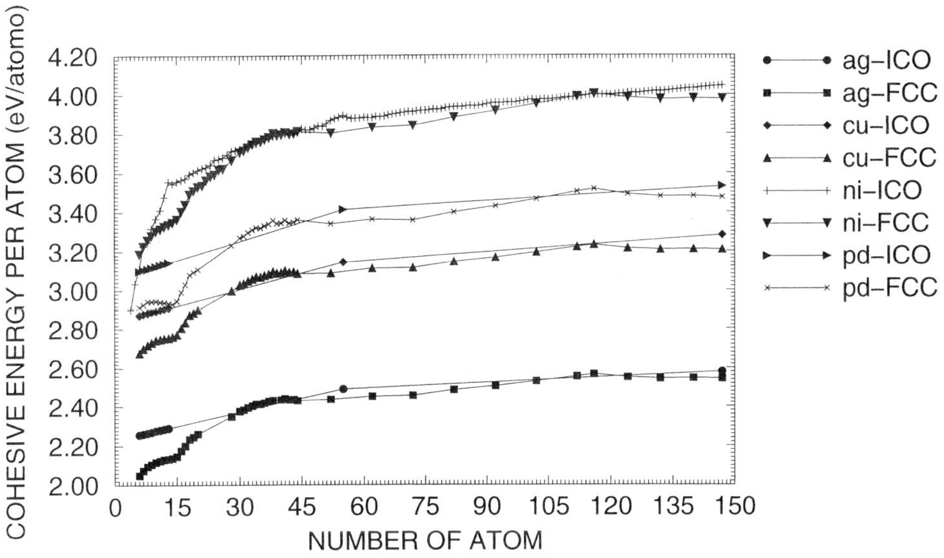

Figure 6. Comparison of the cohesive energy per atom between icosahedral (ICO) clusters[1,2] and *fcc* clusters based on the octahedron as a function of the number of atoms.

Cohesive energy values for icosahedral clusters between 4 and 13 atoms were calculated here in the way described below, for all the structures.

In this work, cohesive energy calculations for Ni, Cu, Pd, and Ag clusters were performed using the values of the parameters as well as the embedding functions obtained by Foiles *et al.*[9] These calculations were carried out for each one of the inequivalent arrays obtained for each structure and were compared between themselves. We considered clusters starting with 4 atoms for different structures, for reasons discussed below. The maximum values in the cohesive energy for each size and for each structure were chosen.

In Fig. 6 we reproduce the results reported for the icosahedron (Ico) in Refs. 1 and 2, and to compare them with the results for the octahedron as a piece of a *fcc* structure for sizes between 4 and 147 atoms. We present results for the whole set of values for Ni. For Ag, Cu and Pd, there are only results for $N = 13, 55, 147$. A rightline from $N = 13$ to $N = 55$ and another from $N = 55$ to $N = 147$ were drawn to have an approximated reference. From Fig. 6 is easy to see that the results show the same behavior for Ni, Cu, Pd, and Ag, *i.e.*, the icosahedron is more stable than the *fcc* clusters except for values around 38 and 116 atoms. The cubo-octahedron clusters tend to be more stable than the icosahedron as the cluster size is increased.[1]

Since the dependence of the cohesive energy on the number of atoms, N, for Ni, Cu, Pd, and Ag, present the same general behavior, and the data for all Ni clusters sizes are known, only the results for nickel are analyzed. The cohesive energy values for all the structures are drawn in Fig. 7, in the range $N = 4$–50. The range is not the same for all the structures. The range for the dodecadeltahedron is up to 20 atoms; 23 for the trigonal prism; 26 for the triangular antiprism; 38 for the decahedron; 48 for the tetrahedron and 56 for the hexahedron. It can be seen from Fig. 3 that up to $N = 15$ atoms all the structures, exception of the *fcc* clusters, tend to follow the icosahedral values. For $N > 15$ dodecadeltahedral, trigonal prism and triangular

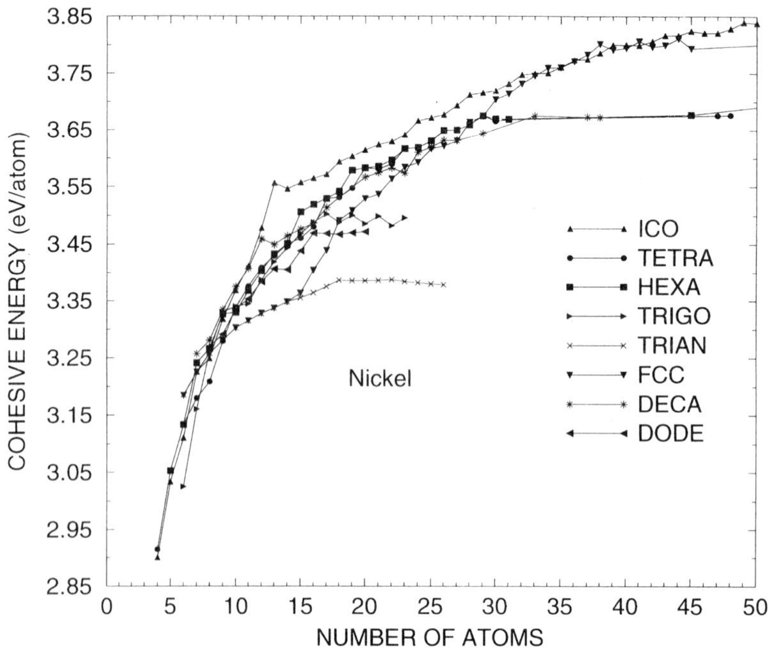

Figure 7. Comparison of the cohesive energy per atom of the different polyhedra used in this work as a function of size, tetrahedron (TETRA), hexahedron (HEXA), trigonal prism (TRIGO), triangular antiprism (TRIAN), decahedron (DECA), and f) dodecadeltahedron (DODEC).

antiprism structures leave this tendency; the same for hexahedral and decahedral structures but for $N > 29$ atoms. That effect is because of the change in the first neighbor distances mentionated above. For small values of N tetrahedral, hexahedral and decahedral structures are the best candidates together. The same property is observed for *fcc* clusters for $N < 10$ and $N > 30$ atoms.

V. Conclusions

We can conclude that the icosahedral clusters are the most stable for clusters with a few atoms, but there are another polyhedral structures that can be used in theoretical calculations, because they are in energetic competition for a few atom clusters. In addition, *fcc* clusters are in competition for sizes around 38 and 116 atoms.

The results here obtained may be improved if relaxation effects are incorporated in calculations by means of steepest-descent strategy, simulation annealing, molecular dynamics or another relaxation technique.

Acknowledgments

This work was partially supported by Consejo Nacional de Ciencia y Tecnología (México) Grants No. G-25851-E and 4920-E9406.

References

1. J. M. Montejano-Carrizales, M. P. Iñiguez, and J. A. Alonso, *J. Cluster Science* **5** (1994) 287.
2. J. M. Montejano-Carrizales, M. P. Iñiguez, J. A. Alonso, and M. J. López, *Phys. Rev. B* **54** (1996) 5961.
3. F. Aguilera-Granja, S. Bouarab, A. Vega, J. A. Alonso and J. M. Montejano-Carrizales, *Solid State Commun.* **104** (1997) 635.
4. F. Aguilera-Granja, J. M. Montejano-Carrizales, J. Guevara, and Ana María Llois, *Solid State Commun.*, in press.
5. F. Aguilera-Granja, J. M. Montejano-Carrizales, and J. L. Morán-López, *Solid State Commun.* **107** (1998) 25.
6. P. Hohenberg and W. Khon, *Phys. Rev. B* **136** (1964) 864.
7. M. S. Daw and M. I. Baskes, *Phys. Rev. B* **29** (1984) 6443.
8. M. S. Daw, *Phys. Rev. B* **39** (1989) 7441.
9. S. M. Foiles, M. I. Baskes, and M. S. Daw, *Phys. Rev. B* **33** (1986) 7893.
10. E. Clementi and C. Roetti, *At. Data Nucl. Data Tables* **14** (1974) 177.
11. A. D. MacLean and R. S. MacLean, *At. Data Nucl. Data Tables* **26** (1981) 197.
12. J. H. Rose, J. R. Smith, F. Fuinea, and J. Ferrante, *Phys. Rev. B* **29** (1985) 2963.

Deformation Effects in the Magnetic Moments of Ni Clusters

J. Hernández-Torres,[1] F. Aguilera-Granja,[1] and A. Vega [2]

[1] *Instituto de Física, "Manuel Sandoval Vallarta"*
Universidad Autónoma de San Luis Potosí
78000 San Luis Potosí, S.L.P.
MEXICO

[2] *Departamento de Física Teórica*
Universidad de Valladolid
E-47011 Valladolid
SPAIN

Abstract

We study the effects of the geometric deformations on the magnetic moments in free Ni_N with different geometrical shapes. Initial geometries of the clusters come from Molecular Dynamics and Monte Carlo calculations based on semi-empirical potentials. We only consider deformations that keep constant the surface area of the clusters. The spin-polarized electronic structure has been calculated within a self-consistent tight binding method considering the $3d$, $4s$, and $4p$ valence electrons in a mean field approximation. The results indicate that the changes in the d component of the magnetic moment as a consequence of the deformation can be explained in terms of the coordination number and interatomic distance, whereas the changes in the sp component can not be explained in terms of those geometrical parameters. Our results are compared with the available experimental data.

I. Introduction

In the last two decades many of the experimental results about free magnetic clusters were performed using the Stern-Gerlach techniques based on the deflection of a cluster beam due to its interaction with an external inhomogeneous magnetic field that allow to determine the average magnetization per atom.[1-6] Those experimental results represent a great step in the understanding of the magnetic properties of small free

Physics of Low Dimensional Systems
Edited by J. L. Morán-López, Kluwer Academic/Plenum Publishers, New York 2001

clusters. A direct comparison with many theoretical calculations for magnetic free standing clusters has been possible and from this comparison a lot of fundamental knowledge was obtained.[7–17]

However, the technological applications of free clusters are really limited since most of the electronic devices are supported on a surface. The surface effects are mainly two: *i*) charge transfer and *ii*) deformations in the cluster structure with respect to that of the free cluster. When clusters are supported on a surface, there is an electronic charge transfer from the cluster to the surface or in the oppossite way. When the aim is to preserve the intrinsic electronic properties of the supported clusters, noble metal substrates like Cu or Ag are used. Then, the electronic interaction between the cluster and the substrate is minimized and therefore the charge transfer is negligeable. However, the modifications of the geometrical structure of the cluster with respect to the free-standing situation are still present, being the main responsible of the changes of the magnetic properties. In this work our aim is to study how small deformations of Ni clusters affect the magnetic moment. As a first step of a more general study, we consider free-standing clusters for which the ground state geometrical structures are available in the literature, and we perform prolate- and oblate-like deformations such that the symmetry in the direction of the main axis and the surface area of each cluster are preserved. For the resulting geometries we calculate the local (orbital- and site-projected) and average magnetic moments. We will compare our results with those obtained experimentally for free-standing Ni clusters[6] in order to see whether or not the deformed geometries can lead to a better agreement than the non deformed ones.

II. Theory

The spin-polarized electronic structure of Ni clusters is determined by solving self-consistently a tight-binding Hamiltonian for the $3d$, $4s$ and $4p$ valence electrons in a mean field approximation. In the usual second quantization notation, this Hamiltonian has the following expression:

$$H = \sum_{i\alpha\sigma} \varepsilon_{i\alpha\sigma} n_{i\alpha\sigma} + \sum_{\substack{\alpha\beta\sigma \\ i\neq j}} t_{ij}^{\alpha\beta} \hat{c}_{i\alpha\sigma}^{\dagger} \hat{c}_{j\beta\sigma}, \tag{1}$$

where $\hat{c}_{i\alpha\sigma}^{\dagger}$ is the operator for the creation of an electron with spin σ and orbital state α at the atomic site i, $\hat{c}_{j\beta\sigma}$ is the anniquilation operator and $n_{i\alpha\sigma}$ is the number operator. The hopping integrals $t_{ij}^{\alpha\beta}$ between orbitals α and β at sites i and j are assumed to be spin-independent and have been fitted to reproduce the band structure of bulk Ni. The variation of the hopping integrals with the interatomic distance r_{ij} is assumed to follow the typical power law $(r_0/r_{ij})^{l+l'+1}$, where r_0 is the bulk equilibrium distance and l, l' are the orbital angular momenta of the $(i\alpha\sigma)$ and $(j\beta\sigma)$ states involved in the hopping process. The spin-dependent diagonal terms account for the electron-electron interaction through a correction shift of the energy levels

$$\varepsilon_{i\alpha\sigma} = \varepsilon_{i\alpha}^{0} + z_{\sigma} \sum_{\beta} \frac{J_{\alpha\beta}}{2} \mu_{i\beta} + \Omega_{i\alpha}. \tag{2}$$

Here, $\varepsilon_{i\alpha}^{0}$ are the bare orbital energies of paramagnetic bulk Ni. The second term is the correction shift due to the spin-polarization of the electrons at site i ($\mu_{i\beta} = \langle n_{i\beta\uparrow}\rangle - \langle n_{i\beta\downarrow}\rangle$). In this term, $J_{\alpha\beta}$ are the exchange integrals and z_{σ} is the sign

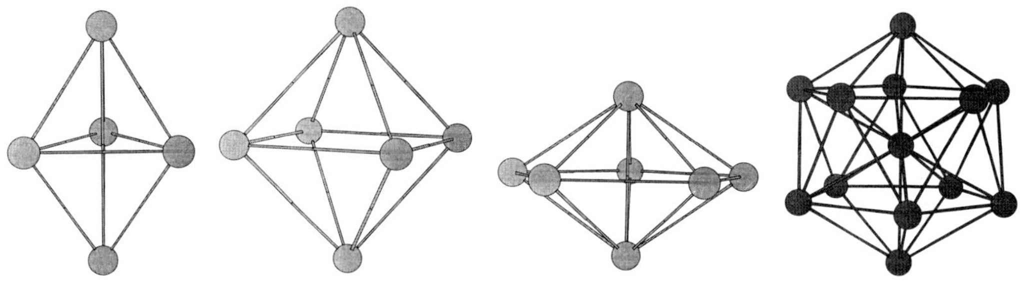

(a) Hexahedral (b) Octahedral (c) Decahedral (d) Icosahedral

Figure 1. Cluster geometries without deformation taken from the Molecular Dynamics calculation.[12]

function ($z_\uparrow = 1$, $z_\downarrow = -1$). The exchange integrals involving s- and p-electrons are neglected and J_{dd} is determined in order to reproduce the bulk magnetic moment. The site- and orbital-dependent self-consistent correction $\Omega_{i\alpha}$ assures the local electronic occupation, fixed in our model by interpolating between the isolated atom and the bulk according to the actual local number of neighbors. The spin-dependent local electronic occupations are self-consistently determined from the local densities of states:

$$\langle n_{i\alpha\sigma} \rangle = \int_{-\infty}^{\varepsilon_F} \mathcal{D}_{i\alpha\sigma}(\varepsilon) d\varepsilon, \tag{3}$$

which are calculated at each iteration by using the recursion method.[12,13] In this way, the distribution of the local magnetic moments ($\mu_i = \sum_\alpha \mu_{i\alpha}$) and the average magnetic moment $[\overline{\mu} = (1/N) \sum_i \mu_i]$ of Ni_N clusters are obtained at the end of the self-consistent cycle.

III. Results

Before presenting the results let us comment on the geometrical structures considered. The geometrical structures used in this work come from different calculations based on Molecular Dynamics (MD) with a Gupta potential[18] or in Monte Carlo (MC) simulations with a Morse potential.[19] For simplicity we only consider closed shell structures. Those geometries coming from MD are: hexahedral ($N = 5$), octahedral ($N = 6$), decahedral ($N = 7$) and icosahedral ($N = 13$). The numbers in the parenthesis correspond to the number of atoms in the cluster. These geometries are illustrated in Fig. 1. And the structures taken from MC are: saturated tetrahedral ($N = 8$), saturated hexahedral ($N = 11$) and centered hexagonal antiprism ($N = 15$). We call saturated thetrahedral ($N = 8$) and saturated hexahedral ($N = 11$) structures to those formed with additional atoms placed in all the faces of the tetrahedra and hexahedra clusters, respectively. These geometries are illustrated in Fig. 2. The deformations studied here are prolate-like and oblate-like, such that the symmetry in the main axis of the cluster as well as the surface area are preserved in order to reduce the number of inequivalent sites within the cluster. The prolate-like (oblate-like) deformations are performed by shrinking (elongating) the interatomic distances of those atoms in the plane(s) orthogonal to the main symmetry axis (we call these atoms the equator atoms). Those interatonic distances out of the plane(s) are fitted such that

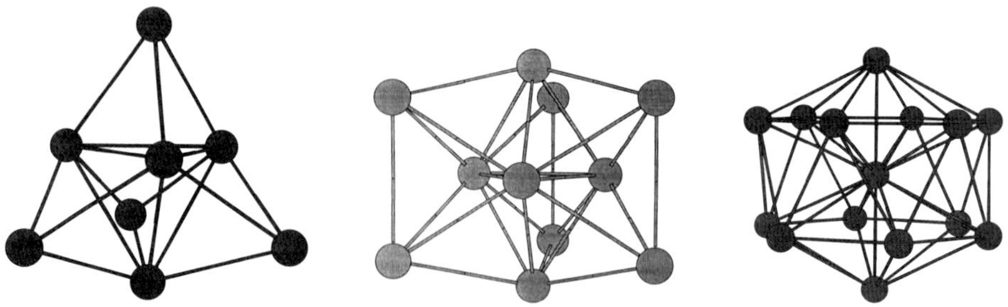

(a) Saturated decahedral (b) Saturated hexahedral (c) Centered hexagonal antiprism

Figure 2. Cluster geometries without deformation taken from the Monte Carlo simulations.[19]

the surface area of the clusters is constant and equal to the one without deformations. The atoms located in the main symmetry axis with the largest distance from the equator plane are called the polar atoms. In the case of the saturated tetrahedral, the prolate-like (oblate-like) deformations are performed by shrinking (elongating) the interatomic bonding of the internal cluster in the same amount, and the external interatomic distances are elongated (shrinked) in such a way that the surface area is also kept constant. The maximum change of the interatomic distance is 15%, however for some clusters this amount is not possible because they lose their shape or they arrive to very short (unphysical) polar distances; in these cases smaller deformation are performed. The deformation performed in the interatomic distances for the MD clusters are hexahedral $\pm15\%$, octahedral $\pm15\%$, decahedral $\pm10\%$ and icosahedral $\pm5\%$, being (+) for the oblate-like and (−) for the prolate-like, respectively. In the case of the MC clusters are saturated tetrahedral $\pm15\%$, saturated hexahedral $\pm15\%$ and centered hexagonal antiprism $\pm4\%$.

In Fig. 3, we present the average magnetic moment per atom for the MD geometries. The *d*- and *sp*-contributions are plotted in Fig. 4. Although in some clusters the average magnetic moments are similar, the local atomic and orbital distribution is different as we will comment. We start discussing the case without deformation. The results present a minimum at $N = 6$ and the dependence of the magnetic moment with the cluster size is in general a non-monotonic decreasing function as it was obtained in a previous work[12] in good qualitative agreement with the experiment.[6]

In the case of the prolate-like deformation the dependence with the cluster size is also non-monotonic and the local minimum at $N = 6$ is also present. For the hexahedral cluster with prolate-like deformation there is a slightly decrease in the average magnetic moment due to the decrease of the local magnetic moment at the equator plane as a consequence of the decrease of the interatomic distance between the equator atoms. For this cluster, the prolate deformation leads to better agreement with the experiment[6] than the non-deformed case. For the octahedral cluster with prolate-like deformation we see a significant variation in the magnetic moment mainly due to the change of orientation of the *sp*-component that flips from an antiparallel (AP) coupling with the *d*-component in the case without deformation to a parallel (P) coupling with the *d*-component (see Fig. 4). In particular the polar atoms show the strongest change. In contrast to the hexahedral, now the prolate deformation leads to a worse agreement with the experiment.[6] In the case of the decahedral and icosahedral clusters

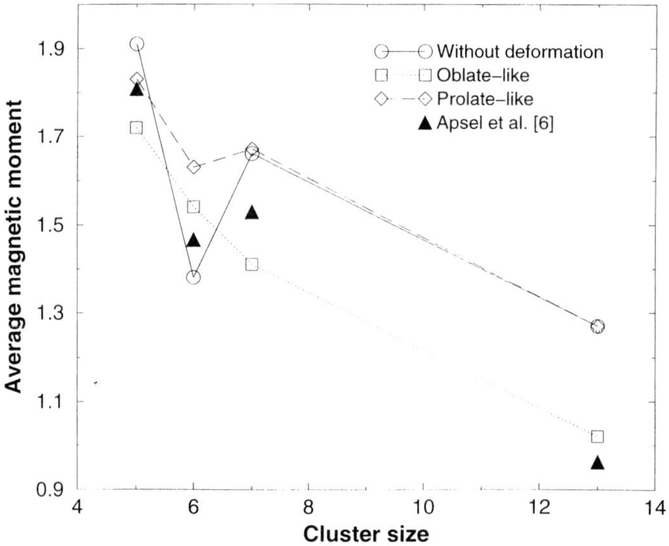

Figure 3. Average magnetic moment per atom (in units of μ_B) as a function of the cluster size for the Molecular Dynamics clusters without deformation, and with prolate-like and oblate-like deformations. The experimental results[6] ($\triangle$) are included for comparison. The lines are for eye guide.

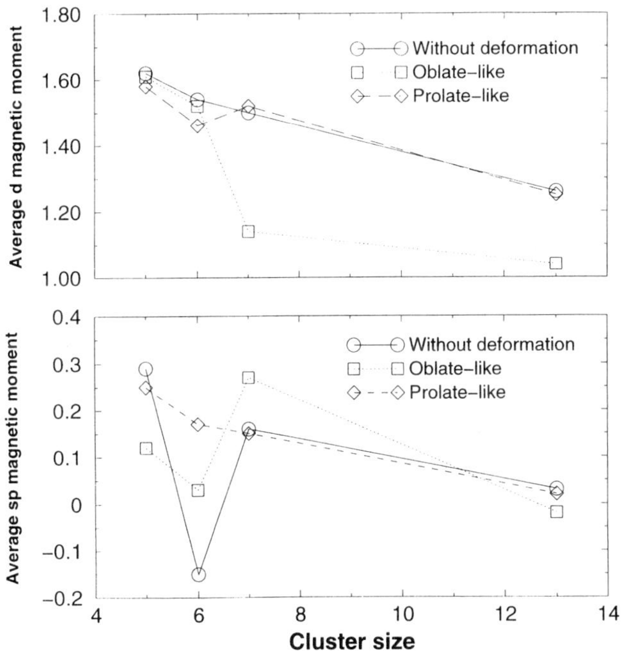

Figure 4. Orbital components of the average magnetic moment (in units of μ_B) as a function of the cluster size for the Molecular Dynamics clusters without deformation, and with prolate-like and oblate-like deformations. In (upper panel) the d-contribution of the magnetic moment, and in (lower panel) the sp-contribution.

with prolate-like deformation, they do not present noticeable variation in the average magnetic moment under the deformation process although the local magnetic moment distribution (in the different sites inside the clusters) is different. In particular the *d*-contribution in the polar atoms increases but this increase is almost canceled by their AP *sp*-contribution; the *sp*-contribution in the polar sites was P with *d* before the deformation.

In the case of the oblate-like clusters the magnetic moment present a monotonic dependence with the cluster size in a completely different behavior than the previous two cases (without deformation and prolate-like) and with the experiment around $N = 6$. For the hexaheadral cluster the magnetic moment decreases due to the reduction of the *sp*-contribution, particularly at the polar atoms as a consequence of the short distance between the polar sites. For the octahedral cluster there is an increase in the magnetic moment due to the variation of the *sp*-component from AP to P with the *d*-component. The polar atoms are the ones displaying the strongest variation. For the decahedral and icosahedral clusters there is a decrease in the magnetic moment due to the reduction of the *d*-component of atoms in the main symmetry axis (polar atoms) due to the deceases of the distance between those atoms. For the icosahedral cluster, the oblate deformation produces a better agreement with the magnetic moment experimentally found for Ni_{13}. In general for the MD clusters the magnetic moment of the oblate-like clusters are smaller than in the prolate-like clusters. Two general trends that are observed for the *d*-contribution are: *i*) the local magnetic moment decreases with increasing the local atomic coordination, *ii*) the local magnetic moment increases with the increase of the interatomic distance. In the case of the *sp*-contribution the behavior is very complex and hard to explain in these terms as it is illustrated in Fig. 4.

In Fig. 5, we present the result for the average magnetic moment per atom of the MC geometries, the *d*- and *sp*-contributions are plotted in Fig. 6. Here the changes as a consequence of the deformations are less significant than for the MD clusters. The dependence of the magnetic moment with cluster size is the expected decreasing function, however we have not enough data to see whether this dependence is a monotonous function or not. In the case of the saturated tetrahedral cluster both type of deformations slightly increase the *d*-contribution of the magnetic moment and in both cases also the *sp*-contribution is strongly reduced with respect to the case without deformation (see Fig. 6). The *sp*-contribution has AP coupling with the *d*-contribution in the prolate-like cluster (in particular for the internal atoms) whereas in the oblate-like clusters the AP coupling is mainly obtained in the external atoms.

For the saturated hexahedral cluster the polar atoms are the ones displaying the largest changes in the *d*-contribution with respect to the clusters without deformation. The magnetic moment increases (decreases) in the prolate-like (oblate-like) case, whereas the atoms in the equator plane do not show appreciable change for any type of deformation. When the average for the total magnetic moment is done the changes in the different components of the local moments almost compensate and the results are very similar to the case without deformation. Nevertheless, here we obtain better agreement with the experiment[6] with the non-deformed geometries. For the centered hexagonal antiprism both type of deformations lead to almost the same magnetic moment although their local atomic distributions are different. The lowest values for the magnetic moments are in the central sites as expected due to the high coordination number. Due to the deformation in the prolate-like (oblate-like) clusters the *d*-magnetic moment of the polar atoms increases (decreases) since the interatomic distance increases (decreases) and the oppossite behavior occurs for the atoms at the

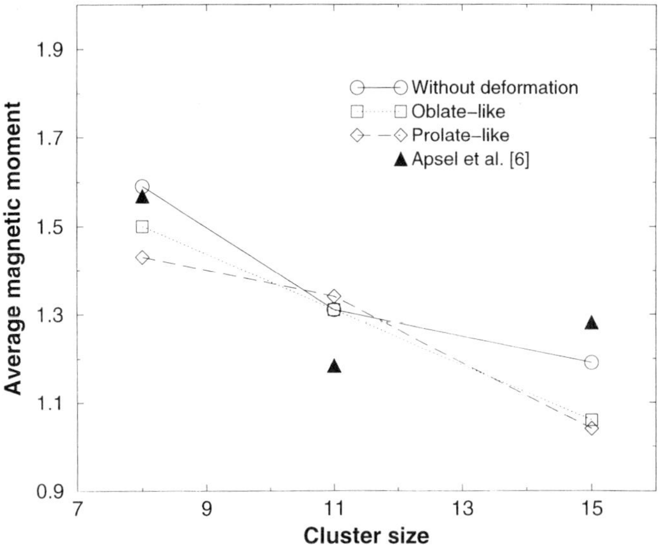

Figure 5. Average magnetic moment per atom (in units of μ_B) as a function of the cluster size for the Monte Carlo clusters without deformation, and with prolate-like and oblate-like deformations. The experimental results[6] (△) are included for comparison. The lines are for eye guide.

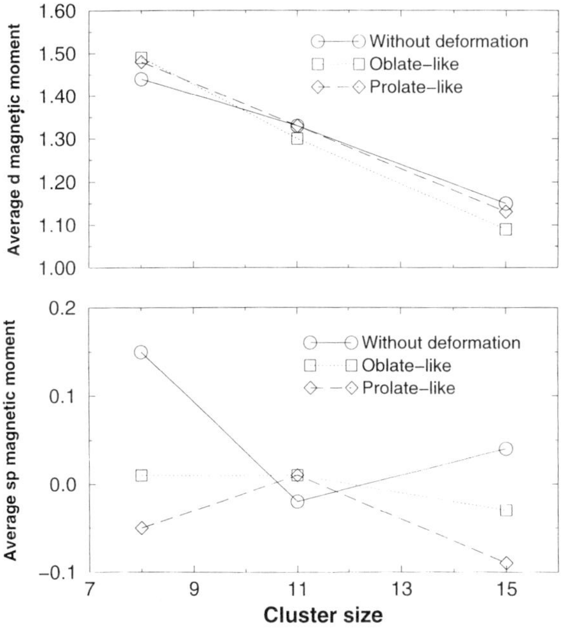

Figure 6. Orbital components of the average magnetic moment (in units of μ_B) as a function of the cluster size for the Monte Carlo clusters without deformation, and with prolate-like and oblate-like deformations. In (upper panel) the d-contribution of the magnetic moment, and in (lower panel) the sp-contribution.

edges. In general the magnetic moments of the clusters with deformations are smaller or equal to those of the clusters without deformation, the only exception being the saturated hexahedral with prolate-like deformation. Moreover the *sp*-contribution for all the MC clusters is smaller than 4%, the only exception is the saturated tetrahedral without deformation where the *sp*-contribution is approximately 10%. As for the saturated hexahedral, the non deformed geometries lead to better agreement with the experiment.[6]

IV. Conclusions

The local and average magnetic moments are strongly related with the geometrical structure and therefore they vary in general under a deformation. The changes in the *d*-component of the clusters with prolate-like and oblate-like deformations can be explained in most cases in terms of simple geometrical parameters such as coordination number and interatomic distances, whereas the *sp*-contribution is very difficult to explain in this way. For the MD clusters in general the oblate-like deformations produce lower magnetic moments than the prolate-like ones, and the oblate-like clusters show a monotonic dependence with cluster size around $N = 6$. For the MC clusters the total average magnetic moment of the prolate-like and oblate-like clusters are very similar and in general its value is equal or smaller than the one obtained without deformation.

Coming back to the problem of free-standing clusters or clusters supported on noble metal substrates our results indicate that the magnetic moments of some clusters may suffer a significant change under deformations as it is the case of icosahedral with prolate-like deformation. However other clusters are practically insensitive although the deformations are considerable like in the saturated hexahedral . The deformations, in some cases, improve the agreement with the experimental results for free-standing clusters.

Ackowledgments

We acknowledge Consejo Nacional de Ciencia y Tecnología, México (Grant No. G-25851-E), DGICYT of Spain (Project PB98-0368-C02) and Junta de Castilla y León, Spain (Project VA70/99).

References

1. D. M. Cox, D. J. Trevor, R. L. Whetten, E. A. Rohlfing, and A. Kaldor, *Phys. Rev. B* **32**, 7290 (1985).
2. W. A. de Herr, P. Milani, and A. Chatelain, *Phys. Rev. Lett.* **65**, 488 (1990).
3. J. P. Bucher, D. C. Douglas, P. Xia, B. Haynes, and L. A. Bloomfield, *Phys. Rev. Lett.* **66**, 3052 (1991).
4. J. P. Bucher, in *Physics and Chemistry of Finite Systems: From Clusters to Crystals*, edited by P. Jena, S. N. Khanna, and B. K. Rao, (Klewer Academic Publishers, Dodrecht, 1992), p. 799.
5. I. M. L. Billas, J. A. Becker, A. Châtelain, and W. A. Heer, *Phys. Rev. Lett.* **71**, 4067 (1993).

6. S. E. Apsel, J. W. Emmert, J. Deng, and L. A. Bloomfield, *Phys. Rev. Lett.* **76**, 1441 (1996).
7. J. P. Jensen and K. H. Bennemann, *Z. Phys. D* **35**, 273 (1995).
8. F. A. Reuse, S. N. Khanna, and S. Bernel, *Phys. Rev. B* **52**, 11650 (1995); F. A. Reuse and S. N. Khanna, *Chem. Phys. Lett.* **234**, 77 (1995).
9. A. N. Andriotis, N. N. Lathiotakis, and M. Menon, *Europhys. Lett.* **36**, 37 (1996).
10. N. Fujima and T. Yamaguchi, *Phys. Rev. B* **54**, 26 (1996).
11. J. Guevara, F. Parisi, A. M. Llois and M. Weissmann, *Phys. Rev. B* **55**, 13283 (1997).
12. S. Bouarab, A. Vega, M. J. López, M. P. Iñiguez, and J. A. Alonso, *Phys. Rev. B* **55**, 13279 (1997).
13. F. Aguilera-Granja, S. Bouarab, M. J. López, A. Vega, J. M. Montejano-Carrizales, M. P. Iñiguez, and J. A. Alonso, *Phys. Rev. B* **57**, 12469 (1998).
14. R. Guirado-López, D. Spanjaard, M. C. Desjonquères, and F. Aguilera-Granja, *J. Magn. Magn. Mater.* **186**, 214 (1998).
15. J. L. Rodríguez-López, F. Aguilera-Granja, A. Vega, and J. A. Alonso, *Eur. Phys. Jour. D* **6**, 235 (1999).
16. J. Guevara, A. M. Llois, F. Aguilera-Granja, and J. M. Montejano-Carrizales, *Solid State Commun.* **111**, 335 (1999).
17. J. A. Franco, A. Vega, and F. Aguilera-Granja, *Phys. Rev. B* **60**, 434 (1999).
18. M. J. López and J. Jellinek, *Phys. Rev. A* **50**, 1445 (1994), and unpublished results.
19. H. J. Hu, L. M. Mei, and H. Li, *Solid State Commun.* **100**, 129 (1996).

Enhancing the Production of Endohedral Fullerenes: a Theoretical Proposal

J. L. Morán-López,[1] J. R. Soto,[2] and A. Calles,[2]

[1]*Instituto de Física, "Manuel Sandoval Vallarta"*
Universidad Autónoma de San Luis Potosí
78000 San Luis Potosí, S.L.P.
MEXICO

[2]*Facultad de Ciencias*
Universidad Nacional Autónoma de México
Apartado Postal 70-646
04510 México D. F.
MEXICO

Abstract

It has been proposed that in a C_{60} molecule containing two kinds of isotopes one could selectively excite one of them to open temporary gates in the carbon cage and induce the penetration of different atoms inside the molecule in order to produce endohedral fullerenes. In the present work we calculate, from first principles, the vibrational excitation spectra substituting one and two ^{12}C atoms for ^{13}C isotopes in the C_{60} molecule. To obtain the force constants between the carbon atoms we performed the second derivatives of the energy calculated within the Hartree-Fock approximation. A comparison of the frequency shifts in the vibrational spectra produced by the isotopes with Raman spectroscopy experiments is also presented.

I. Introduction

Just after the geometry of the C_{60} molecules was elucidated, one of the promising applications of fullerenes, was to encapsulate elements or molecules in their hollow cages. With the exception of La@C_{60}, obtained just after the fullerene discovery,[1] it turned out to be very difficult to encapsulate elements into the fullerenes. In the pioneer experiments in 1985, mass spectroscopy observations of the emitted materials from a laser vaporization of a La-impregnated graphite target, peaks related to the

La@C_{60} and La_2@C_{60} molecules were observed.[2] However the quantities produced of these molecules were too small.

In the early 90s, milligram quantities of other endofullerenes were obtained, through laser[3] and arc[4] vaporization of graphite in a rare gas (He or Ar) atmosphere. Helium or neon can be incorporated inside the C_{60} cage through arc vaporization of carbon electrodes imbedded in these rare gases forming small amounts of He@C_{60} and Ne@C_{60} endofullerenes.[5] Another mechanism to produce K@C_{60} and Rb@C_{60}, among other endofullerenes, is through photofragmentation.[6]

It has been proposed that in a C_{60} molecule containing two kinds of isotopes one could excite selectively one of them to open temporary gates in the carbon cage and induce the production of endohedral fullerenes.[7] Although one can produce deliberately isotopomers by enriching the graphite rods with ^{13}C, due to the natural abundance of this isotope, which is about 1.1%, one may expect that fullerenes produced from natural graphite contain some ^{13}C isotopes.

The presence of this isotope in fullerenes has been observed[8] in Raman spectra obtained from frozen solutions of C_{60}. Other technique that can be used to detect the presence of the ^{13}C isotope is the nuclear magnetic resonance (NMR). This isotope in contrast to ^{12}C atoms carries a nuclear spin of 1/2.

The Nuclear Magnetic Resonance spectroscopy was one of the first techniques used to show conclusively the icosahedral structure of the C_{60} molecule. Such structure, with one isotope substitution, is consistent with a single peak in the NMR spectrum indicating that all carbon sites are equivalent.[9] Once the magnetic dipolar coupling of ^{13}C-^{13}C is measured in NMR experiments, which depends on the inverse cube of the distance between those atoms, the C-C bond distances in C_{60} molecule can be determined.[10] Also through NMR experiments, we believe one could measure the size of the cage deformation provoked by the excitation of the localized modes described in the present work.

The observation of the isotope shift of T_C in M_3C_{60} superconductors produced by carbon isotopes, indicates that the vibration of the carbon atoms plays an important role in the superconductivity mechanism in these materials.[11] Thus, the study of the vibrational spectra of the C_{60} molecule with ^{13}C isotopes inserted in one or two positions in the molecule is of particular interest to elucidate the superconducting mechanism in the M_3C_{60} superconductors.

In the present paper we obtain, from first principles, the modification of the vibrational spectra when one or two atoms of the C_{60} are substituted for ^{13}C isotopes. It is well known, theoretically and experimentally, that the frequency shift goes as $1/\sqrt{m}$. In the present calculation we confirm such dependence and give the exact identity of the modified modes produced by the symmetry breaking.

The paper is organized as follows, in Sec. II we comment the details of the first principle calculation of the force constants for the C_{60} molecule, present the vibrational spectra results and compare them with other *ab initio* calculations and with the experimental data obtained by ir and Raman spectroscopies. In Sec. III the results for the normal mode of vibration of the proposed isotopomers are classified according to the irreducible representations of the respective symmetry group. This identifies, in a very fine way, how the presence of isotopes modifies the whole spectrum, and one can see, through a direct visualization, the normal modes of vibration that deform the molecule due to the isotope larger amplitude of vibration. Finally, in Sec. IV we present our conclusions.

Table I

The bond stretching ϕ_ℓ, the tangential (bond bending on the surface) ϕ_{t1}, and normal to the surface of the molecule, ϕ_{t2}, force constants compared with the results of Quong *et al.*[12] obtained through a calculation using LDA.

R	ϕ_ℓ	$\phi_\ell(Q)$	ϕ_{t1}	$\phi_{t1}(Q)$	ϕ_{t2}	$\phi_{t2}(Q)$
1.403	4.5800	5.0930	2.6550	1.6870	0.7259	0.5370
1.463	2.9175	3.2810	1.9090	1.3750	0.7597	0.7187
2.367	1.0530	0.8500	−0.2141	−0.2180	−0.0254	0.0000
2.482	0.7226	0.6000	−0.6943	−0.4375	0.0709	0.1250
2.866	0.0241	0.0370	0.7041	0.3400	−0.0268	−0.1250
3.598	−0.1134	−0.0850	0.0240	0.0375	−0.0273	−0.0250
3.717	0.1570	0.0350	0.0740	0.0350	−0.1187	−0.1000
4.139	0.0173	−0.0250	0.0158	0.0250	−0.0010	0.0300
4.538	0.0009	0.0000	−0.0264	0.0000	−0.0006	0.0000
4.637	0.0179	0.0300	−0.1835	−0.1250	−0.0184	0.0250
4.862	0.0309	0.0000	0.2051	0.0870	−0.0079	0.0000

II. Force Constants and Vibrational Spectrum of C_{60}

The interatomic forces of the fullerene molecule determine the vibrational spectra. Thus, in order to calculate the vibrational spectra we solved the Hessian matrix in the harmonic approximation using force constants obtained from first principles. We calculated the second derivatives using the central gradient differences of the energy obtained within the Hartree Fock (HF) approximation. The HF calculation was performed with the Gausssian 92 package using a 3-21 G basis.

To simplify the analysis, it is necessary to introduce the following transformation:[12] for each pair of atoms, i and j, one considers the vector that connects them, $\vec{R}_{ji} = \vec{R}_j - \vec{R}_i$, where $\vec{R}_i$ and $\vec{R}_j$ are the vectors from the center of the molecule to atoms i and j, respectively. Then one defines two additional vectors $\vec{R}^i_{t1} = \vec{R}_i \times \vec{R}_{ji} = \vec{R}_i \times R_j$ and $\vec{R}^i_{t2} = \vec{R}_{ji} \times \vec{R}^i_{t1}$. It is important to notice that for near neighbors $\vec{R}_{ji}$ lies on the surface of the molecule, along the bond, $\vec{R}^i_{t1}$ lies also on the surface and is perpendicular to the bond, and $\vec{R}^i_{t2}$ is normal to the surface. We denote the transformed coordinates $ji, t1$, and $t2$.

In Table I we show the results for the force constants as a function of the distance between carbon atoms for the C_{60} molecule in the three directions: along the bond, ϕ_{ji}, the tangential (bond bending on the surface), ϕ_{t1}, and normal to the surface of the molecule, ϕ_{t2}. For comparison we show the results obtained by Quong *et al.*[12] using the Local Density Approximation (LDA). One notices that both calculations have essentially the same distance dependence. The largest force constants correspond to bond stretching, ϕ_{ji}, and the smallest to the bond bending elements normal to the surface ϕ_{t2}.

The vibrational spectrum for the C_{60} molecule containing only ^{12}C atoms obtained using the force constants given in Table I is shown in Table II and is compared to the experimental results[13] and with other *ab initio* calculations.[12,14–17] In the first column we give the classification of each energy level according to the irreducible representation of the I_h group. In the second column we give the values

Table II

The $^{12}C_{60}$ vibrational spectra in cm^{-1}. The symmetry classification is given in the first column. The experimental results[16] are given in the second column. The third column contains our results and the fourth one gives the discrepancy between our results and the experimental results. The next columns contains the theoretical results from other authors,[12,14–17] within various theories and approximations.

sym	Exp. [16]	This work	Dis. %	DFPT [14]	Dis. %	LDFT [12]	Dis. %	LDFT PAO [15]	Dis. %	LDFT Num. [16]	Dis. %	LDFT DG [17]	Dis. %
Ag	498	540	8.43	495	0.60	478	4.02	537	7.83	484	2.81	499	0.20
	1470	1417	3.61	1504	2.31	1499	1.97	1680	14.29	1529	4.01	1525	3.74
T1g	502	608	21.12			580	15.54			566	12.75	579	15.34
	976	963	1.33			788	19.26			825	15.47	830	14.96
	1358	1322	2.65			1252	7.81			1292	4.86	1318	2.95
T2g	566	573	1.24			547	3.36			550	2.83	551	2.65
	865	784	9.36			610	29.48			771	10.87	804	7.05
	914	899	1.64			770	15.75			795	13.02	839	8.21
	1360	1330	2.21			1316	3.24			1360	0.00	1393	2.43
Gg	486	519	6.79			486	0.00			484	0.41	484	0.41
	621	668	7.57			571	8.05			514	17.23	573	7.73
	806	787	2.36			759	5.83			713	11.54	788	2.23
	1076	1091	1.39			1087	1.02			1117	3.81	1122	4.28
	1356	1371	1.11			1296	4.42			1326	2.21	1347	0.66
	1524	1574	3.28			1505	1.25			1528	0.26	1548	1.57
Hg	273	263	3.66	259	5.13	258	5.49	249	8.79	263	3.66	261	4.40
	432	467	8.10	425	1.62	439	1.62	413	4.40	432	0.00	431	0.23
	711	739	3.94	711	0.00	727	2.25	681	4.22	713	0.28	727	2.25
	775	829	6.97	783	1.03	767	1.03	845	9.03	778	0.39	788	1.68
	1101	1168	6.09	1120	1.73	1093	0.73	1209	9.81	1111	0.91	1128	2.45
	1251	1214	2.96	1281	2.40	1244	0.56	1453	16.15	1282	2.48	1297	3.68
	1426	1422	0.28	1450	1.68	1443	1.19	1624	13.88	1469	3.02	1475	3.44
	1578	1636	3.68	1578	0.00	1576	0.13	1726	9.38	1598	1.27	1618	2.53
Au	1143	1176	2.89			850	25.63			947	17.15	972	14.96
T1u	526	556	5.70	527	0.19	547	3.99	494	6.08	533	1.33	535	1.71
	576	628	9.03	586	1.74	570	1.04	643	11.63	548	4.86	591	2.60
	1183	1173	0.85	1218	2.96	1176	0.59	1358	14.79	1214	2.62	1224	3.47
	1429	1400	2.03	1462	2.31	1461	2.24	1641	14.84	1485	3.92	1486	3.99
T2u	356	368	3.37			342	3.93			344	3.37	342	3.93
	680	727	6.91			738	8.53			717	5.44	726	6.76
	1026	1011	1.46			962	6.24			987	3.80	996	2.92
	1201	1188	1.08			1185	1.33			1227	2.16	1234	2.75
	1576	1557	1.21			1539	2.35			1558	1.14	1571	0.32
Gu	400	378	5.50			356	11.00			356	11.00	350	12.50
	760	802	5.53			683	10.13			752	1.05	762	0.26
	924	856	7.36			742	19.70			784	15.15	830	10.17
	970	1000	3.09			957	1.34			977	0.72	984	1.44
	1310	1294	1.22			1298	0.92			1339	2.21	1359	3.74
	1446	1481	2.42			1440	0.41			1467	1.45	1480	2.35
Hu	342	454	32.75			404	18.13			396	15.79	401	17.25
	563	563	0.00			539	4.26			534	5.15	541	3.91
	696	735	5.60			657	5.60			663	4.74	671	3.59
	801	822	2.62			737	7.99			742	7.37	762	4.87
	1117	1237	10.74			1205	7.88			1230	10.12	1248	11.73
	1385	1389	0.29			1320	4.69			1360	1.81	1389	0.29
	1559	1636	4.94			1565	0.38			1588	1.86	1611	3.34
Ave.			4.92		1.69		6.14		10.37		5.18		4.65

obtained experimentally.[13] It is worth to notice that the experimental results correspond to observations of infrared modes in solid C_{60} films deposited on KBr substrates. Due to the nearly ideal molecular solid one expects very small differences with the isolated molecule. In the third and fourth columns we present our results and the discrepancy with the observed values. In the following columns we show the results obtained within the Density Functional Perturbation Theory,[14] Local Density Functional Theory,[12] Local Density Functional/Pseudo Atomic Orbital,[15] Local Density Functional/Numerical,[16] and Local Density Functional/Dgauss.[17] In each case we quote the discrepancy with respect to the experiment. One notices that the results obtained within the DFPT and LDF/PAO, the spectrum has been calculated only partially.

According to group theory the classification of the normal modes of vibration are $\Gamma = 2A_g + 3T_{1g} + 4T_{2g} + 6G_g + 8H_g + A_u + 4T_{1u} + 5T_{2u} + 6G_u + 7H_u$. The calculation that reproduces the best the experimental values is the one carried out by Giannozzi and Baroni;[14] although as mentioned above only some modes have been calculated. We notice that our predictions exceed in more than 10% of the observed values only in two modes, $T_{1g}(1)$ and $H_u(1)$. On the other hand, Quong,[12] Adams,[15] Wang,[16] and Dixon[17] predict 8, 6, 9, and 6 modes that differ more than 10% of the experimental values. In particular all those calculations differ in more than 10% in the $T_{1g}(1)$ and $H_u(1)$ modes.

III. Vibrational Spectra of the isotopomers $^{12}C_{59}{}^{13}C_1$ and $^{12}C_{58}{}^{13}C_2$

The small amounts of He@C_{60} and Ne@C_{60} have been produced[5] by heating C_{60} soot under the atmosphere of those elements at temperatures $\sim 600^0$ C. It is belived that at that temperature the heating process distorts the C-C bonds and opens temporarily a window in the cage. This interpretation is supported by a theoretical calculation,[18] which shows that one can create 9- and 10-membered rings by breaking a C-C bond with a relative small energy cost.

A complementary mechanism to increase the endofullerene production has been proposed, in which the central idea is to excite selectively only one or two carbon atoms in order to build up a larger (distorted) ring with the stretched bonds. The mechanism is as follows: the first step is to produce C_{60} molecules containing two carbon isotopes, $^{A}C_{60-n}\ {}^{B}C_n$ $(A, B = 12, 13, 14)$. Then, one can separate by mass spectroscopy those molecules with one or two atoms of the same isotope and the rest of the other isotope. The next step is to immerse the selected molecules in an atmosphere of the gas to be inserted. Finally one could stretch the bonds related to the minority isotopes by exciting selectively through laser irradiation the localized isotope modes. The deformed bonds would build up a window that may facilitate the incursion of atoms into the hollow cage.

To evaluate the viability of the process, we calculate now the vibrational spectra of molecules with one or two atoms of the ^{13}C isotope. The calculation is carried out with the same force constants and modifying the dynamic matrix with the isotope masses. In the first case, the new symmetry group is C_{1h}. In the second one, when we substitute two ^{12}C atoms for the isotope ^{13}C there are many different possibilities for the position of the pair. We tried various geometries and obtained essentially the same result for all of them. The results, for the vibrational spectra with the group theory classification, of the C_{60} molecule and of the $^{12}C_{59}{}^{13}C$ and $^{12}C_{58}{}^{13}C_2$ isotopomers, are presented in Table III.

Table III

The vibrational spectra in cm^{-1} for the pristine $^{12}C_{60}$ (I_h), $^{12}C_{59}{}^{13}C$ (C_{1h}), and $^{12}C_{58}{}^{13}C_2$ (no group) isotopomers classified according to the respective symmetry group.

$^{12}C_{60}$	sym	$^{13}C_1^{12}C_{59}$	sym	$^{13}C_2^{12}C_{58}$	$^{13}C_1^{12}C_{59}$	sym	$^{13}C_2^{12}C_{58}$	$^{13}C_1^{12}C_{59}$	sym	$^{13}C_2^{12}C_{58}$
263.33	1Hg	262.67	A1	262.15	738.74	A2	738.66	1212.81	A1	1212.64
368.09	1T2u	263.19	A2	262.92	738.76	MIX	738.73	1213.99	A1	1212.88
377.7	1Gu	263.2	A1	263.08	738.76	MIX	738.75	1214.02	MIX	1213.6
454.6	1Hu	263.33	MIX	263.31	781.77	A1	779.66	1214.02	MIX	1214.02
467.26	2Hg	263.33	MIX	263.32	784.11	A2	783.74	1234.21	A2	1232.52
519.18	1Gg	367.32	A1	366.6	784.13	A2	784.11	1236.5	A1	1235.32
540.11	1Ag	368.03	A1	367.93	785.62	A1	785.44	1237.15	A1	1237.03
556.2	1T1u	368.07	A2	368.05	786.66	A2	785.93	1237.16	A2	1237.07
563.53	2Hu	376.81	A1	376.54	786.72	A1	786.69	1237.16	A2	1237.16
572.9	1T2g	377.57	A2	376.86	786.73	A2	786.71	1292.2	A1	1291.1
608.03	1T1g	377.7	MIX	377.69	800.31	A1	799.47	1293.07	A2	1291.48
628.25	2T1u	377.7	MIX	377.69	801.88	A2	801.36	1294.45	A1	1293.39
667.86	2Gg	453.61	A2	452.84	801.92	A1	801.89	1294.46	A2	1294.34
727.07	2T2u	454.01	A1	453.5	801.93	A2	801.92	1319.76	A2	1318.28
734.81	3Hu	454.58	A1	454.32	820.22	A2	818.73	1321.21	A1	1319.7
738.76	3Hg	454.6	A2	454.57	820.41	A1	820.32	1321.94	A2	1321.75
784.13	2T2g	454.6	A2	454.58	821.72	A1	820.49	1328.84	A2	1327.62
786.73	3Gg	465.7	A1	464.96	821.82	A2	821.72	1329.87	A1	1329.18
801.93	2Gu	467.18	A1	466.36	821.82	A2	821.75	1330.41	A2	1330.27
821.82	4Hu	467.21	A2	467.16	828.24	A2	827.46	1368.52	A1	1366.75
829.31	4Hg	467.26	MIX	467.21	828.59	A1	827.79	1369.75	A2	1369.35
856.43	3Gu	467.26	MIX	467.26	828.7	A1	828.06	1371.12	A1	1369.75
899.26	3T2g	517.72	A1	517.52	829.31	MIX	829.12	1371.18	A2	1370.88
962.76	2T1g	519.12	A2	517.85	829.31	MIX	829.24	1386.48	A2	1385.22
999.71	4Gu	519.18	A1	519.15	854.72	A1	853.47	1387.07	A1	1385.31
1011.08	3T2u	519.18	A2	519.17	855.95	A2	855.22	1388.92	A1	1387.85
1091.13	4Gg	539.7	A1	539.29	856.35	A1	856.2	1388.96	A2	1388.92
1168.42	5Hg	554.79	A1	554.31	856.43	A2	856.32	1388.96	A2	1388.95
1173.36	3T1u	556.18	A2	555.27	897.84	A1	896.74	1397.57	A1	1394.12
1176.46	1Au	556.19	A1	556.17	898.86	A2	898.45	1400.17	A1	1399.94
1187.83	4T2u	561.93	A1	561.77	899.26	A2	899.08	1400.18	A2	1400.16
1214.02	6Hg	563.43	A1	562.11	961.2	A2	959.87	1415.44	A1	1413.97
1237.16	5Hu	563.51	A2	563.37	962.21	A1	961.91	1419.58	A1	1419.12
1294.46	5Gu	563.53	A2	563.48	962.76	A2	962.41	1421.21	A2	1420.53
1321.94	3T1g	563.53	A2	563.53	997.39	A2	996.58	1421.96	A1	1420.9
1330.41	4T2g	572.03	A1	571.86	999.21	A1	997.32	1421.99	MIX	1421.71
1371.18	5Gg	572.86	A2	572.21	999.64	A1	999.49	1421.99	MIX	1421.96
1388.96	6Hu	572.9	A2	572.86	999.71	A2	999.55	1479.49	A1	1478.16
1400.19	4T1u	606.75	A2	605.66	1010.14	A1	1009.34	1479.85	A2	1478.88
1416.79	2Ag	608.01	A1	607.84	1010.24	A2	1009.63	1481.42	A1	1480.4
1421.99	7Hg	608.03	A2	607.99	1010.95	A1	1010.43	1481.42	A2	1481.38
1481.42	6Gu	627.42	A1	626.76	1088.22	A2	1086.17	1553.68	A1	1552.66
1557.07	5T2u	628.02	A2	627.77	1090.96	A1	1089.88	1557.06	A1	1554.52
1573.59	6Gg	628.05	A1	627.82	1091.13	MIX	1091.07	1557.07	A2	1557.06
1635.93	7Hu	666.47	A1	666.02	1091.13	MIX	1091.1	1571.28	A1	1570.38
1636.43	8Hg	667.38	A2	666.18	1164.99	A1	1162.64	1572.38	A2	1570.75
		667.84	A1	667.67	1166.38	A2	1164.63	1573.58	A1	1572.88
		667.86	A2	667.81	1168.29	A1	1167.06	1573.59	A2	1573.3
		724.82	A1	721.49	1168.42	MIX	1167.26	1628.74	A1	1625.01
		727.04	A1	726.58	1168.42	MIX	1168.41	1634.56	A2	1630.8
		727.04	A2	727.01	1172.25	A2	1171.72	1635.93	A1	1635.4
		732.65	A1	731.33	1172.48	A1	1171.78	1635.93	MIX	1635.59
		734.28	A2	733.83	1173.36	A1	1173.22	1635.93	MIX	1635.93
		734.63	A1	734.18	1175.9	A2	1175.54	1636.18	A2	1636.11
		734.81	A2	734.62	1186.74	A2	1185.26	1636.2	A1	1636.18
		734.81	A2	734.68	1186.76	A1	1186.54	1636.43	MIX	1636.22
		737.4	A1	737.05	1187.83	A1	1186.98	1636.43	MIX	1636.37
		738.69	A1	738.23	1211.15	A2	1208.42	1636.43	A2	1636.43

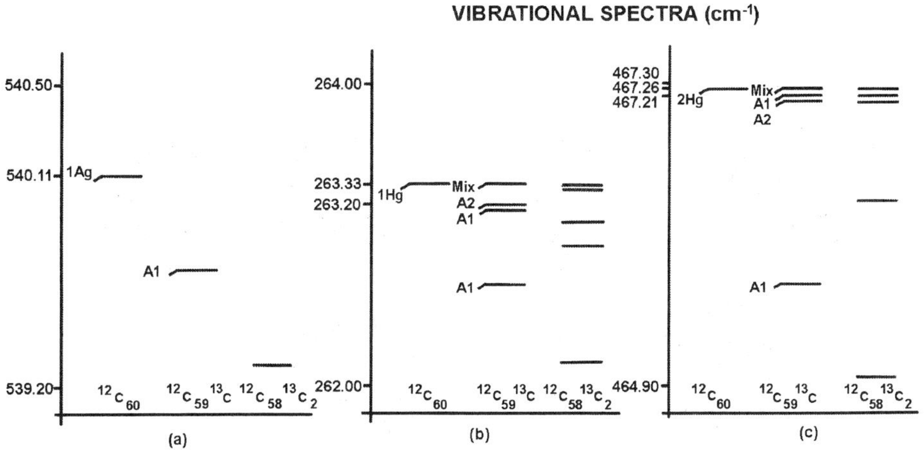

Figure 1. Amplification of the vibrational spectra (cm^{-1}) for a) the $A_g(1)$, b) $H_g(1)$, and c) $H_g(2)$, Raman modes in $^{12}C_{60}$ and the splitted states in the $^{12}C_{59}{}^{13}C$ and $^{12}C_{58}{}^{13}C_2$ isotopomers.

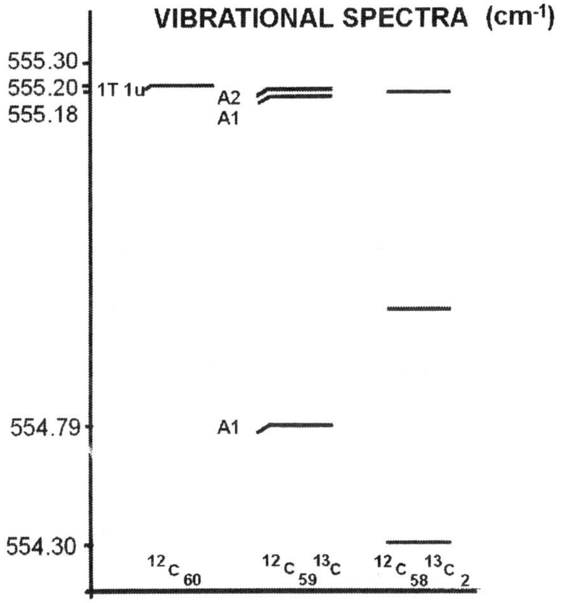

Figure 2. Amplification of the vibrational spectrum (cm^{-1}) for the $T_{1u}(1)$ infrared mode in $^{12}C_{60}$ and the splitted states in the $^{12}C_{59}{}^{13}C$ and $^{12}C_{58}{}^{13}C_2$ isotopomers.

With the help of the symmetry classification it is easy to see which levels, sensitive to Raman or infrared spectroscopies for example, are substantially splitted by the isotope substitution. According to group theory the active Raman and infrared modes, at first order, are $2A_g + 8H_g$ and $4T_{1u}$, respectively. In Figs. 1(a)–1(c) we amplify the vibration spectra for modes which can be observed with Raman spectroscopy. In Fig. 2 we show the splitted modes which can be observed with infrared spectroscopy.

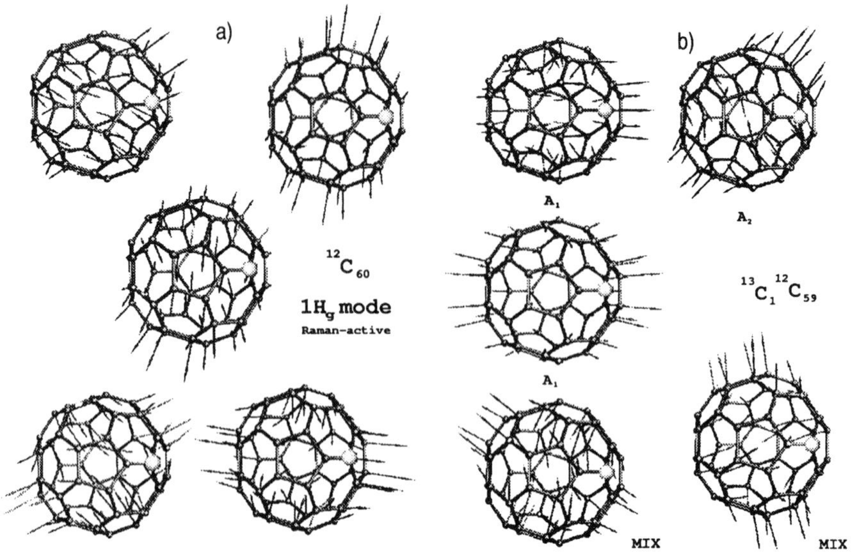

Figure 3. a) The vibration amplitudes for the 5-fold degenerate $H_g(1)$ Raman mode in $^{12}C_{60}$. In this figure the place where the ^{12}C atom is substituted by the isotope is marked with a larger circle. b) The vibration amplitudes for the splitted $H_g(1)$ Raman modes in the $^{12}C_{59}{}^{13}C$ isotopomer.

In addition to the obvious symmetry reduction due to the presence of the ^{13}C isotope, there is a direct isotope effect on the frequencies arising from their dependence on $(k/m)^{1/2}$, where k and m are the force constant and the atomic mass, respectively. In a C_{60} molecule with n ^{13}C isotopes, at first order, the vibrational frequency is expected to be reduced proportional to the difference between $^{12}C_{60}$ and $^{12}C_{60-n}{}^{13}C_n$, that is:

$$\frac{\omega(n)^2}{\omega(0)^2} = 1 - \frac{n}{720},$$

where $\omega(0)$ is an eigenmode frecuency of $^{12}C_{60}$ and 720 au is its atomic mass.

We can compare our results for the frequency shift for the $A_g(2)$ mode with the high resolution Raman spectra obtained by Guha *et al.*[8] In the Raman experiment they obtained 1471, 1470, and 1469 cm^{-1} for the $^{12}C_{60}$, $^{13}C^{12}C_{59}$, and $^{13}C_2{}^{12}C_{58}$ molecules, respectively. From our calculation we obtain 1416.8, 1415.4 and 1414 cm^{-1} in the same order for the respective isotope substitutions. Thus, the shifting in the energy levels, in our calculation, due to the increase in mass is 1.4 cm^{-1} as compared to the 1.0 cm^{-1} predicted from Eq. (1) and as observed in the Raman spectroscopy experiments. This result is quite satisfactory and gives confidence in our calculations.

From Table III and as illustrated in Fig. 1(a) we observe that when we substitute a ^{12}C with one ^{13}C a degeneracy remains. According to group theory that is not correct since there is a reduction of irreducible representations from I_h to C_{1h} groups. This feature, what we call an accidental degeneracy is produced by the fact that in some

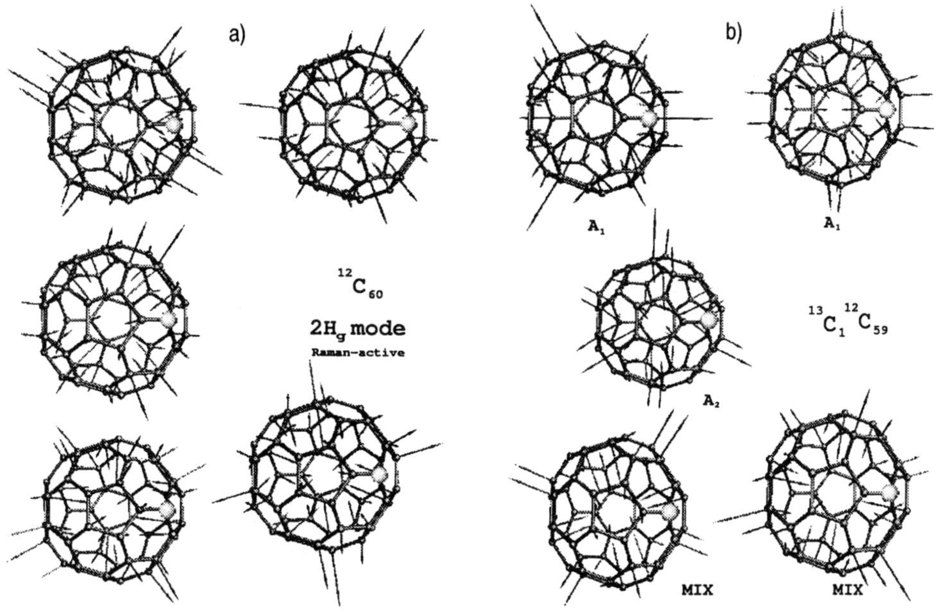

Figure 4. a) The vibration amplitudes for the 5-fold degenerate $H_g(2)$ Raman mode in $^{12}C_{60}$. In this figure the place where the ^{12}C atom is substituted by the isotope is marked with a larger circle. b) The vibration amplitudes for the splitted $H_g(2)$ Raman modes in the $^{12}C_{59}{}^{13}C$ isotopomer.

modes the amplitude of vibration of the isotope is zero, leading therefore to a higher symmetry. We observe that all T modes are frequency splitted with no exception. On the other hand, all the H, $G_u(1)$ and $G_g(4)$ modes show the accidental degeneracy.

In Figs. 3(a) and 3(b) we show the vibration amplitudes of the pristine $^{12}C_{60}$ and of the $^{12}C_{59}{}^{13}C$ isotopomer for the 263.33 cm^{-1} $H_g(1)$, mode and its isotope splitting, respectively. In Fig. 3(a) we present the five degenerate modes and show with a larger circle the position where we locate the isotope. Figure 3(b) contains the four splitted modes and the still degenerate mix mode. Similarly, in Figs. 4(a) and 4(b) we show the vibration amplitudes for the 467.26 cm^{-1}, $H_g(2)$ and its isotope splitting, respectively. From Figs. 3(b) and 4(b), one observes that the amplitude of vibration of these modes in the ^{13}C isotope has zero amplitude. This fact produces a higher symmetry in the molecule, and gives rise to the accidental 2-fold degeneracy that we find in our calculation The localized modes were found also with the help of the visualization.

In Fig. 5 we show for the $^{12}C_{59}{}^{13}C_1$ molecule, three modes of vibration with frequencies 724.82 cm^{-1} ($T_{2u}(2)$), 732.65 cm^{-1} ($H_u(3)$), and 737.4 cm^{-1} ($H_g(3)$). The isotope is represented by a larger circle. One observes very clearly the deformation produced in the molecule by the excitation of those modes.

Finally, in Fig. 6 we show two amplitudes of vibration, in different times, for the original $T_{2u}(2)$ and $H_u(3)$ modes when we substitute two isotope atoms. This shows pictorially the existence of modes that with a proper excitation one could have the possibility to distort the molecule and open larger gates than in the case of a molecule with no isotopes.

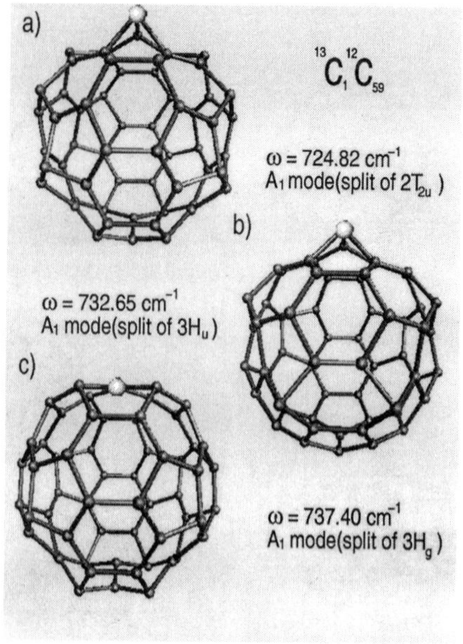

Figure 5. The vibration amplitudes for the modes with frequencies: a) 724.82 cm^{-1} ($T_{2u}(2)$), b) 732.65 cm^{-1} ($H_u(3)$) and c) 737.4 cm^{-1} ($H_g(3)$), in $^{12}C_{60}$ as is seen in the $^{12}C_{59}{}^{13}C$ isotopomer.

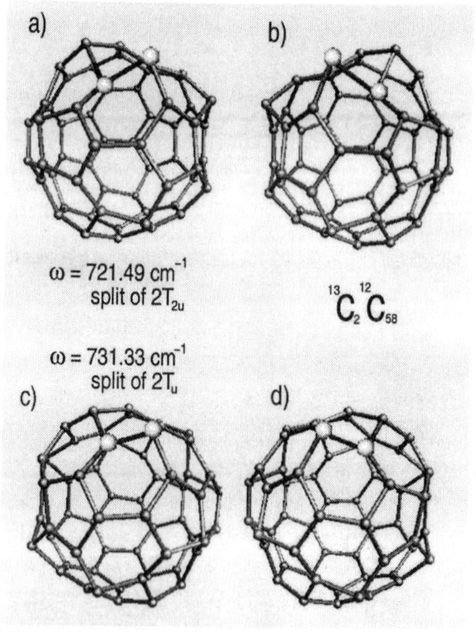

Figure 6. The vibration amplitudes for the modes with two particular frequencies as seen in two different times. Figures a) and b) correspond to the mode 721.49 cm^{-1} ($T_{2u}(2)$), and c) and d) to the mode 731.33 cm^{-1} ($H_u(3)$) in $^{12}C_{60}$ as is seen in the $^{12}C_{58}{}^{13}C^2$ isotopomer.

IV. Concluding Remarks

Motivated by the possibility of improving the methods for the production of larger quantities of endohedral fullerenes, we calculated from first principles the force constants acting in the $^{12}C_{60}$ molecule, and the vibrational spectra of that molecule and its isotopomers when one or two atoms are replaced by ^{13}C isotopes. Based on the results obtained and in view of recent experiments[8] of isotopically resolved Raman spectra of C_{60}, we feel that the proposal of selective excitation is a viable technique for the production of endohecdral fullerenes.

The results for the force constants as well as for the vibrational spectra agree well with other calculations and with the experimental data. We presented in a very fine detail the effects produced by breaking the symmetry through the isotope substitution of one and two ^{13}C isotopes in the C_{60} molecule. We studied in detail the modes that are sensitive to Raman and infrared spectroscopies. In particular, we observe that with one isotope, surprisingly enough, still remains a low degeneracy in the energy levels. As it is observed from the visualized amplitudes, this is produced by the fact that the isotope is still; situation that gives rise to a recovery of a higher symmetry in those modes.

We observed that the isotope substitution provides indeed the possibility of distorting the C-C bonds connecting the isotope with its neighbors through the selective excitation of the localized modes. This situation was illustrated in Figs. 5 and 6, where the larger gates formed by the distorted carbon rings in the C_{60} molecule, were shown schematically.

We expect that this calculation provokes the curiosity of some experimentalists to validate the predictions presented here.

Acknowledgments

This work was partially supported by Consejo Nacional de Ciencia y Tecnología (México), Grant No. G-25851-E.

References

1. H. W. Kroto, J.R. Heath, S. C. O'Brien, R. F. Kurl, and R.E. Smalley, *Nature* **318**, 162 (1985).
2. J. R. Heath, S. C. O´Brien, Q. Zhang, Y. Liu, R. F. Curl, H. W. Kroto, F. K. Tittel, and R. E. Smalley, *J. Am. Chem. Soc.* **107,** 7779 (1985).
3. Y. Chai, T. Guo, C. M. Jin, R. E. Hanfler, L. P. Felipe Chibante, J. Fare, L. H. Wang, J. M. Alford, and R. E. Smalley, *J. Phys. Chem.* **95**, 7564 (1991).
4. R. D. Johnson, D. S. Bethune, and C. S. Yannoni, *Account of Chem. Rec.* **25,** 169 (1992).
5. M. Saunders, H. A. Jimenez-Vasquez, R. J. Cross, and R. J. Poreda, *Science* **259**, 1428 (1993).
6. R. F. Curl, *Carbon* **30**, 1149 (1992).
7. J.L. Morán-L'opez, J.M. Cabrera-Trujillo, and J. Dorantes-Dávila, *Solid State Commun.* **96**, 451 (1995).
8. S. Guha, J. Menendez, J. B. Page, G. B. Adams, G. S. Spencer, J. P. Lehman, P. Giannozzi, and S. Baroni, *Phys. Rev. Lett.* **72**, 3359 (1994).

9. R. D. Johnson, G. Meijer, J. R. Salem, and D. S. Bethune, *J. Am. Chem. Soc.* **113**, 3619 (1991).
10. C. S. Yannnoni, P. P. Berrier, D. S. Bethune, G. Meijer, and J. R. Salem, *J. Am. Chem. Soc.* **113**, 3190 (1991).
11. A. P. Ramirez, A. R. Kortan, M. P. Rosseinsky, S. J. Duclos, A. M. Mujsce, R. C. Haddon, D. W. Murphy, A. V. Makhija, S. M. Zaburark and, K. B. Lyons, *Phys. Rev. Lett.* **68**, 1058, (1992).
12. A. A. Quong, M. R. Pederson, and J. L. Feldman, *Solid State Commun.* **87**, 535 (1993).
13. K. A. Wang, A. M. Rao, P. C. Eklund, M. S. Dresselhaus, and G. Dresselhus, *Phys. Rev. B* **48**, 11375 (1993).
14. P. Giannozzi and S.J. Baroni, *J. Chem. Phys.*, **100**, 8537 (1994).
15. G. B. Adams, J. B. Page, O. F. Sankey, J. Sinha, J. Menéndez, and D. R. Huffman, *Phys. Rev. B* **44**, 4052 (1991).
16. X. Q. Wang, C. Z. Wang, and K. M. Ho, *Phys. Rev. B* **48**, 1884 (1993).
17. D. A. Dixon, B. E. Chase, G. Fitzgerald, and N. Matsuzawa, *J. Phys. Chem.* **1995**, 4486 (1995).
18. R.L. Murry and G.E. Scuseira, *Science* **263**, 791 (1994).

Orbital Magnetism in Low Dimensional Systems: Surfaces, Thin Films and Clusters

J. Dorantes-Dávila,[1] R. A. Guirado-López,[1] and G. M. Pastor[2]

[1] *Instituto de Física, "Manuel Sandoval Vallarta"*
Universidad Autónoma de San Luis Potosí
Alvaro Obregón 64
78000 San Luis Potosí, S.L.P.
MEXICO

[2] *Laboratoire de Physique Quantique*
UMR 5626 du CNRS
Université Paul Sabatier
118 Route de Narbonne
F-31062 Toulouse
FRANCE

Abstract

The local and average orbital magnetic moments $\langle L \rangle$ of several low-dimensional systems are determined in the framework of a self-consistent tight-binding theory. For transition metal surfaces, results for the local orbital magnetic moments $L_\delta(i)$ at different layers i and magnetization directions δ are discussed. It is shown that $L_\delta(i)$ is significantly enhanced at the surface atoms as compared to the corresponding bulk moment $L_\delta(\text{bulk})$. $\langle L \rangle$ depends strongly on the local coordination number and is generally larger the more open the surface is and decreases abruptly as we move from the uppermost layer ($i = 1$) to the second layer ($i = 2$). After some oscillations, convergence to bulk values is reached. The orbital moments at pure surfaces are compared with results for deposited films by considering Co on Pd(111) as a representative example. The role of $\langle L \rangle$ on the magneto-anisotropic behavior of thin Co/Pd(111) films is also discussed. For Ni clusters, a remarkable enhancement of the average orbital moment per atom is observed, which for the very small sizes can be up to an order of magnitude larger than the corresponding bulk value [*e.g.*, $\langle L(\text{Ni}_7) \rangle \simeq 0.5\,\mu_B$ while

Physics of Low Dimensional Systems
Edited by J. L. Morán-López, Kluwer Academic/Plenum Publishers, New York 2001

L(Ni-bulk) $\simeq 0.05\,\mu_B$]. The relation between the enhancement of the local orbital moments $L(i)$ and the reduction of local coordination number is discussed, in particular by comparison with results for TM surfaces. The transition from atomic to bulk-like behavior is determined. For large clusters ($N = 100$–150 atoms) we observe bulk-like quenching at inner atoms and enhanced local moments at the cluster surface. The importance of orbital contributions to the total moments $\bar{\mu}_N$ is quantified. In addition, $\langle L_N \rangle$ amounts to 20–35% of $\bar{\mu}_N$ and is therefore crucial for comparing theoretical results with experiment

I. Introduction

In atoms, Hund's rules predict maximum orbital angular momentum L compatible with maximum spin multiplicity. In contrast, in magnetic transition-metal (TM) solids the orbital moments are almost completely quenched as a result of electron delocalization and band formation. Therefore, interesting orbital contributions to the magnetic behavior are expected in systems like surfaces, thin films and clusters, which are characterized by a reduced dimensionality and local coordination number. Previous investigations of orbital magnetism in low-dimensional systems—in the way from the atom to the solid—show that the local and average orbital moments $\langle L \rangle$ are very sensitive to the local atomic environment.[1–4] In fact, L_δ is originated by spin-orbit (SO) interactions and consequently it is much more sensitive to the details of the local electronic structure. In spite of the remarkable research activity on the subject, a systematic understanding of the environment dependence of the orbital magnetic moment in low-dimensional systems is still lacking. It is the main purpose of this paper to investigate the orbital magnetism of several low-dimensional transition metal (TM) systems (*i.e.* surfaces, thin films and clusters).

II. Theoretical Method

Our theoretical approach is based on a d-electron Hamiltonian which can be written as[4,5]

$$H = H_0 + H_{\mathrm{C}} + H_{\mathrm{SO}} \; . \tag{1}$$

The inter-atomic hopping term H_0 is given by

$$H_0 = \sum_{\substack{\langle i,j \rangle \\ \alpha,\beta,\sigma}} t_{ij}^{\alpha\beta}\, \hat{c}_{i\alpha\sigma}^{\dagger} \hat{c}_{j\beta\sigma} \; , \tag{2}$$

where $\hat{c}_{i\alpha\sigma}^{\dagger}$ ($\hat{c}_{i\alpha\sigma}$) refers to the creation (annihilation) operator of an electron with spin σ at the d orbital α of atomic site i, and $t_{ij}^{\alpha\beta}$ refer to the corresponding hopping integrals.

The Coulomb interaction term H_{C} takes into account the effects of redistributions of the spin- and orbital-polarized density in the unrestricted Hartree-Fock approximation. Orbital polarization (OP) contributions are included as proposed by Hjortstam *et al.*[3,6]

$$H_C = \sum_{i\alpha\sigma} \Delta\varepsilon_{i\sigma}\, \hat{n}_{i\alpha\sigma} \; , \tag{2}$$

where $\hat{n}_{i\alpha\sigma} = \hat{c}^{\dagger}_{i\alpha\sigma}\hat{c}_{i\alpha\sigma}$ is the electron number operator and $\Delta\varepsilon_{i\sigma} = U\Delta n(i) - \sigma J S_{\delta}(i) - B\sum_{im\sigma} m\langle L^z_{i\sigma}\rangle$ is the site- and spin-dependent shift of the d level $\varepsilon_{i\sigma} = \varepsilon^0_d + \Delta\varepsilon_{i\sigma}$. ε^0_d stands for the d-orbital energy in the paramagnetic bulk [$\Delta n(i) = n(i) - n_d(\text{bulk})$]. m refers to the d-electron magnetic quantum number, and B to the Racah coefficient (see Ref. 3).

The average intra-atomic direct Coulomb-repulsion integral is denoted by U and the average exchange integral is denoted by J. Notice that the spin-quantization axis δ is taken to be parallel to the magnetization direction.

The third term in Eq. (1) is the spin-orbit (SO) interaction, which is treated in the usual intra-atomic single-site approximation:[7]

$$H_{\text{SO}} = -\xi \sum_{i,\alpha\sigma,\beta\sigma'} (\vec{L}_i \cdot \vec{S}_i)_{\alpha\sigma,\beta\sigma'}\, \hat{c}^{+}_{i\alpha\sigma}\, \hat{c}_{i\beta\sigma'} \,. \tag{2}$$

Here, $(\vec{L}_i \cdot \vec{S}_i)_{\alpha\sigma,\beta\sigma'}$ refer to the intra-atomic matrix elements of $\vec{L}\cdot\vec{S}$ which couple the up and down spin-manifolds and which depend on the relative orientation between magnetization direction and the lattice structure.

The local densities of electronic states (DOS) $\rho^{\delta}_{im\sigma}(\varepsilon)$ are determined self-consistently for each orientation δ of the spin magnetization $\vec{S}$ with respect to the structure.

The local orbital moments $\vec{L}_{\delta}(i) = [L_{x\delta}(i), L_{y\delta}(i), L_{z\delta}(i)]$ depend on the cluster atom i ($1 \leq i \leq N$) and on δ ($L_{\delta}(i) = |\vec{L}_{\delta}(i)|$). Each component $L_{\alpha\delta}(i)$ is obtained by integration of $\rho^{\delta}_{im\sigma}(\varepsilon)$, choosing α as the orbital quantization direction.

III. Results and Discussion

The parameters used in the calculations are specified as follows. The two-center d-electron hopping integrals are given by the canonical expression in terms of the corresponding bulk d-band widths. The intra-atomic Coulomb exchange integral J yields the proper magnetic moment and exchange splitting in the solid ($J = 0.67$ eV for Fe, $J = 0.76$ eV for Co, and $J = 0.51$ eV for Ni). From atomic Hartree-Fock calculations we estimate the Racah coefficient $B = 0.1$ eV. In addition, the role of OP contributions and the sensitivity of the results on this parameter are analyzed by varying B. The SO coupling constants are taken form Ref. 7 ($\xi = 73$ meV for Fe, $\xi = 88$ meV for Co, and $\xi = 110$ meV for Ni).

In the case of TM surfaces, we find an important enhancement of the orbital moments at the uppermost surface layer. For Fe at the (001) surface, we obtain $\langle L\rangle(\text{surface})/L(\text{bulk}) \simeq 2$, and for the (110) surface $\langle L\rangle(\text{surface})/L(\text{bulk}) \simeq 1.2$. Moreover, in the case of the (110) surface the anisotropy within the surface plane ΔL_{xy} is of the same order of magnitude and sometimes even larger that the off-plane ΔL_{xz}. This is related to the remarkable magneto-anisotropic behavior found in this low-symmetry surface.[5] For the Co (0001) surface, we obtain an enhancement of the orbital moment at the uppermost layer [about 25% for the x direction (within the plane) and 15% for the z direction (perpendicular to the plane)] which is quantitatively similar to that of the compact (110) Fe surface. Another characteristic of Co is the important anisotropy between the x and z directions, $\Delta L_{xz} = L_x - L_z$, which is much larger than in Fe and which is also layer dependent. For Ni surfaces, as in previous cases, L_{δ} is enhanced most significantly at the surface layer. The largest part of this increase is lost already at the second layer. In addition, the ordering of $L_{\delta}(1)$ for different surfaces depends on the magnetization direction δ: $L_z(1)$ is larger at the (111) surface than at the (001), while for $L_x(1)$ the opposite occurs. We found that $L_\delta(i)$ is in general larger at open surfaces than at closer ones and converges rapidly towards the bulk limit already a few layers below the surface ($i \geq 6$),[4] and that a significant anisotropy of the orbital moments is found for different orientations of the magnetization which anticipates the expected spin-orbit contributions to the magneto-crystalline anisotropy energies.

The orbital magnetic moments $\langle L_\delta \rangle$ ($\delta = x, z$) at the interface of Co thin films on Pd(111) also reflect the magneto-anisotropic behavior found in these systems. Indeed, a change of sign in $L_x(i) - L_z(i)$ at the interface layer occurs which is in agreement with the perpendicular contribution of the Co-Pd interface and the 2nd order perturbation theory.[7] Morover, film-substrate hybridizations causes an enhancement of L_δ at the Co atoms ($L_\delta \simeq 0.17$–$0.23\,\mu_B$) as compared to the corresponding bulk moment ($L_\delta(\text{bulk}) \simeq 0.11\,\mu_B$). The obtained environment dependence of $L_\delta(i)$ must be the result of more subtle changes in the electronic structure after deposition. The layer dependence is quite different from that obtained at the pure Co surface. For instance, in Co_4/Pd(111) there is no sharp decrease of $L_\delta(i)$ at the second layer and that $L_\delta(i)$ is also enhanced at deeper layers, even at the fourth layer which is in contact with the Pd substrate. Concerning the anisotropy in L_δ we observe that it is largest at the surface of the film [$L_x(1) = 0.234\,\mu_B$ and $L_z(1) = 0.203\,\mu_B$] and that it decreases as we approach to the interface. Notice the change of sign in $L_x(i) - L_z(i)$ at the interface layer $i = 4$, which is also obtained in other Co_n/Pd(111) films for $n = 1$–4. It is also worth noting that a small spin polarization $2S_z(5) = 0.37\,\mu_B$ is induced at the Pd layer $i = 5$ which is in contact with the Co film.

In the case of free clusters, very little is known about the local atomic environment dependence of orbital magnetism, particularly from the point of view of theory, which has been so far concerned with the dominant spin contributions.[8,9] This is quite remarkable, since a size-dependent enhancement of $\langle L \rangle$ could have direct consequences on the results for the average magnetic moments per atom, and for the comparison with experiment.[10,11]

For Ni clusters, we observe that the reduction of system size causes a remarkable enhancement of $\langle L_\delta \rangle$ with respect to the bulk. Values about eight times larger than $L_b = 0.05\,\mu_B$ are not uncommon for $N \leq 13$. However, comparison with the atomic result $L_a = 2\,\mu_B$ shows that the largest part of the quenching of L takes place already at the smallest clusters, as soon as full rotational symmetry is lost. For example, for Ni_3 (triangle) we obtain $\langle L_z \rangle = 0.40\,\mu_B$. One observes that $\langle L_\delta \rangle$ decreases with increasing N showing some oscillations in the case of *fcc* clusters as bulk-like quenching is approached. We also note that an important enhancement of $\langle L_\delta \rangle$, about 100%, is still present even for the largest considered sizes ($\langle L_\delta \rangle / L_b = 1.8$ for $N = 165$). There are several contributions for the enhancement of $\langle L_\delta \rangle$. First, the increase of the local spin polarizations $S(i)$, which induces larger orbital moments by means of the spin-orbit interactions. Second, orbital polarization contributions amplify this effect thereby reducing the Coulomb repulsion energy. Third, the larger d-electron DOS, and the presence of degeneracies in the single-particle spectrum, favor a more effective spin-orbit mixing, which enhances $\langle L_\delta(i) \rangle$ even in situations where the spin polarizations are saturated ($2\langle S_z \rangle \simeq 10 - n_d$). Finally, there are less predictable effects related to the details of the electronic structure and its dependence on cluster geometry (*e.g.*, the presence of high-symmetry axes, changes in bond-length, etc.) which may affect the orbital moments depending on the d-band filling. In general, we obtain that the enhancement of $\langle L_\delta \rangle$ is driven by the cluster surface, as it is the case for the spin moments.

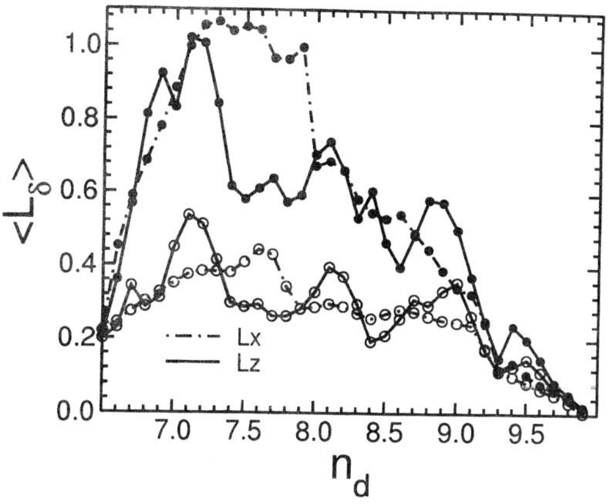

Figure 1. Average orbital magnetic moment per atom $\langle L_\delta \rangle$ in a seven-atom pentagonal bipyramid transition-metal cluster as a function of d-band filling n_d. The magnetization direction δ is taken along a principal C_n symmetry axis ($\delta = z$, full lines) or along a NN bond perpendicular to z ($\delta = x$, dashed lines). Full (open) circles refer to calculations with $B = 0.1$ eV ($b = 0$). Results are given for bulk NN bond length.

Previous calculations of spin-only contributions based on local spin density approximation[9] have been compared with the average magnetizations per atom $\bar{\mu}_N$ derived from Stern-Gerlach experiments, under the assumption of superparamagnetic relaxation.[11] Despite some quantitative differences among the results obtained by different groups, they all agree that the theoretical ground-state spin magnetizations underestimate systematically the experimental $\bar{\mu}_N$ by about 0.3–0.6 μ_B for $N \leq 13$. Our results show that $\langle L \rangle$ is an important contribution to the cluster magnetic moment (20–40%). In fact, taking into account the enhancement of the orbital moments, which in Ni align parallel to the spin moments, the largest part of the discrepancy between theory and experiment is removed.

The trends for other TM's and the effect of *sp*-*d* charge transfers can be inferred from Fig. 1, where the average orbital moment per atom $\langle L_\delta \rangle$ in a pentagonal bipyramid is given as a function of d-band filling n_d ($N = 7$) for $B = 0.1$ (full circles) and $B = 0$ (open circles). In the considered range of n_d, $\langle S_\delta \rangle$ increases approximately linearly with increasing number of d holes, as we move from Ni ($n_d \simeq 9$) to Fe ($n_d \simeq 7$), since the self-consistent calculations yield nearly saturated spin moments. This favors an increase of $\langle L_\delta \rangle$ with decreasing n_d. In contrast to $\langle S_\delta \rangle$, $\langle L_\delta \rangle$ shows a very rich band-filling dependence with strong oscillations. Changes in the sign of the anisotropy $\Delta L = L_z - L_x$ are observed, which is consistent with similar oscillations of the magnetic anisotropy energy.[12] From Fig. 1, we observe that easonable changes in the value of B do not affect the main conclusions. For instance, for $B = 0$ we obtain a very similar size dependence of $\langle L_\delta \rangle$, although with somewhat reduced quantitative values. $\langle L_\delta \rangle|_{B=0}$ is usually about 20–40% smaller than $\langle L_\delta \rangle|_{B=0.1\text{eV}}$. In this case the enhancement with respect to L_b is smaller but still significant. For example, for $B = 0$, and $N = 7$, 13 and 135, $\langle L_\delta \rangle = 0.38\,\mu_B$, $0.14\,\mu_B$, and $0.078\,\mu_B$, respectively ($d = d_b$). The orbital magnetic moments and in particular the enhancement with respect to L_b are therefore important for all magnetic TM clusters. Rhodium clusters deserve a special mention in this context, since in Rh_N the spin moments are smaller and the SO couplings stronger than in $3d$ TM's. Our calculations show that the orbital contributions to the total magnetic moment are also relevant in this case.

Acknowledgments

This work has been supported in part by CONACyT (Mexico) grant 32085E. The *Laboratoire de Physique Quantique* is *Unité Mixte de Recherche* of the CNRS.

References

1. J. Trygg, B. Johansson, O. Eriksson, and J. M. Wills, *Phys. Rev. Lett.* **75**, 2871 (1995); M. Tischer, O. Hjortstam, D. Arvanitis, J. Hunter Dunn, F. May, K. Baberschke, J. Trygg, J. M. Willis, B. Johansson, and O. Eriksson, *Phys. Rev. Lett.* **75**, 1602 (1995); **75**, 1602 (1995); D. Weller, J. Stöhr, R. Nakajima, A. Carl, M. G. Samant, C. Chappert, R. Mégy, P. Beauvillain, P. Veillet, and G. A. Held, *Phys. Rev. Lett.* **75**, 3752 (1995); **75**, 3752 (1995); A. N. Anisimov, M. Farle, P. Poulopoulos, W. Platow, and K. Baberschke, *Phys. Rev. Lett.* **82**, 2390 (1999); H. A. Dürr, G. van der Laan, J. Vogel, G. Panaccione, N. B. Brookes, E. Dudzik, and R. McGrath, *Phys. Rev. B* **58**, R11853 (1998).
2. H. A. Dürr, S. S. Dhesi, E. Dudzik, D. Knabben, G. van der Laan, J. B. Goedkoop, and F. U. Hillebrecht, *Phys. Rev. B* **59**, R701 (1999); K. W. Edmonds, C. Binns, S. H. Baker, S. C. Thornton, C. Norris, J. B. Goedkoop, M. Finazzi, and N. B. Brookes, *Phys. Rev. B* **60**, 472 (1999).
3. O. Hjortstam, J. Trygg, J. M. Wills, B. Johansson, and O. Eriksson, *Phys. Rev. B* **53**, 9204 (1996).
4. J. L. Rodríguez-López, J. Dorantes-Dávila, and G. M. Pastor, *Phys. Rev. B* **57**, 1040 (1998).
5. J. Dorantes-Dávila and G. M. Pastor, *Phys. Rev. Lett.* **77**, 4450 (1996).
6. The OP term is given by $H_{\rm OP} = -B\sum_{im\sigma} m\langle L^z_{i\sigma}\rangle \hat{n}_{im\sigma}$, where i refers to the atom, m to the d-electron magnetic quantum number, σ to the spin, and B to the Racah coefficient (see Ref. 3).
7. P. Bruno, "Magnetismus von Festkörpern und Grenzflächen", *Ferienkurse des Forschungszentrums Jülich* (KFA Jülich, 1993), ISBN 3-89336-110-3, Ch. 24.
8. See, for instance, G. M. Pastor and K. H. Bennemann, in *Metal Clusters*, edited by W. Ekardt (Wiley, New York, 1999), p. 211.
9. M. Castro, Ch. Jamorski, and D. Salahub, *Chem. Phys. Lett.* **271**, 133 (1997); B. V. Reddy, S. K. Nayak, S. N. Khanna, B. K. Rao, and P. Jena, *J. Phys. Chem. A*, **102**, 1748 (1998); F. A. Reuse and S. Khanna, *Eur. Phys. J. D* **6**, 77 (1999).
10. I. M. L. Billas, A. Châtelain, and W. A. de Heer, *Science* **265**, 1662 (1994).
11. S. E. Apsel, J. W. Emmert, J. Deng, and L. A. Bloomfield, *Phys. Rev. Lett.* **76**, 1441 (1996).
12. G. M. Pastor, J. Dorantes-Dávila, S. Pick, and H. Dreyssé, *Phys. Rev. Lett.* **75**, 326 (1995).

Electronic and Structural Properties of Rh_N Clusters: a Molecular Dynamics Study

R. Guirado-López

Instituto de Física, "Manuel Sandoval Vallarta"
Universidad Autónoma de San Luis Potosí
78000 San Luis Potosí, S.L.P.
MEXICO
e-mail:guirado@dec1.ifisica.uaslp.mx

Abstract

A tight-binding total-energy expression is used to study the structure and electronic properties of relatively large Rh_N clusters ($13 \leq N \leq 165$). The clusters are built by adding successive atomic shells around a central atom and the radii of these shells are allowed to relax independently. In all cases, the relaxed structure is far from being homothetical to the unrelaxed one. For small Rh_N clusters ($N \leq 75$) multiple magnetic solutions are obtained and in all cases the local magnetic moments $M(i)$ present a remarkable size and environment dependence. The relation between the observed average magnetization and the cluster geometry is also analyzed.

I. Introduction

As is well known, in cluster structures the bond lengths are expected to be smaller than those of the bulk due to the reduced number of neighbors, this reduction being more pronounced in small particles and becoming less significant with increasing cluster size. In general, for the intermediate sizes one may expect to find a non uniform relaxation profile within the structure since the local coordination number is rapidly changing as we move from the inner atoms to those located at the surface of the particle. This fact allows the existence of different interatomic forces that could drive the geometry of the cluster into another specific crystalline arrangement or amorphous phase.

Physics of Low Dimensional Systems
Edited by J. L. Morán-López, Kluwer Academic/Plenum Publishers, New York 2001

Experimentally, information about the structure of small particles can be obtained using either photoelectron spectroscopy,[1] high-resolution transmission electron microscopy[2] or chemical probe methods.[3] From the point of view of theory, a minimization of the cluster total energy is required, and in this case, the knowledge of the ion-ion repulsive potential and the electronic energy, which govern the dynamics of the atoms, as well as of the electron-electron interactions, which control the charge transfer among the different sites and are responsible for the formation of local magnetic moments, are of fundamental importance. This type of calculation which implies formidable computational requirements are necessary since the magnetic properties and the cluster geometry are interrelated. In this sense, the development of theoretical models allowing the description of the electronic and magnetic properties of cluster structures having a number of atoms comparable to the ones found in the experimental situation is of fundamental importance.

A major step in molecular dynamics was originally proposed by Car and Parrinello.[4] They introduced a new method to minimize the total energy of a solid determined using *ab-initio* calculations based on the local density approximation. As it has been shown previously, the Car-Parrinello molecular dynamics can be easily applied to the tight-binding total energy expression, with the relevant electron-electron interactions treated in the Hartree-Fock approximation (HFA).[5] In this case, the electronic coordinates required to perform molecular dynamics are then defined with respect to the basis set consisting of atomic orbitals that replaces the originally proposed *ab-initio* basis set. This is actually the approach introduced by Khan and Broughton to treat silicon clusters.[6]

Here, we report about the complex internal relaxations that are present in Rh_N clusters, as well as on the size and structural dependence of their magnetic properties, by using a tight-binding molecular dynamics scheme incorporating an extended multi-band Hubbard Hamiltonian for the treatment of magnetic effects. Specifically, we will consider only three dimensional arrangements, which are compact portions of an *fcc* lattice. We will focus on the dependence of the magnetic and structural properties on relevant variables such as electron-electron interactions, bond length, cluster size and structure.

The paper is organized as follows. The model Hamiltonian, its treatment within the HFA, and the implementation of the molecular dynamics within the present tight-binding method is summarized in Sec. II. In Sec. III, we show the obtained numerical results for the structural and magnetic properties of *fcc* Rh_N clusters as a function of the cluster size. Finally, in Sec. IV, the summary and conclusions are given.

II. Theoretical Method

The semiempirical model used here has been described in detail elsewhere,[5] thus we only summarize its main points and discuss the choice of parameters. We use a tight-binding Hubbard Hamiltonian for the d-band in the rotationally invariant form in orbital space,[7] expressed in the basis of real d-orbitals of symmetry t_{2g} (xy, yz, zx) and e_g (x^2-y^2, $3z^2-r^2$). In this basis the most important matrix elements of the on-site Coulomb interaction, *i.e.*, involving one or two orbitals, are: intraorbital $(U+2J)$ and interorbital (U) Coulomb and exchange (J) integrals. The electronic structure of the clusters is determined by solving this model with the interactions treated in the HFA. Using the basis of $4d$-atomic spin-orbitals $|i\lambda\sigma\rangle$ centered at each site i, one finds[5]

$$H_{\rm HFA} = \sum_{i\lambda,\sigma} \epsilon_{i\lambda\sigma} n_{i\lambda\sigma} + \sum_{i,\lambda,\mu\neq\lambda,\sigma} h_{i,\lambda\mu,\sigma} a^{\dagger}_{i\lambda\sigma} a_{i\mu\sigma} + \sum_{i\lambda,j\mu,i\neq j,\sigma} t_{i\lambda,j\mu} a^{\dagger}_{i\lambda\sigma} a_{j\mu\sigma}, \qquad (1)$$

in the usual notation. The on-site energy levels, $\epsilon_{i\lambda\sigma}$, are given in terms of the spin-orbital occupation numbers $\langle n_{i\lambda\sigma}\rangle$ and electron-electron interactions as follows

$$\epsilon_{i\lambda\sigma} = \epsilon^o_d + \Big(U - \frac{1}{2}J\Big)(N_i - N_a) - \frac{1}{2}(U-3J)(\langle n_{i\lambda}\rangle - n_a)$$
$$-\frac{1}{2}\xi_\sigma J M_i - \frac{1}{2}\xi_\sigma (U+J)\langle m_{i\lambda}\rangle, \qquad (2)$$

where $\xi_\sigma = \pm 1$ for $\sigma = \uparrow,\downarrow$, and $i(j)$ stands for the neighbors of site i. Here, we have introduced the occupation number $\langle n_{i\lambda}\rangle = \langle n_{i\lambda\uparrow}\rangle + \langle n_{i\lambda\downarrow}\rangle$, and the magnetic moment $\langle m_{i\lambda}\rangle = \langle n_{i\lambda\uparrow}\rangle - \langle n_{i\lambda\downarrow}\rangle$ of each orbital λ. N_i and M_i are the total number of electrons and the local magnetic moment at atom i, respectively.

The second term in Eq. (1) corresponds to the intrasite interorbital Fock terms ($\bar{\sigma} = -\sigma$),

$$h_{i,\lambda\mu,\sigma} = (J-U)\langle a^{\dagger}_{i\mu,\sigma} a_{i\lambda,\sigma}\rangle + 2J\langle a^{\dagger}_{i\mu,\bar{\sigma}} a_{i\lambda,\bar{\sigma}}\rangle. \qquad (3)$$

Finally, the hopping integrals $t_{i\lambda,j\mu}$ are obtained in the Slater-Koster scheme[8] from the parameters $(dd\sigma)$, $(dd\pi)$ and $(dd\delta)$. We have carried out a least mean square fit of the levels which correspond to a wave function of pure d character calculated within the LDA formalism using only two parameters, $(dd\sigma)$ and $(dd\pi)$, since $(dd\delta)$ is very small and can be neglected. As usual, these parameters are assumed to vary exponentially with the distance R_{ij} between the two considered atoms.

The bond length and the magnetic states of the clusters are determined by minimizing the total energy,

$$E^d(N) = \langle H_{\rm HFA}\rangle - \langle H_{\rm int,HFA}\rangle + E_{\rm rep}, \qquad (4)$$

where the repulsive energy $E_{\rm rep}$ is of the Born-Mayer type, *i.e.*, it is described by a sum of pair interaction energies $A(R_{ij})$ between first nearest neighbors. The corresponding parameters of the model are obtained from bulk properties: cohesive energy, bulk modulus, bandwidth and equilibrium distance.

II.1. Minimization of the Total Energy by Molecular Dynamics

From Eq. (4) it can be seen that the total ground state energy E can be considered as a function of both: *i*) the set of ion coordinates $\{\mathbf{R}_{ij}\}$ and *ii*) the set of components $\{c^n_{i\lambda\sigma}\}$ of the occupied eigenstates in the atomic orbital basis set, called electronic coordinates. Thus, the total energy has to be minimized with respect to all these variables in order to find the ground state of the cluster. This is carried out using the Born-Oppenheimer approximation, *i.e.*, at any position of the ions $\{\mathbf{R}_i\}$, E should be at its minimum with respect to the electronic coordinates. In principle, this can be achieved by a self-consistent diagonalization of the Hamiltonian at each atomic configuration $\{\mathbf{R}_i\}$. Then the force acting on each ion can be computed, and the positions $\{\mathbf{R}_i\}$ are updated according to classical dynamics. However, this procedure would be very time consuming and we have used instead the fictitious Lagrangian method as proposed by Car and Parrinello.[4] This method provides an equation of motion for the electronic coordinates $\{c^n_{i\lambda\sigma}\}$ thus avoiding the self-consistent diagonalization of the

Table I

Hopping parameters $(dd\sigma)_o$ and $(dd\pi)_o$, and electron-electron interaction parameters: U and J (all in eV) estimated for bulk Rh.

	$(dd\sigma)_o$	$(dd\pi)_o$	U	J
Rh	−0.8939	0.3577	1.2	0.52

Hamiltonian at each step. The electronic coordinates are treated as position variables of classical particles with fictitious mass μ. Then the dynamics of the ions with mass M and of these fictitious particles is governed by a classical Lagrangian,[5] with constraint equations arising from the orthogonality of occupied eigenstates. The equations of motion are [the first (second) time derivative is denoted by a single (double) dot]:

$$M\ddot{R}_{i\alpha} = -\frac{\partial E_{\rm tot}(\{c_{i\lambda\sigma}^{(n)}\}, \{\mathbf{R}_i\})}{\partial R_{i\alpha}}, \tag{5}$$

for the ions, with $\alpha = x, y, z$, and

$$\mu\ddot{c}_{i\lambda\sigma}^{(n)} = -\frac{\partial E_{\rm tot}(\{c_{i\lambda\sigma}^{(n)}\}, \{\mathbf{R}_i\})}{\partial c_{i\lambda\sigma}^{(n)}} + \sum_{n'} \Lambda_\sigma^{nn'} c_{i\lambda\sigma}^{(n')}, \tag{6}$$

for fictitious particles (here n and n' are occupied electronic states). The quantities $\Lambda_\sigma^{nn'}$ are Lagrange multipliers given by

$$\Lambda_\sigma^{nn'} = \sum_{i\lambda}\left[\frac{1}{2}\frac{\partial E_{\rm tot}}{\partial c_{i\lambda\sigma}^{(n)}} c_{i\lambda\sigma}^{(n')} + \frac{1}{2} c_{i\lambda\sigma}^{(n)} \frac{\partial E_{\rm tot}}{\partial c_{i\lambda\sigma}^{(n)}} - \mu\dot{c}_{i\lambda\sigma}^{(n)}\dot{c}_{i\lambda\sigma}^{(n')}\right]. \tag{7}$$

The above equations of motion are integrated numerically with a time step Δt using the Verlet algorithm.[9]

III. Results and Discussion

We have performed explicit calculations of the ground state for *fcc* Rh_N clusters to determine their binding energy, electronic structure and equilibrium bond-lengths. In this respect, it is important to say that small Rh_N clusters ($N \leq$ 60–90) are often magnetic in their ground state.[10] In this work, we will consider first the atomic relaxations present in non-magnetic Rh particles the size of which is increased by adding successive shells of neighbors around a central atom. Then we will investigate the influence of a spin polarization on the ground state geometry. The tight-binding parameters and the values of the Coulomb (U) and exchange (J) interactions as defined by Eq. (2) are given in Tables I and II.

First we show in Fig. 1 our results for the relaxation profile of non magnetic *fcc* Rh_N clusters as a function of the size of the particles, obtained using the model presented in Sec. II but neglecting the exchange integral. The relaxation of the distance between the j and $j-1$ shells of neighbors in the cluster, $\Delta R_{j,j-1}/R_{j,j-1}^b$, is defined

Table II

Electronic filling N_a, equilibrium interatomic distance in the bulk R_o, hopping integral exponent q, and the parameters of the pair repulsive potential, A_o and p, for bulk *fcc* Rh.

	N_a	R_o(Å)	q	A_o(meV)	p
Rh	8.0	2.689	3.47	236.0	10.42

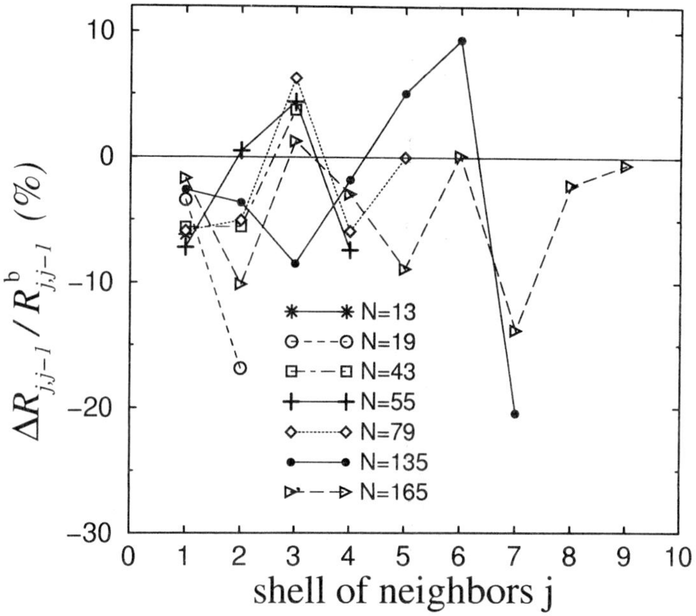

Figure 1. The relaxation of the intershell spacing between the j and $j-1$ shells of neighbors, $\Delta R_{j,j-1}/R^b_{j,j-1}$, for *fcc* Rh_N clusters as a function of size.

as

$$\frac{\Delta R_{j,j-1}}{R^b_{j,j-1}} = \frac{(R_{0,j} - R_{0,j-1}) - (R_{0,j} - R_{0,j-1})_b}{(R_{0,j} - R_{0,j-1})_b}, \tag{8}$$

where $(R_{0,j} - R_{0,j-1})_b = R^b_{j,j-1}$ refers to the distance between the j and $j-1$ shells of neighbors defined for the bulk structure and $R_{0,j}$ is the radius of shell j. Thus, a negative (positive) value of $\Delta R_{j,j-1}/R^b_{j,j-1}$ means a contraction (expansion) of the intershell spacing.

As is well known, metallic clusters have lattice parameters that decrease in general as the number of atoms in these clusters decreases, a fact that has been explained as a surface effect. Our results show that all atomic shells relax inwards leading to an overall contraction of the cluster. However, the relaxed structure is not homothetical to the unrelaxed one. This is illustrated in Fig. 1 where we observe that a complex pattern of intershell distances occurs in the particles upon optimization, which differs considerably from the ideal reference structure. In general, we find a contraction of intershell spacings with respect to the bulk, although for some of them sizeable expansive relaxations occur, as is the case of Rh_{135} where the intershell spacing between the 6 and 5 shells of neighbors presents a 10% expansion.

Experimental results as well as theoretical calculations have shown that transition metal surfaces generally exhibit a relaxation relative to the bulk. While this relaxation is almost negligible on the most closed-packed surfaces, on open surfaces it shows damped oscillations around the bulk interlayer spacing when going into the crystal with a noticeable contraction of the first layer.[11,12] For instance, experiments on (110) transition metal surfaces have shown that the outermost layer relaxes inwards, the second layer slightly outwards and the next layer again inwards. The external shells of the clusters considered in our calculations are too small to exhibit all the characteristics of infinite surfaces, since the edge and corner atoms of the particles should play an important role in determining the morphology of the cluster. However, we can see that the non uniform relaxation profile shown in Fig. 1 is qualitatively similar to the results obtained at surfaces of transition metals. In constrast, it is important to note that in small particles, the outermost layers do not always present the largest contraction (see the results for $N = 43, 79$, and 165). Although the contribution of *sp* electrons has been omitted in our model, we expect that the rather complex size dependence obtained for the inner-shell relaxations should reflect the realistic situation in these kinds of systems.

It is also interesting to precise the energy gained in a non uniform relaxation with respect to a uniform one. For this purpose, we have performed also non magnetic ($J = 0$) electronic structure calculations for Rh_{19}, Rh_{43}, Rh_{55}, and Rh_{79} clusters but now assuming a uniform relaxation process. This gives contractions of the interatomic spacing (with respect to the nearest neighbor distance in the *fcc* Rh bulk) of 5.46, 4.61, 3.97 and 3.49 % for N = 19, 43, 55 and 79, respectively. The energy differences between our molecular dynamics results and those corresponding to the homothetical relaxation for a given size N, $\Delta E^d(N) = |E^d_{\mathrm{MD}}(N) - E^d_{\mathrm{h}}(N)|$, are 0.33, 0.59, 1.12, and 1.14 eV for N = 19, 43, 55 and 79, respectively, *i.e.*, at most 0.02 eV/at.

It is well known that the lattice constants of bulk transition metals are larger in magnetic that in non-magnetic states.[13] It is therefore of interest to investigate the dependence of the relaxation profile obtained in Fig. 1 on the total magnetization of the clusters. Within our model, we have performed also spin-polarized calculations by means of the full ($U \neq 0, J \neq 0$) extended multiband Hubbard Hamiltonian treated in the Hartree-Fock approximation.[5] Moreover, we have explored the existence of multiple magnetic solutions by changing our initial spin-polarized electronic configuration in our self-consistent diagonalization process, a procedure that could be equivalent to different choices of the input potential in the local spin density (LSD) calculations. These multiple magnetic solutions correspond to local minima of the total energy as a function of the magnetic moment of the system for a given geometry, among which the one that gives the lowest total energy is regarded as the ground state of the cluster and the rest with higher energies are only metastable states. In *fcc* Rh_N clusters, we have found stable magnetic solutions up to $N = 43$ with average magnetic moments $\langle M \rangle$ of 1.38, 1.05, and 0.51 μ_B/at for Rh_{13}, Rh_{19}, and Rh_{43}, respectively, in good agreement with the experimental results[10] and other theoretical calculations.[14,15] For Rh_{55} and Rh_{79} several magnetic solutions have also been obtained however they are all less stable than the non-magnetic state. Finally, for larger sizes (N = 135 and 165), no magnetic solution (neither stable or metastable) could be obtained.

In Fig. 2, we show as a typical example the relaxation profile for *fcc* Rh_{55} and Rh_{79} obtained in their non magnetic state, together with the one found in their

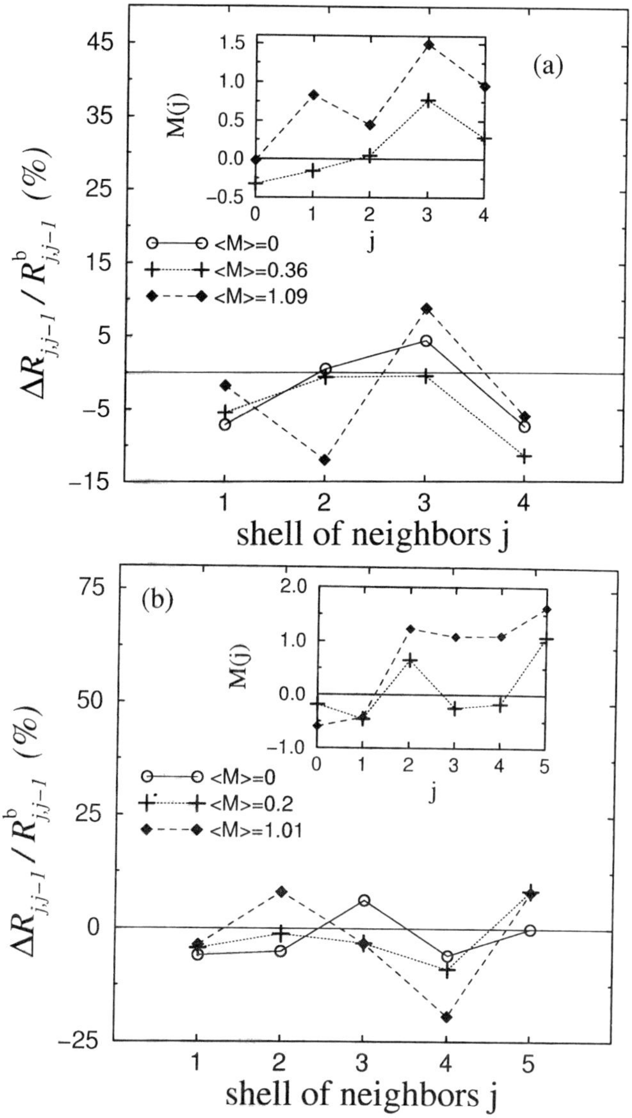

Figure 2. A comparison of the relaxation of the intershell spacing between the j and $j-1$ shells of neighbors, $\Delta R_{j,j-1}/R^b_{j,j-1}$, in the *fcc* (a) Rh_{55} and (b) Rh_{79} clusters for the non magnetic solution as well as for solutions with the lowest, and highest value of the average magnetization $\langle M \rangle$. In the insets, the distribution of the local magnetic moments for the highest and lowest spin solutions in the clusters.

metastable highest and lowest spin solutions. We note that, for both clusters, small magnetizations ($\langle M \rangle \sim 0.3\,\mu_B/\text{at}$) are enough to modify significantly the position of the internal shells. Moreover, we observe that this perturbation is particularly important when the particle has large local magnetic moments.

Concerning the distribution of the local magnetic moments, we can see that they present also an oscillatory behavior as we move from the inner atoms to those located at the surface (see the inset of Figs. 2a and 2b) in agreement with previous calculations.[16] This can be understood by recalling the strong sensitivity of the magnetic properties to the local environment and to the details of the electronic structure around the Fermi level, which is characteristic of systems showing weak non saturated itinerant magnetism. This is the reason why there is no simple relation between the magnitude of the local moments and their magnetic order (see the inset of Figs. 2a and 2b) with the expansions or contractions of the internal shells. These results clearly show how the magnetic properties of Rh_N clusters depend sensitively on their geometry and also how the magnetic ordering can affect the atomic configuration of a given structure. As a consequence, in order to locate the minimum-energy configuration in magnetic transition metal clusters it is necessary to perform a simultaneous variation of all the parameters on which the total energy depends, a procedure that is not common for clusters of these sizes.

Strictly speaking, a complete molecular dynamics scheme should be associated to a complete relaxation algorithm, including radial as well as angular coordinates in the minimization process, however, our model does not include this possibility. Thus it would be interesting to remove the symmetry constraint in order to follow possible structural transformation.

IV. Conclusions

In conclusion, the magnetic and structural properties of *fcc* Rh_N clusters have been determined by using a tight-binding molecular dynamics scheme incorporating an extended multiband Hubbard Hamiltonian which allows the treatment of magnetic effects. This study has revealed a variety of interesting behaviors for the internal positions of the atoms as a function of the cluster size, and how these structural arrangements are influenced by the presence of magnetic order which occurs in small Rh_N clusters. In particular, we have found a highly non uniform relaxation profile of the intershell spacings which may contract as well as expand. However, the net effect is, as usually obtained, a contraction of the cluster diameter. Finally, it is important to precise that our methodology is capable of treating simultaneously both magnetic as well as geometrical degrees of freedom, a procedure that at the moment is not possible for clusters of these sizes.

Acknowledgments

R.G.L. would like to acknowledge the financial support by CONACyT (México) grant J32084-E.

References

1. M. Pellarin, B. Baguenard, J. L. Vialle, J. Lerme, M. Broyer, J. Miller, and A. Perez, *Chem. Phys. Lett.* **217**, 349 (1994).
2. M. Respaud, J. M. Broto, K. Rakoto, A. R. Fert, L. Thomas, B. Barbara, M. Verelst, E. Snoeck, P. Lecante, A. Mosset, J.Osuna, T. Ould Ely, C. Amiens, and B. Chaudret, *Phys. Rev. B* **57**, 2925 (1998).
3. E. K. Parks, B. J. Winter, T. D. Klots, and S. J. Riley, *J. Chem. Phys.* **94**, 1882 (1991); E. K. Parks, L. Zhu, J. Ho, and S. J. Riley, *Z. Phys. C* **26**, 41 (1993).

4. R. Car and M. Parrinello, *Phys. Rev. Lett.* **55**, 2471 (1985); R. Car and M. Parrinello, *Phys. Rev. Lett.* **60**, 204 (1988).
5. B. Piveteau, M. C. Desjonquères, A. M. Oleś, and D. Spanjaard, *Phys. Rev. B* **53**, 9251 (1996).
6. F. S. Khan and J. Q. Broughton, *Phys. Rev. B* **39**, 3688 (1989); F. S. Khan and J. Q. Broughton, *Phys. Rev. B* **43**, 11 754 (1991).
7. A. M. Oleś, *Phys. Rev. B* **28**, 327 (1983).
8. J. C. Slater and G. F. Koster, *Phys. Rev.* **94**, 1498 (1954).
9. L. Verlet, *Phys. Rev.* **159**, 98 (1967).
10. A. J. Cox, J. G. Louderback, S. E. Apsel, and L. Bloomfield, *Phys. Rev. B* **49**, 12 295 (1994).
11. F. Jona and P. M. Marcus, *The structure of surfaces II*, edited by J. F. van der Veen and M. A. van Hove, Springer Series in Surface Sciences Vol. 11 (Springer-Verlag, Berlin, 1988).
12. T.S. Luo and B. Legrand, *Phys. Rev. B* **38**, 1728 (1988).
13. V. L. Moruzzi, J. F. Janak, and A. R. Williams, *Calculated Electronic Properties of Metals* (Pergamon, New York, 1978).
14. B. V. Reddy, S. N. Khanna, and B. I. Dunlap, *Phys. Rev. Lett.* **70**, 3323 (1993); Y. Jinlong, F. Toigo, W. Kelin, and Z. Manhong, Phys. Rev. B **50**, 7173 (1994).
15. P. Villaseñor-González, J. Dorantes-Dávila, H. Dreyssé, and G. M. Pastor, *Phys. Rev. B* **55**, 15 084 (1997).
16. R. Guirado-López, D. Spanjaard, and M. C. Desjonquères, *Phys. Rev. B* **57**, 6305 (1998).

Silicon Nanostructures Grown by Vapor Deposition on HOPG

Paul Scheier,* Björn Marsen, Manuel Lonfat,
Wolf-Dieter Schneider,† and Klaus Sattler

Department of Physics and Astronomy,
University of Hawaii at Manoa,
505 Correa Road,
Honolulu, Hawaii 96822
U.S.A.

Abstract

Silicon nanostructures such as small clusters, superclusters, elongated chains, and tube- or rod-like structures with an average diameter of a few nanometers, have been synthesized by magnetron sputtering on cleaved highly oriented pyrolytic graphite (HOPG). Scanning tunneling microscopy (STM) exhibits that flat, defect-poor areas of the HOPG surface are covered with almost uniformly sized spherical structures of 0.6 ± 0.2 nm, 5.1 ± 1.2 nm, and 15.4 ± 3 nm diameter. Surface regions with defects such as foldings, pits and craters descending a few layers into the graphite surface, are sparsely covered with silicon. In such defect rich regions most of the deposited material is found to be attached to the monatomic step edges forming the crater rims. The average diameter of the silicon nanoparticles that are attached to these steps is $2\,\text{nm} \pm 0.5$ nm. A simulation of the growth process, *i.e.*, deposition of silicon atoms onto a surface with built-in defects, and subsequent surface diffusion and aggregation of the adatoms, reproduces convincingly most of the Si nanostructures observed in the STM topographs.

I. Introduction

Clusters deposited on well-defined surfaces permit to build new materials with novel properties.[1] The current urge for an ever decreasing size of the components in the microelectronics industry displays this particularly relevant for silicon clusters.[2] Their electronic and optical properties are especially sensitive to their size and structure.[3–13]

Since the earliest study on silicon clusters by Honig[14] several experimental investigations on silicon clusters have been performed,[15–30] where a few were STM studies.[23,26–30] Kuk *et al.*[23] deposited Si_{10}-clusters on Au(001) and observed a wide variety of different cluster images even though size-selected clusters were deposited. McComb *et al.*[26] observed a site-specific variation in the electronic characteristics of Si clusters, which were deposited without size-selection but observed with atomic resolution. Dinh *et al.*,[27,28] in the context of an investigation of the optical properties of passivated Si nanostructures, synthesized Si-nanocrystals by laser ablation and by thermal evaporation in an Ar buffer gas and determined the size distribution of a monolayer of these nanostructures on HOPG with an STM. Size-selected Si_{30} and Si_{39} clusters were imaged with a low temperature STM on Ag(111).[29] Manipulation experiments and the appearance of the clusters in the images indicated soft-landing of the clusters. Recently, again in an STM-study, Marsen and Sattler[30] succeeded in creating fullerene-structured nanowires of silicon by magnetron sputtering on HOPG substrates. The present STM study intends to investigate in more detail the submonolayer and monolayer growth regimes of Si-nanostructures on defect-poor and defect-rich HOPG surfaces.

II. Experimental

The synthesis of Si-nanostructures was performed in a high vacuum chamber with a base pressure of $p < 10^{-6}$ Pa. This chamber was connected via vacuum locks to an analysis chamber (base pressure $p < 10^{-8}$ Pa) equipped with a Nanoscope II scanning tunneling microscope (STM) from Digital Instruments. For the synthesis of the Si-nanostructures a magnetron sputter source (MightyMak, Thin Film Products) was used. In an argon atmosphere of 600 Pa at a discharge voltage of 600 V and a typical Ar ion current of 0.2 A, a Si deposition rate of 0.3 nm/sec was obtained. A quartz crystal micro-balance mounted in a distance of 10 cm from the Si target monitored the flux during deposition. The cleaved HOPG substrate (7×7 mm^2), used to collect the sputtered Si, was mounted onto a stainless steel cylinder. This sample holder was placed in a copper block (equipped with heating and cooling facilities) 5 cm in front of the sputter source. A manually operated shutter was placed between the sputter source and the substrate holder during precleaning of the Si target and it served to control the Si arrival fluences. The average size of the Si clusters synthesized by this technique could be varied by changing the sputter parameters, increasing (or decreasing) the source to substrate distance, or a combination of all these parameters.[30] In the present experiments typical exposure times were varied from a few seconds to about a minute, yielding isolated clusters or cluster films of 1 to 3 monolayer (ML) thickness on HOPG, respectively. After deposition, the sample was transfered *in situ* into the STM chamber in order to characterize the deposited silicon nano-structures under stringent ultrahigh vacuum (UHV) conditions. All STM-topographs presented in this work were taken with Pt/Ir tips on the same sample. Large regions or surfaces that exhibit strong variations of the z-dimension (like deep craters) were recorded in constant current mode whereas all other images were recorded in constant height mode. The bias voltage between tip and sample is taken with respect to the latter. Tunneling resistances in the range between 100 MΩ and 6 GΩ yield identical images. Very similar images have been obtained from other samples prepared under the same experimental conditions.

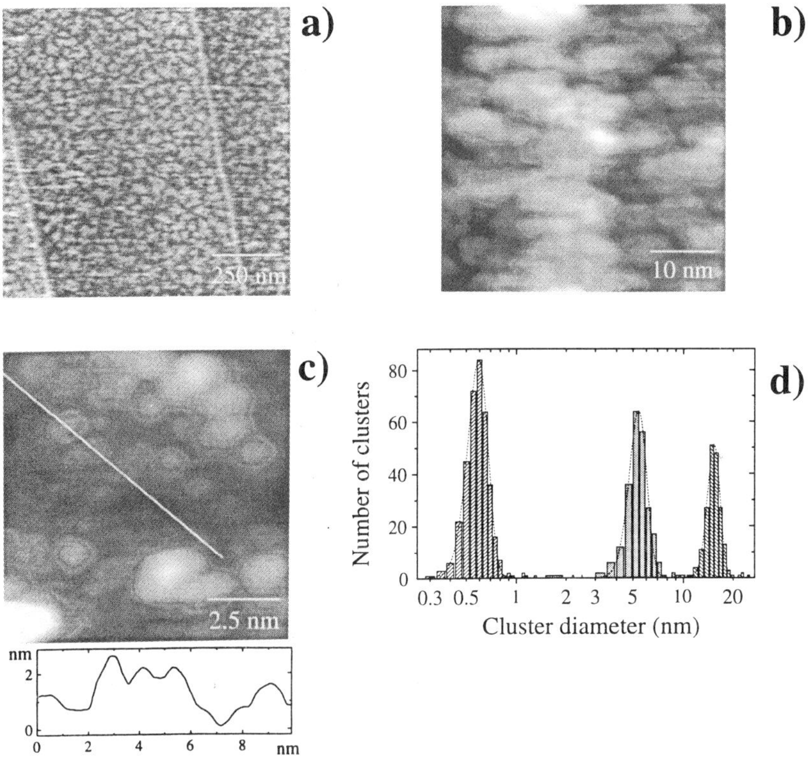

Figure 1. (a)–(c) Room-temperature constant-current topographs of a HOPG surface area at more than 1 ML coverage with silicon clusters. Image size: (a) $1.1 \times 1.1\,\mu\text{m}^2$, (b) $44 \times 44\ \text{nm}^2$, (c) $10 \times 10\ \text{nm}^2$; tunneling parameters: (a) $U = 1.0$ V, $I_t = 0.32$ nA; (b,c) $U = 2.5$ V, $I_t = 0.46$ nA. (d) Line-scan along the white line indicated in (c). (e) Size distribution of the silicon nanoclusters determined from the STM images (a)–(c).

III. Experimental Results

Figure 1(a) shows a $1.1 \times 1.1\,\mu\text{m}^2$ sized area of an HOPG surface covered with about 3 ML of silicon clusters. Two step edges of the HOPG substrate are clearly visible in the image due to the dense decoration with a chain of clusters. Round structures with an average diameter of 16 nm are displayed. Fig. 1(b) and 1(c) show a $44 \times 44\ \text{nm}^2$ and a $10 \times 10\ \text{nm}^2$ area, taken across the left step in the bottom of Fig. 1(a). The bias voltage between tip and sample was increased from 1 V (in Fig. 1a) to 2.5 V. Thereby much smaller structures could be resolved. These images reveal round Si-structures in the size range from 1 to several nanometers. A cross section, indicated by a white line in Fig. 1(c) and shown below reveals that the smallest round structures are semi-spherical with a diameter (FWHM) of about 1 nm. Due to the convolution of tip and object geometries the clusters appear larger as in reality. To correct for this effect we evaluated the tip dimensions on the widths of monatomic steps of pure HOPG yielding a tip contribution of 0.3 nm. Fig. 1(d) shows the corrected size distribution of about 1000 Si-clusters obtained from an analysis of Figs. 1(a)–1(c). It follows that all observed nanostructures fall into three relatively narrow size ranges. The smallest structures have an average diameter of 0.6 ± 0.2 nm, containing up to 10 Si

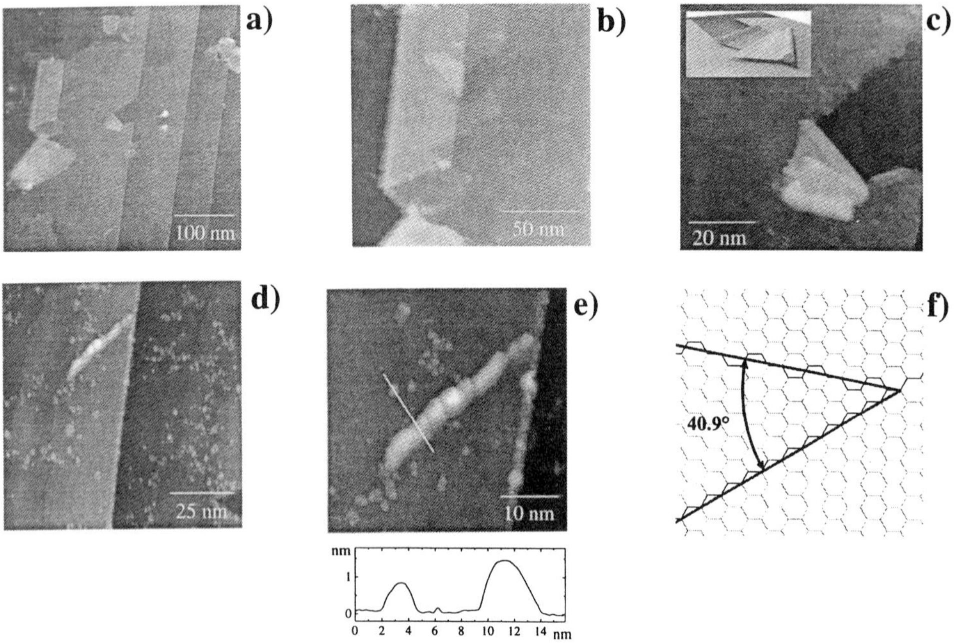

Figure 2. Room-temperature constant-current (a), (b) and constant height (c)–(e) topographs of a HOPG surface area at about 0.1 ML coverage with silicon clusters. Image size: (a) 400×400 nm^2, (b) 150×150 nm^2, (c) 80×80 nm^2, (d) 94×94 nm^2, (e) 41×41 nm^2; the insert in (c) represents a schematic view of the corresponding topograph; tunneling parameters, (a), (d)–(e) $U = 1.96$ V, $I_t = 0.32$ nA; (b) $U = 1.33$ V, $I_t = 0.32$ nA; (c) $U = 1.21$ V, $I_t = 0.32$ nA; (f) Section of the HOPG hexagonal surface lattice showing the angle between armchair and zig-zag directions.

atoms.[3–13,29] Larger aggregates formed by these smallest clusters exhibit diameters of 5.1 ± 1.2 nm and the largest superclusters have sizes in the range of 15.4 ± 3 nm. This observation indicates that the small clusters of 1nm diameter constitute building blocks for the larger aggregates. These findings confirm similar observations made in a recent Atomic Force Microscopy (AFM) study of Si nanocrystals synthesized by thermal evaporation in an Ar buffer gas and collected on HOPG, where the gathering of the Si-nanoclusters at step edges as well as their self-assembly into superclusters has been noted.[31]

Figure 2(a) shows an STM topograph of a 400×400 nm^2 sized area of HOPG taken at a lateral distance of several m from the region shown in Fig. 1. Three step edges are crossing the image from the bottom to the top. The two uppermost layers of graphite are partially folded back on their left side, a phenomenon already well known from earlier STM studies of HOPG.[32,33] Figure 2(b) is a close-up of such a folded region taken at the left side of Fig. 2(a). This image reveals that just a few clusters are attached to the graphite in this region. Figure 2(c) displays a high resolution scan and a schematic model of a doubly folded section of a graphitic layer.

In contrast to the observations made in Fig. 1 the silicon coverage at this new position with a higher density of defects is significantly smaller (about 0.1 ML) and the step edges are less densely decorated although the flux of silicon atoms is expected to be homogeneous over much larger surface areas. In the lower part of the uppermost terrace an elongated Si-structure is visible. A close-up of a 94×94 nm^2 area of this region reveals a chain of silicon clusters at an angle of 41.3° with respect to the step edge. A combination of arm-chair and zig-zag directions in the 2D-graphite hexagonal network yields an angle close to this value as is illustrated in Fig. 2(f). We conclude that the arrangement of the carbon surface atoms in this crystallographic direction provides favorable binding sites for such a chain-like structure. An closer look at this structure (Fig. 2e) shows that the segments of this cluster chain have an average thickness of 3.1 ± 0.3 nm (see line scan) with lengths varying from 2.3 nm to 7.5 nm (uncorrected values). In addition, on the two terraces shown in Figs. 2(b) and 2(c), uniformly sized (1.1 ± 0.1 nm) Si-clusters are distributed randomly. Most of these small silicon clusters form distinct loosely packed groups (only 10% of the small clusters have no neighbors).

Figures 3(a)–3(f) show constant current images of surface areas containing craters and pits[34] with depth down to 10 ML's. Every step edge of the descending terraces is decorated with a chain of silicon clusters. The average diameter of all cluster chains in this area is 3.1 ± 1.1 nm [see Fig. 2(e), line scan] and thus seem not to depend on the width of the terraces limited by the steps. This value corresponds well to the average diameter of the cluster chain observed in Fig. 2. Almost no silicon clusters are found on the flat terraces. Only in the upper left corner of Fig. 3(a) the density of the silicon clusters is large enough to cover more than just the steps. Images taken of adjacent surface regions in this direction exhibit structures which are identical to the ones found in Fig. 1.

In Figs. 3(b), 3(c), 3(f), 3(g), 3(i) and 3(j) flat islands with diameters between 5 and 20 nm are visible on the larger terraces. A high resolution image of one of these islands is shown in Fig. 3(h). A periodic lattice identical to the one of graphite is observed which allows us to identify these small islands as genuine graphite nanoflakes. We note that we were able to obtain this pattern only on very few HOPG flakes indicating a shift and/or rotation of these flakes with respect to the underlying graphite layers.

Figure 3(i) and 3(j) display a smooth elongated structure which we interpret as a silicon nanorod or nanotube.[30] The close-up of the left termination of this nanostructure shows two bamboo like laces. The fact that the step is two ML high (see Fig. 3j) might suggest that the smooth structure is the result of the folding of a graphitic layer. However, the complete missing of a layer termination on the right end of the structure and the bamboo laces contradict this assumption and indicate that this nanostructure is made of silicon. The perfect match of this nanorod to a straight step leads to the hypothesis that the step operates as a mold for the adsorbed silicon atoms and clusters.

Finally we note that in the STM images shown in Figs. 1–3 the respective line scans clearly reveal Si-step decoration of the *upper* step edge (see Fig. 3c).

We summarize our main experimental observations on the growth of Si nanostructures on HOPG as follows. The average silicon coverage varies by a factor of more than 10 between surface regions of different defect density, separated by only 0.1 mm. The diameter of the clusters formed onto defect-poor, flat surface regions is about 0.6 nm while clusters attached to step edges or defects have diameters of about 2 nm and, occasionally, are found to be fused into rod- or tube-like structures. In a coverage range between 0.5 and 5 ML these small clusters often form superclusters. The density and size of the clusters attached to the step edges forming HOPG nanopits are independent of the width of their "feeding" terraces. The rims of nano-sized graphite islands on HOPG are practically free of silicon decoration.

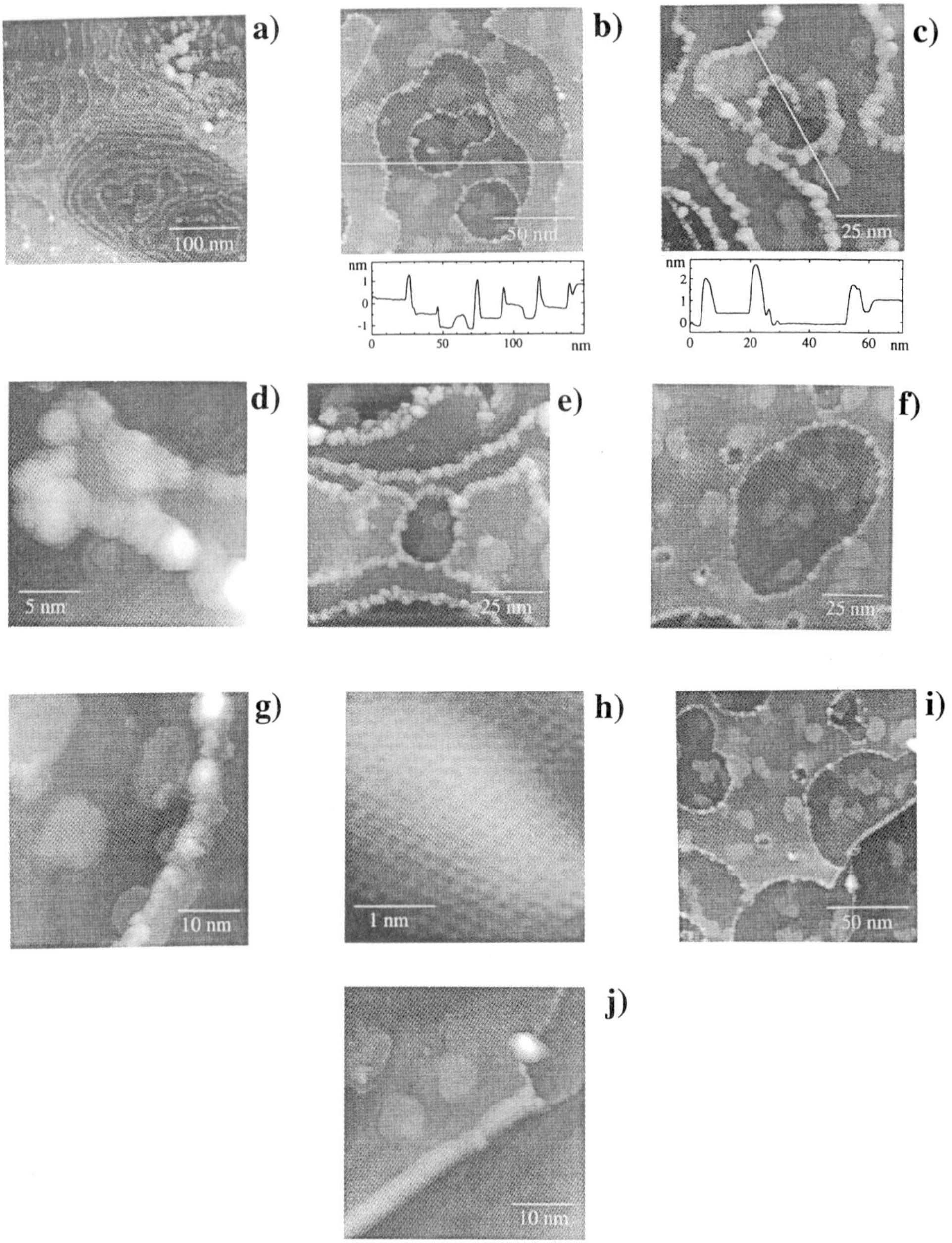

Figure 3. Room-temperature constant-current (a), (b), (i) and constant height (c)–(h), (j) topographs of a HOPG surface area with nanopits of various depth at about 0.5 ML coverage with silicon clusters. Image size: (a) 350×350 nm^2, (b), (i) 150×150 nm^2, (c), (f) 100×100 nm^2; (d) 20×20 nm^2; (e) 86×86 nm^2; (g), (j) 40×40 nm^2; tunneling parameters: (a) $U = -1.5$ V, $I_t = 0.38$ nA; (b) $U = 1.1$ V, $I_t = 4.2$ nA; (c) $U = -1.3$ V, $I_t = 0.38$ nA; (d) $U = -1.4$ V, $I_t = 0.38$ nA; (e) $U = 1.6$ V, $I_t = 0.26$ nA; (f) $U = 0.73$ V, $I_t = 0.51$ nA; (g) $U = 1.2$ V, $I_t = 0.51$ nA. (h) constant-height topograph of a small island in the center of (f) showing the graphite surface lattice with atomic resolution. Image size: 3×3 nm^2, tunneling parameters: $U = 1.2$ V, $I_t = 0.68$ nA. (i), (j) $U = 0.66$ V, $I_t = 1.3$ nA.

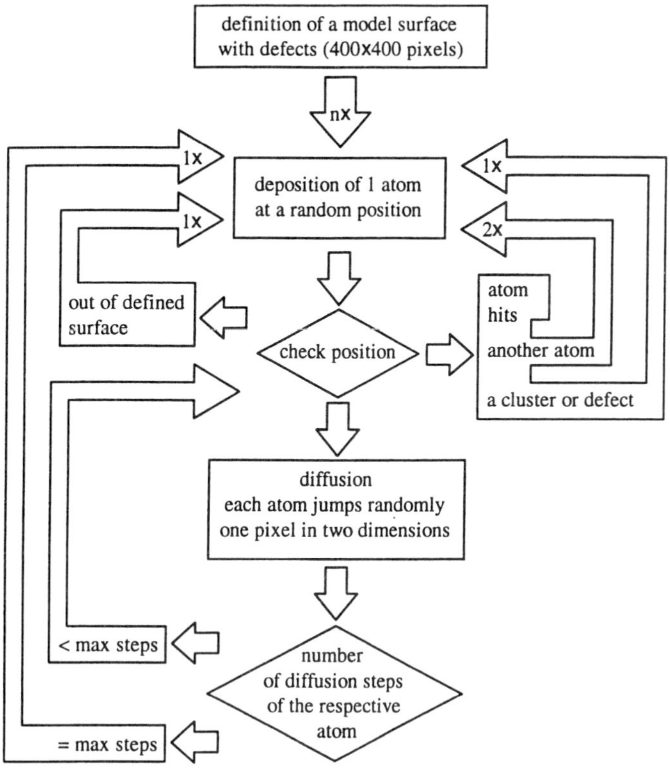

Figure 4. Flow-diagram of a two dimensional growth simulation. n is the number of single atoms which are attached to the surface at all times. If an atom desorbs, diffuses out of the defined surface area or fuses to a cluster or defect it is replaced by a new incoming atom. n is proportional to the flux of impinging atoms. Maxsteps is the number of diffusion steps a single atom can perform before it desorbs from the surface.

IV. Simulation

In order to rationalize the above observations we simulate the growth process, *i.e.*, adsorption, surface diffusion, and clustering of the silicon atoms on HOPG within a simple two-dimensional model sketched in the flow diagram shown in Fig. 4.

In a first step the topography of the surface is defined. Within a two-dimensional array of 400×400 pixels the location of a step or defect is assigned to a value of 1 (black pixel) and all other positions are set to zero (white pixel). Figure 5(a) displays a typical example of such a model surface corresponding closely to the experimental STM image shown in Fig. 3(a). Then the number n of Si atoms which hit the surface at random positions in a given time interval per unit surface area is chosen. Furthermore, the number of diffusion jumps of a single atom before desorption is considered (maxsteps) representing the residence time at the surface.

All atoms which are adsorbed on the surface move randomly along the surface. A collision of an atom with a step or defect immobilizes immediately this atom. Collision with another adsorbed atom leads to the formation of the smallest cluster, a dimer, etc. The probability for the desorption of a cluster is set to zero. The diffusion of the clusters is also set to zero. Therefore, once a cluster is formed on a flat, defect-poor

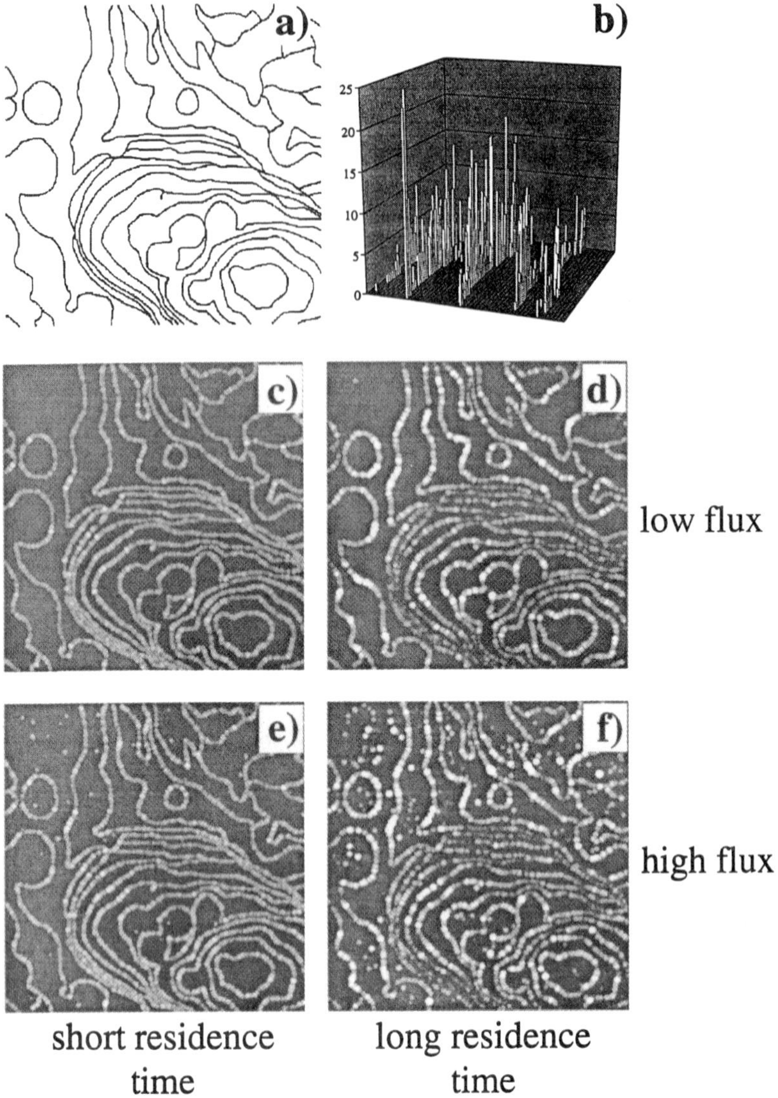

Figure 5. Simulation of the Si-cluster growth (see text). (a) model surface with defects (black lines). (b) bar diagram of the number of accumulated atoms at a few step edges. (c,d) growth pattern after low-flux deposition at short (c) and long (d) residence times of the Si-atoms at the surface. (e), (f) growth pattern after high-flux deposition at short (e) and long (f) residence times. Note the excellent agreement of the results of these simulations with the observed growth patterns shown in Figs. 3.

region its migration to a defect is excluded. and consequently a new defect within a terace is created. Figure 5(b) shows the number of atoms attached to each pixel of a surface area exhibiting 5 step edges in terms of a bar diagram where the height of each column represents the number of atoms at this location at the end of a simulation.

For a direct comparison of the simulated results with the STM images semi-spherical clusters were plotted where the cube of the radius is proportional to the number of atoms within the cluster. Figures 5(c)–5(f) display simulated growth patterns for a surface exhibiting the characteristic topography of Fig. 3(a). The total number of adsorbed atoms is one million for all four simulations. A flux n of 2 atoms and of 1000 atoms per time interval per unit surface area were used in the case of images 5(c) and 5(d) and of 5(e) and 5(f), respectively. In 5(c) and 5(e) the residence time was short (maxsteps=1) whereas in 5(d) and 5(f) an infinite residence time was assumed (maxsteps $= \infty$). For the low flux of atoms clusters are formed almost exclusively along the steps and defects (Figs. 5c and 5d). By contrast, at high flux clusters are formed also on the terraces (Figs. 5e and 5f). At short residence times clusters are very uniform in size (Figs. 5c and 5d) whereas for infinite residence time their size depends strongly on the size of the feeding terrace.

V. Discussion

Comparison of the experimental STM image of Fig. 3(a) with the results of the simulation displayed in Fig. 5 clearly suggests that our simulation captures the essential physics of the growth process. After adsorption of single silicon atoms on the HOPG surface these adatoms move randomly by thermal diffusion along the surface. Collisions among them lead to the growth of Si-clusters. Step edges with unsaturated or dangling bonds constitute preferred nucleation sites and exhibit an effective Schwoebel-Ehrlich[35] barrier for interlayer diffusion of Si-adatoms. In surface regions with increased defect density the probability for the diffusing Si atoms to encounter, at a given residence time, a defect like a step is higher than to collide with another Si atom and to form a cluster on a terrace.

The areas which exhibit pits and craters also contain small HOPG islands or flakes which are practically free of adsorbed silicon particles. For example Fig. 3(c) reveals that only one out of 20 flakes has a cluster attached to its edge. We attribute this observation to two effects. (*i*) A vanishing Schwoebel-Ehrlich barrier on the step edge of these very small HOPG islands allowing for interlayer diffusion. At a critical island size such an effect has been invoked to be responsible for "landsliding" on small Cu-islands.[36,37] (*ii*) Bond-weakening of the HOPG nanoflakes towards the adsorbed Si-atoms due to weak coupling of these flakes to the underlying graphite surface. In the case of Pt on HOPG the perfect stacking of the graphite layers has been shown to be important for optimal bonding.[38,39]

VI. Summary and Conclusions

Silicon nanoparticles were synthesized using magnetron sputtering deposition onto cleaved HOPG. The resulting Si nanostructures were investigated with STM. On defect-poor, flat regions of the HOPG surface Si-clusters with a mean diameter of about 0.6 nm and narrow size distribution were found. On defect-rich surface regions nearly excusively step edge decoration with Si-clusters was observed. A simple two-dimensional simulation of the Si cluster growth successfully describes most of the experimental observations, *e.g.*, the gathering of clusters on step edges and the formation of clusters and superclusters on the terraces. In view of the present results magnetron sputtering might provide an interesting alternative route towards the production of Si nanostructures with potential applications in a future silicon nanotechnology.

Acknowledgments

P. S. gratefully acknowledges an APART grant from the Austrain Academy of Sciences and WDS thanks the Swiss National Science Foundation for financial support.

References

* Permanent address: Institut für Ionenphysik, Universit"at Innsbruck, A-6020 Innsbruck, AUSTRIA, Fax: +43 512 507-2932, e-mail: Paul.Scheier@uibk.ac.at.

† Permanent address: Institut de Physique de la Matiere Condensée, Universite de Lausanne, CH-1015 Lausanne, SWITZERLAND.

1. "Cluster Assembled Materials", *Materials Science Forum*, Vol. 232, edited by K. Sattler, (Trans. Tech. Publ., Switzerland, 1996).
2. M. A. Duncan and D. H. Rouvray, *Scientific American*, Dec. 1989, p. 69.
3. A. P. Alivisatos, *Science* **271**, 933 (1996).
4. J. Shi, S. Gider, K. Babcock, and D. D. Awschalom, *Science* **271**, 937 (1996).
5. E. Kaxiras, *Phys. Rev. Lett.* **64**, 551 (1990)
6. W. Andreonia and G. Pastore, *Phys. Rev. B* **41**, 10243 (1990).
7. C. H. Patterson and R. P. Messmer, *Phys. Rev. B* **42**, 7530 (1990).
8. J. R. Chelikowski, K. M. Glassford, and J. C. Phillips, *Phys. Rev. B* **44**, 1538 (1991).
9. E. Kaxiras and K. Jackson, *Phys. Rev. Lett.* **71**, 727 (1993).
10. U. Röthlisberger, W. Andreoni, and M. Parrinello, *Phys. Rev. Lett.* **72**, 665 (1994).
11. J. C. Grossman and L. Mit, *Phys. Rev. Lett.* **74**, 1323 (1995).
12. M. Menon and E. Richter, *Phys. Rev. Lett.* **83**, 792 (1999).
13. J. Pan and M. V. Ramakrishna, *Phys. Rev. B* **50**, 15431 (1994).
14. R. E.Honig, *J. Chem. Phys.* **22**, 1610 (1954).
15. T. T. Tsong, *Appl. Phys. Lett.* **45**, 1149 (1984).
16. L. A. Bloomfield, R. R. Freeman, and W. L. Brown, *Phys. Rev. Lett.* **54**, 2246 (1985).
17. W. L. Brown, R. R. Freeman, K. Raghavachari, and M. Schlüter, *Science* **235**, 860 (1987).
18. M. F. Jarrold, *Science* **252**, 1085 (1991).
19. E. C. Honea, A. Ogura, C. A. Murray, K. Raghavachari, W. O. Sprenger, M. F. Jarrold, and W. L. Brown, *Nature* **366**, 42 (1993).
20. W. L. Wilson, P. F. Szajowski, and L. E. Brus, *Science* **262**, 1242 (1993).
21. C. Deleruer, M. lanno, G. Allan, E. Martin, I. Mihalcescu, J. C. Vial, R. Romestain, F. Muller, and A. Bsiesy, *Phys. Rev. Lett.* **75**, 2228 (1995).
22. A. A. Shvartsburg, M. F. Jarrold, B. Liu, Z.-Y. Lu, C.-Z. Wang, abd K.-M. Ho, *Phys. Rev. Lett.* **81**, 4616 (1998).
23. Y. Kuk, M. F. Jarrold, P. J. Silverman, J. E. Bower, and W. L. Brown, *Phys. Rev. B* **39** (1989) 11168
24. J. M. Alford, R. T. Laaksonen, and R. E. Smalley, *J. Chem. Phys.* **94**, 2618 (1996).
25. M. F. Jarrold and V. A. Constant, *Phys. Rev. Lett.* **67**, 2994 (1991).
26. D. W. McComb, B. A. Collings, R. A. Wolkow, D. J. Moffat, C. D. Mac Pherson, D. M. Rayner, P. A. Hackett, and J. E. Hulse, *Chem. Phys. Lett.* **251**, 8(1996).
27. L. N. Dinh, L. L. Chase, M. Balooch, L. J. Terminello, and F. Wooten, *Appl. Phys. Lett.* **65**, 3111 (1994).

28. L. N. Dinh, L. L. Chase, M. Balooch, W. Siekhaus, F. Wooten, *Phys. Rev. B* **54**, 5029 (1996).
29. S. Messerli, S. Schintke, K. Morgenstern, A. Sanchez, U. Heiz, and W.-D. Schneider, *Surf. Sci.*, to be published
30. B. Marsen and K. Sattler, *Phys. Rev. B* **60**, (1999).
31. T. van Buuren, L. N. Dinh, L. L. Chase, W. J. Siekhaus, and L. J. Terminello, *Phys. Rev. Lett.* **80**, 3803 (1998).
32. H.-V. Roy, C. Kallinger and K. Sattler, *Surf. Sci.* **407**, 1 (1998).
33. H.-V. Roy, C. Kallinger, B. Marsen, and K. Sattler, *J. Appl. Phys.* **83**, 4659 (1998).
34. G. Bräuchle, S. Richard-Schneider, D. Illig, R. D. Beck, H. Schreiber, and M. M. Kappes, *Nucl. Instr. Meth. Phys. Res. B* **112**, 105 (1996).
35. K. Morgenstern, G. Rosenfeld, E. Laegsgaard, F. Besenbacher, and G. Comsa, *Phys. Rev. Lett.* **80**, (1998) 556 and References therein.
36. M. Giesen, G. Schulze Icking-Konert, and H. Ibach, *Phys. Rev. Lett.* **80**, 552 (1998).
37. M. Giesen, G. Schulze Icking-Konert, and H. Ibach, *Phys. Rev. Lett.* **82**, 3101 (1999).
38. U. Müller, K. Sattler, J. Xhie, N. Venkateswaran, and G. Raina, *Z. Phys. D* **19**, 319 (1991).
39. J. Xhie, K. Sattler, M. Ge, and N. Venkateswaran, *Phys. Rev. B* **47**, 15835 (1993).

NanoMet: From Lab to Market

M. Kasper

Matter Engineering AG
5610 Wohlen
SWITZERLAND

Abstract

How do you measure the success of a scientist's idea? Certainly a valuable criterion is its potential to open a new market or conquer an existing one. For basic research the much more "down-to-earth" economic success is especially challenging due to the large gap between laboratory experiment and practical application. A project that straddled this gap is NanoMet, a nanoparticle measurement technique for on-line characterization of submicron sized particles in gas suspension. It has evolved from inventions and laboratory work carried out in the group of H.C. Siegmann. NanoMet comes with two on-line sensors where particles are electrically charged and subsequently precipitated on a measurement filter with current amplifier. The sensors differ in the charging principle; in the diffusion charger (referred to as DC) positive ions from a corona discharge diffuse onto the particles. Therefore, the filter current is proportional to the total scattering surface of the particle ensemble in gas/particle interactions. In the other sensor (referred to as PAS, photoelectric aerosol sensor) aerosol particles are illuminated by ultraviolet light and photoelectrically charged. The electrons are then quickly removed from the gas by an electric filter. As photoemission involves absorption of a photon by the particle bulk material and emission of an electron through the particle surface, the resulting charge on the particles is proportional to the active surface and a material coefficient. This material coefficient is especially large for particles from fossil fuel combustion. Simultaneous operation of the two sensors yields both the active surface (DC sensor) and the active surface times material coefficient (PAS). Division of the readings provides the material coefficient which turns out to be characteristic of the particle source. Thanks to a pocket size diluter with tunable dilution ratio the measurable concentration range stretches from (vehicle) raw emissions to ambient air/occupational exposure measurements. With NanoMet one has an instrument at hand which solves the unsatisfactory situation that the same particles are measured and legally limited by different, often not even correlating methods, only depending on the actual legal situation. For vehicle exhaust emissions legal limits have been significantly lowered recently and will be even tighter in the near future. In fact, they will be so low that legal methods (gravimetric analysis) are not sensitive enough to decide whether a vehicle meets the limits or not. This situation has induced a growing demand in more sensitive new techniques such as NanoMet to efficiently survey emissions of existing vehicle fleets. NanoMet is a portable, easy-to-use instrument for applications in the field as well as in the laboratory. This paper reports results and experiences with a unit from the first series and discusses applications and accessories beyond the present state.

I. Introduction

The results of the VERT project[1] have demonstrated that particle emissions from diesel engines can be reduced to toxicologically tolerable limits only by means of particulate traps. Once in use, however, the filtration efficiency of such traps, as well as the total particle emissions of the vehicle have to be periodically assessed. In order to minimize interruptions of vehicle availability for the owner, efficiency tests are to be carried out "in the field" without the need to dismount the trap from the vehicle, and the duration of the test itself should be kept as short as possible. Furthermore, the test method must be sensitive both at high and low particle concentrations as are found upstream and downstream of the trap, respectively, and it should be easy to operate for the user.

One of the merits of the VERT project is that particle measurement was not restricted to the legal methods but extended to any available laboratory technique. From this large pool of methods, it were not the legal ones that fulfilled all above criteria, as they are either not accurate enough or may not be applied under field conditions. Instead, a combination of a diluter and at least two on-line sensors—devices which had originally been designed for laboratory use—turned out to be the most suitable choice.

This result lead to the VERT follow-up project "NanoMet", the goal of which is to integrate the on-line sensors and diluter in a portable, easy-to-use instrument for trap survey in the field. This paper reviews the working principles of the system and reports results and experiences with a unit from the first series.

II. Setup and Working Principles

II.1. NanoMet

Figure 1 shows a principal view of the NanoMet set up. A detailed description of it has been given by Matter,[2] which will be briefly reviewed here. NanoMet consists of a heatable diluter, a particle classifying unit with a centrifuge and a diffusion battery, and two on-line sensors. A cyclone at the sampling inlet keeps coarse particles out of the system. The time resolution of the system is determined by the rotation frequency of the diluter disk and the response time of the sensors. The current version achieves a lower limit of 1 second and for the near future a value of 0.1 seconds is planned. Reproducibility of measurements lies within a margin of a few percent.

The NanoMet modules can be used as stand alone devices or in combination to best fit the user's needs. For example, ambient air measurements or occupational exposure assessment do not require the diluter due to low particle concentrations, but the use of both sensor types may be desirable. In another setup with high particle concentrations the user may wish to use the heatable diluter but complement or even replace the NanoMet sensors by his own technology.

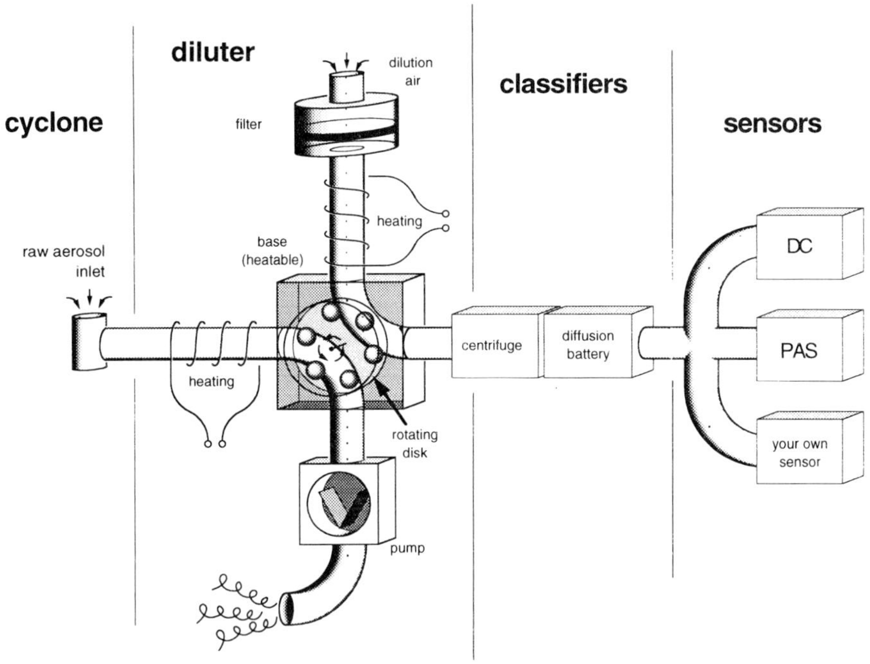

Figure 1. Schematic drawing of the NanoMet set up. Aerosol enters the cyclone and flows through the heated diluter and the classifiers to the sensors. Details are given in the text.

II.2. Diluter

The pocket size diluter consists of a steel base through which two aerosol ducts are drilled, one for the undiluted aerosol as sampled from the source, the other for particle free air.[3] On one side of the base the channels are milled open. A rotating disk pressed against the polished surface of the base seals the orifices off against the environment. Small cavities in the back of the disk transport well-defined volumes from the raw aerosol orifice on the primary side into the particle free air on the secondary side from where the dilute aerosol flows to the sensors. The dilution ratio is determined by the rotation frequency of the disk and the flow rate on the dilute aerosol side. The orifices in the base are designed in such a way that primary and secondary channel are never in direct contact. Thus pressure fluctuations on one side cannot cause disturbance on the other.

Steel base and dilution air can be heated up to 150°C. This is useful when the raw aerosol is hot and dilution is to take place before the aerosol cools down to ambient temperature. In the measurement of tailpipe emissions this separation of dilution from cooling process prevents the condensation of volatiles, especially water and sulfuric acid. Selection of a lower heating temperature provokes condensation of some species while more volatile substances remain in gas phase. This feature can be used to distinguish between aerosol constituents that have different boiling points.

Dilution is not needed if particle concentrations in the raw aerosol are low such as downstream from a particle trap. In this case the diluter is easily bypassed with a valve. A separate unit holds the power supply, a pump for the raw aerosol, the

Table I

Selected properties of the diluter.

	diluter
dimensions of supply unit	258 × 148 × 312 mm 1/2–19" plug–in unit
dimensions of diluter head	360 × 180 × 90 mm
length of connecting hoses	up to 3 m
raw aerosol flow	1 l/min
dilution air flow	0.5 . . . 5.0 l/min, determined by connected sensors,
disk rotation frequency	0.15 . . . 3.0 Hz
dilution ratio	1 : 30 . . . 1 : 600 (disk with 10 cavities) 1 : 150 . . . 1 : 3000 (disk with 2 cavities)
heating temperature	OFF/80°C/120°C/150°C

controls for the diluter disk motor and heating. Supply unit and diluter head are interconnected by shock and heat protected cables and hoses. The most important technical specifications are listed in Table I.

II.3. Classification

Three classifying elements are used in NanoMet. A cyclone with cut size 2.5 m in the raw aerosol channel prevents coarse particles from entering the diluter. Behind the diluter a centrifuge removes particles larger than 200 nm. Its cutoff can be varied towards larger diameters by tuning the rotation frequency. The third classifier is a multi-stage diffusion battery located after the centrifuge with cutoff at 4 different mobility diameters. Size classification facilitates the measurement of size spectra; in combination with the DC sensor reading the number concentration per size class can be calculated.

In the particle centrifuge, an aerosol stream is exposed to the centrifugal field of a rotating drum forming a solid body vortex. In the flow field, a circular gap between an outer and an inner cylinder, the particles are forced to the outward. Thus regions of enriched and diluted particle concentrations will develop. Unlike in older designs reported in the literature, where the separated particles were collected on a foil cladding the inner surface for later determination of the size distribution, the classification serves to produce an aerosol flow fraction reduced of particles larger than the size class of interest, which is fed to the measuring devices downstream.

The centrifuge has two adjustable parameters: speed (centrifugal force) and throughput (residual time). So it is a freely tunable particle classifier. When operated, it also acts as a ventilator, this is why the cut size is not coupled with pressure drop. Though for the application in NanoMet the centrifuge's significance will be that of an adjustable preseparator, it can be also used for classifying particles according

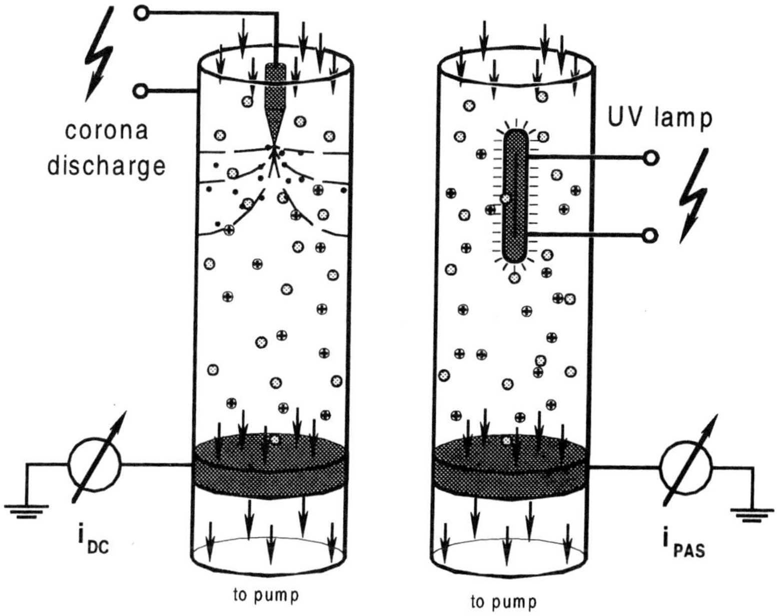

$$i_{DC} \propto \sum S_{active} \qquad i_{PAS} \propto \sum S_{active} \cdot \phi$$

i_{DC}, i_{PAS}	filter current of DC sensor / PAS
S_{active}	active surface of a single particle
ϕ	material coefficient of the aerosol
Σ	sum over all particles

Figure 2. Working principles of the diffusion charging sensor (DC, left) and the photoelectric aerosol sensor (PAS, right). The measured current is proportional to the total active surface of the aerosol. The PAS signal further depends on a material coefficient.

to their aerodynamic diameter down to 50 nm or below. After having tested a first design which shows the feasibility but has grade efficiency curves of unsatisfactory steepness, a new outlay is now under investigation.

For the size classification of particles according to their mobility diameter, a diffusion battery is applied. It is of tube type with layers of wire gauze. The adjustable parameters are the number of gauze layers and the volume flow. The present design comprises a series of layer packages with exits in between through which the reduced gas stream is fed to the connected sensors. Cut sizes of 15, 35, 75 and 190 nm could be realized. Even though the penetration curves are rather flat, they can be clearly distinguished for the given diameters.

II.4. Sensors

NanoMet comes with two on-line sensors where particles are electrically charged and subsequently precipitated on a filter (Fig. 2). The electric current flowing from the filter is amplified and measured. The sensors differ in the charging principle; in the diffusion charger (referred to as DC) positive ions from a corona discharge diffuse

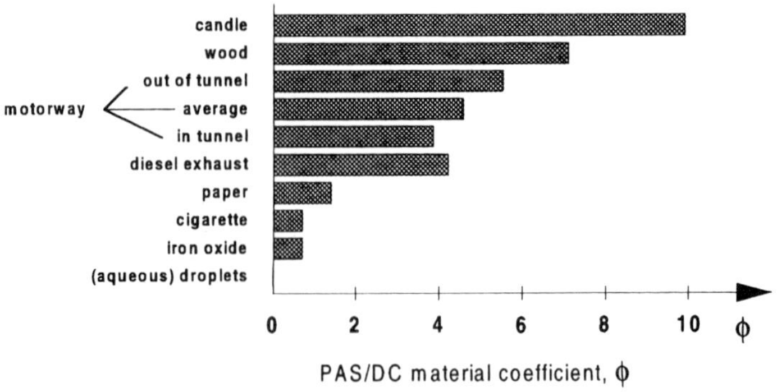

Figure 3. Material coefficients of particles from different sources.

onto the particles. Therefore, the filter current is proportional to the total active surface of the aerosol, ΣS_{active}. A detailed derivation of this concept is given in the next section. In the photoelectric aerosol sensor (referred to as PAS) aerosol particles are illuminated by ultraviolet light and photoelectrically charged. The electrons are then quickly removed from the gas by an electric filter. As photoemission involves absorption of a photon by the particle bulk material and emission of an electron through the particle surface, the resulting charge on the particles is proportional to the active surface, ΣS_{active}, and a material coefficient, ϕ. This material coefficient is especially large for solid particles from fossil fuel combustion.

Simultaneous operation of the two NanoMet sensors yields both the active surface (DC sensor) and the active surface times material coefficient (PAS). Division of the readings provides the material coefficient which turns out to be characteristic of the particle source. Figure 3 shows a list of particle sources and the respective material coefficients. A very high photoelectric activity is observed for soot from candles and wood fire while cigarette smoke and iron oxide exhibit little such activity. Droplets of water or condensed organic materials do not undergo photoemission at all; their material coefficient is zero.

The measurable particle size range of both sensors is in principle unlimited towards the lower end. At the upper end, the PAS cuts at approximately 1 micron because the emitted electrons will be drawn back to the surface of the charged particle. Only if the particle is smaller than the mean free path of the electron in the gas ($\sim$ 1 micron) the electrons will fail to find the particle and be lost in the gas. Thus only submicron particles remain charged and contribute to the signal. For diffusion charging with the DC sensor no upper size limit exists, but the cyclone, centrifuge and diffusion battery keep coarse grains out of the system. The most important properties of the sensors are listed in Table II.

III. Theoretical Background and Calibration

III.1. The Active Surface

In the DC sensor ions from the corona discharge diffuse between the neutral gas molecules and eventually reach the surface of an aerosol particle. Thus the DC measures the integral attachment cross section of ions colliding with the aerosol particles. As a sticking coefficient of one can be assumed for ions, this equals the collision cross

Table II

Selected properties of the NanoMet sensors.

	DC sensor	PAS
dimensions	258 × 148 × 312 mm 1/2–19" plug–in unit	258 × 148 × 312 mm 1/2–19" plug–in unit
physical property being measured	active surface, independent of material	active surface, material specific (carbonaceous soot)
measurement range (filter current)	0–1000 fA	0–5000 fA
resolution corresponding to ...	1 fA = 10^{-15} A 1 $\mu m^2/cm^3$	1 fA = 10^{-15} A 0.125 $\mu g/m^3$ EC (only diesel exhaust)
particle size range	1 nm ... 2.5 μm (integal)	1nm ... 1 μm(integal)
aerosol flow	1.5 l/min	2 l/min
time resolution	$<$ 1 sec	$\sim$ 1 sec

section or collision frequency. If the ion concentration is kept low enough to avoid multiple charging, the collision frequencies of ions and neutral atoms are alike, otherwise Coulomb repulsion will decrease the ion attachment rate. Furthermore, the aerosol particles have to be larger than about 10 nm because only in this case the image force can be neglected. For smaller particles the image force will increase the attachment probability.[4]

If these conditions are fulfilled, the DC sensor measures the integral collision cross section or attachment cross section, respectively. This is an important quantity by itself, as it determines adsorption kinetics and thereby has significant influence on chemical reactions between particles and the surrounding gas phase or gas phase reactions, where the particles serve as catalyst. It also determines particle growth by attachment of material from the gas phase. One could say it is the fraction of geometric surface which is directly accessible from outside. For this reason the term 'active surface', S_{active}, is used. A related designation, 'Fuchs-Surface', has been introduced in connection with the Epiphaniometer.[5]

III.2. Mobility and Active Surface

Another important physical property of aerosol particles is their mobility, b, which determines the velocity ν a particle obtains as response to the action of an external drag force: $\nu = bF$. The mobility is measured with a differential mobility analyzer (DMA) where it is inversely proportional to the applied voltage. Furthermore, b is related to the diffusion constant, $D = bkT$, where k is Boltzmann's constant and T the temperature. D, in turn, determines the probability that the particle will diffuse to the walls of a vessel containing the aerosol.

The two concepts of microscopic active surface, S_{active}, and macroscopic mobility, b, are connected by an astonishingly simple relation, $S_{\text{active}} b = \text{const}$, which holds over the whole diameter range within an uncertainty of a few percent independently of particle size, shape or material.[6] This is plausible as the drag force, F, is also due to collisions with carrier gas molecules. The collisions determine the momentum transfer and thereby also the drag force. Thus, inverse proportionality of the DC signal with the diffusion coefficient is established via active surface and mobility. The reading of the DC sensor may then be considered a direct measure of the amount of particles deposited in a tubing system such as the bronchial epithelium of the human lungs.

III.3. Calibration of the Sensors

Calibration of the DC sensor is now very easy. Test aerosol is fed into a DMA where a mobility class is selected. This monodisperse aerosol flow is split in two and guided to a condensation nucleus counter (CNC) and the DC sensor. Without further calibration, the DMA voltage determines particle mobility, b, the CNC measures their number, N, and the DC sensor captures the total active surface of the monodisperse aerosol, $S_{\text{active}} N$. The only parameter remaining to be tuned is the amplification factor in the DC sensor. This concludes the calibration process.

The reading of the PAS differs from that of the DC sensor by a multiplicative material coefficient, ϕ. This requires a test aerosol of constant and reproducible photoelectric properties. Only recently has this condition been fulfilled by a new aerosol generator.[7] Using this test aerosol allows one to calibrate the PAS against the DC sensor by tuning the PAS/DC ratio, ϕ, to the desired fixed value.

IV. Aerosol Characterization

IV.1. Solid Particles and Condensates

The term aerosol designating particles in gas suspension is a very general expression for a highly complex system. Aerosols are far from being stable. The particles diffuse through the gas on random paths, eventually colliding with each other. As surface forces are very strong, the particles will stick and a single, larger particle emerges from the collision. In the aerosol ensemble coagulation and agglomeration make the particle concentration decrease as time passes while the average diameter increases. At the same time, particles interact with the surrounding gas molecules in chemical reactions, vapors condense to form droplets or volatile particles evaporate. Moreover, ambient aerosol particles are individuals; not two of them are alike. Even so well-defined an aerosol source as a diesel engine releases to the environment a large variety of particles in the submicron size range. A major part of mass and number is concentrated in solid carbonaceous soot particles. Ashes and minerals from engine wear or additives make up for the rest of the solid granules. But this is only one half of the nanoworld.

Water and inorganic acids, mostly sulfuric acid, condense to droplets which often exceed the solid particles by far both in number and in mass. This inhomogeneous mixture of solid soot, salts and volatiles fills the air that we breathe. It is obvious that water droplets in the human lung will not yield the same effect as carbonaceous soot or minerals. If, on the other hand, the emission of particles is to be prevented the source must be clearly identified. In the diesel engine carbonaceous soot is a result of the engine's working principle whereas sulfuric acid originates from sulfur in the fuel. A

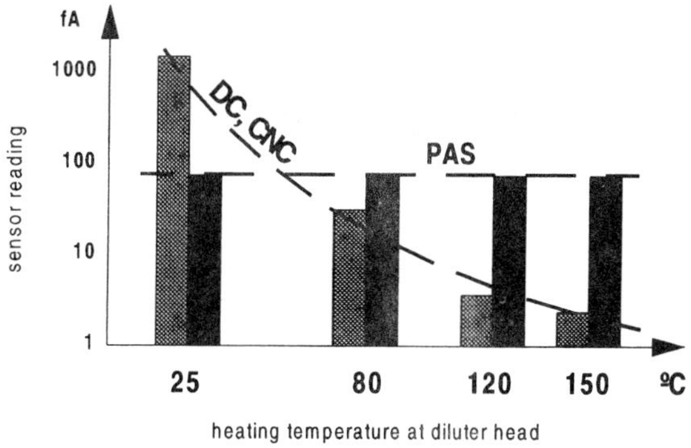

Figure 4. Droplets are detected by the DC sensor (or a Condensation Nucleus Counter, CNC) but not by the PAS. As the diluter head is heated to higher temperature less volatiles condense. The DC signal decreases while the PAS displays a constant value because solid soot particles are not affected by heating.

measurement technique that characterizes ambient or emission aerosol must therefore be able to distinguish particles with different physical and chemical properties, and it should be fast enough to allow for changes in aerosol composition during sampling.

NanoMet characterizes particles in two steps. First, droplets are separated from solid particles using the heatable diluter. Then the material coefficient, ϕ, of the dilute mixture is compared to known or calibrated values. The following examples illustrate our method. The NanoMet approach to treat droplets as different species aims at preventing them from ever condensing at all. This is accomplished by heating the diluter head and dilution air to 150°C, which sets the process of aerosol dilution prior to cooling of the mixture to room temperature. Thereby the order of the two processes—dilution and cooling—is reversed, as opposed to the dilution tunnel where conditions are most favorable for the formation of droplets. Furthermore, two lower heating temperatures can be chosen (120°C, 80°C) at which some of the less volatile compounds may condense while others remain in gas phase. This allows one to further characterize the volatile components of the aerosol, *e.g.*, weakly acid water vapor will not condense at 120°C while vapor of more concentrated sulfuric acid will. The principle is illustrated in Fig. 4. As temperature in the heated diluter head is increased, the reading of the "overall surface" DC sensor decreases. The PAS detects an almost constant concentration of solid carbonaceous soot particles at any temperature.

IV.2. Source Apportionment

An application of the material coefficient measurement of NanoMet is shown in Fig. 5. Emissions of a diesel engine operated with two types of fuel were characterized using SMPS and NanoMet. With standard fuel the SMPS spectrum shows the usual log-normal distribution peaking at 70 nm. The NanoMet reading is equal for both sensors, which can be interpreted as "all particles" (DC) are "soot particles" (PAS). In the second case the fuel was doped with some organo-metallic compound. The effects of this additive on the emission spectrum measured with a scanning mobility particle sizer (SMPS) are twofold: *a*) The number of soot particles is reduced by half an order of magnitude. This result is reflected in a reduction of the PAS signal to a third of its previous value. *b*) The additive generates new small particles in high numbers which

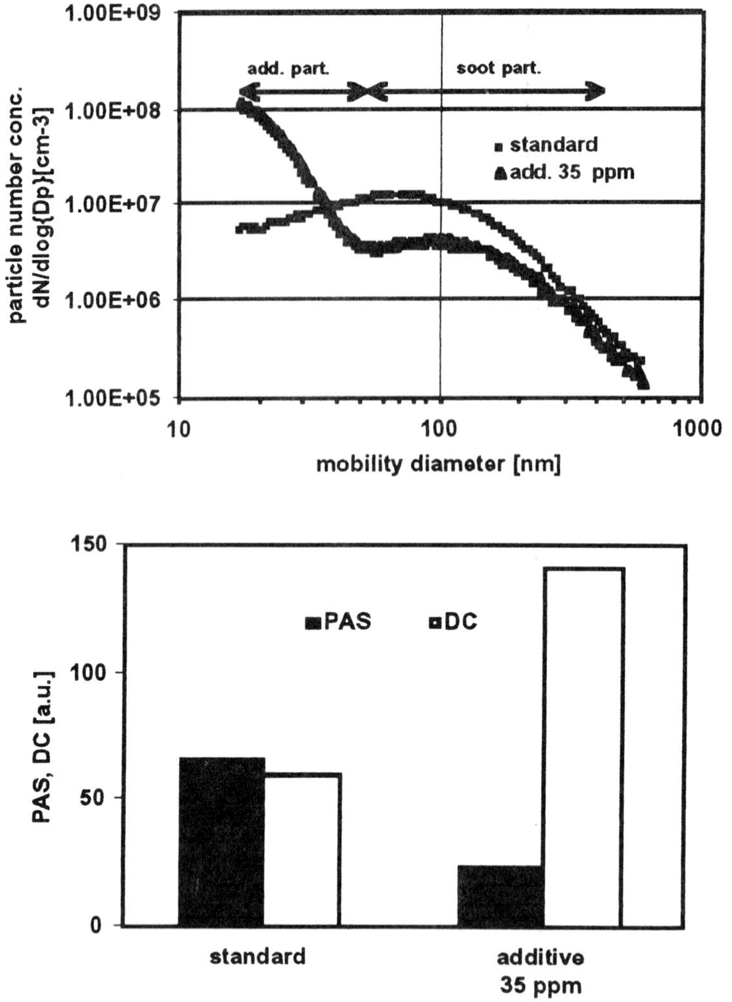

Figure 5. Emission measurements with scanning mobility particle sizer (SMPS, upper) and NanoMet (lower) at a diesel engine operated with standard fuel and cerium doped fuel. Due to the additive the number of soot particles is decreased while the total particle number increases (SMPS spectra). This result is also reflected in the PAS and DC signals.

are suspected not to be soot particles. NanoMet confirms this suggestion by an almost tripled DC reading. Thus the relevant information (soot reduction; increase of total particle emissions) is obtained in a very straightforward manner without the need for additional measurements or chemical analysis.

V. Applications

Owing to the high sensitivity of the PAS and DC sensor, combined with the diluter, the whole concentration range between emission and ambient air/occupational exposure measurement can be covered. With NanoMet one has an instrument at hand which solves the unsatisfactory situation that the same particles are measured and legally limited by different, often not even correlating methods, only depending on the actual legal situation. Numerous ambient air and emission measurements have been carried out during the last years where sensors and diluter worked under field conditions. Prototypes of the centrifuge and diffusion battery have yielded encouraging test results and will soon be intergrated as new NanoMet modules.

VI. Summary

The newly developed NanoMet technology that emerged from the results of the VERT project has passed its first field tests. According to the initial goals of the project, NanoMet measures particles *in-situ, i.e.* as aerosol. The time resolution (1 second) and the sensitivity of the instrument facilitate transient measurements in a concentration range stretching from ambient air up to vehicle emissions. NanoMet distinguishes particles of different chemical composition and characterizes their diffusion behavior. Thus, information on source and toxicity of the aerosol is obtained. The instrument fits in a large suitcase and is robust enough to be operated under fairly rough field conditions. The design can be simplified to a point where no special skills in particle measurement are required any more to operate NanoMet.

Acknowledgments

The author wishes to thank A. Mayer and H. C. Siegmann for continuous support and many helpful discussions.

References

1. A. Mayer, U. Matter, G. Scheidegger, J. Czerwinski, M. Wyser, D. Kieser, and J. Weidhofer, "Particulate Traps for Retro-Fitting Construction Site Engines VERT: Final Measurements and Implementation." *SAE Technical Paper Series,* 1999-01-0116, (1999).
2. U. Matter, H. C. Siegmann, and H. Burtscher, "Particle Emissions from Diesel Engines; Measurement of Combustion Exhaust Occupational Exposure". *2nd ETH Zurich Workshop "Nanoparticle Measurement", Proceedings, BUWAL 1998.*
3. Ch. Hueglin, L. Scherrer, and H. Burtscher, *J. Aerosol Sci.* **28**, 1049 (1997).
4. A. Filippov, *J. Aerosol Sci.* **24**, 423 (1993).
5. S. N. Pandis, U. Baltensperger, J. K. Wolfenbarger, and J. H. Seinfeld, *J. Aerosol Sci.* **22**, 417 (1991).
6. A. Keller, M. Fierz, K. Siegmann, H. C. Siegmann, A. Filippov, "Evaluation of the Surface Properties of Nanoparticles", Submitted to *J. Aerosol Sci,* (2000).
7. L. Jing, (1999) "Generation of Combustion Soot Particles for Calibration Purposes". *3rd ETH Zurich Workshop "Nanoparticle Measurement", Proceedings, BUWAL 1999.*

Characterization of Nanoparticles by Aerosol Techniques

H. Burtscher[1] and B. Schleicher[2]

[1] *Fachhochschule Aargau*
University of Applied Science
CH-5210 Windisch
SWITZERLAND

[2] *Norsk Hydro*
Research Center Porsgrunn
P.O. Box 2560
N-3901 Porsgrunn
NORWAY

Abstract

A number of techniques has been developed to study particles suspended in a carrier gas, that is an aerosol. These methods allow for example to determine number concentration, size and mass of the particles, or yield information on morphology and chemistry of the particles. Some are also feasible to study adsorption and desorption processes either by monitoring chemical changes on the particle surface (by photoelectron emission) or by determining the change in particle size, which can be done on a monolayer-scale resolution. Examples are given of how information on shape and structure of the particles can be obtained by measuring the mass *versus* diameter, thus obtaining a fractal-like dimension. These techniques can be very useful to investigate nanometer sized particles, often in real time on a time scale in the order of seconds or even less.

I. Introduction

Usually nanoparticles or clusters are studied in cluster beams or embedded in a matrix or deposited on a solid matrix. The first technique has the advantage of having the particles isolated without interaction with their surrounding, however, the time for observation is very short. In the second method time is available, but there will always be some undesired interference with the supporting solid, which may cause artifacts.

Here an alternative way will be discussed, which is somehow 'in between' the techniques mentioned above; the particles are investigated while suspended in a carrier gas ("aerosol"). If the particle concentration is low enough to avoid rapid agglomeration, particles can be kept in the gas for a long time maintaining their initial shape, they can be transported by the gas flow for example from the source to the analysis-section, and interaction with the surrounding gas is less pronounced than with a solid. To minimize gas-particle interaction, ultraclean noble gas can be used (helium evaporated from the liquid).

Small particles suspended in gas also play an important positive or negative role in many fields, some examples are: A number of nanomaterials are produced via aerosol routes.[1] In ambient air particles are found in concentrations of typically 10^4–10^6 particles/cm^3. Depending on the composition, the particles can be toxic, act as a catalyst for a number of reactions in atmospheric chemistry and have a dominant influence on atmospheric electricity. In clean room technology even extremely low concentrations may have fatal consequences. Most of these particles (at least in number concentration) have sizes below some hundred nanometers. This means that they are smaller than the wavelength of visible light and therefore are invisible. Natural sources are for example sea spray or volcanic emissions. The most important antropogenic source of submicron particles are combustion processes, first of all diesel engines. Particles are produced in the combustion process directly (primary aerosol) or by gas to particle conversion of material emitted in the gas phase, often induced by photochemistry ('photochemical smog').

To study these particles a number of *in-situ* tools have been developed, which allow investigating them in their original environment. Most of these techniques are based on the interaction of the particle with the surrounding gas (diffusion, stationary motion under influence of an external force, particle motion in an accelerated gas flow) or interaction with electromagnetic radiation, first of all light.

In the following, some of these tools will be introduced. Then examples for applications of these techniques to characterize nanometer sized particles in size, structure and chemistry and to monitor dynamic processes, as adsorption of trace gases on the particles will be given.

II. Tools for Particle Analysis

II.1. Particle Analysis by Interaction with the Carrier Gas

The stationary motion of particles under the influence of an external force is described by Stokes law, if the motion is slow enough to be in a laminar regime and if the particle are large compared to the mean free path λ in the gas ($\lambda = 64$ nm in air at normal conditions)

$$\vec{F} = 3\pi\eta d\vec{v}\,, \tag{1}$$

where $\vec{F}$ is the force, η the viscosity, d the particle diameter, and $\vec{v}$ the particle velocity. A linear relation exists between velocity and force, defined by the mobility b

$$b = \frac{1}{3\pi\eta d}\,, \tag{2}$$

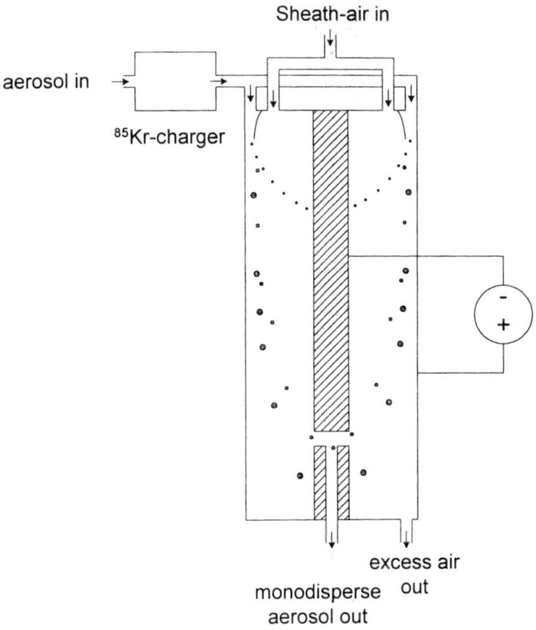

Figure 1. Differential mobility analyzer. Particles are charged by attachment of ions, produced by the ^{85}Kr source.

b linearly depends on $1/d$. The concept of the mobility also holds for particles much smaller than λ, the relation between force and velocity is then given by[2]

$$\vec{F} = \frac{\pi\delta\eta d^2\vec{v}}{\lambda}, \qquad b = \frac{\lambda}{\pi\delta\eta d^2}, \tag{3}$$

δ is a scattering parameter, depending on whether the scattering is diffuse or specular.

Now the mobility depends on $1/d^2$. The mobility of electrically charged particles can easily be measured by their drift in an electrical field, a setup therefore is shown in Fig. 1. Charging usually is done by attachment of ions, produced by a radioactive source (bipolar diffusion charging).

Based on the mobility as measured quantity, the 'mobility diameter' or 'Stokes diameter' d_b is defined as diameter of a sphere, having the measured mobility. In the continuum- and in the molecular range Eqs. (2) and (3) define d_b. In the transition range the relation is given by an empirical approximation (Stokes-Cunningham correction, see for example Hinds[3]). The particle mobility also determines diffusion, the relation between mobility and diffusion constant D is given by the Stokes-Einstein relation

$$D = bkT\,. \tag{4}$$

So far the stationary motion has been considered. In the case of accelerated motion, an inertia term has to be added

$$\vec{F} = \frac{\vec{v}}{b} + m\frac{d\vec{v}}{dt}\,. \tag{5}$$

This equation yields a relaxation time constant τ

$$\tau = mb\,. \tag{6}$$

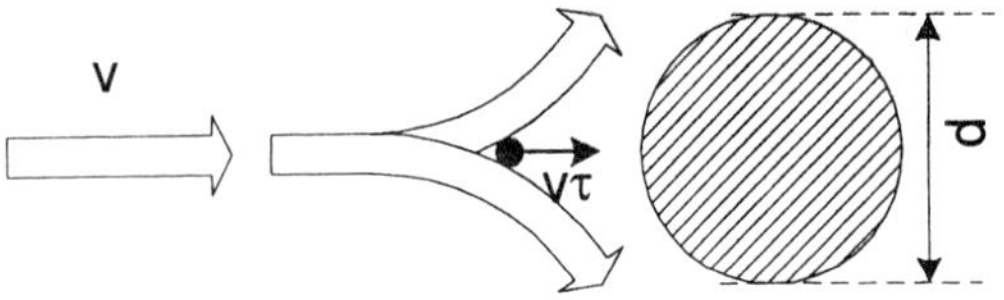

Figure 2. Particle impaction.

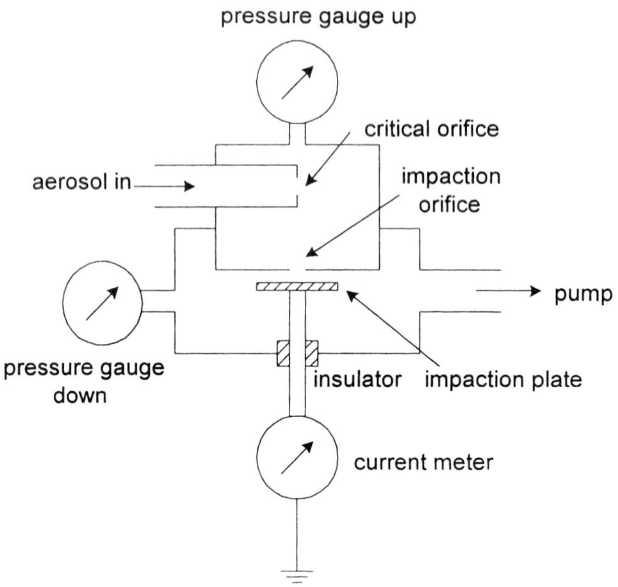

Figure 3. Low pressure impactor. If particles are charged before entering the impactor, the concentration of impacted particles can be measured by measuring the current at the impaction plate. The critical orifice at the entrance determines the flow rate.

The relaxation occurs fast (for a unity-density particle, 100 nm in diameter, $\tau = 8.8 \times 10^{-8}$ s), the assumption of a stationary motion therefore is valid in many cases.

If the gas flow is deflected by an obstacle, τ and the gas velocity v determine if a particle hits the obstacle or follows the gas flow (impaction) (see Fig. 2).

The relaxation time τ depends on particle mobility and mass. Based on τ the 'aerodynamic diameter' d_{ae} is defined as the diameter of a unit-density sphere, having the same τ as the particle under consideration. d_{ae} is measured by impactors, which are found in different designs.[3] Figure 3 shows a design for a low pressure impactor, which is used for very small particles.[4] Measuring mobility and impaction allows to determine the particle mass as function of its diameter and thus yields structure related information, as will be shown in Sec. III.1.

II.2. Attachment of Atoms, Molecules or Ions (Diffusion Charging)

Another technique to study particles is via the diffusion attachment of 'labeled' atoms or molecules. Mainly two labeling techniques are used. First, one can use radioactive species, as done in an instrument called Epiphaniometer,[5] where a lead isotope is applied. The sticking coefficient for lead can be assumed to be one, this means that attachment probability and collision probability are identical. Attached atoms are detected by their radioactive decay, *i.e.* by a counting techniques. This means that this technique has the advantage of being very sensitive, very low concentrations can be measured.

The second technique uses ions. The measurement then becomes very simple. The ions can be produced by an electrical corona discharge, the charge, deposited on the particles can be measured by capturing the particles in a filter mounted isolated and connected to a current meter. If the ion density is kept low enough to avoid multiple charging (no repulsive Coulomb potential), the probability of ion and atom attachment is mainly the same, *i.e.* the image potential can be neglected.

Both techniques yield an integral attachment probability or mass transfer rate, which depends on the particle size and concentration. It is important for example for adsorption of gaseous species, for example toxic material. The attachment probability of a particle of a certain size is nearly inversely proportional to its mobility.[6] This means that mainly the same particle property determines mobility, diffusion and adsorption kinetics. The very simple measurement may therefore yield important information.

If the particle size is known, it can also be used as a simple way to determine the number concentration or vice versa, if the number concentration is measured together with the attachment probability, the mean diameter can be calculated.

II.3. Photoelectric Charging

So far particle characterization according to physical properties has been considered. To obtain chemical information particles can be ionized by irradiation with UV-light (photoelectric charging). The energy of the photons must be between the photoelectric threshold of the particles and the ionization energy of the carrier gas. Due to collisions with the surrounding gas electron energy and direction cannot be measured, what is measured is the integral photoelectric yield. Diffusion of photoelectrons or ions, formed by attachment of a photoelectron to a gas molecule, in the surrounding of the emitting particle will lead to a reattachment of the electron. The probability for reattachment is negligible small for nanoparticles, however, it becomes important for larger particles.

As photoelectron emission is very sensitive to the particle surface, adsorption of trace gases in the sub-monolayer range can easily be detected. Of course, electron emission not only depends on the particle composition, but also on its size. Experiments[7] and calculations[8] show that photoemission has almost the same size dependence as the attachment probability, discussed in Sec. II.2. This means that when looking at the ratio of photoelectric charging and diffusion charging, size and concentration are eliminated, only chemistry remains.

III. Applications

III.1. Determination of the Structure of Agglomerated Aerosol Particles by Combined Mobility and Inertia Analysis

Due to Brownian motion particles suspended in a carrier gas collide and agglomerate. If the particles are liquids or the temperature is high enough to allow sintering between the collided particles, the agglomerates will be mainly spherical. However, if solid particles in a cold region collide, the resulting particle will be an agglomerate which is held together by van-der-Waals forces. The shape and therefore the density of the agglomerates can differ (chain, grape-like shape), mainly dependent on the conditions under which the agglomerates have been formed. To describe the structure of the agglomerate, the so-called fractal-like dimension D_f was introduced. D_f connects the mass m of an agglomerate with its radius of gyration R_g by the equation (see for example Mandelbrot[9])

$$m \propto R_g^{D_f} . \tag{7}$$

In general, the more compact the agglomerate, the larger the fractal-like dimension. Schmidt-Ott[10] could show that the radius of gyration can be replaced with the mobility diameter of the agglomerate d_b for fractal-like dimensions larger or equal to two in the free molecular regime

$$m \propto d_b^{D_f} . \tag{8}$$

III.1.1. Experiment. Particle Production and Agglomeration Process

Nanosized metal particles were produced by an evaporation/nucleation technique: A wire made of the material of interest was heated in a controlled way. Some of the material of the wire evaporates and the metal vapor was subsequently cooled in the gas which streams by. The nucleation process induced thereby leads to the formation of nanosized particles, their size depends on the amount of vapor available for particle formation. By this technique, we produced Pd, Ag, and Co particles. Carbon particles were produced by an electrical discharge between two graphite electrodes facing each other. Also in this process, material evaporates and the gas cools the vapor resulting in particle formation. Particles of oil were produced by spraying a solution of low-volatile oil in alcohol into the gas stream. These droplets are then guided into a heated zone evaporating the solvent which is then removed by a charcoal absorber. As a carrier gas, nitrogen was used with a flow rate of 4 l/min, particle concentration was always about 10^7 cm^{-3}. The particles are then transported either directly to the measurement equipment described below or are first guided through an agglomeration chamber. In this chamber, which is about 15 litres in volume, the residence time of about 4 minutes of the particles is long enough to allow formation of agglomerates from the primary particles.

III.1.2. Measurement of Mobility Diameter and Particle Density

To measure the mobility of the particles, they are charged in a well defined way in a diffusion charger and size selected by a differential mobility analyzer as described previously. Particles in the nanometer size range leaving the DMA carry only one elemental charge. The particles then enter a low-pressure impactor where they are selected according to $\tau = mb$. The impaction behavior of an impactor is described by the Stokes number. For an impactor as described above, the Stokes number is proportional to particle mass and mobility diameter and inverse proportional to the upstream pressure squared.[4] Each impactor has a fixed Stokes number at which 50% of the particles are impacted (Stk_{50}). Impaction efficiency of the monodisperse particles selected by the DMA can be changed by varying the pressure inside the impactor. Knowing the Stokes number at which 50% of the particles are impacted and their

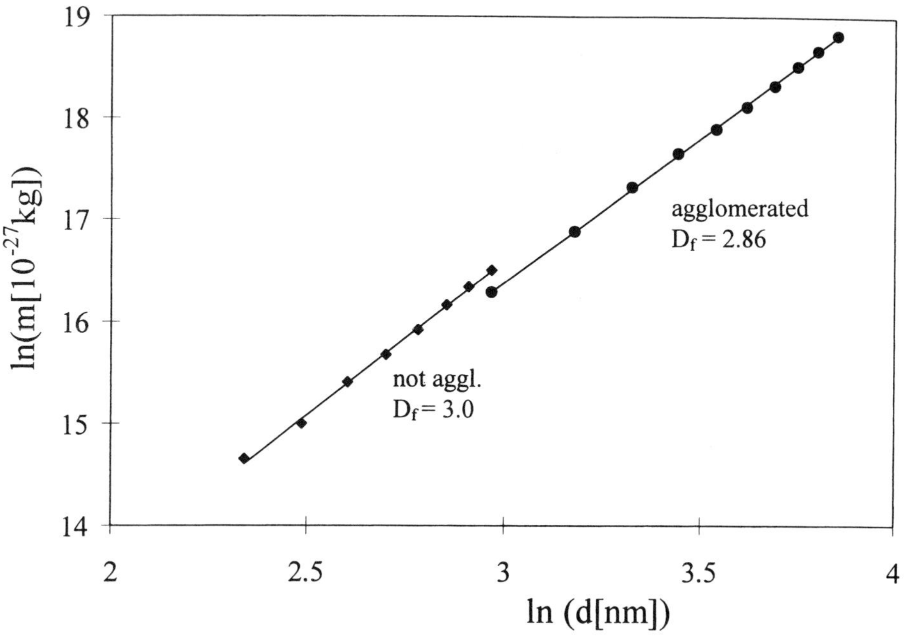

Figure 4. Mass *versus* mobility diameter for agglomerated and not-agglomerated silver particles.

mobility leads directly to particle mass. The fractal-like dimension can then be determined from the slope of a $\log m$ *vs.* $\log d_b$ plot. With this set-up it is possible to measure particles down to radii of 2 nm.

III.1.3. Results

First, the impactor has to be calibrated by determining Stk_{50}. This is done with the oil droplets, which have a well known density and a spherical shape. We obtained with our impactor design $Stk_{50} = 0.19$ with an error of about 10%.[11] Based on this, we could now determine mass and density of other particles. Without the agglomeration chamber we found for Pd and Ag particles a density which is almost independent on mobility radius. This indicates a fractal-like dimension of close to 3. The situation is different for graphite particles which were produced by spark discharge. The density decreases with size and the fractal-like dimension was only 2.3–2.5. This indicates that the graphite particles are agglomerates composed of smaller primary particles immediately after production. The results for Co particles gave also a result less than 3 for the fractal-like dimension. This may be explained by the fact that Co is a magnetic material which could lead to an increased coagulation right after production.

After the agglomeration chamber a lower D_f was found in all cases. As an example Fig. 4 shows the case of silver. The increase in size accompanied by a decrease in D_f if the particles are passed through the agglomeration chamber is obvious. The most pronounced difference, however, is found for Co particles. This again may be explained by the fact Co is a magnetic material. It is known that magnetic particles try to form chain-like agglomerates having a low D_f.[12] Table I, shows a summary of the results for D_f.

Table I

Fractal-like dimension with and without agglomeration chamber.

	D_f Bypassing agglomeration chamber	D_f Passing agglomeration chamber
Pd	(2.7–3.0)	(2.54–2.75)
Ag	(2.9–3.0)	(2.75–2.86)
Co	(2.7–2.8)	(2.2–2.3)
C	(2.3–2.5)	(2.05–2.35)

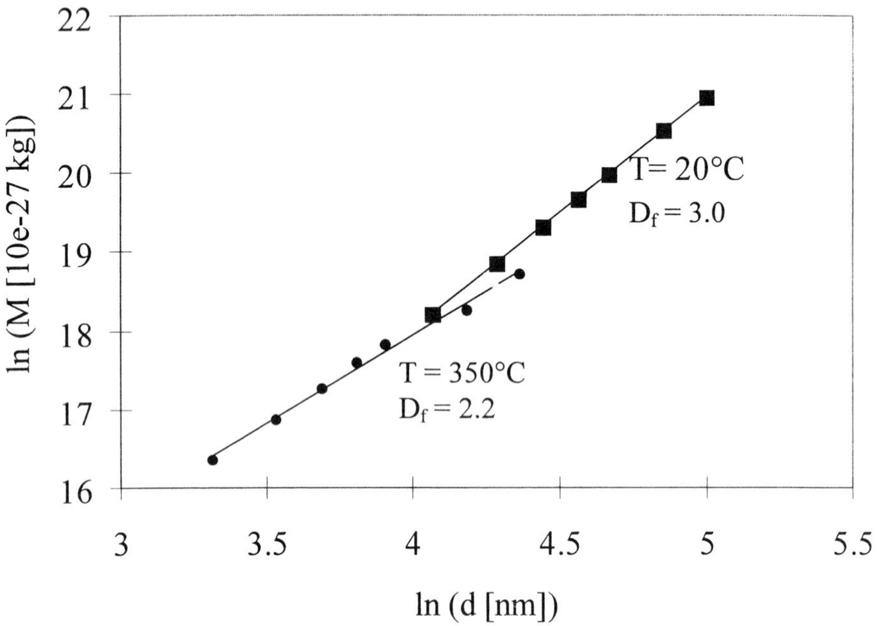

Figure 5. Mass *versus* mobility diameter for particles from a spark ignition engine, which are passed through a heated section at different temperature.

Our experiments give a density of about 2.4 g/cm^3 for Pd and about 2.1 g/cm^3 for the Ag nanoparticles. However, bulk density data for Pd and Ag are higher by a factor of 4 as compared to particle density. Still, we do not know exactly how to explain this difference.

Another example is shown in Fig. 5. Particles emitted by a spark ignition engine have been analyzed in the same way. Before analysis, the particles are passed through a heated tube (thermodesorber), were volatile material can be removed.[13] Heating this tube to 350°C leads to a significant reduction in size and in D_f. This can be

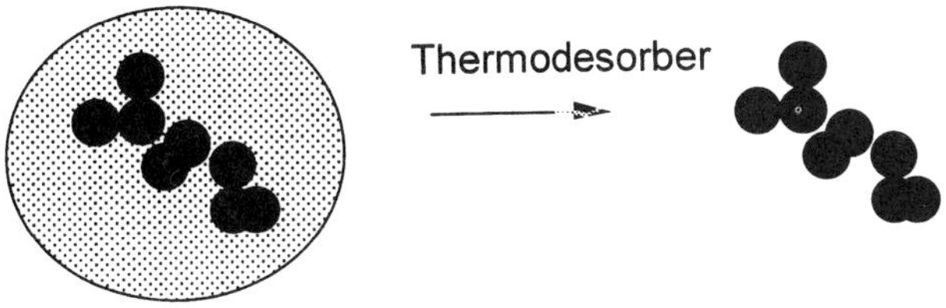

Figure 6. Model for particles from a spark ignition engine. A 'skeleton' formed by the nonvolatile fraction is covered by a liquid fraction, which may be removed by the thermodesorber.

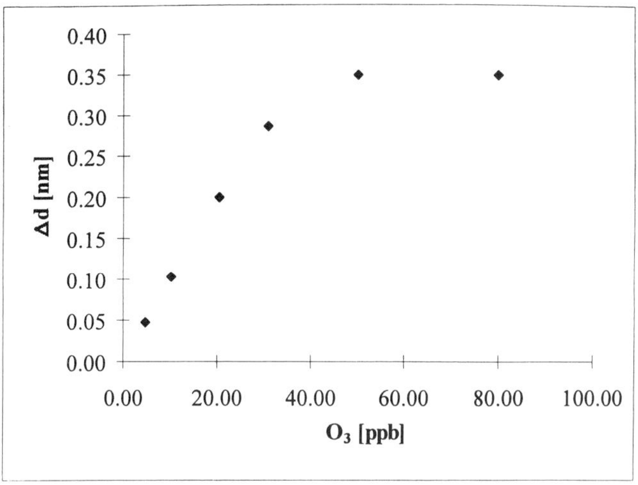

Figure 7. Particle growth by O_3-adsorption.

explained by the high content of volatile material of these particles. This material is at least partly liquid, which leads to coalescence after agglomeration, *i.e.*, the particle always remain spherical, the fractal like dimension is 3. If the volatile fraction is removed, agglomerates of the nonvolatile part become 'visible' (see Fig. 6).

III.2. Monitoring of Adsorption and Desorption Processes

The resolution of differential mobility analysis (DMA) is good enough to measure changes in particle diameter in the order of monolayers. A useful setup is the so-called tandem DMA. One particle size is selected in a first DMA. These particles are guided through a reaction chamber, where for example a trace gas is absorbed. This chamber is followed by a second DMA, measuring the change in particle size occurring in the reaction chamber. Fendel *et al.*[14] studied the interaction of carbon particles with ozone using this method. They found an increase in particle diameter, shown in Fig. 7. This increase is independent of the initial particle size (particles ranging from 5 to 60 nm have been investigated). The reaction time was about 15 s. Obviously the growth comes to an end, when all adsorption sites are occupied. After leaving the particles another 160 s in an O_3 free gas, the particles shrink and are left 0.2 nm smaller than before the adsorption. This shows a carbon oxidation by the ozone, followed by a desorption as CO or CO_2. Weingartner *et al.*[15] performed such measurements to study hygroscopic properties of particles. In the reaction chamber there is a humidifier.

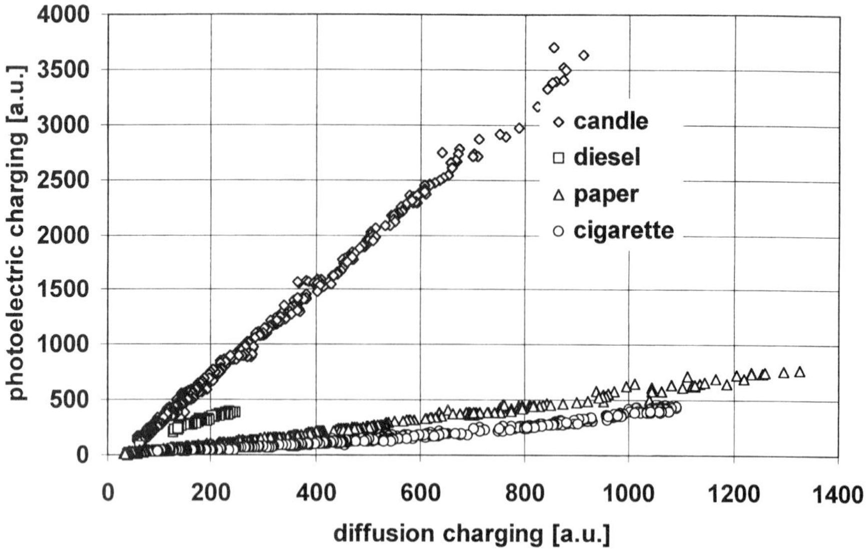

Figure 8. Photoelectric charging *versus* diffusion charging for different combustion sources.[18]

The reaction chamber can also be a thermodesorber. In this case the volatile fraction of a particle can be removed. This is of importance for example for the analysis of combustion particles, where one wants to distinguish the nonvolatile elemental carbon core from organic and inorganic volatiles.[16,17]

III.3. Particle Characterization by Diffusion and Photoelectric Charging

From all particles usually occurring in the atmosphere, those arising form incomplete combustion of organic material, show the highest photoelectric yield, These particles may therefore be selectively detected. Using photoelectric charging and diffusion charging in parallel, even particle from different sources can be distinguished, as shown in Fig. 8. A similar setup was also used by Ammann *et al.*[19] to study volcanic nanoparticles. When sampling in a volcanic plume the dilution changes all the time. To get quantitative results nevertheless, it is of crucial importance to have a normalization, which again can be obtained by looking at the ration of PE and DC. In this case it was found that photoemission is dominated by particles containing CuCl particles. As CuCl belongs to the volatile fraction of the magma, which degasses in an early stage, the presence of CuCl indicates transport of fresh magma. As these measurements can be done with a time resolution of less than one second, it can also be used to monitor dynamic processes as exhaust measurements during engine acceleration.

IV. Conclusions

Particle analysis by aerosol techniques allows to obtain information on particle concentration, size, mass and structure (for example in terms of a fractal dimension). The resolution of the size measurement is good enough to detect particle growth in the order of monolayers, which allows a sensitive monitoring of ad- and desorption processes or also restructuration of particles.

A simple and powerful tool to obtain chemical information is photoelectron emission from the particles. Even if it is not possible to measure energy, direction or spin of emitted electrons, (this information is immediately lost due to collisions with gas molecules) the integral yield and the photothreshold can be very useful. This techniques allows for example to distinguish particles from different sources (in the case of combustion-generated particles diesel engine emissions can clearly be distinguished from cigarette smoke) or to monitor aging processes (in volcanic emissions particles formed from fresh magma and from aged magma can be distinguished) or again to study adsorption or desorption processes, as already fractions of a monolayer can significantly change the photoelectric yield.

References

1. T. Kodas and M. Hampden-Smith, *Aerosol Processing of Materials* (J. Wiley, New York, 1999).
2. W. A. Fuchs, *The Mechanics of Aerosols* (Dover, New York, 1989).
3. W. C. Hinds, *Aerosol Technology* (John Wiley & Sons, New York, 1999).
4. J. Fernandez de la Mora, N. Rao, and P. H. McMurry, *J. Aerosol Sci.* **21**, 889 (1990).
5. S. N. Rogak, U. Baltensperger, and R. C. Flagan, *Aerosol Sci. Technol.* **14**, 447 (1991).
6. A. Keller, M. Fierz, K. Siegmann, and H. C. Siegmann, *J. Aerosol Sci.* (1999), Submited.
7. H. Burtscher, *J. Aerosol Sci.* **23**, 549 (1992).
8. A. V. Filippov, A. Schmidt-Ott, and W. Fendel, *J. Aerosol Sci.* **24**, S501 (1993).
9. B. Mandelbrot, *The Fractal Geometry of Nature* (Freeman, San Francisco, 1982).
10. A. Schmidt-Ott, *Appl. Phys. Lett.* **52**, 954 (1988).
11. B. Schleicher, S. Künzel, and H. Burtscher, *J. Appl. Physics* **78**, 4416 (1995).
12. G. A. Niklasson., A. Torebring, C. Larsson, C. G. Granqvist, and T. Farestam, *Phys. Rev. Lett.* **60**, 1735 (1988).
13. H. Burtscher, S. Künzel, and Ch. Hüglin, *J. Aerosol Sci.* **4**, 389 (1998).
14. W. Fendel, D. Matter, H. Burtscher, and A. Schmidt-Ott, *Atmos. Environ.* **9**, 967 (1995).
15. E. Weingartner, H. Burtscher, and U. Baltensperger, *Atmos. Environ.* **31**, 2311 (1997).
16. D. Steiner and H. Burtscher, *Environ. Sci. Technol.* **23**, 1254 (1994).
17. U. Matter, H. C. Siegmann, and H. Burtscher, *Environ. Sci. Technol.* **33**, 1946 (1999).
18. K. Siegmann, L. Scherrer, H. C. Siegmann, *J. Mol. Struct. (THOCHEM)* **458**, 191 (1999).
19. M. Ammann, L. Scherrer, W. Mller, H. Burtscher, and H. C. Siegmann, *Geophys. Res. Lett.* **19**, 1387 (1992).

Carbon Formation in Combustion

Konstantin Siegmann

Laboratory for Solid State Physics
Swiss Federal Institute of Technology (ETH)
CH-8093 Zürich
SWITZERLAND

Abstract

Laser ionization mass spectroscopy studies and aerosol analyses of combustion byproducts from inside of a laminar, atmospheric pressure diffusion flame burning with argon diluted methane are presented. Large molecules and small carbonaceous particles are found. The molecules are Polycyclic Aromatic Hydrocarbons (PAHs), with molecular masses up to 800 amu. The diameters of the particles range between 2 and 50 nm. Formation and destruction of particles and PAHs can be monitored as a function of the height in the flame. It is found that particles are formed before large PAHs appear. The relation between PAHs and particles is discussed. It is proposed that PAHs are formed on the surface of the particles and evaporate into the gas-phase when synthesis is completed. A mechanism for soot particle formation in diffusion flames is presented.

I. Introduction

A diffusion flame is the simplest combustion device, known to man for over a million of years.[1] It is the flame that forms by burning the fuel in air, and means that oxygen and fuel are not mixed prior to combustion.

Any hydrocarbon fuel will thereby produce carbon particles, that is, soot. The glowing particles are responsible for the yellow luminosity of the flame. Often, those particles are not released into the atmosphere but are oxidized in the combustion process. When the gas flow in the flame is not turbulent, one speaks of a laminar flame. A laminar flame has the advantage of a reaction time scale, which means that the height-above-burner can be transformed into a time spent in the combustion zone.[2] By investigation of flame components as a function of height one can therefore trace the combustion process.

Any hydrocarbon may serve as fuel. We chose to use methane because of its simplicity: Only one carbon atom is present in one formula unit. Larger carbonaceous species are therefore solely the result of carbon-carbon bond formation; decomposition processes of the fuel can not occur. Also, methane is the major component in natural

Physics of Low Dimensional Systems
Edited by J. L. Morán-López, Kluwer Academic/Plenum Publishers, New York 2001

gas and is generally believed to be the cleanest burning hydrocarbon fuel as it produces less CO_2 and more H_2O than other fossil fuels because of its high H/C ratio.[3,4]

A laminar diffusion flame with methane as fuel, as it is mainly used in our investigations, does not emit any hydrocarbons or particles into the environment; all species are burned inside of the flame.

One can group the great number of flame investigations into two categories. The first group consists of methods that are applied *in situ. In situ* methods do not require sampling of gas probes but analyze the species in their natural environment. This is the large field of spectroscopy, mainly using lasers as light sources.[5–8] The advantage of *in situ* methods is that they do not disturb the flame. For the analyses of a single large molecule (*e.g.*, a member of the group of Polycyclic Aromatic Hydrocarbons, PAHs) optical spectroscopy is inapplicable, simply because there are far too many similar molecules present in the flame.[9]

The disadvantage of disturbing the flame by withdrawing samples is compensated by the many powerful techniques that can be applied to analyze the sample.[10] Also, sampling artifacts can be estimated. For example, we find that the height profiles do not vary much if we either withdraw samples in direction of the flame axis or perpendicular to the flame axis. This shows that the sampling depth is small enough compared to flame dimensions.[11]

Mass spectroscopy is often used to analyze the constituents of the sample.[12–14] To do so, the molecules and particles have to be charged. One can either directly analyze the charged flame species,[9] or one can ionize the neutral ingredients.[15,16] Ionization is best carried out using UV-light from a laser, because PAHs efficiently absorb UV-photons. Both, one-photon ionization[17–19] and two-photon ionization[15,16] mass spectroscopy have been realized.

Mass spectroscopy is mainly used to analyze low-pressure premixed flames,[13,14] but there are some reports on the analysis of diffusion flames.[15,18,20] The difficulty associated with atmospheric diffusion flames is that the flame burns at normal pressure, whereas in the mass spectrometer high-vacuum prevails. The usual technique of transferring samples from normal pressure into high vacuum is to employ heated transfer capillaries.[21,22] The disadvantage of this technique is that larger molecules (larger than about 300 amu, *e.g.*, $C_{24}H_{12}$, coronene) and particles are lost in the transfer line, simply because their volatility is much to low, the capillary can not be heated enough. We therefore constructed a new mass spectrometer inlet device that does not rely on heated transfer capillaries.[15]

In order to analyze the gas samples for particles we rely on "traditional" aerosol methods.[10] The features of those methods are summarized below:

1) A Differential Mobility Analyzer (DMA) is used to determine size spectra of the particles from the flame, and it is also used to size select particles with a defined diameter.[23,24]
2) Photoelectric Charging is employed to characterize particle surface properties.[25]
3) By the use of a low pressure impactor densities and fractal-like dimensions of the particles are determined.[26,27]

Combining the results from the mass-spectroscopic and aerosol investigations a model for soot particle formation in diffusion flames shall be developed.

II. Experimental

The experimental setup including burner, extraction system, mass spectrometer and aerosol analyses has been described previously.[11,15,27–29] Briefly, a 70 mm high laminar atmospheric pressure diffusion flame, burning with methane/argon mixtures on a cylindrical Santoro type burner[30] is investigated. Other fuels were also used on the same burner. The argon dilution reduces the amount of soot produced inside of the flame. Typical mixing ratios are: 425 ml/min methane and 335 ml/min argon. The diameter of the flame is 10 mm at the burner mouth. The flame is stabilized by a surrounding airflow.

Two extraction systems for obtaining flame gas samples are employed, a quartz microprobe (outer diameter = 1 mm) and a steel tube (diameter = 8 mm) with a pinhole at the bottom, placed across the flame. Tiny amounts of combustion gases are extracted from the symmetry (middle) axis of the flame. By adjusting and monitoring the underpressure applied (2.0 mbar in the case of the steel tube; 10 mbar for the quartz probe) the amount of flame gases extracted can be determined and controlled. The samples are instantly diluted with an inert gas (nitrogen or argon) within the sampling device to quench all reactions and prevent adsorption/coagulation processes. While the dilution in the steel tube is immediate [depending only on the thickness (0.1 mm) of the tube wall], the flame gases in the quartz probe have to travel about 10 mm before they are diluted. However, the quartz microprobe disturbs the flame only minimally whereas the steel tube causes a strong perturbance downstream of the sampling position. The microprobe samples horizontally whereas the steel tube takes vertical samples. By comparing the two opposed sampling devices it is found that they yield almost the same spectra and height profiles of molecules and particles.[11,28] We conclude therefore that the sampling depth, sampling artifacts and perturbance caused by the two sampling systems are negligible.

The sampling position in the flame is adjusted by moving the burner with a computer driven mechanical motion. It is thus possible to take samples from any point in the flame. The height resolution of the sampling is better than 0.1 mm.

The gaseous samples from the flame are examined either with a Time-of-Flight Mass Spectrometer (TOF-MS) or by classical aerosol techniques, such as a Differential Mobility Analyzer (DMA) combined with photoelectric charging or a low-pressure impactor. The preparation of the gaseous samples differs if either the TOF-MS or aerosol techniques are employed. The most important difference is the dilution ratio in the probe, it is about 1:10 for the mass-spectroscopic investigations but about 1:400 for the aerosol experiments. In the case of the TOF-MS the dilution is determined by the sensitivity of the instrument and can not be increased, whereas aerosol techniques are more sensitive and a high dilution is preferred in order to prevent coagulation of the particles. The sampling system was thoroughly tested for altering the properties of the combustion particles. To do so, a well-characterized test aerosol (*e.g.*, palladium particles produced from a hot palladium wire) was added to the methane fuel. It was found that the whole procedure of sampling and diluting had no effect on the test aerosol, the size distributions have been retained unchanged.[28] Size distributions of particles in the range between 2 and 100 nm in diameter are recorded in the following way: The diluted combustion aerosol is guided through a bipolar, radioactive diffusion charger. The DMA then selects negatively charged particles of an electrical mobility that can be chosen by the applied voltage. The electric mobility is a function of particle size only, since the particles are small enough that multiple charging is improbable. Therefore, a monodisperse fraction of the aerosol leaves the DMA. The aerosol is split into two branches for detection in sensitive electrometers. One of the branches determines the total charge deposited in a filter. This charge is proportional to the amount of particles with the diameter as selected by the DMA. The other branch is equipped with a photoemission stage, where the photoelectric yield is determined.

The particles are illuminated with a low pressure mercury lamp at a wavelength of 185 nm and are thus photoelectrically charged. Comparison of the particle currents in the two branches gives directly the fraction of particles that have emitted one electron. This fraction is referred to as photoelectric yield. The photoelectric yield is a sensitive probe of the surface of the particles; it is mainly determined by the surface coverage with PAHs.[10,25] Alternatively, a low-pressure impactor is used after the DMA. The impactor yields the mass of the impacted particles if their mobility, as measured by the DMA, is known.

The flame species were investigated using two different TOF-MS. The instrument in Zürich[15] was used for molecules having masses below 1700 amu, whereas with the Stuttgart instrument[31] species with masses up to 1000000 amu were studied. Both spectrometers are equipped with UV-lasers for ionization, with the Zürich instrument electron-impact ionization is also feasible. The wavelengths of the light from the three excimer-lasers are 193, 248, and 308 nm, respectively. For equal mass ranges and UV-lasers the Zürich and the Stuttgart instrument yielded the same spectra. The inlet system for introducing combustion gases from atmospheric pressures directly into the high-vaccum of the TOF-MS is used for both instruments. After the flame samples are rapidly diluted, they are pumped fast (within about 1 sec) through a steel tube with a large inner diameter (8 mm) to a pulsed valve. The valve opens for about 1 ms to form a supersonic gas pulse inside the high-vacuum chamber of the TOF-MS. There, the flame species are photo-ionized, accelerated and then mass-selected in a drift tube. Several spectra are accumulated in order to enhance sensitivity. Flame molecules with masses up to 788 amu ($C_{64}H_{20}$)[20] and soot particles up to many 1000 amu could be detected using this setup. For recording PAH height profiles only the Zürich TOF-MS was used. Values proportional to the PAH concentration in the flame were obtained by integrating over the appropriate mass peak.

III. Results

III.1. Laser Ionization Mass Spectroscopy of Flame Components

High-performance time-of-flight mass spectroscopy (TOF-MS)[31] combined with photo-ionization is applied to analyze combustion byproducts from a laminar diffusion flame, burning with argon-diluted methane. The spectra from Figs. 1–4 were taken from the soot and PAH rich part of the flame. The TOF-MS is equipped with three different excimer lasers for ionization, operating at three different wavelengths (193 nm, 248 nm and 308 nm; The corresponding photon energies are: 6.4 eV, 5.0 eV and 4.0 eV). We have recorded mass spectra up to 1000000 amu. With increasing mass, the probability of getting a count in a channel in the high-mass range drops continuously. The mass spectra show a strong dependence on the wavelength of the ionization laser used. Because the ionization potential of PAHs (depending on PAH; generally > 7 eV) and fullerenes (about 8 eV) is larger than the energy of one photon from any laser used, two photons are necessary for ionization of those species. The energy of two photons of 308 nm is not sufficient to ionize small PAHs, whereas fullerenes can be ionized by two 308 nm photons.

Figure 1 shows a spectrum which was taken with 248 nm ionization (5 eV/photon). From 0 to 600 amu, PAHs are found, as shown in the inset. This wavelength ionizes PAHs with the least fragmentation, probably due to a resonant 2-photon-process. The first signal observed belongs to a molecule with a mass of 78 amu, which is assigned to benzene (C_6H_6). The absence of peaks below 78 amu shows that there is

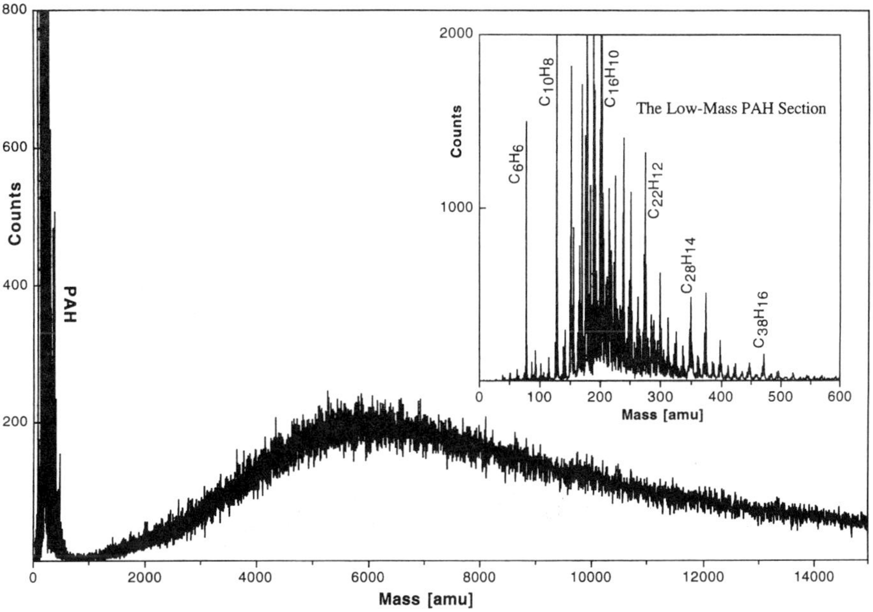

Figure 1. Mass spectrum of flame gases, ionized with with 248 nm UV light. The inset shows an enlargement of the low-mass region.

little fragmentation. From a thorough analysis of the PAHs found conclusions concerning their growth processes were made. For example, it is found that PAHs with the least possible number of hydrogen atoms are formed preferentially.[20] From 1000 to 20000 amu, a broad distribution of soot particles is found, no distinct mass peaks are resolved. Counts at non integral mass numbers are an indication for fragmentation due to the laser irradiation in the TOF-MS. Interesting is the absence of counts at around 1000 amu. With 248 nm ionization PAHs, and in the high mass part, a broad distribution of soot particles is found. The maximum of the soot distribution at about 6000 amu corresponds to carbonaceous particles with a diameter of about 2 nm, if one assumes spherical particles and the specific density of diameter of graphite, 2.26 gcm^{-3}. This diameter is a lower limit because soot particles are less dense than graphite.

If the KrF excimer laser with 248 nm wavelength is focused into the cluster beam using a quartz lens, fullerenes are observed in the mass spectra. Figure 2 shows a typical mass spectrum, using UV light at 193 nm wavelength (6.42 eV/photon) for ionization. PAHs are seen in the low mass part of the spectrum. Below about 500 amu we find strong peaks corresponding to well-known PAHs (*e.g*, $C_{16}H_{10}$, $C_{24}H_{12}$) and smaller peaks of their fragments (*e.g.*, $C_{16}H_9$, $C_{24}H_{11}$) which result from the PAHs by hydrogen abstraction during the laser ionization process. From 500 amu towards larger masses, a broad hill culminating at about 3000 amu is eye-catching. Up to about 3500 amu the hill is overlaid by a structure with individual mass peaks. Distinct series of signals can be observed with spacings of 18 amu, as shown in the inset. There are two separate series showing this pattern. The 18 amu spacing could be attributed to water clusters. Three series would imply two particular molecules serving as cores for more than one hundred attached water molecules. However, this pattern could

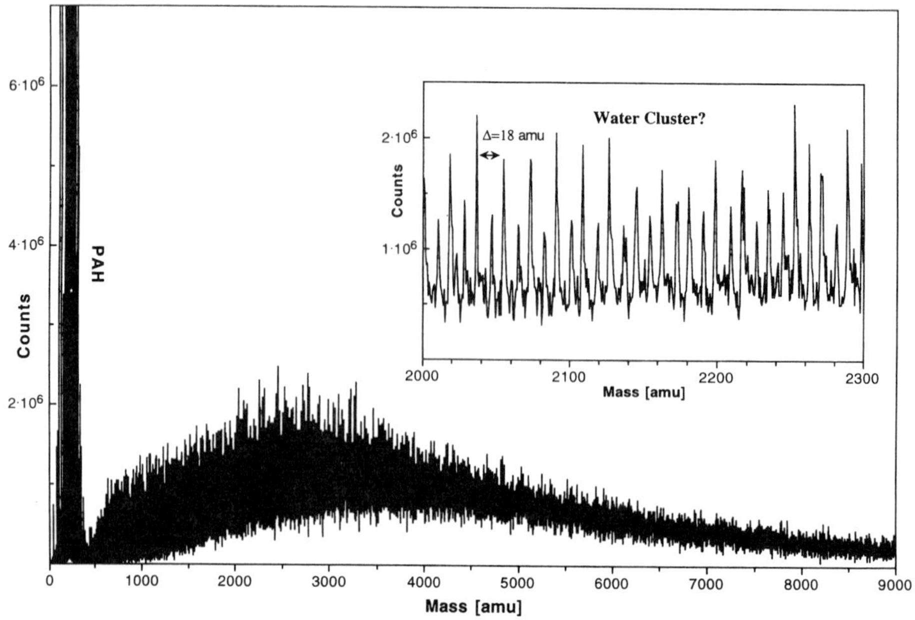

Figure 2. Mass spectrum of flame gases, ionized with with 193 nm UV light. The inset shows an enlargement from the region between 2000 and 2300 amu.

also correspond to multiply charged molecules with a C_3- or C_6-spacing. A broad distribution of soot particles is also observed when 193 nm UV light is employed for ionization.

The third spectrum (Fig. 3) is taken under the same experimental conditions, yet a third laser with a wavelength of 308 nm (4.03 eV/Photon) is employed for ionization. The low-mass part (0–600 amu) shows PAHs and fragments thereof. With 308 nm UV light, the PAHs are strongly fragmented. With 308 nm ionization exclusively fullerenes are seen in the upper mass section of the spectrum. Their distribution displays the usual "magic numbers"; that is, the signals corresponding to C_{60} and C_{70} have a relatively high intensity. Fullerene signals are broad compared to PAH signals.

In the spectrum presented in Fig. 4 C_{60} was added to the flame and the ionizing wavelength was 308 nm. Adding C_{60} to the flame results in a strong peak at the appropriate mass (720 amu), together with smaller peaks which are fragments of the C_{60} added. Due to the laser irradiation units of C_2 are split off the added C_{60}. However, the shape of the C_{60} signal is different from the signal arising from flame species alone. The inset compares the C_{60} peak from the flame species with the peak of the added C_{60}. It can be seen that the flame borne C_{60} appears at slightly higher masses, corresponding to a longer flight time. It therefore seems that not all of the C_{60} observed is formed in the gas phase in the flame. The "delayed" detection of the flame C_{60} could be an indication that fullerenes from the flame are themselves fragments of larger carbonaceous particles. The "delay" of the flame species would then be due to the time needed for fullerene synthesis in the ion-source of the TOF-MS. However, it is unlikely that fullerenes are synthesized from smaller precursor molecules because the density of carbonaceous material in the molecular beam is much too low, and the time available is much too short. We conclude that the fullerene signals originate from evaporation or fragmentation from the surface of small soot particles. An increase in

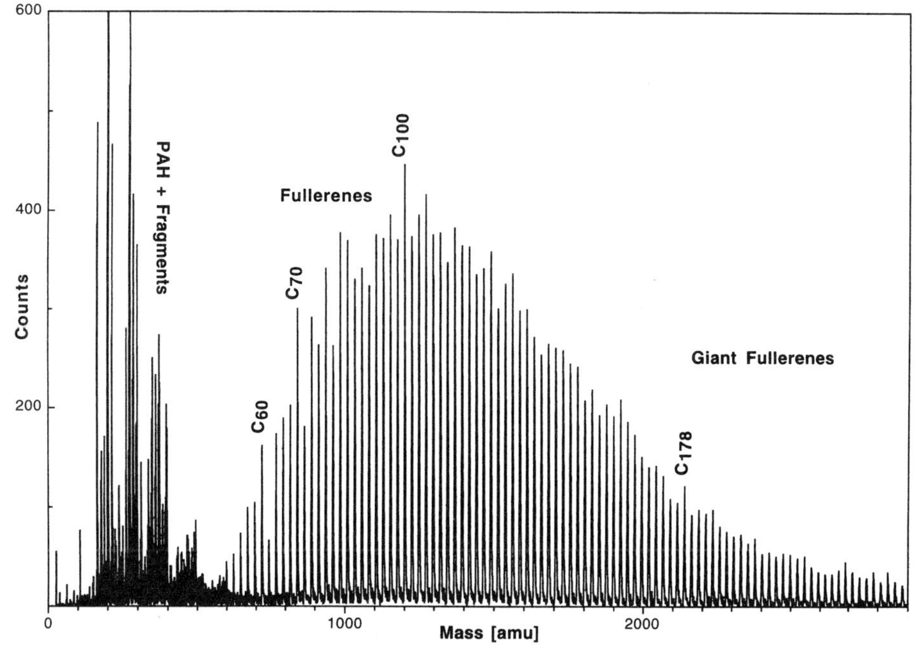

Figure 3. Mass spectrum of flame gases, ionized with with 308 nm UV light.

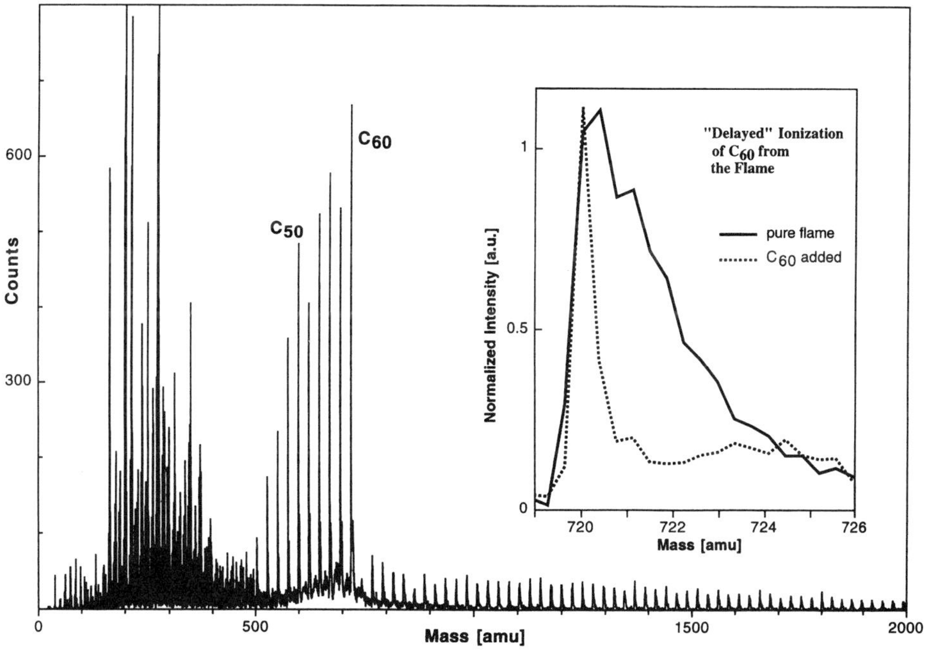

Figure 4. Mass-spectrum of flame gases, ionized with with 308 nm UV light, C_{60} is added to the flame. The inset compares the peak of the added C_{60} with the peak of the C_{60} from the undoped flame.

Table I

The effects of ionization wavelength on the mass spectra of flame components.

	193 nm	248 nm	308 nm
PAH	Some fragmentation (H–abstraction)	Very little fragmentation	Only PAH larger than $C_{14}H_{10}$ are ionized, strong fragmentation
Fullerenes	Are not observed	Are only observed when laser is strongly focused	Are always observed
Soot	Small soot particles and water cluster are observed	Small soot particles are observed	Soot particles are not observed

laser intensity shifts the fullerene distribution to smaller species in accordance with fragmentation as the dominant process. The effects of ionizing wavelength on mass spectra of flame species are summarized in Table I.

III.2. Height Profiles of PAHs and Soot Particles

Height profiles of PAHs and soot particles in the flame will be discussed in this section. Because the flame is laminar, the height above burner (HaB) is also a reaction time scale. Low heights in the flame correspond to short times spent in the combustion zone, whereas larger heights are related to longer reaction times.[2] In Fig. 5 the findings on PAHs and particles in our flame are displayed. On the left side particle size spectra are shown, as obtained with the DMA. They are assigned to the appropriate height in the flame where they were measured. On the right side the maxima in concentrations of selected PAHs (as measured with the TOF-MS) are assigned to their corresponding height in the flame (see also Fig. 7). It is clearly seen that particles are formed before PAH concentrations peak. It can also be seen that larger PAH are generally formed later (higher) in the flame. This is the consequence of their growth process, larger PAHs grow from smaller ones by the addition of small molecules (*e.g.*, acetylene, C_2H_2).[32] Consequently, small PAHs are the precursors of larger ones. We will show in the following that this PAH growth process takes place at the surface of the particles.

A detailed analysis of the height profiles showed that some PAHs are formed faster than others of comparable size. We explain this behavior by postulating an efficient PAH growth process which specifically forms those PAHs. This "reactive dimerization" pathway is described in Ref. 15.

The particle size distributions from the flame (Fig. 5) are approximated by log-normal functions,

$$\frac{dn}{d\ln D} = \frac{n}{\sqrt{2\pi}\sigma}\exp\left[-\frac{(\ln D - \ln\langle D\rangle)^2}{2\sigma^2}\right],$$

which are useful to extract characteristic parameters such as the total particle number density, n, and the geometric mean diameter, $\langle D\rangle$.[23]

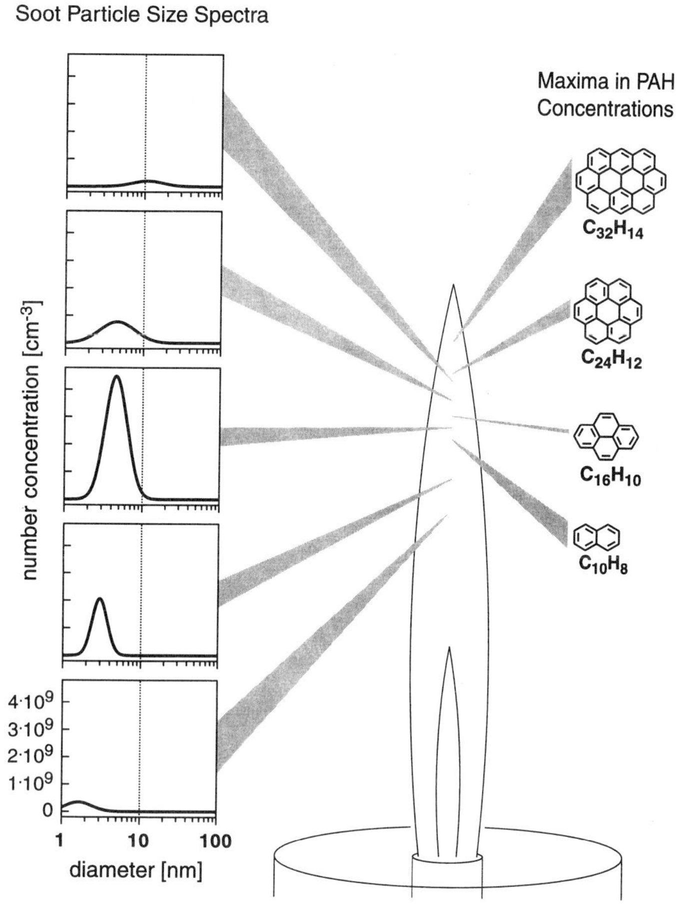

Figure 5. A representation of the methane flame with PAH-maxima and soot particle size distributions, assigned to the height in the flame where they were measured.

These parameters are plotted in Fig. 6 as functions of the height above the burner (HaB) along the flame axis. Up to 52 mm HaB, the particle number grows explosively while the geometric mean diameter grows only to about 5 nm. Above 54 mm HaB, the number density decreases strongly again. This effect is apparently due to coagulation in the flame, as the geometric mean diameter increases rapidly by now. Above 62 mm HaB, the particles start to shrink again strongly while the number density drops more steeply. This indicates the onset of soot particle oxidation. Under the conditions of this experiment, no particles above 66 mm HaB were detected.

The mean diameters of the particle ensemble are far at the lower end of the size range usually encountered in aerosol science. Particles with a mobility corresponding to equivalent spheres of 2–3 nm could be large molecules containing roughly around 500 carbon-atoms. They are with the same size range as the particles observed with mass spectroscopy (see preceding chapter).

To investigate the relationship between soot particles and PAHs, height profiles of the latter are plotted together with the particle number concentration in Fig. 7. It can be seen that the strong increase in particle number starts at a height in the

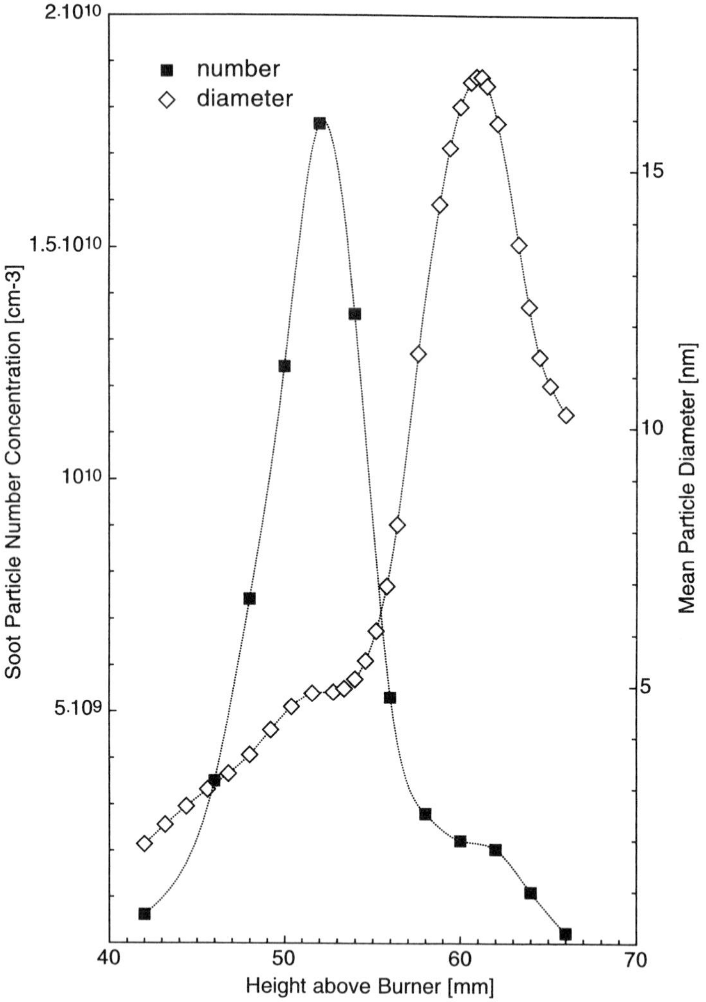

Figure 6. Number concentrations and diameters of particles as a function of height above burner.

flame where the concentration of large PAHs (with molecular weights of more than about 300 amu) is still low. The concentration of the largest PAH shown, ovalene ($C_{32}H_{14}$, 398 amu), increases only where the number concentration of the particles already decreases.

It has been hypothesized by some that large PAHs are the precursors of soot particles.[33] Large PAHs should coagulate forming the first, primary soot particles. PAHs are planar molecules and all of their chemical bonds lie in this plane. Accordingly, the PAHs forming the particle are held together only by van-der-Waals forces. Because such van-der-Waals forces are quite weak and the temperature in the flame is high, very large PAHs are needed to form stable clusters. We have determined, for example, that the PAH perylene ($C_{20}H_{12}$, 252 amu) desorbs within fractions of a second from carbon particles even at temperatures below 100°C.[34] It has been estimated that soot-precursor PAHs must have masses of around or larger than 1000 amu.[14]

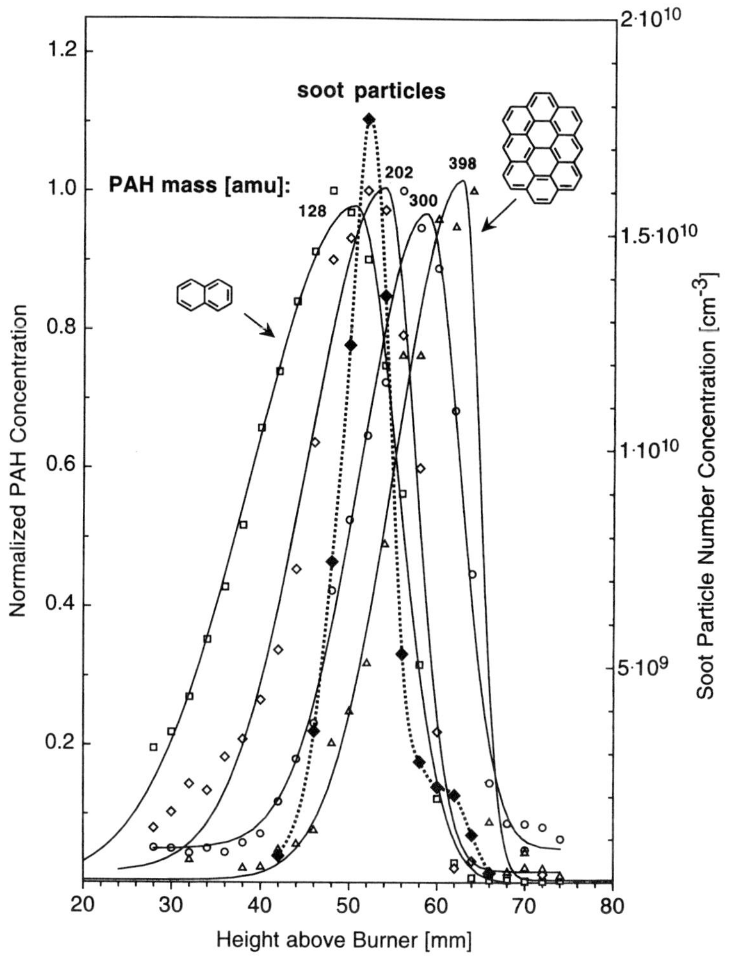

Figure 7. Normalized PAH concentration and particle number concentration as a function of height above burner. (128 amu = $C_{10}H_8$; 202 amu = $C_{16}H_{10}$; 300 amu = $C_{24}H_{12}$; 398 amu = $C_{32}H_{14}$).

The assumption of PAHs being soot precursors is based on the experimental observation that the concentration of PAHs decreases where the first particles appear.[35,36] This observation was made in low pressure premixed flames. From our experiments it can be concluded that this is not the case for atmospheric pressure diffusion flames.

Yet even the reverse is seen in Fig. 7: Large PAHs are formed later than soot particles. We therefore propose that particles are the precursors for PAHs. The PAHs are formed on the surface of the particles and evaporate into the gas phase when synthesis is complete. This mechanism is corroborated from measurements of the photoelectric yield. It was found previously (in many experiments) that PAHs adsorbed at particle surfaces strongly increase their photoelectric yield.[25]

Figure 8 shows the photoelectric yield of particles normalized to the particle surface. Also, the height profile of ovalene ($C_{32}H_{14}$, 398 amu) is displayed. Low in the flame, the yield of the small particles is high. We postulate that PAH precursor

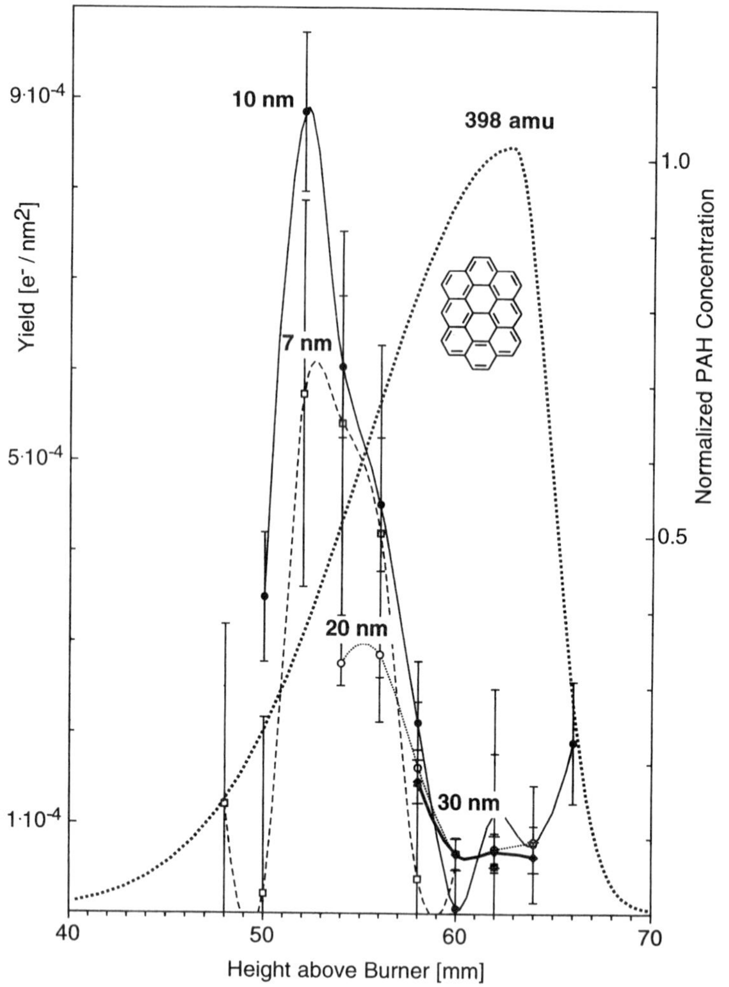

Figure 8. Photoelectric yield of particles with different diameters, normalized to their surface aerea; and height profile of ovalene ($C_{32}H_{14}$ 398 amu).

molecules, still beeing chemically bond to the surface, are responsible for the high photoelectric yield. At a height in the flame where the larger gas phase PAHs appear the yield decreases strongly. We conclude that PAHs are formed on the particle surface and evaporate into the gas-phase when synthesis is complete (that is, when all chemical bonds are saturated) leaving behind a carbonaceous particle of low photoelectric yield. Particles therefore change their properties with time spent in the combustion zone. This can be demonstrated by measuring particle densities with the low-pressure impactor.[27] Figure 9 shows the specific density of particles with a diameter of 37 nm as a function of the height in the flame. (The argon dilution in the flame is smaller than in the previous experiments.) Flames burning with different fuels (methane, CH_4; ethylene, C_2H_4; and propane, C_3H_8) were investigated. Ethylene yields slightly denser particles than the other fuels, probably because in ethylene the carbon-atoms are already sp^2 hybridized. It is seen that with increasing height above burner the

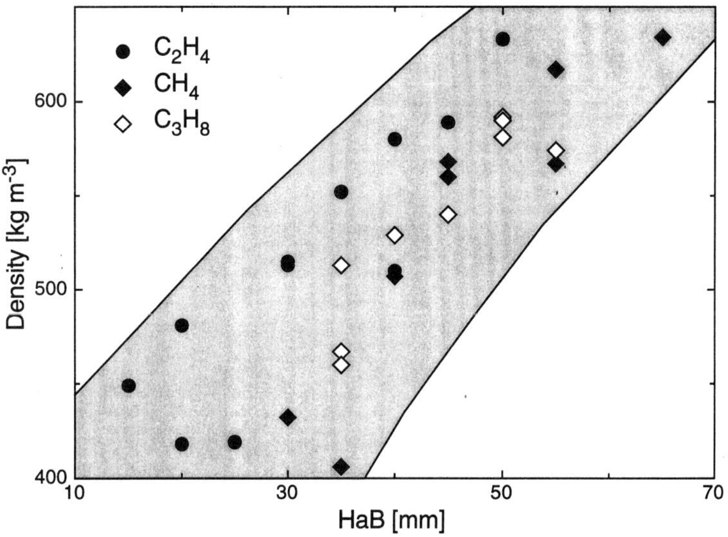

Figure 9. Densities of particles with a diameter of 37 nm as a function of heigt above burner for diffusion flames burning with different fuels (C_2H_4 = ethylene, CH_4 = methane, C_3H_8 = propane).

particles become denser, independent of the fuel used. Two effects are responsible for the increase in density. First, the primary nuclei in the particle can rearrange forming a more compact structure. This effect changes the geometry of the particles and increases their fractal-like dimension, as found and described in Ref. 27. Second, the internal structure of the particles becomes more graphitic with the time spent in the combustion zone, that means that the aromatic rings form increasingly larger and more compact arrays. As this graphitization process proceeds, larger PAHs are formed and released into the gas-phase. PAHs larger than about 1000 amu are bond to the particles by strong van-der-Waals forces and will therefore not be released into the gas phase. This could be the explanation for the minimum observed at 1000 amu, as seen in Fig. 1.

IV. Discussion

A rough picture of soot formation in diffusion flames shall be developed here. This picture is based on the observations demonstrated in the foregoing sections and differs somewhat from the current models.[33] However, studies from D'Alessio *et al.* corroborate our findings[37] (see also the corresponding article in Ref. 33) By using spectroscopic methods they come to similar conclusions, especially what concerns soot inception. They suggest: "... that there is an initial fast polymerization process building bricks of which are aromatic compounds with few condensed rings (no more than 2–3 rings)." They also state that primary soot particles "(consist) of two–ring structures connected together with some non aromatic, some aliphatic bridges. I am thinking of structures of tar which later on become more compact and more aromatic." Also, Dobbins *et al.* state that: "The existence of young, precursor soot particles in flames in an unaggregated, liquid-like state prior to the development of the more commonly observed carbonaceous soot aggregates is well established".[38]

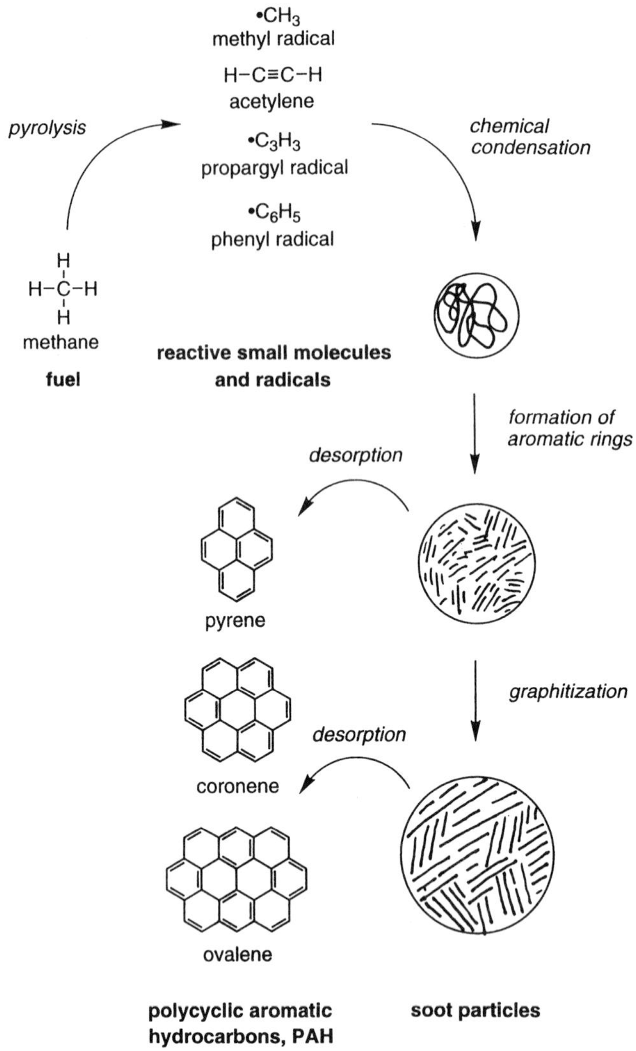

Figure 10. Model of soot formation in an atmospheric pressure methane diffusion flame.

Our findings support these sights, and our picture of carbon formation in diffusion flames is summarized in Fig. 10. First, a small fraction of the fuel (methane) decomposes during oxidation and heating into hydrocarbon radicals which, under fuel rich conditions, form small molecules, particularly acetylene. The latter adds hydrocarbon radicals for growth and the growing unsaturated, radicalic hydrocarbons condense chemically to form the first particles. The primary particles consist of chainlike aggregates which are symbolized as a ball in Fig. 10. Those particles may be more or less liquid and they may contain small aromatic moieties, interconnected by flexible hydrocarbon chains.

During heating of the particles in the flame more and more of their chainlike structure is converted into aromatic rings and compounds, as a consequence of their extraordinary(aromatic) stability. The polyaromatic compounds can detach from the particles when they layat the surface and all chemical bonds are saturated. Those are the PAHs found in the gas phase. PAH synthesis therefore takes place on the particles using their surface as a template. As heating continues the particles form increasingly larger arrays of aromatic rings, becoming more and more graphitic (and therefore denser). Thermal graphitization of carbon-soot is described in Ref. 39. As a consequence the PAHs which desorb into the gas phase are larger. In the same time the particles quickly coagulate and simultaneously pick up small radicals and acetylene for surface growth. The irregular aggregate structure of soot particles is attributed to coagulation. In Fig. 10 the evolution of a spherical primary particle is shown. This picture of soot formation is supported by the mass-spectroscopic experiments. The counts at non-integral mass numbers observed in the broad distribution of soot particles in the high-mass range of the spectrum are an indication of fragmentation inside the TOF-MS. This fragmentation might be the consequence of PAHs desorbing from the particles in the ion source of the mass spectrometer. The desorption process could smear out the individual mass peaks, resulting in the observed, unstructured, hill-like distribution. Also, the observation of fullerenes (Fig. 3) is the result of soot particle heating, as discussed in Sect. I. In order to produce fullerenes, the soot particles have to be irradiated strongly with UV light (see Table I). Consequently, the particles become very hot and fullerenes desorb from their carbonaceous core. The fullerene mass peaks are broad and it has been demonstrated above that this feature is due to their formation from particles in the ion source of the TOF-MS. In analogy to PAHs, fullerenes are formed at the soot particle surface through heating, although to higher temperatures which are not reached in the diffusion flame.

Acknowledgment

This contribution is dedicated to my father, Hans Christoph Siegmann, on the occasion of his 65th birthday celebration.

I wish to thank H. C. Siegmann for many helpful and stimulating discussions. K. Sattler contributed much to the material presented here. I am grateful to H. Hepp and G. Skillas for material from their theses. I wish to thank the group of T. P. Martin for letting us use their instrument. Finally, thanks to all members of the Laboratory of Combustion Aerosol and Suspended Particles for their support.

References

1. C. K. Brain and A. Sillen, *Nature* **336**, 446 (1988).
2. R. E. Mitchell, A. F. Sarofim and L. A. Clomburg, *Combustion and Flame* **37**, 227 (1980).
3. P. McGeer and E. Durbin, *Methane: Fuel of the Future* (Plenum Press, New York, 1982) p. 101.
4. R. Isaacson, *Methane from Community Wastes* (Elsevier, New York, 1991) p. 1.
5. R. Puri, T. F. Richardson, and R. J. Santoro, *Combustion and Flame* **92**, 320 (1993).
6. G. M. Faeth and Ü. Ö. Köylü, *Combust. Sci. and Tech.* **108**, 207 (1995).
7. A. D'Alessio, A. D'Anna, G. Gambi, and P. Minutolo, *J. Aerosol Sci.* **29**, 397 (1998).
8. P. P. Radi, B, Mischler, A. Schlegel, A.P. Tzannis, P. Beaud, and T. Gerber, *Combustion and Flame* **118**, 301 (1999).

9. P. Weilmünster, A. Keller, and K. H. Homann, *Combustion and Flame* **116**, 62 (1999).
10. H. Burtscher, *J. Aerosol Sci.* **23**, 549 (1992).
11. M. Kasper and K. Siegmann, *Combust. Sci. and Tech.* **140**, 333 (1998).
12. G. Bermudez and L. Pfefferle, *Combustion and Flame* **100**, 41 (1995).
13. H. Righter, W. J. Grieco, and J. B. Howard, *Combustion and Flame* **119**, 22 (1999).
14. T. Baum, S. Loeffler, P. Weilmünster, and K. H. Homann, *Phys. Chem.* **96**, 841 (1992).
15. K. Siegmann, H. Hepp, and K. Sattler, *Combust. Sci. and Tech.* **109**, 165 (1995).
16. J. Ahrens, R. Kovacs, E. A. Shafranovskii, and K.H. Homann, *Ber. Bunsenges. Phys. Chem.* **98**, 265 (1994).
17. J. Boyle, L. Pfefferle, J. Lobue, and S. Colson, *Combust. Sci. and Tech.* **70**, 187 (1990).
18. C. S. McEnally, L. D. Pfefferle, *Combust. Sci. and Tech.* **116**, 183 (1996).
19. C. S. McEnally, L. D. Pfefferle, R. K. Mohammed, M. D. Smooke, and M. B. Colket, *Analytical Chemistry* **71**, 364 (1999).
20. K. Siegmann and K. Sattler, *J. Chem. Phys.* **112**(2), 698 (2000).
21. H. J. Heger, R. Zimmermann, R. Dorfner, M. Beckmann, H. Griebel, A. Kettrup, and U. Boesl, *Analytical Chemistry* **71**, 46 (1999).
22. S. Senkan and M. Castaldi, *Combustion and Flame* **107**, 141 (1996).
23. H. Hepp and K. Siegmann, *Combustion and Flame* **115**, 225 (1998).
24. H. Burtscher, D. Matter, and H. C. Siegmann, *Atmospheric Environment* **27A**, 1255 (1992).
25. H. Burtscher, and H. C. Siegmann, *Combust. Sci. and Tech.* **101**, 327 (1994).
26. B. Schleicher, S. Künzel, and H. Burtscher, *J. Appl. Phys.* **78**, 4416 (1995).
27. G. Skillas, H. Burtscher, K. Siegmann, and U. Baltensperger, *J. Colloid and Interface Sci.* **217**, 269 (1999).
28. M. Kasper, K. Siegmann, and K. Sattler, *J. Aerosol. Sci.* **28**(8), 1569 (1997).
29. M. Kasper, K. Sattler, K. Siegmann, U. Matter, and H. C. Siegmann, *J. Aerosol Sci.* **30**, 217 (1999).
30. R. J. Santoro, H. G. Semerjian, and R. A. Dobbins, *Combustion and Flame* **51**, 203 (1983).
31. U. Zimmermann, N. Malinowski, U. Nher, S. Frank, and T. P. Martin, *Z. Phys. D* **31**, 85 (1994).
32. H. Bockhorn, F. Fetting, and H. W. Wenz, *Ber. Bunsenges. Phys. Chem.* **87**, 1067 (1983).
33. H. Bockhorn (Ed), *Soot Formation in Combustion, Mechanisms and Models* (Springer, Berlin-Heidelberg-New York, 1994).
34. Ch. Hüglin, J. Paul, L. Scherrer, and K. Siegmann, *J. Phys. Chem. B* **101**, 9335 (1997).
35. J. T. McKinnon and J. B. Howard, *Twenty-Fourth Symposium (International) on Combustion/The Combustion Institute* 965 (1992).
36. J. T. McKinnon and J. B. Howard, *Combust. Sci. and Tech.* **74**, 175 (1990).
37. P. Minutolo, G. Gambi, A. D'Alessio, and S. Carlucci, *Atmospheric Environment* **33**, 2725 (1999).
38. R. A. Dobbins, R. A. Fletcher, and H.-C. Chang, *Combustion and Flame* **115**, 285 (1998).
39. G. Ertel, H. Knörzinger, and J. Weitkamp, in *Handbook of Heterogeneous Catalysis* (VCH, Weinheim, 1997).

Photoemission Applied to Volcanology: Another Idea of Hans Christoph Siegmann

Giuseppe Faraci

Dipartimento di Fisica
Università di Catania
ITALY

In the late spring of 1985 Hans Christoph Siegmann came to Catania for a seminar. Argument of the seminar was the possible application of the photoemission technique to detect submicron particles in the plume of a volcano, a novel idea of Hans Christoph for eruption control and forecasting.

Catania lies at the foot of the Etna volcano, and in the local university a group of volcanologists looks after Etna.

I was involved by Erio Tosatti who was in Zurich at that time and called me by phone asking if I could organize the seminar with an audience of physicists and volcanologists. With enthusiasm I accepted the plan, since one of my fields of interest is photoemission. Few days after the phone call, I fetched Hans at the Catania-Fontanarossa airport. We did not know each other, so I was keeping in my hands an issue of *Physics Today* to be identified.

The seminar was very interesting and several questions were arised by a number of sceptic people. One of the geologists, Vittorio Scribano, was particularly interested and together we organized a visit to the craters of the volcano who, as often is the case, was emitting ash, smoke and some lava from the main crater.

Five persons left Catania for the top of the mountain after the lunch; Hans, his wife Katrina, Vittorio, my daughter Carla, ten years old, and myself. Vittorio drove his car till the altitude of about 2300 m; we climbed up to the south east crater (about 3050 m) to watch the boiling lava inside the volcano's mouth (See picture 1).

In Fig. 2 is visible the crater cone and fluid lava at 1000°C. A fantastic view. At the sunset we begun our way downhill, overtaking some small lava brooks which in the meantime had crossed our road.

At about nine p.m. we were again in town. Few days after Etna was heavily erupting for about a week.

Physics of Low Dimensional Systems
Edited by J. L. Morán-López, Kluwer Academic/Plenum Publishers, New York 2001

Picture 1. Etna south-east crater. Hans Christoph Siegmann with Giuseppe Faraci (on the right).

Picture 2. Etna Volcano in spectacular performance.

This visit represented the beginning of a joint research mainly conducted by Heinz Burtscher and Markus Ammann; several trips to Etna followed, with very nice results. See the references for further details.

References

1. H. Burtscher, P. Cohn, L. Scherrer, H. C. Siegmann, G. Faraci, A. R. Pennisi, V. Privitera, R. Cristofolini, and V. Scribano, *J. Volcanol. Geotherm. Res.* **33**, 349 (1987).
2. M. Ammann; Ph.D. Thesis, ETH, Zürich 1992, and references therein.

From Nanoparticles to Health Effects

Reinhold Wasserkort

Division of Bioengineering and Environmental Health
Center for Environmental Health Sciences
Massachusetts Institute of Technology
21 Ames St, 16-743, Cambridge, MA 02139
USA

Abstract

Nanoparticles, or ultrafine particles in the submicrometer range, are omnipresent in urban air. The particles originate predominantly from the combustion of fossil fuels from automobiles, home heating and power generation. Human exposure to these particles is therefore almost unavoidable and several health impacts have been associated with such exposure. Among them are the increased risk for lung cancer, mortality and morbidity due to cardiovascular and pulmonary disease and an increase in asthma incidence. This paper aims to provide a short overview of current knowledge of air pollution related health effects stretching from the exposure, via biological targets, to the question of personal susceptibility and risk. Special emphasis will be placed on the hypothesized link between nanoparticles and lung cancer.

I. Introduction

Suspended particulate matter is a nearly ubiquitous urban pollutant. The distribution of particle sizes ranges from smaller than 10 nm to larger than 100 μm. These particles are not distributed evenly over this size range but have been found to occur in a so-called trimodal size distribution, consisting of a nucleation mode (up to ~ 50 nm), accumulation mode (~ 50 nm to $\sim 2\,\mu$m) and coarse particulate fraction (larger than $2\,\mu$m).[1] The predominant source of the particles in the submicrometer range (*i.e.* nanoparticles) is the combustion of fossil fuels. The coarse particle fraction in contrast comprises particles of different origins, such as dust, tire wear, large diesel soot, fly ash, and salt particles.

While suspended particles may have a variety of effects ranging from impaired visibility to an increase of the albedo, the concern about possible human health impacts of particles is a major motivation for the research into nanoparticles.[2] This anthropocentric view has led to additional classifications of the particle sizes which reflect primarily the inhalability of aerosols. While particles with an aerodynamic diameter of more than 10 μm are effectively filtered in the nose the smaller particles may enter the respiratory tract. Therefore particulate matter up to 10 μm (PM10) is also called the inhalable particle fraction and this has therefore long been the size standard for measurements of particle concentrations and exposure assessment.[3] More recently, however, the focus has shifted to particles with smaller diameters, *i.e.* those with aerodynamic diameters of up to 2.5 μm (PM2.5) as this fine particulate fraction was found to be predominantly correlated with observed health effects, rather than the coarse particle fraction (EPA 1999).[4,5] The overview that follows below will primarily focus on the hypothesized link between exposure to nanoparticles and lung cancer. This focus has been chosen because lung cancer is among the leading causes of deaths in western countries and air pollution is often considered a major factor contributing to the extend of lung cancer related deaths. The last chapter, however, will briefly discuss mortality due to cardiovascular disease and asthma as important public health concerns that have also been linked to nanoparticle exposure.

II. Particle Uptake

Upon entry into the respiratory tract particles will follow the air flow but will become deposited along the tracheal, bronchial or alveolar walls depending on the size of the particles. Particles in the range between 2.5 and 10 μm deposit primarily in the upper bronchi, while those with smaller diameters may become deposited anywhere along the airway bifurcations and may disperse into the alveoli, which are the terminal airspaces and are functionally the location of the actual gas exchange.

The physical mechanisms of particle deposition in the airways are basically the same than those in technical systems that are frequently utilized to measure particle sizes: impaction, gravitational settling and diffusion (Fig. 1). The first two modes are important for the larger particles while the ultrafine particle fraction predominantly settles by means of diffusion. At the same time, however, it should be noted that not all particles that enter the respiratory tract are also deposited. Especially the nanoparticles, *i.e.* those particles that are generally not deposited by impaction, can again be exhaled and it has been estimated that the overall deposition efficiency of these particles is less than 20%.[1]

III. Bioavailability and Xenometabolism

The fate of the deposited particulate matter as well as of the material adsorbed to the particle surface depends on the location of the deposition at the inner lining of the respiratory tract (which are epithelial cells) and the biological mechanisms activated in response to this deposition.

If the deposition occurred in the bronchi or bronchioles, an active ciliate mechanism may remove the particles relatively quickly from the respiratory tract (within hours to days). Those particles, however, that where deposited in the alveoli cannot be removed with such an active clearance process as the alveoli are void of any cilia. Here, macrophages (which belong to the cellular guardians of the immune system) may ingest those particles and remove them in a very slow process (up to a year or longer). Ultrafine particles that penetrated the epithelial membranes might become trapped in the interstitial space of the alveoli where only intracellular transport mechanisms are active.[6]

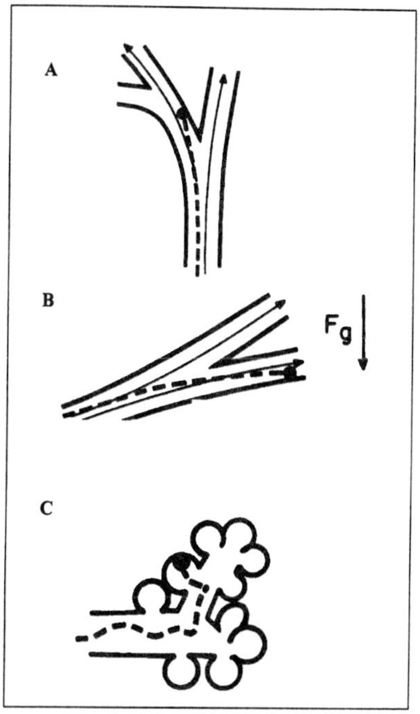

Figure 1. Deposition modes of particles in the lung: (A) impaction; (B) gravitational; (C) diffusion (modified after Raabe[6]).

Particle deposition may lead to mechanical damage of the cells that are affected but also oxidative processes due to particle adsorbates can damage the cell membranes, both of which may cause immediate response such as inflammation, injury or cell death.[7]

After the particles are ingested by macrophages or the particle adsorbates are taken up by epithelial cells they become subject to enzymatic attack. So-called xenometabolic enzymes, with the task of converting chemicals of non-biological origin (hence, xenobiotics) into polar molecules that are more soluble and therefore can be excreted more easily. Many of these enzymes belong to the large family of cytochrome P450 dependent oxygenases, with each member reacting specifically with certain groups of xenobiotic chemicals. CYP1A1 is one of these enzymes that modifies especially polycyclic aromatic hydrocarbons (PAH), a larger class of organic chemicals that is typically adsorbed to the surface of combustion particles.[8,9] However, some of the metabolic intermediates become reactive electrophilic epoxides that may react with other components of the cell, such as proteins and nucleic acids. The best studied example of these intermediates is benz[a]pyrenediolepoxide (BPDE), derived from the hydroxylation of benzo[a]pyrene. BPDE preferentially binds to the 2-amino position of the guanine bases in the DNA, forming a DNA adduct (see Fig. 2).

IV. DNA Adducts

DNA adducts represent a form of DNA damage that in itself is not deleterious but may potentially lead to mutations, *i.e.* to a change of the genetic information and therefore an error in the 'blueprint' of the cell. This could happen while the DNA is

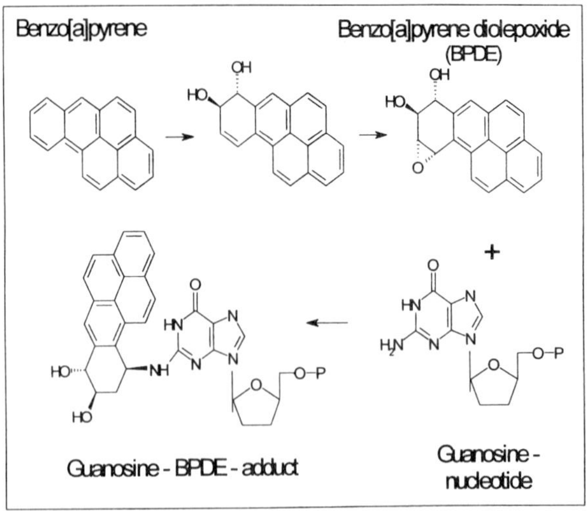

Figure 2. Xenometabolism of benzo[a]pyrene, catalyzed by the enzyme Cyp1A1.BPDE is the ultimate carcinogen that interacts with DNA.

replicated, *i.e.* prior to each cell division, and the covalently modified base (*i.e.* the DNA adduct) leads to misreading, fall-off or other errors by the DNA polymerase delta, the main enzyme responsible for DNA replication in human cells.[10,11] Such mutations are important as they may lead to abnormal cell behavior which in turn may lead to disease. Cancer is but one important example of such diseases and virtually any known cancer derives from cells the genetic information of which has been changed by inherited or somatically acquired mutations. Mutations that are inherited are present in all cells of that person while somatic mutations in general occur only in those cells or tissues in which the mutation originated.

V. DNA Repair

An important question here is whether or not all DNA adducts that occur in a cell will also lead to mutations. As an approximation one can compare the numbers of adducts in human tissues on the one hand and on the other the somatic mutations rates in human cells. Adduct occurrence seems to outnumber the mutation rate by several orders of magnitude. This is expected from taking into account the experimentally determined number of individual adduct species and the fact that many of such adducts coexist at the same time. Basically all techniques capable of detecting DNA adducts will do so only for a few selected species of adducts but will not be able to determine the complete spectrum of adducts. Therefore any determination of adduct load so far detects only a subfraction of adducts.[12,13] An estimation of the total number of adducts/cell could be well above 10 000, which translates to $\sim 1.5 \times 10^{-6}$ adducts/basepair. (based on B[a]P adducts found and extrapolated to consider non-B[a]P adducts as well.) The somatic mutation rate in human cells on the other hand has been estimated at $\sim 5 \times 10^{-11}$ mutations/basepair/cell division.[14] This clearly represents a difference of several orders of magnitude. How can this discrepancy be explained, especially since in experimental systems a link between DNA adducts and particular mutations is found?[15,16]

The most important factor to be considered here is DNA repair. Adducts are recognized and mostly removed by an intricate cellular damage repair system. Among the several repair systems that exist in cells nucleotide excision repair is the one that recognizes PAH adducts.[17] DNA glycosylases identify the damage and other enzymes (endonucleases, DNA polymerases, DNA ligases in conjunction with several accessory proteins) act in concert to excise the damaged base, remove the abasic site and fill the resulting gap complementary to the information from the opposite DNA strand and finally link the replaced nucleotide with the neighboring ones to fully restore the DNA strand.[18]

It could be speculated that incomplete recognition of DNA adducts could still be responsible for mutations due to the PAH adducts. Denissenko *et al.*[19] actually describe that a slow repair of such adducts could be used to explain the strand bias found between transcribed and untranscribed strands in the DNA. An alternative or additional mechanism, however, might be that the DNA repair mechanisms itself is not error free. Most notable are the DNA polymerases responsible for filling the gap after the damaged base has been removed. These enzymes have a comparatively low fidelity and their error rate has been estimated at about 10^{-4}.[20] This means that for every 10 000 nucleotides replaced there would be one that would be an incorrect replacement. In this case the damage would become manifest, *i.e.* create a mutation, as the DNA repair system may not be able to repair itself.

VI. Mutational Spectrometry

As there are different pathways that eventually may lead to a mutation the important question becomes whether or not such pathways can be distinguished from each other. Being able to make such distinctions is more than merely a matter of scientific curiosity as it may help to unravel the real causes for those mutations suspected to lead to a cancer. This is particularly important in cases like lung cancer where both exogenous (DNA adducts) and endogenous pathways (error prone DNA repair) may lead to the oncomutations that give rise to cancer.

A technology that is potentially able to uncover the origin of mutations is mutational spectrometry. This technology is build upon the original observation by Benzer and Freeze in the fifties that any mutagen creates a specific set of mutations that can be as characteristic for this mutagen as a fingerprint is for a criminal suspect.[21,22]

In this concept, mutations that were created by DNA adducts would lead to a different set of mutations than those that result form misincorporation errors of DNA polymerases. Using this approach and a novel methodology to detect rare mutations in human tissue samples Coller *et al.*[23] determined the mutational spectra in a mitochondrial sequence in smokers and non smokers. The results indicated that both spectra were virtually the same and that the variation detected in these spectra were not bigger than the variation found within one person alone. These data suggest that smoking did not contribute to the mutations found in the DNA analyzed in this study. Though this finding is highly suggestive, it does not prove that cigarette smoking is not to be made responsible for the mutations leading to cancer. This is because the oncomutations arise in nuclear DNA rather than mitochondrial DNA. Detecting mutations in the nuclear DNA, however, is inherently more difficult than those in mitochondrial DNA as the latter is present in as many as several thousand copies per cell while there are only two copies of nuclear DNA per cell. Therefore an answer to the question of what is causing the mutations that lead to lung cancer cannot be given with the results available at this time.

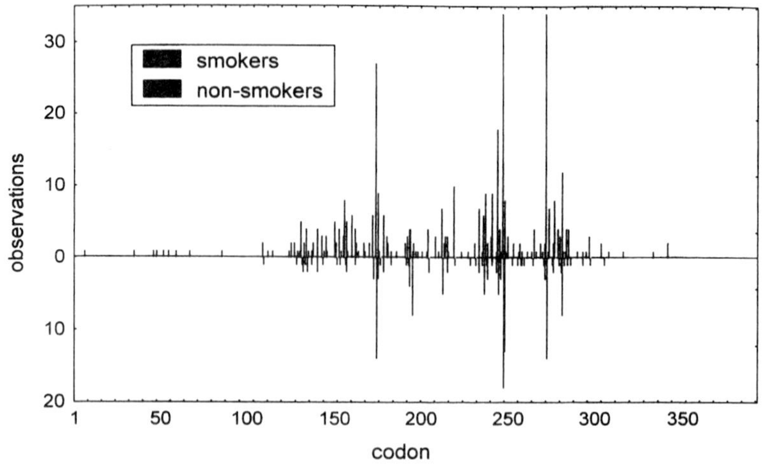

Figure 3. Mutational spectrum of the TP53 gene. Mutations were compiled from 254 smokers (upper panel) and 123 non-smokers (lower panel). Data obtained from the IARC darabase.[25]

Nevertheless, a valuable source of information is present in mutational databases that contain the information form all mutations within a specific gene discovered by different laboratories. An extensive collection of mutations is available for mutations occurring in the tumor suppressor gene TP53. The gene is located in the nuclear DNA (chromosome 17). This is a widely studied gene and it was found that this gene is mutated in more than 50% of all human cancers.[24] Figure 3 shows a compilation of those mutations found in people with lung cancer and for which the smoking status was known.[25]

Comparison of the two spectra indicates that the hotspot mutations, *i.e.* mutations that occur more frequently than by chance are basically identical between smokers and nonsmokers. This would suggest that the pathways leading to such mutations would be the same in both smokers and non smokers. As a consequence it appears rather unlikely that cigarette smoking would lead to those mutations that eventually lead to lung cancer. What is limiting the usefulness of this compilation, however, is that all mutations reported here have been found in people that already developed lung cancer (both in smokers and non smokers). Therefore the similarity of the mutations could also simply be a result of the cancers rather than the cause leading to cancer. To unambiguously determine which mutations, or mutational hotspots, are to be made responsible such spectra have to be compared from healthy and noncancerous lungs of both smokers and non-smokers. Nevertheless, as smoking has been clearly demonstrated to be causally related to lung cancer, the question would be whether or not this link has to be via mutations or whether alternative mechanisms might exist. One plausible but untested possibility would involve an effect of components of the cigarette smoke that may directly (*i.e.* without involving mutations) alter the growth characteristics of lung epithelial cells.[22] Such a mechanism is conceivable also because of the observation that cessation of smoking will reduce the risk of getting lung cancer.[26]

VII. Mutations and Cancer

While understanding the origins of onco mutations is clearly important this alone may not be sufficient to determine the individual risk of getting lung cancer. Lung cancer is, like basically all other cancers as well, a multistep disease in which several mutations have to be acquired in particular genes in order to reach the malignant, and thus fatal, state.[27] Genes that are important in this multistep model of cancer are the so-called tumor suppressor genes, an example of which is the TP53 gene already mentioned earlier. The product of the TP53 gene is the p53 protein which is a multi-functional transcription factor involved in the control of cell cycle progression, DNA integrity and cell survival in cells exposed to DNA damging agents,[28] Other important genes are *e.g.* those that are involved in cell-to cell signaling or those that determine the cellular growth rate.

As human cells contain two copies per autosomal gene a mutation that knocks out one of the important genes may not have any immediate consequences, as the second copy of the gene is still functional. However, if a second mutation knocks out this remaining copy as well, the function of this gene can no longer be executed. Assuming that two of such mutations would have knocked out the gatekeeper TP53 gene, the cell in which theses mutations happened would have lost its ability to commit suicide. Such a suicide is a genetically programmed 'last exit' (apoptosis) for cells that have acquired too much damage and therefore cannot function anymore normally. Having a single cell die is therefore a better strategy for the whole tissue than coping with a malignent cell. The loss of a gatekeeper gene could be compared to the loss of the CEO in an organization such that the important "stay alive and grow or it is time to go out of business" decision cannot be made anymore and the program is stalled in its default mode that leads to further growth of the cell. If in such a cell (which has lost the CEO and is no longer able to option for the 'last exit') any additional mutations that may disrupt the normal cell signaling, or provide the cell with growth stimuli could potentially lead to a more rapid growth of this cell than it normally should, the cell may get out of balance and convert to a rapidly growing cell cluster within the surrounding normal tissue. A carcinoma would be born.

Within this model (Fig. 4) it becomes clear that any person who would already have inherited one or even more of those mutations required to lead to that particular cancer would have be a 'person at risk'.[29] The probability of acquiring the additional mutations necessary for the completion of this cascade within a lifetime would be much bigger than when ALL of these mutations would have to be acquired irrespective of which mutational pathway causes the individual mutations.

A consequence of this model therefore is that the "risk" of a person to acquire a particular cancer would largely depend on the genetic predisposition of that person rather than on, *e.g.* the exposure to potentially carcinogenic substances like PAH adsorbed to nanoparticles. However, it is conceivable that the latter would contribute to the total mutational load and thereby accelerate the carcinogenic process even though it would not be the sole determinant.

VIII. Polymorphisms

Additional risk factors, which also depend on the genetic predisposition of a person, would come from genetic polymorphisms that may render certain proteins or enzymes either more or less efficient compared to the normally functioning enzyme.

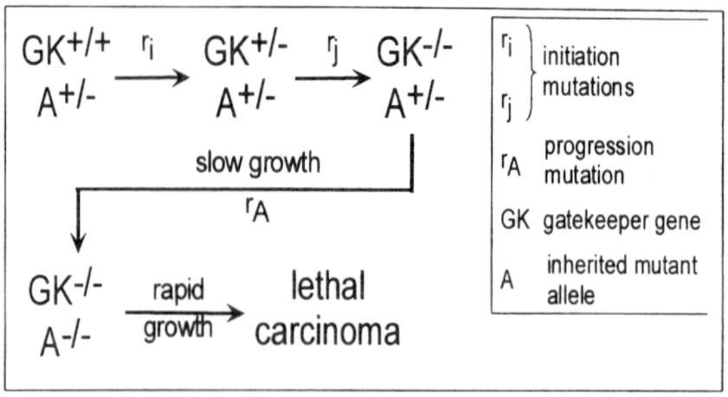

Figure 4. Multi-step model of cancer. Modified from Tomita-Mitchell *et al.*,[30] and Herrero *et al.*[29] This model was originally designed for colon cancer. The exact number of mutations required to lead to lung cancer is not known.

Such polymorphisms have been shown to occur, *e.g.*, in the xenometabolising enzymes mentioned above and the susceptibility of the affected people to cigarette smoke was higher than in people without this polymorphism. Other important enzymes would, *e.g.*, include the DNA repair enzymes and the DNA polymerases as the efficiency of such enzymes may very well influence the overall mutation rates. Any polymorphism in one or more of such genes may cause a person to be more or less susceptible to either a particular disease or to particular exposures and thereby put this person into the 'at risk' category. Research aimed at identifying the kinds of polymorphisms most important to certain diseases is currently under way in many laboratories.

IX. Nanoparticles and Non-Cancerous Diseases

Several epidemiological studies in the last decade have suggested a correlation between air pollution, especially the fine particulate fraction, and an increased daily mortality due to cardiovascular disease.[31] In a long term follow-up study conducted in six US American cities air pollution levels were monitored and the effects on hospitalization rate, mortality and pulmonary disease were monitored.[32,33] As a result of this study increases in daily mortality due to cardiovascular and pulmonary diseases were found to be correlated to particulate air pollution levels.

The connecting mechanisms behind this correlation, however, has remained largely elusive. An interesting hypothesis was developed by Seaton *et al.*[34] According to this hypothesis, particles deposited in the airways cause subacute levels of inflammation due to their irritant impact on epithelial cells. This inflammation might be particularly triggered by nanoparticles that are retained in the interstitial space of the alveoli, which is void of any particle removal mechanism. Whether the mechanical properties of the particles alone might be responsible or rather the oxidative lung damage caused by particle associated reactive chemicals is unclear. Nevertheless, the immunological response to this inflammation would then be similar to the response to any other injury as well: the blood coagulability increases due to elevated levels of fibrinogen and factor VII in the blood, as well as sequestration of hemoglobin and red blood cells. In people that are already at risk (like the elderly and people with preexisting diseases) this slightly higher level of coagulability may then lead to cardiovascular episodes which, in the worst case, may be fatal. Fatalities would then occur predominantly at days with increased levels of particle pollution. Even though this model appears plausible it has so far not been possible to verify its validity experimentally.[35]

Asthma is another public health concern as its prevalence has increased dramatically during the last decades.[36] And this increase has also been observed in children of young age. Among the diverse range of factors particle pollution has also been discussed.[37,38] Several observations, however, cast doubt on the assumption that particulates are primarily responsible for asthma. In a study that compared air pollution and asthma prevalence in Germany before and after the unification in 1989 this allergic disease was negatively correlated with air pollution levels.[39] Furthermore with declining levels of air pollution the asthma prevalence nevertheless appeared to increase and this increase was largely ascribed to changes in life style, resulting in larger exposure to common allergens such as house dust mites and a lack of early sensitization.[39] Also, on a global scale asthma has been observed more frequently in correlation with a western lifestyle than in conjunction with air pollution levels. Air pollution has been found generally much worse in countries like India, Taiwan, Uzbekistan, China and Eastern Europe, but childhood asthma predominantly is a problem in countries of the west like Britain, Australia, New Zealand and Ireland.[40,41]

In addition, most incidences of asthma have been reported with a very high degree of heritability that may be as high as 75%.[42] Unless experimental and epidemiological evidence may demonstrate a plausible link between asthma and nanoparticles, this link has to be considered as speculative and so far as little substantiated.

X. Conclusion

The ubiquitous presence of nanoparticles in our atmosphere raises concern about the possible public health impacts of theses particles. Since lung cancer is the leading cause of cancer related death in the West, and cigarette smoking as an important form of air pollution has been strongly associated with lung cancer it is frequently believed that the carcinogen exposure from combustion particles is causing lung cancer. However, considering present experimental evidence, this link has not been conclusively proven. Particle exposure might be considered as an external risk factor only for those people that already have a genetic precondition for cancer. For this risk group, however, particle exposure could be a significant risk factor.

Increased mortality due to cardiovascular disease presumably affects a comparatively small number of people, particularly elderly and those with respiratory preconditions. Nevertheless, it may present a direct effect of particulate exposure. A mechanisms for this link has not been experimentally demonstrated.

According to current observations combustion particles are rather unlikely to be directly responsible for the prevalence of asthma.

References

1. P. J. Lioy and J. Zhang, "Air pollution", *Air pollutants and the respiratory tract*, edited by D. L. Swift and W. M. Foster, (Basel, Marcel Dekker, New York, 1999) p. 1.
2. R. Wasserkort, Ph.D. Thesis, ETH Zürich, 1995.
3. C. A. d. Pope and D. W. Dockery, *Am. Rev. Respir. Dis.* **145**, 1123 (1992).
4. D. W. Dockery, A. C. d. Pope, X. Xu, J. D. Spengler, J. H. Ware, M. E. Fay, B. G. Ferris, Jr., and F. E. Speizer, *N. Engl. J. Med.* **329**, 1753 (1993).
5. EPA, *Environ. Health Perspect.* **106**, A373-4 (1998).

6. O. G. Raabe, "Respiratory exposure to air pollutants", *Air pollutants and the respiratory tract*, edited by D. L. Swift and W. M. Foster, (Basel, Marcel Dekker, New York, 1999) p. 39.
7. W. S. Beckett, "Detecting respiratory tract responses to air pollutants", *Air pollutants and the respiratory tract*, edited by D. L. Swift and W. M. Foster, (Basel, Marcel Dekker, New York, 1999).
8. R. Barale, L. Giromini, G. Ghelardini, C. Scapoli, N. Loprieno, M. Pala, F. Valerio, and I. Barrai, *Mutat. Res.* **249**, 227 (1991).
9. J. L. Durant, A. L. Lafleur, W. F. Busby, Jr., L. L. Donhoffner, B. W. Penman, and C. L. Crespi, *Mutat. Res.* **446**, 1 (1999).
10. F. P. Guengerich, M. S. Kim, M. Muller, and L. G. Lowe, *Recent Results Cancer Res.* **143**, 49 (1997).
11. G. J. Latham, A. G. McNees, B. De Corte, C. M. Harris, T. M. Harris, M. O'-Donnell, and R. S. Lloyd, *Chem. Res. Toxicol.* **9**, 1167 (1996).
12. C. M. Dale and R. C. Garner, *Food Chem. Toxicol.* **34**, 905 (1996).
13. K. Hemminki, C. Dickey, S. Karlsson, D. Bell, Y. Hsu, W. Y. Tsai, L. A. Mooney, K. Savela, and F. P. Perera, *Carcinogenesis* **18**, 345 (1997).
14. S. A. Grist, M. McCarron, A. Kutlaca, D. R. Turner, and A. A. Morley, *Mutat. Res.* **266**, 189 (1992).
15. M. J. Mass, A. J. Jeffers, J. A. Ross, G. Nelson, A. J. Galati, G. D. Stoner, and S. Nesnow, *Mol. Carcinog.* **8**, 186 (1993).
16. Ross and Nesnow, *Mutat. Res.* **421**, 155 (1999).
17. D. Gunz, M. T. Hess, and H. Naegeli, *J. Biol. Chem.* **271**, 25089 (1996).
18. A. Sancar and M. S. Tang, *Photochem. Photobiol.* **57**, 905 (1993).
19. M. F. Denissenko, A. Pao, G. P. Pfeifer, and M. Tang, *Oncogene* **16**, 1241 (1998).
20. W. P. Osheroff, H. K. Jung, W. A. Beard, S. H. Wilson, and T. A. Kunkel, *J. Biol. Chem.* **274**, 3642 (1999).
21. S. Benzer and E. Freeze, *Proc. Natl. Acad. Sci. USA* **44**, 112 (1958).
22. H. A. Coller and W. G. Thilly, *Environ. Sci. Technol.* **28**, 478A (1994).
23. H. A. Coller, K. Khrapko, A. Torres, M. W. Frampton, M. J. Utell, and W. G. Thilly, *Cancer Res.* **58**, 1268 (1998).
24. C. C. Harris, *J. Investig. Dermatol. Symp. Proc.* **1**, 115 (1996).
25. P. Hainaut, T. Hernandez, A. Robinson, P. Rodriguez-Tome, T. Flores, M. Hollstein, C. C. Harris, and R. Montesano, *Nucleic Acids Res.* **26**, 205 (1998).
26. H. Pohlabeln, K. H. Jockel, and K. M. Muller, *Lung Cancer* **18**, 223 (1997).
27. P. Amitage and R. Doll, *Brit. J. Cancer* **8**, 1 (1954).
28. T. Unger, M. M. Nau, S. Segal, and J. D. Minna, *EMBO J.* **11**, 1383 (1992).
29. P. Herrero-Jimenez, G. Thilly, P. J. Southam, A. Tomita-Mitchell, S. Morgenthaler, E. E. Furth, and W. G. Thilly, *Mutat. Res.* **400**, 553 (1998).
30. A. Tomita-Mitchell, B. P. Muniappan, P. Herrero-Jimenez, H. Zarbl, and W. G. Thilly, *Gene* **223**, 381 (1998).
31. D. W. Dockery and C. A. Pope, *Annu. Rev. Public Health* **15**, 107 (1994).
32. F. E. Speizer, *Environ. Health Perspect.* **79**, 61 (1989).
33. F. E. Speizer, M. E. Fay, D. W. Dockery, and B. G. Ferris, Jr., *Am. Rev. Respir. Dis.* **140**, S49-55 (1989).
34. A. Seaton, W. MacNee, K. Donaldson, and D. Godden, *Lancet* **345**, 176 (1995).
35. A. Seaton, A. Soutar, V. Crawford, R. Elton, S. McNerlan, J. Cherrie, M. Watt, R. Agius, and R. Stout, *Thorax* **54**, 1027 (1999).
36. S. T. Holgate, *Nature* **402** Suppl., B2 (1999).
37. P. J. Barnes, *Postgrad. Med. J.* **70**, 319 (1994).

38. E. von Mutius, *Pediatr. Pulmonol. Suppl.* **16**, 86 (1997).
39. E. von Mutius, S. K. Weiland, C. Fritzsch, H. Duhme, and U. Keil, *Lancet* **351**, 862 (1998).
40. W. Cookson, *Nature* **402** Suppl., B5 (1999).
41. H. Williams, C. Robertson, A. Stewart, N. Ait-Khaled, G. Anabwani, R. Anderson, I. Asher, R. Beasley, B. Bjorksten, M. Burr, T. Clayton, J. Crane, P. Ellwood, U. Keil, C. Lai, J. Mallol, F. Martinez, E. Mitchell, S. Montefort, N. Pearce, J. Shah, B. Sibbald, D. Strachan, E. von Mutius, and S. K. Weiland, *J. Allergy. Clin. Immunol.* **103**, 125 (1999).
42. D. L. Duffy, N. G. Martin, D. Battistutta, J. L. Hopper, and J. D. Mathews, *Am. Rev. Respir. Dis.* **142**, 1351 (1990).

Electroencephalograms in Epilepsy: Complexity Analysis and Seizure Prediction within the Framework of Lyapunov Theory

H. R. Moser,[1,*] B. Weber,[2] H. G. Wieser,[2] and P. F. Meier[1]

[1] *Physics Institute*
University of Zürich
CH-8057 Zürich
SWITZERLAND
[*] *e-mail: moser@physik.unizh.ch*

[2] *Department of Neurology*
University Hospital
CH-8091 Zürich
SWITZERLAND

Abstract

Epileptic seizures are defined as the clinical manifestation of excessive and hypersynchronous activity of neurons in the cerebral cortex and represent one of the most frequent malfunctions of the human central nervous system. Therefore, the search for precursors and predictors of a seizure is of utmost clinical relevance and may even guide us to a deeper understanding of the seizure generating mechanisms. We extract chaos-indicators such as Lyapunov exponents and Kolmogorov entropies from different types of electroencephalograms (EEGs). We concentrate on EEGs that originate from intracranially implanted electrodes (semi-invasive and fully invasive recording techniques), which provides particularly "clean" signals in terms of noise-level and stationarity. Among the analytical methods we tested up to now, we find that the spectral density of the local expansion exponents is best suited to predict the onset of a forthcoming seizure. We also evaluate the time-evolution of the dissipation in the EEGs: it exhibits strongly significant variations that clearly relate to the time relative to a seizure onset. We mainly address ourselves to hidden properties in these signals,

Physics of Low Dimensional Systems
Edited by J. L. Morán-López, Kluwer Academic/Plenum Publishers, New York 2001

e.g., changes that indicate a seizure cannot be detected by a visual inspection. Further, we investigate interictal EEGs (i.e., far away from a seizure) in order to characterize their more general properties, such as the convergence of the reconstructed quantities with respect to the number of phase space dimensions. Finally, we discuss our results within the general context of complex dynamical systems.

I. Introduction

The human brain may be considered to be a complex system in the natural sense of this term, i.e., it is made up of enormously many constituents and sub-constituents that interact in a complicated way. This general point of view immediately raises the question whether the accumulated knowledge about complex laboratory systems still applies, or whether biological systems are different in every respect. In this contribution we shall exemplify that certain general findings within the "theory of complexity" prove true also in the context of brain research.

We focus on a quantitative characterization of electroencephalograms (EEGs), since they still offer one of the most important tools to study the many facets of neuronal dynamics. The human EEG has been investigated using a wide range of different time series analysis techniques (for a review see Ref. 1) To far extent this means the extraction of time-independent characteristic quantities, preferably dynamical invariants. In recent years, nonlinear time series analyses have been applied in various areas of EEG research. It has been shown that chaos-indicators such as Lyapunov exponents (LEs) (see, e.g., Ref. 2) and correlation dimensions (e.g., Ref. 3) show characteristic changes as a function of certain physiological and pathological brain states.

These chaos-indicators, particularly the Kolmogorov entropy (see below), are generally considered to be complexity measures. Since a measured or calculated chaotic trajectory admits prediction only over a short time-interval, the underlying dynamics is "hard to understand". In oversimplified terms, the natural imagination that "complex" means complicated then is recovered as a determinable quantity, namely the average rate of information loss. (Apart from the links between attractor topologies and LEs, in the context of EEGs we preferably look at the dynamical aspects of this subject matter.) Further below (in Sec. III.3) we shall discuss what this means with regard to the number of phase space dimensions.

Complexity measures have also been used to analyze EEGs recorded from epileptic patients. A general finding is that the analysis of seizure activity results in strongly decreased correlation dimensions[4] and reduced largest LEs.[5] It has also been shown that the correlation dimensions of interictal EEG recordings (i.e., far away from a seizure) have the power to localize the seizure onset zone in so-called temporal lobe epilepsy.[6,7] Further, Lehnertz and Elger have shown that drops in the correlation dimensions of intracerebral EEG recordings seem to be more prominent a few minutes prior to seizures when compared with interictal EEGs.[8,9]

The aim of the present study is to examine our EEG recordings by means of "Lyapunov theory", where we concentrate on both preictal changes (i.e., EEG-changes a few minutes prior to a seizure onset) as well as on more general quantitative properties of the EEGs under consideration. In our view, this has not yet been carried out in detail. At present, we are constrained to analyze the EEGs from only a few individuals, which is explained further below. Nevertheless, our methods permit to detect the preictal changes in visually very similar signals, although the relative incidence of an observed kind of change stays unknown at this stage. Further, it has been known for a long time that seizure prediction cannot be based only on the density of visible epilepiform patterns in the EEG: a thorough investigation of the practicable computational methods is desirable.

Besides an investigation of the advantages and shortcomings of the different extracted quantities, we address the question whether our obtained complexity measures and dissipations may lead us to practical applications, *e.g.*, warning systems for a forthcoming seizure. This then would even strongly enlarge the importance of EEG changes prior to epileptic seizures.

II. Data Acquisition

In the present contribution we exemplify our methods by means of EEG recordings originating from intracranially placed electrodes in a patient suffering from mesial temporal lobe epilepsy. We analyze the signals from bilaterally implanted so-called foramen ovale (FO) electrodes, *i.e.*, an electrode on either temporal lobe of the patient (left, right). Further, we present a result based on an 8×8 subdural grid electrode (10 mm spacing between neighboring contacts). The grid is placed over the left frontal cortex in a patient suffering from frontal lobe epilepsy.

The semi-invasive FO electrodes[10] yield excellent intracranial recordings without the need of a burr hole and are mainly applied in the presurgical evaluation of patients with a potential mesial temporal seizure onset. Each electrode consists of 10 contacts that we use to obtain 9 bipolar signals of neighboring contacts. This modified version of the previously used FO electrodes was especially designed for our studies in order to suffice the requirements of the analyses. The placement of a grid electrode requires craniotomy, *i.e.*, the standardized surgical opening of the skull, and thus provides the most invasive recording technique investigated in this work.

On the hardware side, the FO recordings were performed using a Telefactor beehive telemonitoring system. The data were band pass filtered (0.1–70 Hz) and, after 12 bit A/D conversion, digitally stored with a sampling frequency of 200 Hz. The subdural recording was stored with a sampling rate close to 184 Hz, 0.53–85 Hz band pass filtering, and again a 12 bit A/D precision. Clearly these sampling rates discard the high frequencies that, expectedly, are also there. However, to our present knowledge, this is of lesser importance. Part of the reason for this may be seen in the comparatively slow conduction mechanisms in the brain.

The commonly used scalp recordings from the surface of the skin (not shown here) are non-invasive and therefore clearly the most convenient ones. But, unfortunately, the quality of the scalp-EEG data is attenuated due to the large distance between the electrodes and the source of the brain electric activity. On the other hand, FO (and significantly more so the subdural grid electrodes) require neurosurgical procedures, but they yield an outstanding quality of EEG recordings.

III. Data Analysis

III.1. Methods

EEGs may be interpreted as an overall activity, *i.e.*, an integral output of an enormously complex underlying neuronal dynamics. Therefore it appears highly questionable whether there exists an analysis method that traces an EEG back to the signal-generating origin. In view of this difficulty, it does not astonish that most analyses concentrate on a different goal, namely the characterization of an EEG in terms of time-converged indicators such as fractal dimensions or LEs, that may be attributed to different brain states of a person (at different times, *i.e.*, within separate time-windows), to distinct areas in the brain, or to different individuals.

One of the crucial points in this type of analysis is the question whether the dominating features of an EEG can be described by means of a sufficiently small number of variables (generally the number of phase space dimensions) that numerically still can be handled. Besides the neuronal complexity, it is the presence of noise that renders this question notoriously delicate. The amount and type (as well as the origin) of the noise in an EEG is not known, at least to far extent. Various filtering methods may help us to end up with converged characteristic quantities, but this alters also the underlying "true" signal. Further, part of the noise may well be relevant to the physiological brain state of a person, and so it is even questionable whether we should try to get rid of it at any cost.

This contribution is thought to be mainly of methodological interest, rather than to be adequate to the many facets of epilepsy: we investigate EEGs from only two patients that suffer from mesial temporal lobe epilepsy and from frontal lobe epilepsy, respectively. Further, on the here presented computational level we have not yet investigated a large number of patients, which prevents us from drawing statistical conclusions. This has several reasons. Firstly, a 10-polar FO electrode has been newly constructed for this study (see above), and so there are not that many data available at present. Secondly, we observed that close to a seizure onset the extracted quantities strongly depend on the pathological specifications of the patients: therefore, single case studies are not necessarily less relevant than a statistics involving a broad population of individuals. Finally and most importantly, not every measurement yields sufficiently long stationary EEG segments as we intended to use in this rather analytical approach.

Regarding the complex structure of an EEG, it appears to be desirable to combine as many as possible different analysis methods, since they generally focus on different aspects of the underlying dynamics. Thus, one hopes that additional approaches help to assemble a more complete picture. *e.g.*, the well-known power spectrum proved to be specific to various brain states such as sleep stages, while surrogate data techniques mainly tackle the question of determinism *versus* stochasticity (see, *e.g.* Refs. 11–14). There is a multitude of contributions that determine attractor dimensions, many of the pertinent references may be found in Refs. 13 and 15. More recently, a characterization of isolated, local events by means of wavelet theory gained interest, in a context of EEGs in epilepsy, see Ref. 16.

However, there is not so much literature dealing with LEs extracted from physiological time series, some of the "exceptions" may be found in Refs. 2, 5, and 17. Particularly few contributions present a whole Lyapunov spectrum, *i.e.*, as many LEs as phase space dimensions, see, *e.g.* Ref. 2. One of our goals is a contribution to bridge this gap. We calculate Lyapunov spectra as well as related quantities, namely dissipations and Kolmogorov entropies,[18] and spectra of local expansion exponents.[19] Concerning seizure prediction, the latter proved amazingly discriminative with respect to different time-windows of a preictal signal.

The LEs (and therefore also all the related quantities listed above) have been calculated according to the method of Benettin *et al.*,[20] and we use the well-known Gram-Schmidt orthogonalization procedure in order to obtain also the exponents other than the principal (*i.e.*, largest) one. This method is based on a direct evaluation of distances, namely at every mesh-point of a signal it measures to what extent initially nearby trajectories are separated after a time-step. The essentials of our calculation (as well as many of the numerical pitfalls) are outlined in Ref. 21.

Since in the present applications we never experienced that the LEs depend on the initial conditions (*i.e.*, all the points of a visually stationary EEG segment seem to belong to the same "basin of attraction"), we consider the Kolmogorov entropy to be the sum of the positive LEs. Note that this quantity comprises all those directions in phase space that in a time average are attended with an expansion, and so it is actually the hallmark of what we call complexity. In many contexts (but not always), a definition that characterizes the dynamical aspects suits the situation better than an attractor's topology or dimension (*e.g.*, in the case of Hamiltonian chaos an attractor does not even exist). The dissipation, then, is the sum of all the d LEs λ_i, d being the number of phase space dimensions.

A major part of our analyses is based on spectra of local expansion exponents. Essentially, these spectra are calculated the following way. The standard method to evaluate LEs[20] requires a reorthonormalization of d direction vectors at time-points t^n that are equidistantly separated by an interval τ (in our case the sampling interval). These vectors are the directions along which a local separation or contraction of nearby trajectories is measured. The corresponding local expansion exponents γ_i^n then are summed up in d histograms $N(\gamma_i)(i = 1, \ldots, d)$, where N is the spectral intensity. Note that the γ_i^n are the quantities that in a time average lead to the LEs λ_i. The resolution of a spectrum is given by the width of a column in the histogram and must be adapted to the size of the occurring structures. This way a given input signal results in a set of d spectra that reveal statistical properties of the local flow as a function of time, although these spectra are time-independent quantities. In Ref. 19 we illustrate their properties by means of greatly different model systems.

III.2. Seizure Prediction

We proceed now to our first example. We analyze the preictal (30–0 minutes before a seizure onset) and ictal EEGs from a patient suffering from a clearly unilateral left mesial temporal lobe epilepsy. The patient underwent a presurgical evaluation including long-term FO EEG monitoring. The EEGs were recorded using a 10-polar version of the FO electrode, see above. Since we perform multichannel reconstruction, the electrode fixes the number of phase space dimensions d, which is followed by important consequences that we shall discuss further below.

Both electrodes (left, right) lead to 9 bipolar channels of neighboring contacts each. This avoids the problem to define a reference potential (such as, *e.g.*, the common average of all the available contacts). During time-windows of 90 s, *i.e.*, 18000 data vectors, we assume the signals to be stationary. This was confirmed by visual inspection in order to avoid obvious artifacts and unequal occurrence of epileptiform patterns, and also by separately analyzing the first and second half of the time series. Additional filtering (besides the further above mentioned band pass filter on the hardware side) is carried out by means of a convolution with a Gaussian function (FWHM = 4 mesh-intervals). This type of filter is simple and well-defined, and it introduces no additional free parameters apart from the strength of filtering. All the 9 signals on either side then are analyzed simultaneously in order to span the two phase spaces (no delay-embedding).

Figure 1 displays five spectra of local expansion exponents originating from different time-windows according to the caption (primary epileptogenic area, which in this

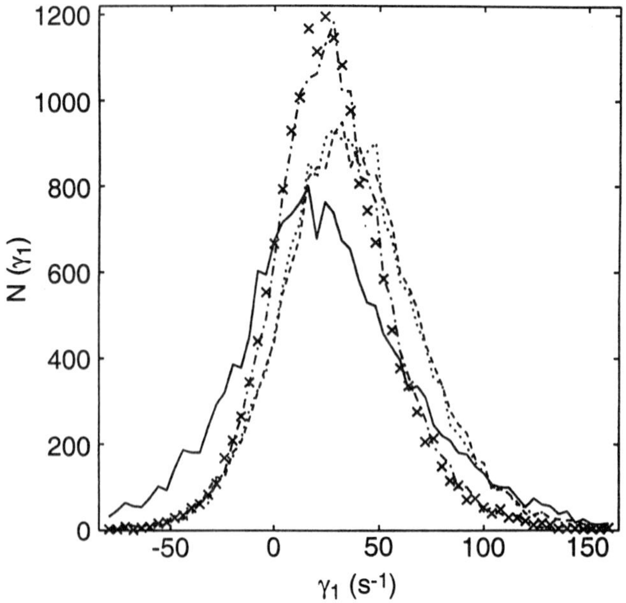

Figure 1. Five spectra of local expansion exponents originating from different time-windows of the left FO electrode (primary epileptogenic area). (– – –): around 28 min before seizure onset; (···): around 14 min before onset; (– · – · – ·): around 4 min before onset; (x x x): the interval 95–5 seconds before onset; solid: within the ictal phase, *i.e.*, during the seizure. The spectra are made up of those local exponents that in a time average lead to the largest respective Lyapunov exponents. At fixed resolution, the spectral intensity $N(\gamma_1)$ is considered to be a dimensionless number of events.

patient is the left mesial temporal lobe). Surely the computational conditions for all the spectra must either be kept constant or organized automatically, in order to avoid arbitrariness. Each time-interval leads to $d = 9$ spectra, but we present only the ones that relate to the principal, *i.e.*, largest respective LE λ_1. The time-windows most distant to the seizure onset (some 28 and 14 minutes before the onset) lead to almost identical spectra within the statistical (Poissonian) uncertainty, *i.e.*, for EEGs with the same global properties our extracted quantities are reproducible, although the actually measured EEG of course never repeats itself. A few minutes before the seizure onset, however, the spectral shapes become remarkably different, although all the four preictal spectra originate from visually very similar time series. Hence this example indeed exhibits a preictal change that cannot be observed by a visual inspection of epileptiform patterns in the EEG.

The solid spectrum in Fig. 1 originates from the ictal phase, where the EEG shows a clearly visible increase of regularity. However, it was not possible to find a 90 s interval of acceptable stationarity. Therefore we used a segment of only 45 s and, for the sake of correspondence with the other curves, we magnified the ictal spectrum by a factor of two. This is necessarily attended with an enlarged (off Poissonian) statistical uncertainty in the spectrum.

Figure 2 presents the very same situation (with the same time-windows) for the non-primary epileptogenic area, *i.e.*, the right mesial temporal lobe. Just the ictal spectrum is missing due to a strong non-stationarity of the corresponding signal. Again

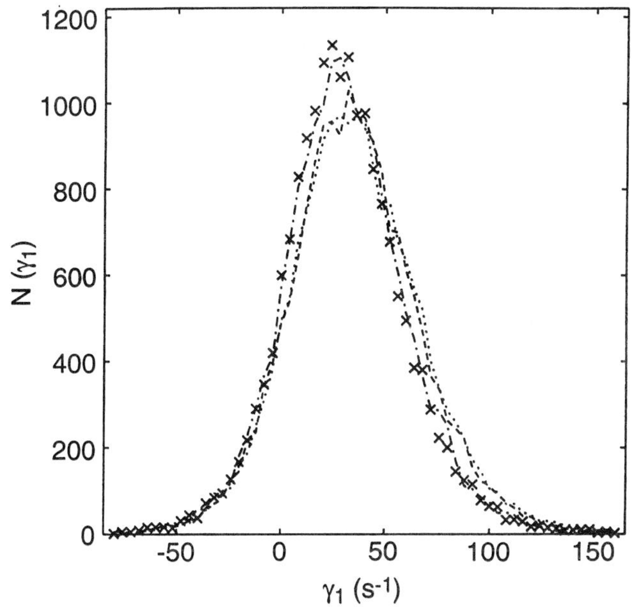

Figure 2. As Fig. 1, but the data originate from the non-primary epileptogenic area (right FO electrode). Compared to the left-hand side, the solid curve (ictal phase) has been omitted because of a strong non-stationarity of the corresponding signal.

we note the agreement between the two spectra extracted from the EEG segments most distant to the seizure onset. We observe a statistically significant preictal change that, expectedly, is less pronounced compared to the primary epileptogenic area.

Moreover, Figs. 1 and 2 reveal that we have to compromise between the length of the time-windows and their stationarity, since the EEG alters towards a seizure onset. This is even true if we find long pieces that are not disrupted by obvious non-stationarities: in such cases we may address the length of the time-windows as the resolution of the observed preictal changes.

Now we go over to other indicators that relate to the LEs, namely to the further above defined Kolmogorov entropy and to the dissipation. However, we stay with the same seizure event. A first remark concerns a mathematical property of these quantities. The LEs, and therefore also their sum, are dynamical invariants, *i.e.*, they stand any (*e.g.*, nonlinear) transformation into a new set of variables (which are the recorded channels in our case). This is not true for the above spectra of local expansion exponents, see Ref. 19. However, since the electrodes stay fixed during the whole EEG recording time, this is of minor importance. The spectra of local expansion exponents offer lots of very detailed possibilities for a comparison, while the LEs are quantities of more generality. This then holds also for the Kolmogorov entropy and for the dissipation.

Since most of our results are based on multichannel reconstruction, the above considerations have a specific implication: the more or less conductive environment of the implanted electrodes is followed by a mixing of the individual channels. However, at least in principle, a dynamical invariant should be unaffected by this circumstance.

Figure 3 depicts the Kolmogorov entropy as well as the dissipation as a function of time for both temporal lobes of our patient. As in Figs. 1 and 2, we use 90 s windows

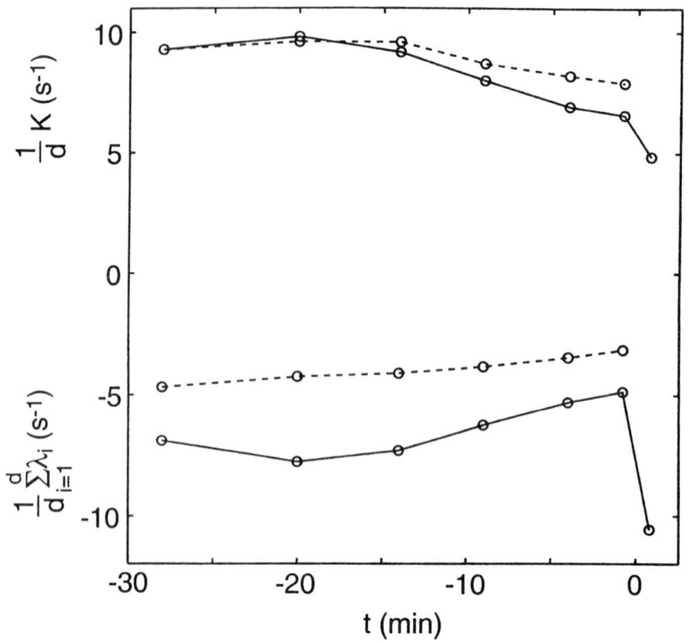

Figure 3. Kolmogorov entropy (positive values) and dissipation (negative values) as a function of time. The zero on the time axis refers to the seizure onset. (—): primary epileptogenic area, left side; (– – –): non-primary epileptogenic area, right side. The normalization factor $1/d$ accounts for different numbers of phase space dimensions: this way the results from different electrodes become comparable. In the present case there is $d = 9$, *i.e.*, 9 bipolar channels from either FO electrode.

out of the same 30 min total interval (and again the non-primary epileptogenic area does not permit to obtain a result from the ictal phase). Both plotted quantities have been normalized by a factor of $1/d$, where d is still 9. The reason lies in the "arbitrariness" of the phase space dimensionality d: the presence of noise admits an arbitrarily large d, and we aim at a representation where the results from other (*i.e.*, forthcoming) electrodes will be comparable with the ones presented here.

The Kolmogorov entropies in Fig. 3 show a noticeable decrease towards the seizure onset. Expectedly, this is more pronounced at the primary epileptogenic area. Beyond the seizure onset, this reduction becomes even stronger, in accordance with visible properties of the signal. We think that this behavior is in agreement with dimensional analyses in Refs. 6, 8, 9, and 22.

The dissipation as a function of time exhibits an interesting property of the recorded signal: since the sum of the d LEs increases towards the seizure onset, the underlying "dynamical law" must become less dissipative. On more speculative grounds we may argue that the dissipation acts as a stabilizing element that gradually gets lost towards a seizure onset, and then, during the seizure, gets restored or even overcompensated.

We are aware of the inaccuracies that may occur in the calculation of the LEs, particularly the negative ones. Part of the reason lies in the difficulties attended with the search of sufficiently many and sufficiently "good" nearest neighbors for every point of the input signal, a problem that has incessantly been discussed since the

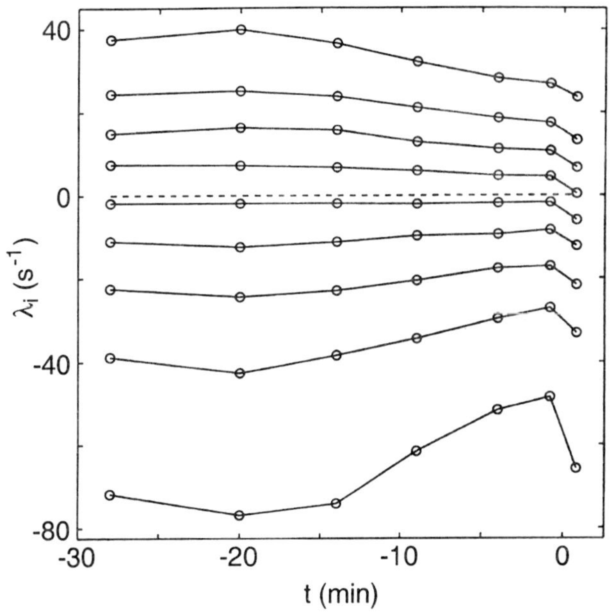

Figure 4. Lyapunov spectrum, *i.e.*, the set of all the d Lyapunov exponents, represented as a function of time. This plot corresponds to Fig. 3 with all respects, *i.e.*, it emerges from the very same calculations concerning the primary epileptogenic area. Note that the negative exponents increase towards the seizure onset, while the transition into the ictal phase is followed by a drop in all the individual Lyapunov exponents.

pioneering treatises such as Ref. 23. In Ref. 21 we outline our experiences on this matter with particular emphasis on EEG applications. Nevertheless we think that the errors to far extent should be common to all the individual calculations in Fig. 3, and so the plotted trends should survive a moderate alteration of the chosen methods.

Finally, Fig. 4 relates to Fig. 3 with all respects, *i.e.*, it refers to the very same calculations. However, instead of plotting the sum of LEs, we present them individually (*i.e.*, we display the Lyapunov spectrum as a function of time). Again we stress that such results strongly depend on the recording technique, particularly with regard to the noise-level and the number of phase space dimensions. In addition, the type and strength of filtering should also be seen as a part of the signal that necessarily enters the results (predominantly the spread of a Lyapunov spectrum). Since Fig. 4 resolves the individual LEs, we may address it as a type of data reduction of lesser extent, whereas Fig. 3 gives more integral quantities that in all probability are closer to a physical or physiological interpretation.

III.3 Determinism and Dimensionality

Up to now, we focused on seizure prediction, and the analyses were based on multichannel EEGs recorded with bilaterally implanted FO electrodes. Below we aim at a more general characterization of the EEG, and so we leave the topic of seizure prediction in its narrower sense. We concentrate on a 64 channels interictal signal obtained from an 8×8 subdural grid electrode implanted over the left frontal lobe in a patient suffering from frontal lobe epilepsy.

Besides the number of available channels, such signals turn out to be particularly "well-behaving" in terms of stationarity, and, as far as we can judge this with our methods, in terms of the noise-level. However, the semi-invasive recordings (that we used for the Figs. 1 through 4) should not be seen as a lower standard than the fully invasive ones, since clinically the FO electrodes require less invasive surgical measures.

Since the grid-data are particularly well suited to perform general research on the nature of brain electric activity, *i.e.*, beyond the context of seizure prediction, we investigated this type of EEG in quite some detail. We renounce to present all of our findings, but we would like to briefly list what we summarized in Ref. 21. The spatial resolution within the not too large area of the grid has been quantified. Further, the comparison with the delay-embedding technique[24] has been performed in terms of spectra of local expansion exponents, as well as by means of LEs: we found interesting consequences of the different kinds of noise-sensitivity. We plotted spectra of local expansion exponents referring also to LEs other than the principal one, and we compared them to the corresponding results of a surface-EEG. Finally, we discussed the effects attended with a time reversed calculation of LEs. Expectedly, for a measured EEG such a numerical test fails, while for simulated time series of marginal complexity our algorithm stands the expansion of the phase volume, even in cases with a rather strong dissipation.

Up to this point, we suppressed an issue of utmost importance, namely the choice of the number of phase space dimensions d. The presence of noise renders the measured signal infinitely dimensional, and so it appears questionable whether a calculation in a low-dimensional phase space makes sense at all. In a context of seizure prediction, one might even adopt the point of view that the preictal changes simply should be detected, no matter how one gets there. This, however, is just one of our purposes.

We now persue a quite obvious procedure, namely the reconstruction of dynamical invariants for different d. If we achieve convergence upon an increase of d, we conclude that the given signal "seemingly" can be represented in d dimensions, as we shall discuss below. We exemplify this procedure by means of Lyapunov spectra, where we represent their sets of the d LEs as spectral densities.

Figure 5 displays two Lyapunov spectra based on 16 and 40 phase space dimensions, respectively. We utilize an integer number of rows in the 8×8 grid electrode (see above), which facilitates a systematic study of the effects attended with an increase in d.

The solid curves in Fig. 5 are made up of the reciprocal distances between adjacent LEs. This yields $d - 1$ values that, on the abscissa, are placed at the center position of these distances. The solid line then connects the points. Evidently these curves, in particular the 40-dimensional one, look rather toothed, since a minute variation in the small distances gets strongly amplified in their reciprocal values. The dashed curves represent a smoothening procedure that particularly well suits the present situation: we collect several LEs (increasingly many towards the center of the spectra) in order to obtain larger distances. The respective reciprocal values then are correspondingly weighted and placed at the "center of mass" of the LEs used at a time. Note that this is not a fit and still represents the outcome of our calculation.

Several plots in the sense of the ones in Fig. 5 reveal that we achieve convergence in the range of, say, 10 up to 12 phase space dimensions. We cannot commit ourselves to an exact number, since the "minimum d" depends on the applied criteria. *e.g.*, we may focus on the largest and smallest LEs only, or rather on the spectral shapes. Retrospectively we may comment that all of our figures should be seen in the light

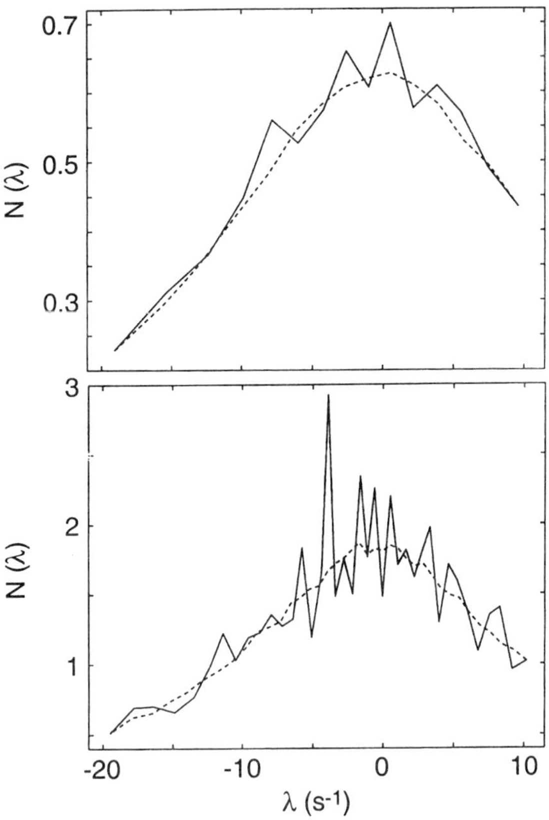

Figure 5. Lyapunov spectra, *i.e.*, the set of all the d Lyapunov exponents, represented as a spectral density (see text). This time, we use the signal originating from the 8×8 grid electrode. Top: rows 1 and 2 in the grid, *i.e.*, $d = 16$. Bottom: rows 1 through 5, *i.e.*, $d = 40$. Just as, *e.g.*, in Fig. 1, we prefer to see the intensity $N(\lambda)$ as a dimensionless number of events, although its construction implies units of seconds. This representation is inconvenient for small d, but for increasingly large d it exhibits the convergence of the Lyapunov spectra.

of this result. However, we are always aware that different EEG recording techniques are attended with different noise-levels: in most cases the minimum d will exceed the above estimation.

In spite of an approximate knowledge of the minimum d, the number of phase space dimensions of a deterministic share in the unfiltered EEG stays unknown. Therefore, we cannot distinguish between the so-called spurious exponents and the true ones. We think it is fair to consider the Lyapunov spectrum as a whole, rather than to inquire the meaning of the individual LEs.

We think, our finding that the minimum d is in the order of 10 deserves some special comments. Clearly the EEG is quite an integral quantity that mainly covers some overall aspects of the underlying neuronal dynamics, as we stated further above. Moreover, there is almost no access to the properties of the high frequencies (that we necessarily have to filter away). Nevertheless, it astonishes that these signals emerging from complex and strongly coupled systems in general (*i.e.*, also in contexts of non-biological systems) turn out to be rather low-dimensional. This means that merely a few degrees of freedom are active in the sense that they contribute to the number of phase space dimensions. In view of the large number of neurons and their complicated interactions, we think it is amazing that the EEG seems to be no exception with this respect.

Within solid state physics, there are comparatively well-controlled laboratory systems where the phase space dimensionality is also "lower than expected". We think of electron spin-resonance experiments, where an anisotropic sample is located in a static magnetic field, and at the same time a microwave field is irradiated. Then, the absorbed microwave power as a function of time serves as the measured signal. Although under certain conditions this signal appears to be strongly chaotic, the number of phase space dimensions d is not too large, see, *e.g.* Ref. 19. In such cases, the spin dynamics is said to take place in collective modes, which to far extent accounts for the observations. It is a known fact that similar effects exist in neuronal dynamics, too: epileptic seizures are attended with a highly synchronous activity of the affected neuronal assemblies. This, then, is followed by a strongly increased regularity in the EEG. (On the other hand, a low attractor dimension does not necessarily mean a small number of phase space dimensions, as is readily confirmed by means of dynamical systems that are given by a set of coupled differential equations.) We may speculate that related synchronization effects happen also in the healthy brain. At present, however, our knowledge on this matter is rather poor.

Further, we may ask how Fig. 5 relates to surrogate data techniques, since they generally aim at the question of determinism *versus* stochasticity. Do plots in the spirit of Fig. 5 mean that the underlying signals are deterministic? We adopt the following point of view. The overall amount of information in the signals gets truncated by, *e.g.*, the filtering procedure and by the limited sampling resolution. In particular, it may happen that the observable part of a non-deterministic signal mimics deterministic properties with almost every respect. Moreover, further above we mentioned the difficulties attended with the search of the nearest neighbors: our methods fulfil the definition of the LEs only approximately. We end up with signals and methods that are consistent with a strong deterministic share in the signals, although this does not answer the question of the "true" nature of the EEG.

Finally, we would like to mention that the knowledge of the minimum d in the EEG is followed by a remarkable consequence. Namely, this allows us to compute the corresponding Kolmogorov entropy K, and its inverse then is the time interval where the trajectory can be predicted. The 16-dimensional Lyapunov spectrum in Fig. 5 is clearly converged, thus yielding an upper limit for K, namely $K = 34.6\ \mathrm{s}^{-1}$. Hence, we arrive at a lower limit for the prediction interval, namely $K^{-1} \approx 0.029$ s, which encloses several sampling intervals. This, however, applies only to the filtered signal, never to the true EEG (which contains also the high frequencies that are notoriously hard to separate from the noise). Nevertheless, at every time point we may inspect whether the predicted signal fits the actually measured one, in order to detect a physiological alteration in the EEG. Obviously this cannot be carried out readily, these considerations rather constitute a quite ambitious future project.

IV. Summary

To sum up, we analyzed two different types of intracranially measured EEGs, which comprises semi-invasive and fully invasive recordings. We laid particular emphasis on bilaterally implanted foramen ovale electrodes in a patient suffering from mesial temporal lobe epilepsy. The heart of our chosen methods lies in the divergence rate of initially nearby trajectories, *i.e.*, we extracted LEs, entire Lyapunov spectra, spectra of local expansion exponents, Kolmogorov entropies, and dissipations. We investigated the time-evolution of these quantities when a seizure onset is approached and even passed.

From our results (as well as from others that are not shown here) we draw the following conclusions. Far away from a seizure onset we observe little time dependence in the extracted quantities. Thus, we are confident that our results are reproducible for different time series with the same global properties. Hence, a change in the plotted indicators can significantly be attributed to the physiological properties of the EEG.

The preictal changes are strongly present, but, except a few minutes prior to a seizure onset, they are qualitatively different in the patients we investigated up to now (here we presented only the analyses from a single case study). We expect further differences in other patients, and so a preictal change should be seen as a characteristic property of an individual. Further, we note that the preictal changes are qualitatively the same ones in both temporal lobes of the patients.

The spectral densities of the local expansion exponents are particularly well suited to detect EEG changes, since they offer more possibilities for a comparison than indicators consisting of a plain number. These spectra reveal that immediately before a seizure onset the changes in some way are opposite to the ones attended with the transition into the ictal phase. This holds also for the dissipations (but not for the Kolmogorov entropies), at least in the patients we investigated up to now.

Further, we examined how our methods apply and relate to the more general properties of the EEG, *i.e.*, beyond the context of seizure prediction. On this score, we investigated a particularly "clean" signal originating from an 8×8 subdural grid electrode. In this contribution, we concentrated on the convergence of the Lyapunov spectra with respect to the size of the spanned phase space. This, then, led us to the well-known problem of determinism *versus* stochasticity, which we discussed also in the general context of "complexity".

What about practical applications? In our opinion, the peculiar time-evolution of the dissipation might offer a chance for a physical or even physiological interpretation. This, however, exceeds the scope of our rather methodological investigations we performed up to now. Concerning clinical applications, recall that the preictal changes are found to take place on a "useful" time scale of several minutes, thus leaving enough time for a possible intervention. In particular, we think of an implanted device that provides a warning system for seizure onsets that are a few minutes ahead. An even more ambitious version of this idea consists of a device that, by means of electrical stimulation or instillation of antiepileptic drugs, is able to prevent seizures. However, at present these plans are clearly beyond our concrete intentions and possibilities.

Acknowledgments

We gratefully acknowledge the support by the Swiss National Science Foundation.

References

1. F. Lopes da Silva, in *Electroencephalography*, edited by E. Niedermeyer, and F. Lopes da Silva (Williams & Wilkins, Baltimore, 1993), p. 1097.
2. D. Gallez and A. Babloyantz, *Biol. Cybern.* **64**, 381 (1991).
3. J. Röschke and J. B. Aldenhoff, *Biol. Cybern.* **64**, 307 (1991).

4. A. Babloyantz and A. Destexhe, *Proc. Nat. Acad. Sci. USA* **83**, 3513 (1986).
5. L. D. Iasemidis, J. C. Sackellares, H. P. Zaveri, and W. J. Williams, *Brain Topography* **2**, 187 (1990).
6. K. Lehnertz and C. E. Elger, *Electroenceph. Clin. Neurophysiol.* **95**, 108 (1995).
7. B. Weber, K. Lehnertz, C. E. Elger, and H. G. Wieser, *Epilepsia* **39**, 922 (1998).
8. K. Lehnertz and C. E. Elger, *Phys. Rev. Lett.* **80**, 5019 (1998).
9. C. E. Elger and K. Lehnertz, *Eur. J. Neurosci.* **10**, 786 (1998).
10. H. G. Wieser and S. Moser, *J. Epilepsy* **1**, 13 (1988).
11. M. C. Casdagli *et al., Physica D* **99**, 381 (1996).
12. J. Theiler *et al., Physica D* **58**, 77 (1992).
13. J. Theiler and P. E. Rapp, *Electroenceph. Clin. Neurophysiol.* **98**, 213 (1996).
14. P. Achermann *et al., Eur. J. Neurosci.* **6**, 497 (1994).
15. W. S. Pritchard and D. W. Duke, *Int. J. Neurosci.* **67**, 31 (1992).
16. S. Blanco *et al., Phys. Rev. E* **57**, 932 (1998).
17. J. Röschke, J. Fell, and P. Beckmann, *Electroenceph. Clin. Neurophysiol.* **86**, 348 (1993).
18. J.-P. Eckmann and D. Ruelle, *Rev. Mod. Phys.* **57**, 617 (1985).
19. H. R. Moser, P. F. Meier, and F. Waldner, *Phys. Rev. B* **47**, 217 (1993).
20. G. Benettin, L. Galgani, and J.-M. Strelcyn, *Phys. Rev. A* **14**, 2338 (1976).
21. H. R. Moser, B. Weber, H. G. Wieser, and P. F. Meier, *Physica D* **130**, 291 (1999).
22. C. E. Elger and K. Lehnertz, in *Epileptic Seizures and Syndromes*, edited by P. Wolf, (John Libbey and Company Ltd., London, Paris, 1994), p. 541.
23. J.-P. Eckmann, S. O. Kamphorst, D. Ruelle, and S. Ciliberto, *Phys. Rev. A* **34**, 4971 (1986).
24. F. Takens, in *Dynamical Systems and Turbulence, Lecture Notes in Mathematics*, edited by D. A. Rand and L. S. Young, (Springer, Berlin, 1981) p. 366.

Sulfur Nanowires Elaboration and Structural Characterization

E. Carvajal,[1] P. Santiago,[2] and D. Mendoza[1]

[1] *Instituto de Investigaciones en Materiales*
Universidad Nacional Autónoma de México
Apartado Postal 70-360, 04510 México, DF
MEXICO

[2] *Instituto Nacional de Investigaciones Nucleares*
Salazar, 52045 Estado de México
MEXICO

Abstract

We have fabricated sulfur nanowires by immersion of a nanoporous anodic alumina template in a S-CS_2 solution. Structural properties of these nanowires were investigated by means of various techniques as scanning, transmission and high resolution electron microscopy (SEM, TEM, HRTEM) and differential scanning calorimetry (DSC). We found that the porous template imposes a preferential direction on the growing of the wires and, at the same time, there were several curly wires when they were released from the template. Nevertheless, the crystalline structure differs from the corresponding bulk sulfur, and preliminary DSC characterization shows a melting behavior that differs also from the one observed on bulk sulfur.

I. Introduction

Sulfur is one of the most studied chemical elements, because its technological applications. It plays a relevant role in many industrial processes, has a remarkable environmental impact, and it has a diverse physical behavior. As a bulk material crystallizes in different structures and it is possible to build up a great variety of sulfur molecules, being the orthorhombic-α the STP form of the cyclo-octasulfur, the most stable molecule that occurs in all phases. The α-sulfur transforms into monoclinic β-sulfur, at 94.4°C, before it reaches the melting point that changes from 112°C to 119.6°C going from one to another structural phase. In fact, there are many reported values of the melting point, because the freezing point of sulfur depends on the pressure and temperature history of the melt; anyway, the *natural* melting point (the freezing point of an equilibrated melt) is considered 114.6° C. Obtained from the melted sulfur, there are another monoclinic allotrope: the γ-sulfur, which has a higher density but decomposes at room temperature. These three allotropes are the best well characterized solid forms of bulk sulfur.[1,2]

Among their transport properties it is known that the thermal conductivity of sulfur decreases when the temperature is increased, ranging from 11 W/m·deg at 4.2 K to 0.15 W/m·deg at 373 K; it is one of the best thermal insulators. Sulfur is an electrical insulator too. Some authors propose that the electrical conductivity has two contributions, due to electrons and holes with mobilities of $\sim 10^{-4}$ cm^2/V·s and $\sim$10 cm^2/V·s respectively.[1] Nevertheless, it is possible to observe a different electrical behavior in sulfur, because it transforms to a metallic phase at 90 GPa (room temperature) and even it transforms to a superconducting phase;[3] both cases are accompanied by a crystallographic change. Previous theoretical work has predicted that sulfur becomes a superconductor in a *bcc* phase with a T_c of 15 K above a pressure of 550 GPa, but it was not predicted the superconductivity in the lower pressure metallic phases.[4] This is just an example of the large scale changes in physical properties induced by pressure.

On the other hand, the synthesis and study of materials with nanometric dimensions[5] or in geometrically confined structures[6] have recently attracted great interest, due to the fact that many bulk properties of the material can change under those conditions. These new properties open the possibility of having other applications of these novel materials. In particular, having in mind the possibility of build up an unidimensional system, the novel observations in the physical properties of bulk sulfur and that the behavior of nanoscopic systems are usually far from those of bulk materials, we have prepared sulfur nanowires to study some of its physical properties.

II. Experimental Section

The first step to prepare the sulfur nanowires was to make a nanoporous alumina template. To generate the template we anodized an aluminium sheet (Merck, purity 99.95%) in a sulfuric acid solution. As mentioned elsewhere[7] this kind of alumina template has an hexagonal array of parallel nanochannels, at least with a short range order, and the separation and diameter of these nanochannels depends on the acid concentration and the applied voltage used in the anodization process. The electrolyte nature and the bath temperature are another two parameters that influence the growing velocity of the nanochannels.[8] With this information we used a 20% wt. H_2SO_4 aqueous solution and a 20 V voltage to make the electrochemical anodization of the aluminium samples. Keeping the temperature near 0°C for a period of time of about four hours we successfully made anodic templates with pore diameters of $\sim$20 nm and some microns in length.[9]

Once we separated these templates from the remaining aluminium sheet we put them in the same acid solution for two hours to remove the impervious layer of the template. After this, the templates was washed with deionized water and dried in air, and they were ready to be used as the matrix where we kept the sulfur wires. To make the wires we used a S-CS_2 solution (10 g in 40 ml) where the templates remains immersed by a period of 15 days in a flask. When we took off the hosting templates from the S-CS_2 solution, we payed special attention on preventing some sulfur precipitation over the template surface, to avoid any sulfur crystal which may hide the wires or interfere other experiments on the wires.

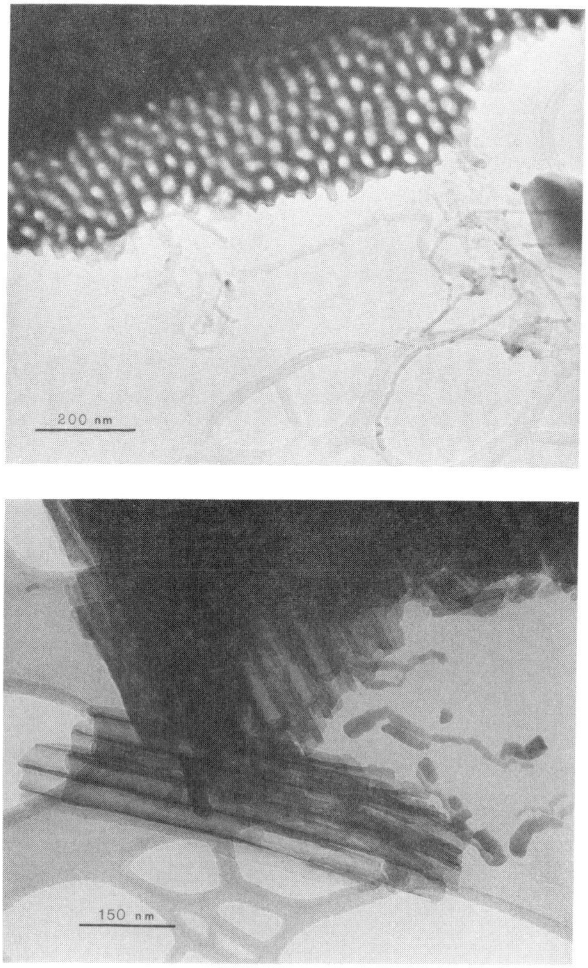

Figure 1. TEM alumina template and released sulfur nanowires.

III. Results and Discussions

The most interesting feature of the nanoscopic systems is the modification of their physical properties, from the corresponding bulk material, owing to the reduction of some spatial dimension and the parallel increase of the surface to volume ratio. This is the reason that motivate us to have some control over the diameter of the nanoporés in the prepared anodized templates, because it represents the barrier on the transversal dimension of our wires. By SEM first,[9] and TEM later, we obtained images that confirmed the presence of pores on the template. The set of pores are over all the surface; they are ordered and have almost uniform dimensions (Fig. 1).

To make the structural study of the wires we released them from the hosting template by grinding it in an agate mortar; then we added deionized water to this powder and dropped a sample on a TEM grid. In this way it was possible to observe both the template and the sulfur nanowires. We found a lot of curly wires and some straight very long sulfur wires. The longest of these wires seems to be almost monocrystalline such as it was preferentially piled up plane after plane, perpendicular to the axis wire.

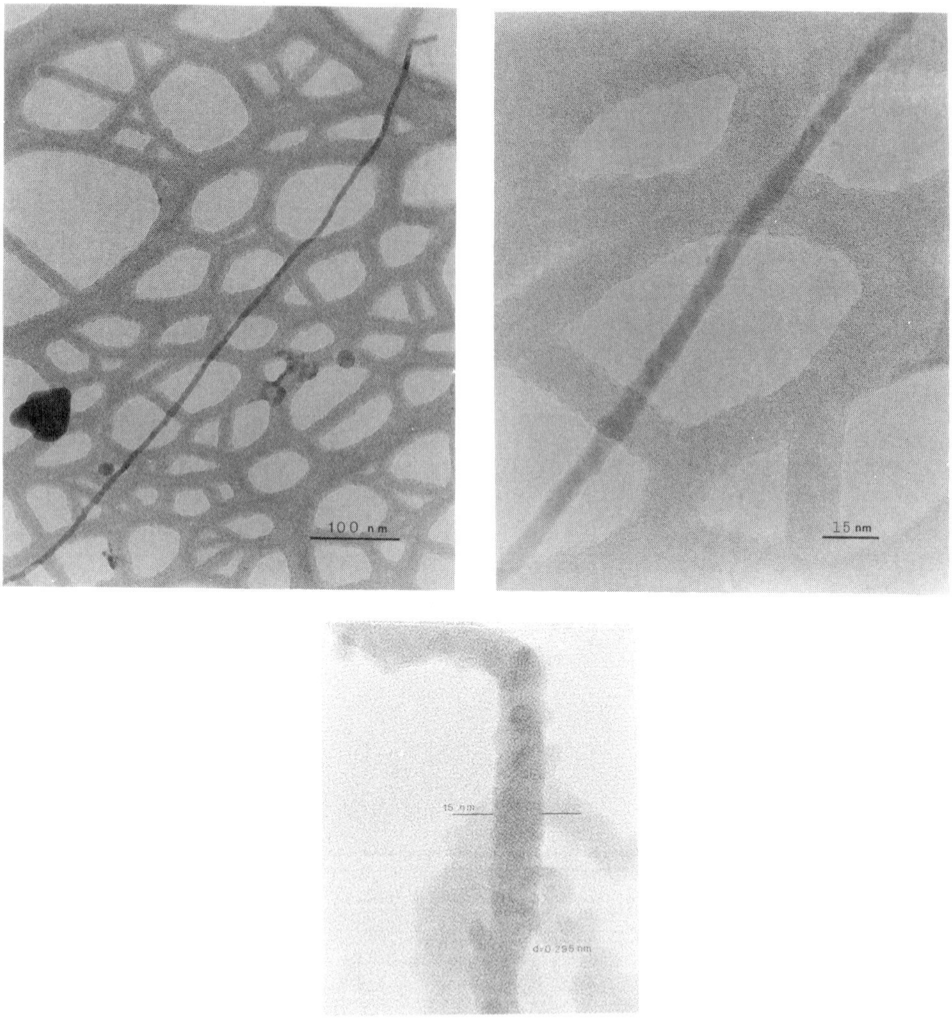

Figure 2. (a) TEM sulfur nanowire micrograph, (b) HRTEM micrograph where the lattice fringes contrast is shown, and (c) a digitized image with its FFT.

Anyway, it does not matter which kind of wire we observe, the electron diffraction patterns make us think that the wires crystallize on a monoclinic phase, because the lattice parameter values are near the corresponding to the β-sulfur phase.

The necessity of the TEM study is not just because we want to know the crystallographic structure acquired by the wires as they grow up in the pores, but we need to know how they are growing and how much of the pores space was filled up. Our initial concern was to study the effect of the confinement on the wires behavior and, to measure their conduction properties, we need wires that fill completely the hollow template space. This is to keep them in there until we finish our measuresments.

The samples were characterized by HRTEM and electron diffraction in a Jeol JEM2010 microscope operated at 200 KV and equipped with an energy dispersive spectrometer. Figure 2(a) shows a straight sulfur nanowire with a diameter of 15 nm

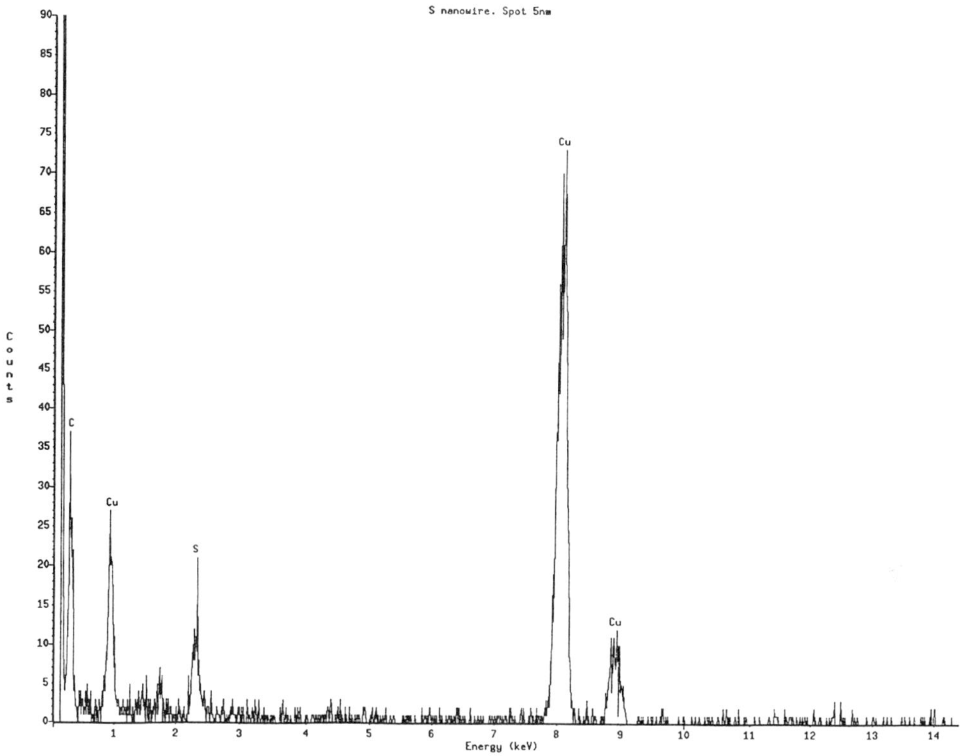

Figure 3. EDS spectrum corresponding from a five-nm zone of a nanowire.

and more than 1000 nm of length. Figure 2(b) is a magnification of the nanowire where it is possible to see a lattice fringes contrast. A digitized image is shown in Fig. 2(c) in order to get the lattice spacing by using a Fast Fourier Transform technique (FFT). The lattice parameter found is $d = 0.295$ nm which does not correspond with any reflection reported for a sulfur structure in the literature. The FFT image is the typical for a nanowire structure, which agrees with the experimental diffraction pattern obtained for a nanowire. However, in order to get the proper characterization, we are working on nano-beam techniques. A chemical elemental microanalysis was also performed by a 5 nm nanowire zone; this is shown in Fig. 3. The microanalysis spectrum shows sulfur, but there is not any other element. There is also another kind of sulfur nanowires, which bends (Fig. 4). At the bending zones, it is possible to see defects like stacking faults, which give the enough energy to the nanowire to be bended. It is also possible to see that some nanowires are twisted on themselves.

Finally, differential scanning calorimetry measurements was made for the template, the bulk sulfur used as the precursor and the nanowires embedded into the template. The heat flow data *vs.* temperature obtained are shown in Fig. 5. These results show that the characteristic temperatures associated with the transformations of the sulfur nanowires embedded into the nanoporous alumina template, and those associated with bulk sulfur are different. At first look, these results may be related to the confinement of the sulfur into the nanopores, or more interestingly, to the intrinsic nature of the nanowires. More detailed experiments are in progress to study this interesting phenomena, which will be published elsewhere.

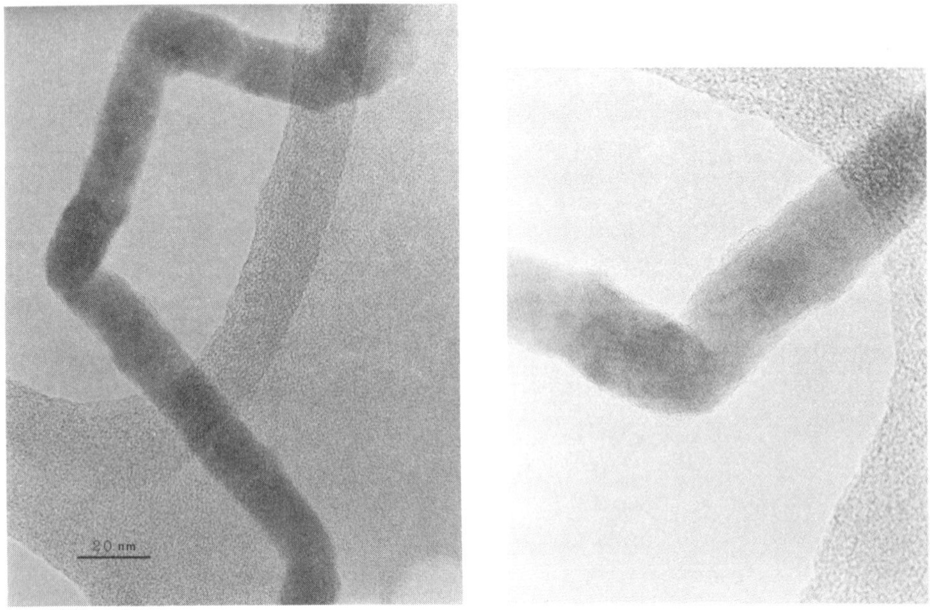

Figure 4. (a) HRTEM micrograph of a bending nanowire, and (b) a digitized image which shows the stacking faults and a rotation of the nanowire on itself.

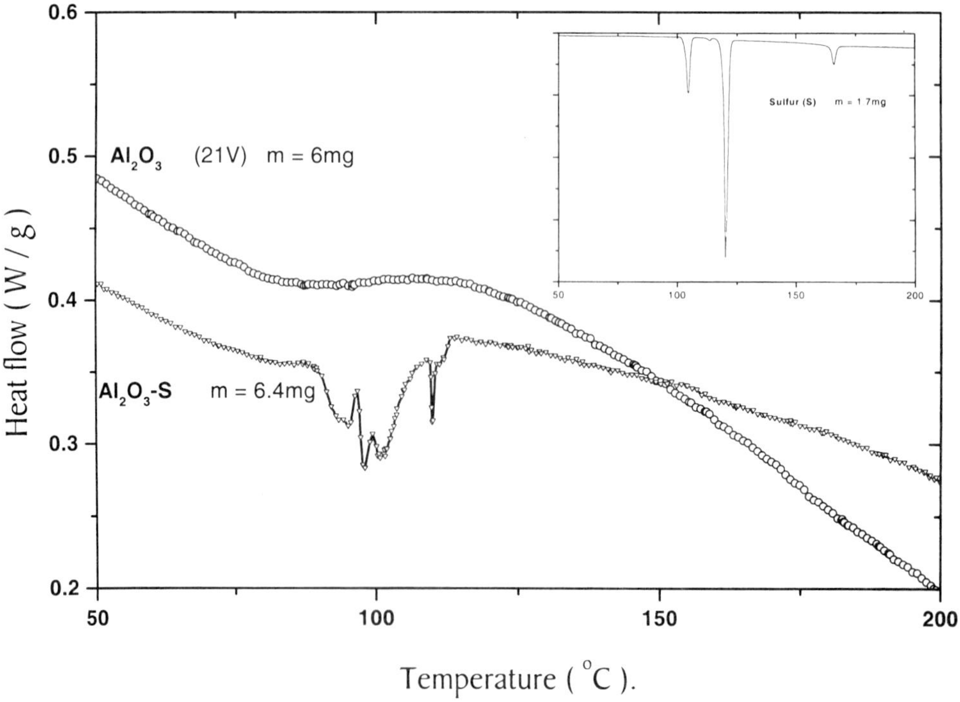

Figure 5. DSC measurements for bulk sulfur, the template and the nanowires in the nanochannels.

IV. Conclusions

Sulfur nanowires have been synthesized at room temperature by a template approach. The chemical microanalysis does not show the presence of any other element except S along different zones of the nanowire. However this is not a definitive conclusion about the nature of these nanowires even when there is no chemical evidence to consider a chemical reaction during the nanowires formation. On the other hand, the preliminary electron diffraction study does not show agreement with any single sulfur phase. We are performing an exhaustive nano-beam diffraction study in order to find the sulfur nanowire structure.

Differential scanning calorimetry measurements disagree also with the bulk sulfur data reported in the literature; we found that the characteristic temperatures associated with the bulk sulfur transformations are different to those for the wires. But this information is not conclusive because the phenomenon may be associated with the time passed between the wires preparation and the DSC measurement.

Acknowledgments

This work was supported by Conacyt-México under grant 32306-E. We thank to C. Vázquez for the DSC measurements, E. Caballero for preparing photographs and to R. Escudero for fruitfull discussions. E. Carvajal recognize the financial support provided by CONACYT, DGEP-UNAM and the IWPLD Organizer Committee.

References

1. B. Meyer, *Chem. Rev.* **76**, 367 (1979).
2. W. N. Tuller (ed.), *The sulphur data book* (McGraw-Hill, New York 1954) p. 4, 16.
3. V. V. Struzhkin, R. J. Hemley, H. Mao, and Y. A. Timofeev, *Nature,* **390**, 382 (1997).
4. O. Zakharov and M. L. Cohen, *Phys. Rev. B,* **52**, 12572 (1995).
5. G. Schmid and L. F. Chi, *Adv. Mater.* **10**, 515 (1998).
6. C. R. Martin, *Science,* **266**, 1961 (1994).
7. T. Kyotani, L. F. Tsai, and A. Tomita, *Chem. Mater.* **8**, 2109 (1996).
8. R. L. Penn and J. F. Banfield, *Science,* **281** 969 (1998).
9. P. Santiago *et al.*, *Journal of Materials Research.* (Submitted)

Nanodots and Nanowires of Silicon

K. Sattler

Department of Physics and Astronomy
University of Hawaii
2505 Correa Road
Honolulu, HI 96822
USA

Abstract

When the size of silicon is reduced towards the nanometer range, new properties emerge due to a dramatic change in bonding conditions, and due to electron and hole state quantization. Bulk silicon is characterized by diamond-type crystal structure, with sp^3-hybridization and 4-fold coordination. With decreasing size, silicon undergoes a phase change to a more close-packed atomic arrangement, which characterizes atomic and electronic structures of small Si clusters. In order to study size-dependent properties of silicon nanoparticles, we apply scanning tunneling microscopy (STM) and scanning tunneling spectroscopy (STS). For silicon clusters we determine the energy gap as a function of size. We show that pristine silicon particles show a major transition in their electronic properties at about 15 Å. We find that by vapor-condensation in UHV, silicon is also able to form quasi-one-dimensional structures. Nanowires with diameters from 3 nm to 7 nm, more than 100 nm long were produced. Considering the calculated free energies and band gaps for several possible wire structures we suggest that silicon nanowires tend to grow with a fullerene-type Si_{24}-based core structure.

I. Introduction

Silicon is at the center of today's modern electronics consumer industry. From cars to washing machines, almost all consumer goods contain silicon microprocessors in some form. Since there is a trend towards ever smaller electronic devices, microelectronics may eventually go over to nanoelectronics[1–4] and may be the key technology in the 21st century. New nanosize-specific functions are expected to exceed those of currently manufactured materials and instruments.

During the last two decades extensive work has been performed on semiconductor nanostructures. In particular, the electronic band structure of semiconductor quantum dots is of interest.[5] Also the carrier transport and tunneling through dots,[6,7] arrays

Physics of Low Dimensional Systems
Edited by J. L. Morán-López, Kluwer Academic/Plenum Publishers, New York 2001

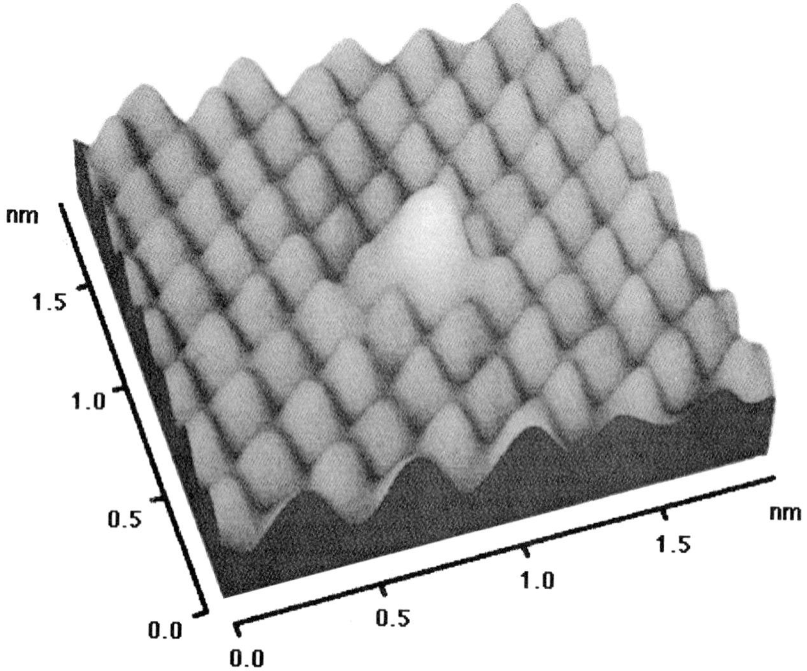

Figure 1. STM image of a 2.5 Å silicon cluster on HOPG.

of dots,[8] and nanogranular semiconductors[9] are important issues. Cluster-assembled materials[10] are new granular materials with quantum-confinement determining their properties. Recently, quantum wires became the focus of increasing interest.[11–15] Both, wires and dots may show quantum confinement effects when their diameters are small enough. However, the surface of these structures may determine their overall properties.

We have studied nanodots and nanowires using scanning tunneling microscopy and spectroscopy. We studied the size-dependence of the energy gap of silicon particles by application of voltage-dependent STM. Both, nanodots and nanowires were grown by quenching the atomic vapor. It is surprising that nanowires grow under such conditions as the diamond structure of silicon does not show a preferential direction.

II. Experimental

The nanodots and nanowires were grown on highly oriented pyrolytic graphite (HOPG) upon submonolayer deposition of silicon. Starting from a base pressure of $\sim 10^{-8}$ torr, the deposition was done by dc magnetron sputtering of silicon at an argon pressure of 5 to 8 mtorr. Once the Si atoms arrive on the HOPG substrate, they are quickly thermalized. With this technique the clusters form after surface diffusion by quasi-free growth on the inert substrate. A 2.5 Å, silicon cluster is shown in Fig. 1.

After the deposition the samples were transferred to the STM chamber (1×10^{-10} torr). Since local STS of clusters is not feasible at room temperature due to the thermal drift, voltage-dependent STM[16] was our STS method of choice. Once a Si cluster was selected, a series of STM images at different bias voltages was recorded. If the cluster has a gap and the bias voltage is tuned to be within the gap, there are no states available for tunneling. Accordingly, the cluster is "invisible" until the bias voltage is high enough to allow tunneling into the "conduction band" or out of the "valence band" of the cluster. In constant-current mode this is reflected by the apparent height Δz of the clusters as it varies with the bias voltage. The equivalent measurement in constant-height mode is the difference ΔI in tunneling current on the substrate and on the cluster.

Clusters in the size range from a few Angstroms to a few nanometers were analyzed on the substrates. Samples produced at a discharge current of 50 mA or below had well-separated clusters on the smooth HOPG substrate. At higher currents, the coverage was higher and coalescence of clusters was observed. For longer exposure and higher currents we have observed the formation of nanoscale silicon wires[11] on the substrate.

III. Silicon Nanodots

Theoretically, small silicon clusters were first proposed to be fragments of the crystalline bulk.[17,18] Such clusters have many dangling orbitals and very low average coordination numbers. In further studies crystalline structures were shown not to be the ground states but either to correspond to high-energy local minima.[19–22] Tomanek and Schlueter[23,24] have used a local-density functional (LDF) method, Pacchioni and Koutecky[20] and Balasubramian[25] have performed configuration interaction (CI) calculations, and Ballone *et al.*[26] have used simulated annealing techniques to study small silicon clusters.

Silicon clusters, $(Si)_n$, with just a few atoms ($n < 10$) are theoretically predicted to have compact atomic arrangement. The close-packed structure is typical for metallic rather than covalent systems. As bulk silicon is a semiconductor (band gap 1.1 eV), a major change in the electronic properties can therefore be expected with size reduction. The critical size n^* for a transition from covalent to metallic bonding was theoretically estimated, but with very different values: $n^* = 100$–1000 (Ref. 24) and $n^* = 50$.[27]

Medium-sized silicon clusters, with up to several hundred atoms per cluster, are little understood. Various atomic structures have been proposed.[27–32] Due to the relatively large number of atoms, semiempirical techniques such as the interatomic force field method[33] were used instead of *ab initio* techniques. In this size range there seems to be a major structural transformation. Jarrold *et al.*[34] found that clusters up to ~ 27 atoms have a prolate shape while larger clusters have more spherical shapes. Also, abrupt changes in properties of medium-sized Si clusters have been observed in photoionization measurements, at $n^* = 20$–30.[35]

Since the discovery of room-temperature visible photoluminescence in silicon nanocrystals[36] and porous silicon,[37] the size-dependence of the energy gap of Si nanostructures has been discussed extensively.[38–41] The quantum size effect, resulting in a blue-shift of the energy gap with decreasing size, was widely believed to be at the heart of the novel optical properties.[36,42–45] The energy gap was found to increase significantly with decreasing size, typically between 1.3 and 2.5 eV for particles of 5 to 1 nm in size.[45] We note that the surfaces of these clusters had been passivated with hydrogen or oxygen. For very small, pristine Si clusters band gaps below the bulk gap of 1.1 eV[23,24,46,47] and even zero gaps are predicted.[28,47] What a difference the surface passivation makes was illustrated recently in a calculation of pristine silico clusters

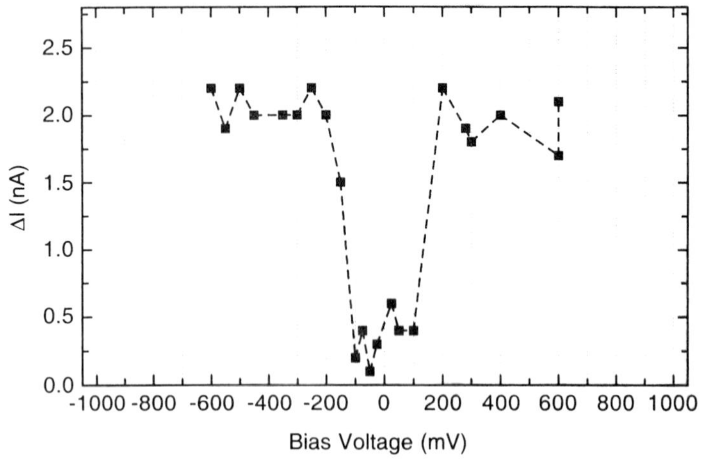

Figure 2. $\Delta I(V)$ plot for a 2.5 Å silicon cluster.

and their hydrogenated counterparts.[48] The pristine clusters show a density of states with zero energy gap between HOMO and LUMO. The passivated clusters however show wide band gaps up to 3.4 eV. We show experimentally that the properties of pristine silicon clusters differ significantly from those of passivated clusters.

An STM image of a 2.5 Å cluster is shown in Fig. 1. The atomic structure of graphite is visible. No other silicon cluster is in the vicinity. Also, the graphite lattice around the cluster is not distorted. It indicates that the interaction between cluster and substrate is small. In fact we were able to move such clusters on the substrate by using the STM tip without changing the cluster's shape.

The $\Delta I(V)$ curve for the 2.5 A-cluster is shown in Fig. 2. ΔI is a measure for the density of states (DOS) of the cluster. A reduction of ΔI means that fewer states are available for electrons to tunnel into or out of the cluster. The value of ΔI is close to zero in a range from -100 mV to $+100$ mV. Interpolating between the data we determine a band gap of 250 meV. ΔI does not vanish completely which indicates that there is a small density of states still present. The observed gap is symmetric around zero bias voltage. This shows that there is no charge transfer between cluster and substrate. Charge transfer would lead to a contact potential, which would produce a shift of the energy levels of the cluster relative to those of the substrate.

In Fig. 3 we show the $\Delta I(V)$ curve for a 8.5 A-cluster. Again, a band gap with steep edges is displayed. ΔI in this case is found to be zero in the gap region since the cluster completely vanishes in the STM image. There are no accessible states in the cluster and the tunneling electrons pass through the cluster without interaction. Accordingly only the substrate is visible in the STM image. The gap is nearly symmetric around zero bias. Outside of the gap the ΔI values fluctuate due to quality differences of the STM images.

In Fig. 4 the energy gap of the analyzed clusters is plotted versus the cluster size. In the range between 15 A and 40 A only clusters with zero gap are observed. For smaller clusters, zero gaps are found as well but non-zero gaps predominate. Below 15 Å the gaps tend to increase with decreasing cluster size. The largest gap recorded is 450 meV, for clusters with 5 A and 8.5 A. There is significant scatter of the data points in the small size range. Clusters of similar size can have very different energy gaps. For example at 8 ± 1 A we find a zero-gap cluster and one with a 250 meV gap.

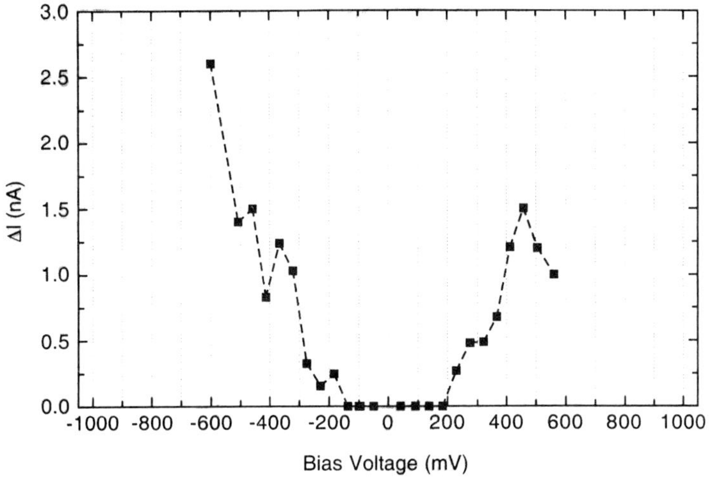

Figure 3. $\Delta I(V)$ plot for a 8.5 Å silicon cluster.

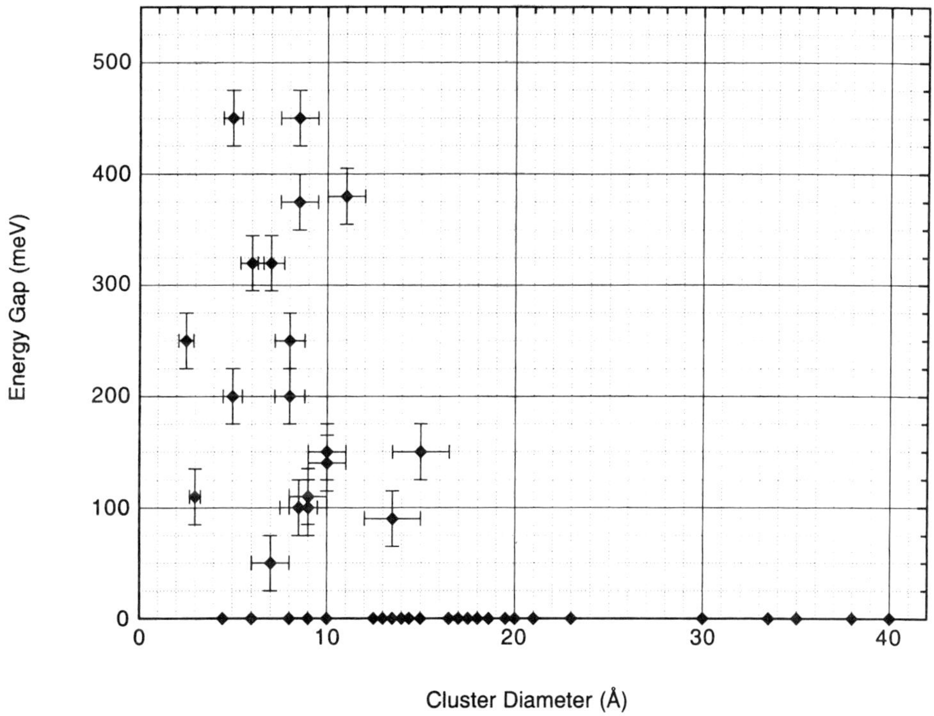

Figure 4. Energy gap values for silicon clusters determined by STS.

We note that the energy gap strongly differs for small clusters of similar size. In our experiment we do not determine the number of atoms in a cluster but rather its diameter. In the uncertainty window of $\pm 10\%$ there are clusters with slightly different number of atoms. Such spread in the number of atoms per cluster and the presence of various isomers explains the observed scatter of gap data.

The results are surprising at first. Bulk silicon has a band gap of 1.1 eV and one may expect it to be much higher for small particle sizes due to the quantum size effect. Our results show that the gaps of unpassivated and passivated silicon clusters are entirely different.

For unpassivated silicon clusters, dangling bonds are present at the surface. It is a consequence of the strong directional properties of the sp^3-hybrids and the missing neighbors at the surface. Since they are partially filled the surface states associated with the dangling bonds are located in the energy gap around the Fermi level. This has been well demonstrated for the surface of bulk silicon using a number of methods including tunneling spectroscopy.[49] These experiments show that surface states mostly fill the band gap of the unpassivated silicon surface. For clusters of silicon, with unpassivated surface, a similar situation can be expected. In fact, a recent calculation of Si_{29}, Si_{87}, and Si_{357} shows density of states with zero energy gap between HOMO and LUMO.[48] The passivated clusters, $Si_{29}H_{36}$, $Si_{87}H_{76}$, and $Si_{357}H_{204}$, however show the broad band gaps of 3.44 eV, 2.77 eV, and 1.99 eV, respectively, just as expected for passivated particles in this size range. Evidently the zero gap of the pristine particles originates from the surface states.

In our experiment we observe zero band gaps for all clusters with sizes larger than 15 Å. We assume that the dangling bond states make the surface regions of the clusters conductive. Electrons tunneling between STM tip and sample rather pass through the conducting cluster surface than through the insulating core. Therefore zero band gaps are observed for large silicon clusters. This is independent of the particle diameter within our probed size range.

At $\sim$ 15 Å, a major change in the cluster properties occurs. The energy gap suddenly opens up. We associate this with the covalent-metallic transformation suggested previously for silicon. Accordingly, large silicon clusters are sp^3-hybridized with strong covalent bonds and rigid bond angles giving rise to high surface state density. Around a critical size n^* the nature of the bonds gradually changes to hybridization other than sp^3- or even to the atomic s- and p-configurations. Below n^* the clusters are not fragments of the bulk anymore but rather have close-packed structures resembling those of metals. Some of the silicon clusters exhibit zero-energy gaps even below n^*, as seen in Fig. 4. It shows that the transition at n^* does not occur for all silicon clusters. Some clusters have sp^3-coordinated bonds even at very small size. They coexist with the compact clusters as structural isomers.

In summary, we have measured the energy gap of pristine silicon clusters with diameters between 2.5 Å and 40 Å, using scanning tunneling spectroscopy. We have observed two size ranges with significantly different electronic properties. Energy gaps up to 450 meV exist for clusters of 15 Å or smaller, while all the larger clusters show zero energy gaps. This observation leads us to conclude that a major transition in the cluster structures occurs at a size of about 15 Å.

IV. Nanowires of Silicon

In recent years, efforts have been made to produce nanometer scale silicon wires in a controlled manner, using common semiconductor processing steps.[50,51] Individual nanowires of silicon have been produced by a number of methods, for example by natural masking,[52] lithography,[53] wet-chemical etching[54,55] and a combination of laser ablation cluster formation and vapor-liquid-solid growth.[56] In our laboratory, we have produced silicon nanowires by quenching Si vapor on solid supports.[11] The structures

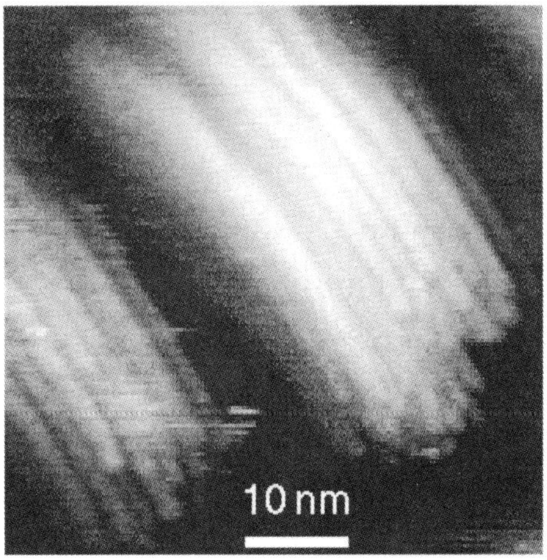

Figure 5. Bundles of vapor-grown silicon nanowires, imaged by STM.

are analyzed by STM. We address the question why quasi-1D structures may grow for silicon by suggesting several wire structures. These are investigated by Self-Consistent-Field Molecular Orbit calculations to find the most stable one.

An STM image of silicon nanowires is shown in Fig. 5, where 20–30 wires are grouped in bundles. The cylindrical shapes of the wires can be seen. The nanowires are about 50 nm long, the diameters are 5±0.5 nm. We have observed similar nanowires on several samples. This shows that the wires prefer to be alligned parallel in bundles in order to saturate their dangling surface bonds. Besides wires in bundles we occasionally found individual wires.

In order to understand why silicon vapor may grow in form of wires we construct several linear polyhedric networks:

a) Si_{12}-cage polymer structure (12 atoms per unit cell)
b) Si_{15}-cage polymer structure (10 atoms per unit cell)
c) Si_{20}-cage polymer structure (based on the Ih dodecahedron, 30 atoms per unit cell)
d) Si_{24}-cage polymer structure (based on the D6d icosahedron, 36 atoms per unit cell)

The suggested structures are shown in Fig. 6. All models have in common a stacking of Si cages, in the center of which lies the wire axis. While these lattices deviate significantly from bulk diamond, tetrahedral configuration of the Si atoms is maintained.

In structure a) (C_{3v} symmetry), the axis of the wire passes through the centers of buckeled Si_6-rings. Two adjacent rings are connected by three bonds and form a Si_{12} cage. This cage represents the unit cell, which is repeated every 6.31 Å. The surface of the wire consists of buckled hexagons.

Structure b) (C_{5v} symmetry) consists of planar pentagons, joined through five outward oriented interstitial atoms. Two pentagons together with the interstitial atoms form a cage. The unit cell of this structure contains 10 atoms and the repeat distance is 3.84 Å. The surface of the wire consists of buckled hexagons.

Structure c) (C_{5v} symmetry) is built from Si_{20}-cages. These dodecahedra (Ih symmetry) are the smallest possible fullerene structures, consisting of 12 pentagons. In the wire structure, two adjacent cages share one pentagon. The 30 atoms of one and a half such cages build the unit cell, which is repeated every 9.89 A. Pentagons make up the surface net of this wire.

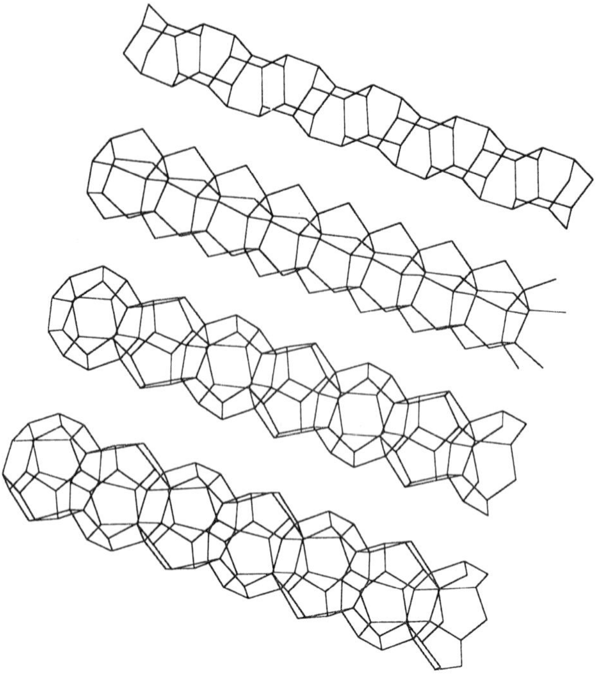

Figure 6. Four models of nanowire core structures, as polymers of Si_{12}-, Si_{15}-,Si_{20}-, and Si_{24}-cages.

Structure d) (C_{6v} symmetry) is similar to structure c), only that Si_{24}-cages (D_{6d} symmetry) are used as building blocks. These units contain 12 pentagons and two hexagons. The hexagons are shared by two adjacent cages and are concentric to the wire axis. The unit cell consists of 36 atoms and the lattice parameter is 10.03 Å. The surface of the wire consists of pentagons.

In order to find the most stable of the proposed structures we calculated their binding energies and HOMO-LUMO gaps.[11] We employ the PM3 Self-Consistent Field Molecular Orbit (SCF-MO) theory derived by Stewart.[57] In Fig. 7, the binding energies are plotted for wires of various lengths. We find that the atoms in longer wires are bound more strongly. The binding energies saturate at relatively small cluster sizes, for example at about $n = 30$ (3.6 eV) for structure a). Clearly, the diamond-type wires [a) and b)] have binding energies which are much smaller than the fullerene-type wires [c) and d)]. The Si_{24} fullerene has a symmetry axis specified for preferential addition of further cages. Among the four considered configurations, structure d) also has the highest binding energy per atom and the largest energy gap. If energetics is responsible for the wire formation in its early stage, then structure d) should grow preferentially. Once the Si_{24}-based polymer has formed, the wire may continue to add layer by layer, further increasing its diameter.

Figure 8 shows two views of the charge density distribution of the Si_{24} cluster as obtained from the PM3 analysis. The two hexagons in the otherwise pentagonal network are on opposite sides of the Si_{24} cluster with fullerene structure. This makes the cluster's surface anisotropic for the addition of further atoms along the built-in axis.

In summary, nanodots and nanowires of silicon were produced by quenching the atomic vapor on solid supports in UHV. For nanodots, an electronic transition is ob-

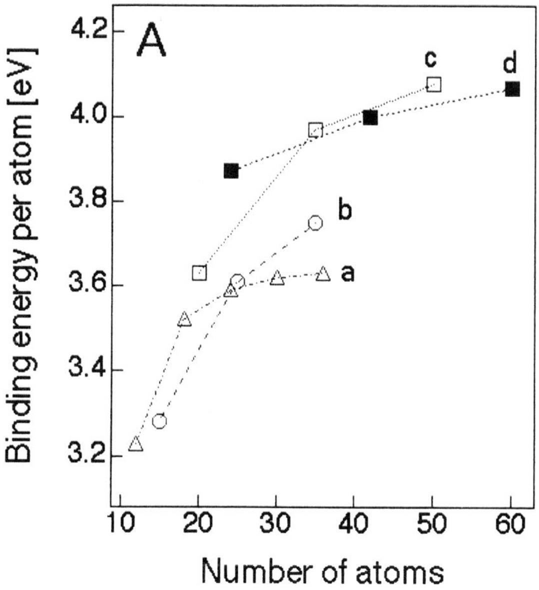

Figure 7. Binding energies for silicon nanowires obtained from PM3 calculations.

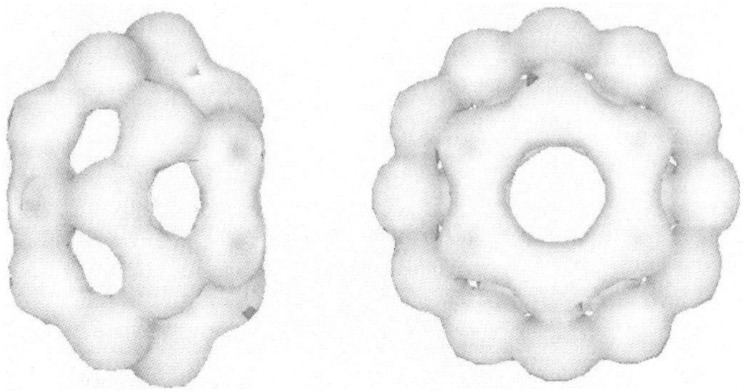

Figure 8. Charge density isosurface of the fullerenic Si_{24}.

served at about 15 Å which is related to a transition from diamond-like to compact structures. For silicon nanowires, diamond-like and fullerene-like structures are considered. Calculations of binding energies indicate that fullerene-based structures have the highest probability to form.

References

1. M. Van Rossum, *Mater. Sci. Eng.* **20**, 128 (1993).
2. F. A. Buot, *Phys. Reports* **234**, 1 (1993).
3. K. F. Goser, C. Pacha, and M. L. Rossmann, *Proc. IEEE* **85**, 558 (1997).
4. H. L. Hartnagel, R. Richter, and A. Grub, *Electronics and Communications Engineering Journal* **3**, 119 (1991).
5. G. W. Bryant, *Phys. Rev. B* **40**, 1620 (1989).

6. N. Blanc, P. Gueret, and R. Germann, *Physica B* **189**, 135 (1993).
7. M. Lonfat, B. Marsen, and K. Sattler, *Chem. Phys. Lett.* **313**, 539-543 (1999).
8. G. Chen, G. Klimeck, I. I. I. Goddard, and A. William, *Phys. Rev. B* **50**, 8035 (1994).
9. T. A. Burr, A. A. Seraphin, and K. D. Kolenbrander, *Phys. Rev. B* **56**, 4818 (1997).
10. *Cluster Assembled Materials*; Vol. 232, edited by K. Sattler (Trans. Tech. Publications, Zuerich, 1997).
11. B. Marsen and K. Sattler, *Phys. Rev. B* **60** (1999).
12. N. T. Bagraev, E. T. Chaikina, and A. M. Malyarenko, *Solid-State Electron* **42**, 1199 (1998).
13. B. Li, D. Yu, and S.-L. Zhang, *Phys. Rev. B* **59**, 1645 (1999).
14. V. Ng, H. Ahmed, and T. Shimada, *Appl. Phys. Lett.* **73**, 972 (1998).
15. J. Westwater, D. P. Gosain, and S. Usui, *Phys. Stat. Solidi A* **165**, 37 (1998).
16. M. F. Crommie, C. P. Lutz, and D. M. Eigler, *Phys. Rev. B* **48**, 2851 (1993).
17. J. C. Phillips, *J. Chem. Phys.* **83**, 3330 (1985).
18. S. Saito, S. Ohnishi, and S. Sugano, *Phys. Rev. B* **33**, 7036 (1986).
19. K. Raghavachari, *J. Chem. Phys.* **84**, 5672 (1986).
20. G. Pacchioni and J. Koutecky, *J. Chem. Phys.* **84**, 3301 (1986).
21. R. Biswas and D. R. Hamann, *Phys. Rev. B.* **36**, 6434 (1987).
22. U. Roethlisberger, W. Andreoni, and M. Parrinello, *Phys. Rev. Lett.* **72**, 665 (1994).
23. D. Tomanek and M. A. Schlueter, *Phys. Rev. Lett.* **56**, 1055 (1986).
24. D. Tomanek and M. A. Schlueter, *Phys. Rev. B* **36**, 1208 (1987).
25. K. Balasubramanian, *Chem. Phys. Lett.* **135**, 283 (1987).
26. P. Ballone, W. Andreoni, R. Car, and M. Parrinello, *Phys. Rev. Lett.* **60**, 271 (1988).
27. J. R. Chelikowsky, *Phys. Rev. Lett.* **60**, 2669 (1988).
28. X. G. Gong, *Phys. Rev. B* **52**, 14677 (1995).
29. K. M. Ho, A. A. Shvartsburg, B. Pan, Z.-Y. Lu, C.-Z. Wang, J. G. Wacker, J. L. Fye, and M. F. Jarrold, *Nature* **392**, 582 (1998).
30. D. A. Jelski, B. L. Swift, and T. T. Rantala, *J. Chem. Phys.* **95**, 8552 (1991).
31. E. Kaxiras and K. Jackson, *Phys. Rev. Lett.* **71**, 727 (1993).
32. A. M. Mazzone, *Phys. Rev. B* **54**, 5970 (1996).
33. J. R. Chelikowsky, K. Glassford, and J. C. Phillips, *Phys. Rev. B* **44**, 1538 (1991).
34. M. F. Jarrold and V. A. Constant, *Phys. Rev. Lett.* **67**, 2994 (1991).
35. K. Fuke, K. Tsukamoto, and F. Misaizu, *J. Chem. Phys.* **99**, 7807 (1993).
36. H. Takagi, H. Ogawa, Y. Yamazaki, A. Ishizaki, and T. Nakagiri, *Appl. Phys. Lett.* **56**, 2379 (1990).
37. L. T. Canham, *Appl. Phys. Lett.* **57**, 1046 (1990).
38. L.-W. Wang and A. Zunger, *J. Phys. Chem.* **98**, 2158 (1994).
39. H. Yorikawa, H. Uchida, and S. Muramatsu, *J. Appl. Phys.* **79**, 3619 (1996).
40. A. Kux and M. B. Chorin, *Phys. Rev. B* **51**, 17535 (1995).
41. M. Hirao and T. Uda, *Surf. Sci.* **306**, 87 (1994).
42. L. E. Brus, P. F. Szajowski, W. L. Wilson, T. D. Harris, S. Schuppler, and P. H. Citrin, *J. Am. Chem. Soc.* **117**, 2915 (1995).
43. L. N. Dinh, L. L. Chase, M. Balooch, W. J. Siekhaus, and F. Wooten, *Phys. Rev. B* **54**, 5029 (1996).
44. S. S. Iyer and Y. H. Xie, *Science* **260**, 40 (1993).

45. T. van Buuren, L. N. Dinh, L. L. Chase, W. J. Siekhaus, and L. J. Terminello, *Phys. Rev. Lett.* **80**, 3803 (1998).
46. A. Sieck, D. Porezag, and K. Jackson, *Phys. Rev. A* **56**, 4890 (1997).
47. E. Kaxiras, in Cluster Assembled Materials; Vol. 232, edited by K. Sattler (TransTech Publications, Zurich, 1996), p. 67.
48. P. Melinon, P. Keghelian, B. Prevel, A. Perez, G. Guiraud, J. LeBrusq, J. Lerme, M. Pellarin, and M. Broyer, *J. Chem. Phys.* **107**, 10278 (1997).
49. J. A. Stroscio, R. M. Feenstra, and A. P. Fein, *Phys. Rev. Lett.* **57**, 2579 (1986).
50. H. I. Liu, D. K. Biegelsen, F. A. Ponce, N. M. Johnson, and R. F. W. Pease, *Appl. Phys. Lett.* **64**, 1383 (1994).
51. H. Namatsu, Y. Takahashi, M. Nagase, and K. Murase, *J. Vac. Sci. Technol. B* **13**, 2166 (1995).
52. M. Gotza, B. Saint-Cricq, and P.-H. Jouneau, *Microelectron. Eng.* **27**, 129 (1995).
53. A. G. Nassiopoulos, S. Grigoropoulos, and D. Papadimitriou, *Thin Solid Films* **297**, 176 (1997).
54. Y. Shi, J. L. Liu, F. Wang, Y. Lu, R. Zhang, S. L. Gu, P. Han, L. Q. Hu, Y. D. Zheng, C. Y. Lin, and D. A. Du, *J. Vac. Sci. Technol. A* **14**, 1194 (1996).
55. J. L. Liu, Y. Shi, and Y. D. Zheng, *J. Vac. Sci. Technol. B* **13**, 2137 (1995).
56. A. M. Morales and C. M. Lieber, *Science* **279**, 208 (1998).
57. J. J. P. Stewart, *J. Comput. Chem.* **10**, 221 (1989).

Electronic Properties of AFM-Defined Semiconductor Nanostructures: Quantum Wires and Single Electron Transistors

S. Lüscher,[1] R. Held,[1] A. Fuhrer,[1] T. Heinzel,[1]
K. Ensslin,[1] M. Bichler,[2] and W. Wegscheider,[2]

[1] *Solid State Physics Laboratory*
ETH Zürich
8093 Zürich
SWITZERLAND

[2] *Walter Schottky Institut*
TU München, 85748 Garching,
GERMANY
and
Institut für Angewandte und Experimentelle Physik
Universität Regensburg,
93040 Regensburg
GERMANY

Abstract

The electron gas in AlGaAs heterostructures can be depleted below regions which are patterned with an atomic force microscope. This leads to laterally insulating regions across which in-plane gate voltages can be applied in order to tune the electronic nanostructures. In long quantum wires we find that the wire potential can be laterally shifted in real space while keeping the general transport characteristics of the wire unchanged. With an additional front gate electrode the wire potential and location can be changed over large parameter ranges. The wire walls are found to be steep and smooth on the length scales relevant for electronic transport. In single electron transistors fabricated with the same technology we find that the charge on the dot can be tuned by up to 70 electrons. Because of the steep walls the electronic dot shape at the Fermi energy resembles the lithographic pattern with high accuracy.

Physics of Low Dimensional Systems
Edited by J. L. Morán-López, Kluwer Academic/Plenum Publishers, New York 2001

I. Introduction

Electronic nanostructures imbeded in AlGaAs heterostructures have revealed a rich variety of physical phenomena.[1] Usually the fabrication of these nanostructures is based on electron beam lithography in combination with a pattern transfer technique such as etching or ion bombardment.[2] The potential of scanning probe techniques[3] for the fabrication of nanostructures has long been recognized and realized in a variety of approaches.[4] Local anodic oxidation[5] has proven particularly useful for the fabrication of electronically active nanostructures on AlGaAs heterostructures.[6] The distinctive features of such nanostructures are:

1.- highly tunable in-plane gates via laterally patterned oxide lines,
2.- the possible combination of in-plane and conventional top gates,
3.- steep and smooth edges of nanostructures in combination with short depletion length.

In this publication we present a summary of results obtained on quantum wires and single electron transistors (SET) fabricated with local anodic oxidation. The quantum wires can be shifted in real space by suitable lateral gate voltages while leaving the wire potential as a whole basically unchanged. The wires can be fabricated and tuned over a wide range of parameters namely from short quantum point contacts showing conductance quantization up to 40 μm long wires with almost single mode occupancy. The single electron transistors fabricated with the same technology display pronounced Coulomb oscillations as a function of several in-plane and top gate voltages.

II. Fabrication

The conducting tip of an AFM is negatively biased with respect to the two-dimensional electron gas (2DEG) embedded in an AlGaAs-GaAs heterostructure. By controlling the humidity, the tip-sample voltage and distance, as well as writing speed, fine oxid lines can be patterned on the sample surface. The electron gas has to be no deeper than 40 nm below the sample surface such that it can be locally depleted below the oxidized lines. In our case we use high-quality 2DEGs located 37 nm below the sample surface. A 10 μm wide Hall bar was defined by optical lithography and wet chemical etching. In a final fabrication step, the Hall bar was covered by a homogeneous Ti/Au top gate electrode. The sample was mounted in the mixing chamber of a ^{3}He/^{4}He-dilution refrigerator. Under these conditions and with the top gate grounded, the 2DEG has a sheet density of 4.3×10^{15} m^{-2} and a mobility of 102 m^2/Vs. The Drude scattering length as extracted from the $B = 0$-resistance is $\ell_D = 11.0$ μm.

The insets of Figs. 1 and 3 display typical AFM defined oxide lines for quantum wires and quantum dots. Lateral gate voltages can be applied across the laterally insulating regions in order to tune the respective nanostructure in its electronic width and carrier density.

III. Quantum Wires

We first present results on a rather long quantum wire ($L = 40$ μm) with a lithographic width of about 150 nm. An AFM topography is shown in the upper right corner of Fig. 1. The resistance of the wire indicates that about 8 lateral modes are occupied if all gates are grounded. For high quality quantum wires two questions are relevant: How smooth is the wire edge and how wide is the electronic width of the wire

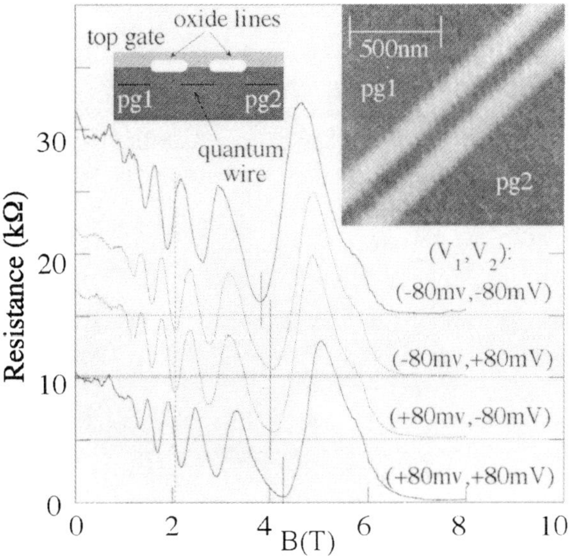

Figure 1. Magnetoresistance for different planar gate voltages (V_1,V_2). Curves are vertically offset by 5 kΩ each. The wire can be tuned in both shape and position by V_1 and V_2. The vertical lines mark the positions of filling factor 4. The two dotted curves correspond to situations where the wire potential is basically the same (positions of minima line up) but the electrons reside in different spatial locations. Inset in upper right corner: Surface topography of the sample, as measured with an atomic force microscope. Two oxide lines appear as bright lines, since the oxidized regions are elevated 8 nm above the heterostructure suface (gray regions). The oxide lines define a wire (lithographic width 150 nm), sandwiched between 2 planar gate electrodes, pg1 and pg2. Inset in upper left corner: Cross section of the sample across the wire. The 2DEG is depleted underneath the oxide lines, thus defining the quantum wire and the planar gates. The structure is covered by a homogeneous top gate electrode. (From Ref. 7).

compared to the lithographic width? The first question can in principle be answered by using a method applied by Thornton *et al.*[8] A rough wire edge is expected to lead to a characteristic resistance maximum once an external magnetic field forces the electrons on classical cyclotron orbits which are roughly twice as large as the electronic wire diameter. We had to make our wires as long as 40 μm to see a faint feature of this effect (see Fig. 1, $B \approx 0.4$ T) from which we estimate that more than 95% of the electrons are scattered specularly at the wire edges. The electronic width of the wire can also be estimated by assuming a parabolic wire potential and following the deviation of the magnetic field positions of the minima in the magnetoresistance from a 1/B periodicity.[9] The lateral depletion length is defined as the difference between the lithographic and electronic wire width devided by two (the wire has two edges). With this approach we find a lateral depletion length of 15 ± 5 nm. This value can be further reduced by appropriate lateral gate voltages. The extremely short lateral depletion length in combination with the smooth wire edge forms the basis for the interesting electronic properties of our AFM defined nanostructures.

Figure 1 shows magnetoresistance traces for a quantum wire which can be tuned by voltages V_i, applied to two planar gates pgi ($i = 1, 2$), as well as with a top gate voltage V_{tg}. By using the model of Berggren *et al.*[9] to fit the positions of the

Shubnikov-de Haas minima, we obtain the one-dimensional electron density n_{1D} and the electronic wire width w. The four magnetoresistance traces in Fig. 1 were taken for different planar gate voltages (V_1, V_2) with the top gate grounded. By changing V_1 and/or V_2, n_{1D} and w can be tuned from, $n_{1D} = 5.5 \times 10^8$ m^{-1}, $w = 116$ nm, for $V_1 = V_2 = -80$ mV to $n_{1D} = 6.7 \times 10^8$ m^{-1}, $w = 141$ nm, for $V_1 = V_2 = +80$ mV. Our scan range is limited to $\mid V_i \mid \leq 80$ mV, since a leakage current across the oxide lines is detected for higher $\mid V_i \mid$. Within this range, the wire contains 8 spin-degenerate one-dimensional subbands. We find that the two middle traces in Fig. 1 which are obtained for an antisymmetric set of gate voltages ($V_1 = -V_2 = 80$ mV and $-V_1 = V_2 = 80$ mV) reveal minima at exactly the same magnetic fields within experimental accuracy. This means that the wire potential itself depends only on the 1D carrier density n_{1D} and remains basically unchanged as the wire potential is laterally shifted through the host crystal by the in-plane gate voltages.

The resistance at small magnetic fields for each of the four traces in Fig. 1 displays reproducible fluctuations. A variety of experiments on conductance fluctuations in one- and two-dimensional systems has been performed, acompanied by substantial theoretical work.[10–12] These fluctuations are known as "universal conductance fluctuations",[13] although true universality can be disturbed by various effects.[14–17] In order to tune the interference pattern one has to change an experimental parameter which is usually the Fermi energy[18] of the system or an external magnetic field.[19–21] Because the impurity configuration is never exactly known, experiment and theory have to be compared on a statistical level.

In order to obtain a better understanding how the fluctuations can be interpreted we present in Fig. 2 a conductance map as a function of the two in-plane gate voltages. The diagonal axis defined by $V_1 = V_2$ corresponds to a change in Fermi energy or n_{1D}, while the center of the wire potential remains at the same location. The axis given by the diagonal $V_1 = -V_2$ corresponds to constant n_{1D} and w. Both axes have been calibrated by measuring and fitting $R(B)$ for various (V_1, V_2). Reproducible fluctuations in G are observed as a function of both n_{1D} and δy. They are almost completely smeared out at $T \approx 850$ mK indicating a phase coherent origin.

The conductance fluctuations as a function of n_{1D} can be analyzed in the statistical framework as done by others and we find resonable agreement with data in the literature. The new feature in our experiment is the observation of conductance fluctuations as a function of a spatial shift of the system while the other system parameters (wire width, carrier density) remain unchanged. Obviously a detailed understanding of this situation requires a sophisticated calculation taking into account the random potential distribution in the vicinity of the wire potential. Here we will proceed with some qualitative arguments without the necessity of a statistical analysis of the conductance fluctuations. The background potential in semiconductor heterostructures has been calculated for quantum wires and point contacts.[22] We expect that on average, the number of scattering potentials inside the wire changes by one if the wire is displaced by $\Delta y = d^2/2L$, where L is the length of the wire and $1/d^2$ is the relevant density of scattering potentials. The factor of 1/2 takes into account that potential bumps may enter the wire on one side as well as exit it on the other. We find $\Delta y = 2.7$ nm for $d = \ell_q$ and $\Delta y = 400$ nm for $d = \ell_e$. This suggests that the observed quasi-period in δy is caused by individual small-angle scattering potential bumps the wire passes on its way through the crystal. Their density is comparable to $1/\ell_q^2$, where ℓ_q is the quantum scattering length. For our sample we determine ℓ_q to be about 460 nm from the magnetic field dependence of the Shubnikov-de Haas oscillations. In a simple picture, such a potential bump causes an indentation at the

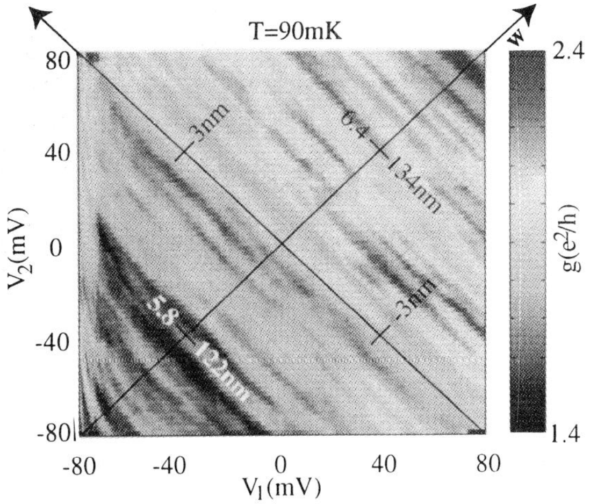

Figure 2. (a) (color) Conductance G at $T = 90$ mK and for $V_{tg} = 0$, $B = 0$ as a function of (V_1,V_2). The data have been obtained by sweeping V_1 and changing V_2 in steps of 2 mV. The two diagonal axes correspond to the electronic wire width w as well as to the electron density n_{1D} in the wire, measured in units of 10^8 m^{-1}, and to the wire displacement δy (see text). From Ref. 7.

wire edge as it enters the wire, at which electrons can scatter. As the wire is further displaced, the potential bump is at some point completely surrounded by conduction electrons, and the wire boundary becomes straight again. Obviously this simplified picture needs more refinement, *e.g.* the screening length plays an important role as well. Our interpretation is solely based on geometrical arguments and the assumption, that a potential bump entering or leaving the wire changes the interference pattern of the electron trajectories and thus leads to a feature in the resistance trace. This picture gives us a handle to interprete conductance fluctuation without the necessity to involve a complicated statistical analysis. It is obvious how this method of shifting electronic structures by suitable lateral electric fields can also be applied to quantum dots and other nanostructures. It can be an important tool to distinguish impurity induced features in transport experiments from "real" effects by intentionally fabricated potential shapes.

IV. Single Electron Transistors

Single electron transistors have been realized in a variety of ways.[1] Here we discuss results were quantum dots in AlGaAs/GaAs heterostructures are fabricated by AFM-defined nano-lithography as described before. The inset in Fig. 3 shows such a quantum dot. The bright lines are written with the AFM, the electron gas is depleted below. The point contact openings are tuned by the voltages applied to contacts PC1 and PC2, respectively. The charge on the island is controlled by voltages applied to the plunger gates I and II. The principle of how such an SET can be tuned by in-plane gates as well as by the homogeneous top gate[23] gives a lot of flexibilty for the operation.

The conductance trace in Fig. 3(a) shows Coulomb peaks of very high quality. The point contacts are almost pinched off such that the system is decoupled from the outside world. The width of the Coulomb peaks is in this case only determined by temperature. By fitting the experimental lineshape with the appropriate theoret-

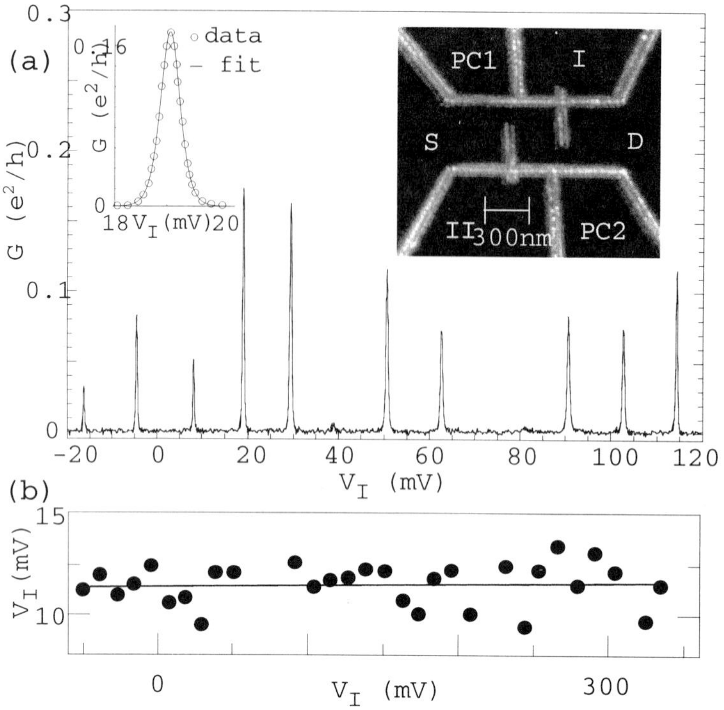

Figure 3. (a) Conductance G as a function of V_{I}. Pronounced Coulomb oscillations are observed. Right inset: AFM picture of the dot. The oxid lines (bright) define the dot as well as its coupling to drain and source through the tunneling contacts tunable via the planar gates PC1 and PC2. Gates I and II are used to tune the occupation number of the dot. (b) Peak spacing ΔV_{I} as a function of V_{I}. The solid line is a linear fit to the experimental data. Because the line is almost horizontal this hints at small shape deformations. From Ref. 25.

ical expression[24] (inset in the upper left corner of Fig. 3a) we find in the best case a temperature of 90 mK in agreement with the base temperature of our dilution refrigerator.

The distance between Coulomb peaks on the voltage axes is determined by *i*) the Coulomb charging energy, *ii*) the lever arm converting voltage into energy, and *iii*) the energy spectrum in the dot possibly influenced or even dominated by interaction effects. Only the last quantity is the really interesting one. In order to extract the influence of the energy spectrum on the Coulomb peak spacing one usually plots the peak spacing as a function of the tuning voltage (see Fig. 3b). The linear fit (solid line) is fairly horizontal indicating that the capacitance between the dot and the tuning electrode changes only very little over the investigated range of parameters. This is in pronounced contrast to observations on top gate-defined quantum dots, where the lever arm can change by up to 30%. Yet it is another confirmation that the AFM-defined quantum dots are very rigid and that their electronic shape resembles pretty much the lithographic pattern.

Another feature is apparent from Fig. 3(a): The Coulomb peaks are not equidistantly spaced. The smallness of the dot (electronic area 170 nm $\times$ 170 nm) leads to confinement energies which are about 1/3 of the Coulomb charging energy. This allows us to study level statistics and spin effects with a very high resolution.[25]

V. Conclusions

Semiconductor nanostructures are fabricated by AFM nano-lithography. We show that high quality quantum wires and single electron transistors can be realized. These structures have the additional degree of freedom that their properties can be tuned by in-plane electric fields. This allows us to shift a wire potential with respect to the host crystal lattice leading to the detection of individual potential humps as they enter and leave the wire potential. High-quality single electron transistors reveal a pronounced influence of the energy level spectrum in the dot on the spacing of the Coulomb peaks. This powerful method of AFM lithography lends itself for the patterning of semiconductor nanostructures and can most likely also be extended to superconducting, magnetic and other nanostructures.

References

1. For a review on quantum dots see L. P. Kouwenhoven, C. M. Marcus, P. L. McEuen, S. Tarucha, R. M. Westervelt, and N. S. Wingreen, "Electron Transport in Quantum Dots", in Mesoscopic Electron Transport, *Proceedings of a NATO Advanced Study Institute*, edited by L. P. Kouwenhoven, G. Schön, and L. L. Sohn, (Kluwer, Dodrecht, Netherlands, 1997), Ser. E, Vol. 345, p. 105.
2. For a review see H. Smith and H. G. Craighead, *Physics Today* 24, February 1990
3. For a review, see *Technology of Proximal Probe Lithography*, edited by C. R. K. Marrian, (SPIE Optical Engineering Press, Bellingham, WA, 1993); R. Wiesendanger, *Appl. Surf. Sci.* **54**, 271 (1992).
4. M. A. McCord and R. F. W. Pease, *Appl. Phys. Lett.* **50**, 269 (1987); Y. Kim and C. M. Lieber, *Science* **257**, 375 (1992); L. L. Sohn and R. L. Willett, *Appl. Phys. Lett.* **67**, 1552 (1995); M. Wendel, S. Kuühn, H. Lorenz, J. P. Kotthaus, and M. Holland, *Appl. Phys. Lett.* **65**, 1775 (1994); D. M. Eigler and E. K. Schweizer, *Nature (London)* **344**, 524 (1990); H. J. Mamin, P. H. Guethner, and D. Rugar, *Phys. Rev. Lett.* **65**, 2418 (1990); H. J. Mamin, S. Chiang, H. Birk, P. H. Guethner, and D. Rugar, *J. Vac. Sci. Technol. B* **9**, 1398 (1991); E. S. Snow and P. M. Campbell, *Appl. Phys. Lett.* **64**, 1933 (1994); E. S. Snow and P. M. Campbell, *Science* **270**, 1639 (1995); R. S. Becker, J. A. Golavchenko, and B. S. Swartzentruber, *Nature (London)* **325**, 419 (1987); J. A. Dagata, J. Schneir, H. H. Harary, and C. J. Evans, *Appl. Phys. Lett.* **56**, 2001 (1990); R. S. Becker, G. S. Higashi, Y. J. Chabal, and A. J. Becker, *Phys. Rev. Lett.* **65**, 1917 (1990); E. S. Snow, P. M. Campbell, and P. J. McMarr, *Appl. Phys. Lett.* **63**, 749 (1993); P. M. Campbell, E. S. Snow, and P. J. McMarr, *Appl. Phys. Lett.* **66**, 1388 (1995); H. Sugimura, T. Uchida, N. Kitamura, and H. Masuhara, *Appl. Phys. Lett.* **63**, 1288 (1993); D. Wang, L. Tsau, and K. L. Wang, *Appl. Phys. Lett.* **67**, 1295 (1995); E. S. Snow, D. Park, and P. M. Campbell, *Appl. Phys. Lett.* **69**, 269 (1996); E. S. Snow, D. Park, and P. M. Campbell, *Superlattices Microstruct.* **20**, 545 (1996); S. Sasa, T. Ikeda, C. Dohno, and M. Inoue, *Jpn. J. Appl. Phys.*, to be published; K. Matsumoto, M. Ishii, K. Segana, Y. Oka, B. J. Vartanian, and J. S. Harris, *Appl. Phys. Lett.* **68**, 34 (1996); H. W. Schuhmacher, U. F. Keyser, U. Zeitler, R. J. Haug, and K. Eberl, *Appl. Phys. Lett.* **75**, 1107 (1999); G. Binnig, M. Despont, U. Drechsler, W. Häberle, M. Lutwyche, P. Vettiger, H. J. Mamin, B. W. Chui, and T. W. Kenny, *Appl. Phys. Lett.* **74**, 1329 (1999); B. Klehn, S. Skaberna, and U. Kunze, *Superl. and Microstr.* **25**, 473 (1999); B. Klehn and U. Kunze, *J. Appl. Phys.* **85**, 3897 (1999).

5. E. S. Snow and P. J. McMarr, *Appl. Phys. Lett.* **66**, 1388 (1995); B. Irmer, M. Kehrle, H. Lorenz, and J. P. Kotthaus, *Semicond. Sci. Technol.* **13**, 79 (1998); E. S. Snow, D. Park, and P. M. Campbell, *Appl. Phys. Lett.* **69**, 269 (1996); E. S. Snow, D. Park, and P. M. Campbell, *Superlatt. Microstruct.* **20**, 545 (1996); K. Matsumoto, M. Ishii, K. Segana, Y. Oka, B. J. Vartanian, and J. S. Harris, *Appl. Phys. Lett.* **68**, 34 (1996); J. Shirakashi, K. Matsumoto, N. Miura, and M. Konagai, *Appl. Phys. Lett.* **72**, 1893 (1998); B. Irmer, M. Kehrle, H. Lorenz, and J. P. Kotthaus, *Appl. Phys. Lett.* **71**, 1733 (1997); S. C. Minne, H. T. Soh, P. Flueckiger, and C. F. Quate, *Appl. Phys. Lett.* **66**, 703 (1995); M. Ishii and K. Matsumoto, *Jpn. J. Appl. Phys., Part 1* **34**, 1329 (1995); R. Garcia, M. Calleja, and H. Rohrer, *J. Appl. Phys.* **86**, 1898 (1999).
6. R. Held, T. Heinzel, P. Studerus, K. Ensslin, and M. Holland, *Appl. Phys. Lett.* **71**, 2689 (1997); R. Held, T. Vancura, T. Heinzel, K. Ensslin, M. Holland, and W. Wegscheider, *Appl. Phys. Lett.* **73**, 262 (1998); R. Held, S. Lüscher, T. Heinzel, K. Ensslin, and W. Wegscheider, *Appl. Phys. Lett.* **75**, 1134 (1999); T. Heinzel, R. Held, S. Lüscher, T. Vancura, K. Ensslin, T. Blomqvist, I. Zozoulenko, and W. Wegscheider, *Advances in Solid State Physics (Festköperprobleme)* **39**, edited by B. Kramer, (Vieweg, Braunschweig 1999) p. 161.
7. T. Heinzel, S. Lüscher, K. Ensslin, W. Wegscheider, and M. Bichler, *Phys. Rev. B* **61**, R13353 (2000).
8. T. J. Thornton, M. L. Roukes, A. Scherer, and B. P. Van de Gaag, *Phys. Rev. Lett.* **63**, 2128 (1989).
9. K.-F. Berggren, G. Roos, and H. van Houten, *Phys. Rev. B* **37**, 10118 (1988).
10. S. Feng, P. A. Lee, and A. D. Stone, *Phys. Rev. Lett.* **56**, 1960 (1986).
11. M. Cahay, M. McLennan, and S. Datta, *Phys. Rev. B* **37**, 10125 (1988).
12. B. L. Altshuler, P. A. Lee, and R. A. Webb, "Mesoscopic Phenomena in Solids", *Modern Problems in Condensed Matter Sciences* **30**, edited by V. M. Agronovich and A. A. Maradudin, (North Holland, Amsterdam, 1991).
13. P. A. Lee, A. D. Stone, and H. Fukuyama, *Phys. Rev. B* **35**, 1039 (1987).
14. T. Onishi, M. Kawabe, K. Ishibashi, J. P. Bird, Y. Aoyagi, T. Sugano, and Y. Ochiai, *Phys. Rev. B* **48**, 12353 (1993).
15. K. Nikolic and A. MacKinnon, *Phys. Rev. B* **50**, 11008 (1994).
16. R. Harris and H. Guo, *Phys. Rev. B* **51**, 5491 (1995).
17. N. Zhu, H. Guo, and R. Harris, *Phys. Rev. Lett.* **77**, 1825 (1996).
18. J. C. Licini, D. J. Bishop, M. A. Kastner, and J. Melngailis, *Phys. Rev. Lett.* **55**, 2987 (1985).
19. C. P. Umbach, S. Washburn, R. B. Laibowitz, and R. A. Webb, *Phys. Rev. B* **30**, 4048 (1984).
20. T. J. Thornton, M. Pepper, H. Ahmed, G. J. Davies, and D. Andrews, *Phys. Rev. B* **36**, 4514 (1987).
21. A. K. Geim, P. C. Main, P. H. Beton, L. Eaves, S. P. Beaumont, and C. D. W. Wilkinson, *Phys. Rev. Lett.* **69**, 1248 (1992).
22. J. A. Nixon and J. H. Davies, *Phys. Rev. B* **41**, 7929 (1991); J. A. Nixon, J. H. Davies, and H. U. Baranger, *Phys. Rev. B* **43**, 12638 (1991).
23. S. Lüscher, A. Fuhrer, R. Held, T. Heinzel, K. Ensslin, and W. Wegscheide *Appl. Phys. Lett.* **75**, 2452 (1999).
24. C. W. J. Beenakker, *Phys. Rev. B* **44**, 1646 (1991).
25. S. Lüscher, T. Heinzel, K. Ensslin, W. Wegscheider, and M. Bichler, submitted

Properties of the Thue-Morse Chain

M. Noguez and R. A. Barrio

Instituto de Física
Universidad Nacional Autónoma de México
Apartado Postal 20-364
01000 México D. F.
MEXICO

Abstract

In recent years there has been a great progress in the theory of elementary excitations in non-periodic systems. In particular, electronic states in quasicrystals are peculiar because they seem to be neither localized, nor extended, but critical. That is the case of the Fibonacci chain. Of course, localization of states due to disorder in one dimension is ill-defined, since a chain is infinitely unstable facing disorder. However, there are other non-periodic one-dimensional systems that are constructed by inflation rules that are not quasiperiodic, and definitely neither disordered. A good example of these interesting systems is the Thue-Morse chain. The electronic tight-binding spectra of finite Thue-Morse chains show bands of perfectly extended states, together with a fractal distribution of gaps, whose edges present truly localized states. In this work we examine this spectrum exactly, in the limit of infinite chains, where there are dramatic changes to this picture.

I. Introduction

One dimensional non-periodic sequences are very useful objects in a wide variety of fields, as pure mathematics, computation, cryptography, language theory and linguistics, cellular automata and topological dynamics, just to mention a few. At first glance, one should think that these have no useful applications in physics, but one would be wrong. One dimensional systems are good models for real systems and are usually quite tractable mathematically. Their importance has increased enormously since new materials and technologies have been recently developed to realize one-dimensional arrays of atoms, or systems that behave very close to these. Examples are numerous, as quantum wires, superlattices made with layers of atoms that change in one direction, and solids with one-dimensional quasicrystalline order, as Al-Pd, Al_{80}-Cu_{20}-Mn_{15}, and many other alloys.

Physics of Low Dimensional Systems
Edited by J. L. Morán-López, Kluwer Academic/Plenum Publishers, New York 2001

A useful way to investigate the physical properties of these arrays is to find the states of a simple model Hamiltonian for quasiparticle excitations, which in the limit case of a periodic chain are the well known Bloch states, or plane waves that span the whole lattice. In a random chain these states are no longer possible, and all states become localized in one dimension. We shall be interested in a peculiar chain, called the Thue-Morse sequence, that is non-periodic, but non-random either.

The most popular example of a chain with these characteristics is the Fibonacci sequence,[1] which is a one-dimensional quasicrystal. This chain is constructed in a deterministic way by giving a rule for growing. This rule can be expressed in various ways, as a recursion formula, an inflation relation or by concatenation matching rules. It is obvious that one needs at least two kinds of objects to define the sequence, say A and B. The Fibonacci inflation rule is $A \to AB$ and $B \to A$, which produces a sequence of A's and B's that is never repeated, but that is quasicrystalline, since the inflation scaling length is irrational and equal to the "Golden Ratio" $\sigma = (1+\sqrt{5})/2$. The Thue-Morse chain can be defined by an inflation rule as well, but the sequence is not quasicrystalline, since the inflation scaling is 2. However, the sequence is non-periodic.

The spectra of excitations of quasicrystalline sequences have been subject of numerous studies recently[2] and their main features are: 1) There is a series of bands and gaps disposed in a fractal way. 2) The infinite chain presents only extended states, but the wave function is self-similar. 3) Any imperfection in the sequence produces a weak localization of practically all the states, which are called critical states, that form a Cantor set. These defects include surface atoms in finite chains.

The purpose of this paper is to show that the spectra of excitations in the Thue-Morse chain is different from the quasicrystalline case. In fact, we believe that the properties of the Thue-Morse chain are rather unique and interesting, due to its peculiar symmetry. There have been former works, either experimental,[3] numerical[4] and analytical[5] that prove the persistence of extended states, even in finite chains.

II. The Thue-Morse Chain

The Thue-Morse chain was first treated by Prouhet[6] when attempting to form a language with two letters with the restriction that no words should contain the same letter three consecutive times. It owes its name to A. Thue[7] and M. Morse,[8] who rediscovered it while studying thoroughly aperiodic sequences and problems in topological dynamics, respectively.

There are various ways of uniquely define the sequence, for instance, by an inflation rule similar to the Fibonacci chain:

$$\begin{aligned} A &\to AB, \\ B &\to BA. \end{aligned}$$

After n applications of the rule, one defines a chain S_n with 2^n sites, since one doubles the number of sites in each operation. For instance, if one uses as a starting elements $\epsilon_0 = A = -1$ and $\epsilon_1 = B = 1$, then

$$\epsilon_{2n} = \epsilon_n \quad \text{and} \quad \epsilon_{2n+1} = -\epsilon_n .$$

Analogously, if one defines the complement of a chain S_n as another chain S_n^c, in which one exchanges all A elements by B's, the next chain is exactly the concatenation

$$S_{n+1} = S_n + S_n^c .$$

Figure 1. The 4th generation of a Thue-Morse sequence of white and black balls.

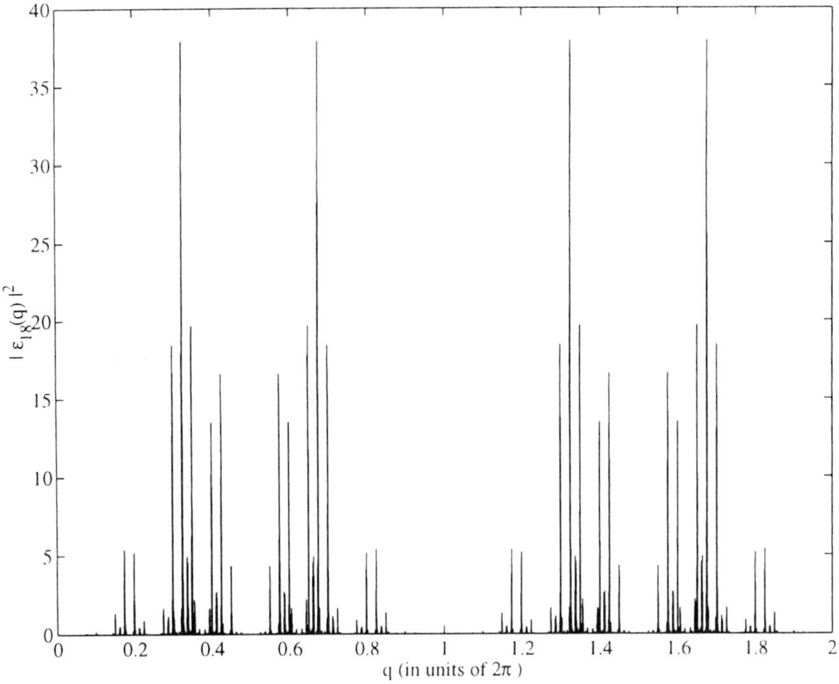

Figure 2. The static structure factor for the 18th generation of a Thue-Morse sequence.

In Fig. 1 we illustrate a Thue-Morse chain of generation $n = 4$, starting with a white atom on the left. Observe that chains of even n present full mirror symmetry, and the odd generations have anti-mirror symmetry. This fact is of great importance in understanding the behaviour of the eigenfunctions, as we shall see later. The Fourier transform, defined as

$$\epsilon_n(q) = \sum_{j=0}^{2^n-1} \epsilon_j e^{-i2\pi jq}, \tag{1}$$

produces an static structure factor that is singularly continuous:

$$S_n(q) = |\epsilon_n(q)|^2 = 2^n \prod_{j=0}^{n-1} \left[1 - \cos\left(2\pi 2^j q\right)\right]. \tag{2}$$

In Fig. 2 the structure factor for a chain of generation $n = 18$ is shown. Observe that there is a period of 2π and the spectrum is symmetric around $(2n+1)\pi$ and dense, since the amplitude of the peaks in between the gaps becomes increasingly small.

So far, we have exhibited only the so called diagonal problem, in which the sites are of two species. Analogously, one can define a Thue-Morse sequence of bonds.

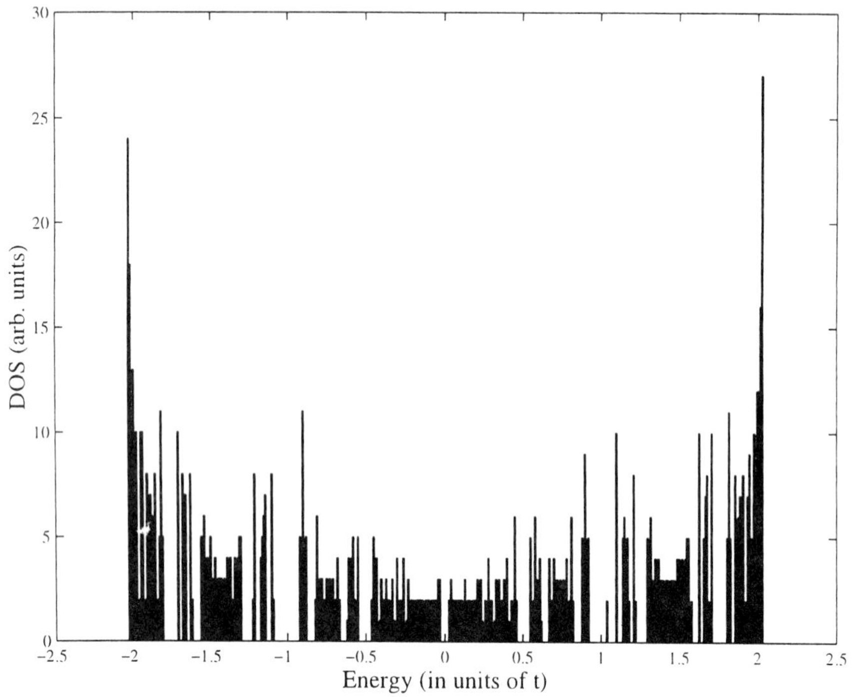

Figure 3. Total density of states for a chain with 16 384 sites, corresponding to site disorder in which $\varepsilon = -\varepsilon_c = 0.5$ and $t_i = 1$.

III. Model Hamiltonian

To study the excitation spectra in the Thue-Morse chain, we use a general form of a first-neighbours tight-binding Hamiltonian,

$$H = \sum_i \{(\varepsilon_i - 2\alpha t_i)|i\rangle\langle i| + t_i(|i\rangle\langle i+1| + cc)\} . \quad (3)$$

This is a convenient form since one can treat various interesting cases. For example, a s-band electron problem corresponds to $\alpha = 0$, in the site problem the self-energies ε_i vary from site to site, and in the bond problem, they are constant but the hopping integrals t_i follow the sequence with two different values, say t_L and t_S. A classical phonon Hamiltonian is obtained by setting $\alpha = 1$, and one can deal with mass or spring disorder at will. Also, the magnon case can be treated with the pertinent quasi-boson operators.

There are various ways to solve this problem and to obtain the density of states (DOS). The least imaginative one is to diagonalize the sparse tridiagonal matrix Eq. (3) by using one of the many efficient numerical algorithms in the literature. This procedure is easy but is restricted to chains with at most 10^5 sites. An example of such a calculation is give in Fig. 3 for a chain of generation $n = 14$.

Observe that the eigenvalues follow a general weight that corresponds to the linear chain. However, there is a complicated, but not fractal disposition of bands and gaps. There is a gap in the center of the spectrum that is surrounded by a true band of extended states. Here we first notice the drawbacks of dealing with finite chains,

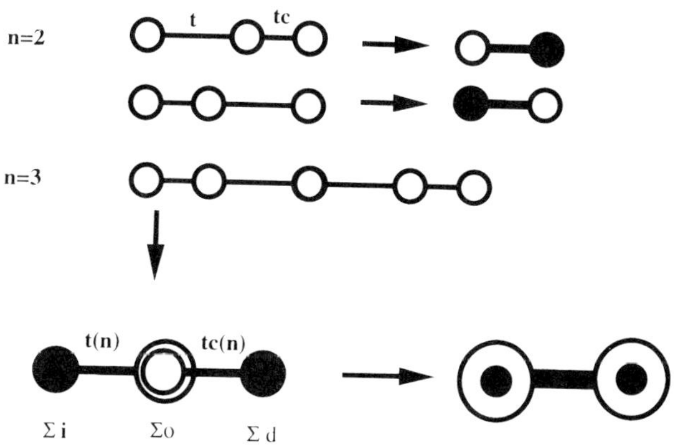

Figure 4. Scheme to renormalize the central sites in the Thue-Morse chain.

because one finds that the wave functions, outside of the central bands, are critical or localized, which is not true for a defect-less infinite chain, as we shall show below.

There is a second numerical method that allows to compute DOS for very long chains (10^{24}) with very little effort. This is based on a renormalization (decimation) procedure, similar to the one first reported for the Fibonacci sequence.[9] In the next section we shall illustrate this method in detail for the bond problem.

IV. Renormalization Procedure

Consider bond disorder. First we write the equations of motion obeyed by the resolvent Green's function of the Hamiltonian Eq. (3), for generation $n = 1$:

$$G = (EI - H)^{-1} = \begin{pmatrix} E - \Sigma & t \\ t^c & E - \Sigma \end{pmatrix}, \tag{4}$$

where the constant Σ gives the origin of energies, and there are two sorts of bonds. An equivalent system for the complement chain can be written by just interchanging $t \leftrightarrow t^c$. The problem to solve a chain of any generation can be reduced to this 2×2 problem if one renormalizes the coordinates of all intermediate sites. This can be done step by step, as illustrated in Fig. 4.

In every step one has to consider the central site with care, and find recursive expressions for the diagonal terms on the central atom $[\Sigma_0(n)]$, the atom on the left $[\Sigma_i(n)]$, on the right $[\Sigma_d(n)]$, as well as for the non-diagonal terms $[t(n)]$ and the same quantities for the complement chain. Explicitly, the needed recursion formulae are:

$$\begin{aligned} \Sigma_0(n) &= \Sigma_i(n-1) + \Sigma_d^c(n-1) - \Sigma_0(n-1), \\ \Sigma_i(n) &= \Sigma_i(n-1) + \frac{t^2(n-1)}{E - \Sigma_0(n)}, \\ \Sigma_d(n) &= \Sigma_d^c(n-1) + \frac{t_c^2(n-1)}{E - \Sigma_0(n)}, \\ t(n) &= \frac{t(n-1)t^c(n-1)}{E - \Sigma_0(n)}. \end{aligned} \tag{5}$$

The case of diagonal disorder can be treated in a similar fashion. Note that this case is more symmetric than the bond case, since one needs only two sites (instead of three sites and two bonds) to start the sequence. Therefore, there is no need for a central site self-energy. The recursion formulae for this case are different for odd or even generations, since that renormalized chains have to reflect the fact that odd chains possess anti-mirror symmetry and even generations produce mirror symmetry. For n odd,

$$t(n) = \frac{t(n-1)t_c(n-1)t(0)}{[E-\Sigma^c(n-1)][E-\Sigma(n-1)]-t(0)^2},$$

$$\Sigma(n) = \Sigma(n-1) + \frac{t^2(n-1)[E-\Sigma(n-1)]}{[E-\Sigma^c(n-1)][E-\Sigma(n-1)]-t(0)^2}. \tag{6}$$

To write down the equations for the complement, one exchanges all quantities by their complement. Observe that in this case $t(n) = t_c(n)$. For n even,

$$t(n) = \frac{t(n-1)t_c(n-1)t(0)}{[E-\Sigma^c(n-1)]^2-t(0)^2},$$

$$\Sigma(n) = \Sigma(n-1) + \frac{t^2(n-1)[E-\Sigma^c(n-1)]}{[E-\Sigma^c(n-1)]^2-t(0)^2}. \tag{7}$$

The recursion formulae can be written in terms of the quantities belonging to two generations before, so one recovers all symmetries and it is more convenient for the computer calculation. Other problems, as the phonon, or the magnon recursion relations can be obtained in similar way. Notice that the quantity that one obtains is the Green's function for either the central site, or the surface sites, that is, one obtains local responses. One can generalize this method to obtain total densities of states if one writes down recursion expressions for the imaginary part of the Green's functions, since the total DOS is defined as $\rho(E) = -\text{Im Tr}\, G(E)/\pi$. To illustrate the results from this renormalization method for very long chains, in Fig. 5(a) we show the averaged local DOS obtained by adding up the local densities in the four local configurations of sites (bonds) in the chain. Observe that there is a continuous band around $E = 0$. The bands obtained with this renormalization procedure for the diagonal disorder case are well reproduced (compare Fig. 5b with Fig. 3), but the relative weights are distorted because of locality. The phonon bands for spring disorder are shown in Fig. 3(c) and show that there are one-dimensional sound waves at low frequency with a perfectly defined velocity of sound, comparable with the ordered chain.

The symmetry of the lattice allows to calculate several properties of the spectra in an exact form, and we shall deal with this immediately.

V. Exact Solution for the Site Problem

Observe in Fig. 5 that the band edges of the spectrum are different from the ones expected in a perfectly ordered chain. Take for instance the site problem, in which we assume (without loss of generality) that $\Sigma_0 = -\Sigma_0^c = \Sigma$, and $t = 1$. In a perfect crystal of alternate sites, there will be two bands, whose exact edges are situated at $\pm\Sigma$ and $\pm\sqrt{\Sigma^2+4}$. In the Thue-Morse chain, these band edges can be calculated exactly by taking advantage of the mirror (anti-mirror) symmetry present in chains of any generation. In the limit $\Sigma \to \infty$ there are monomers (with a state at the center of

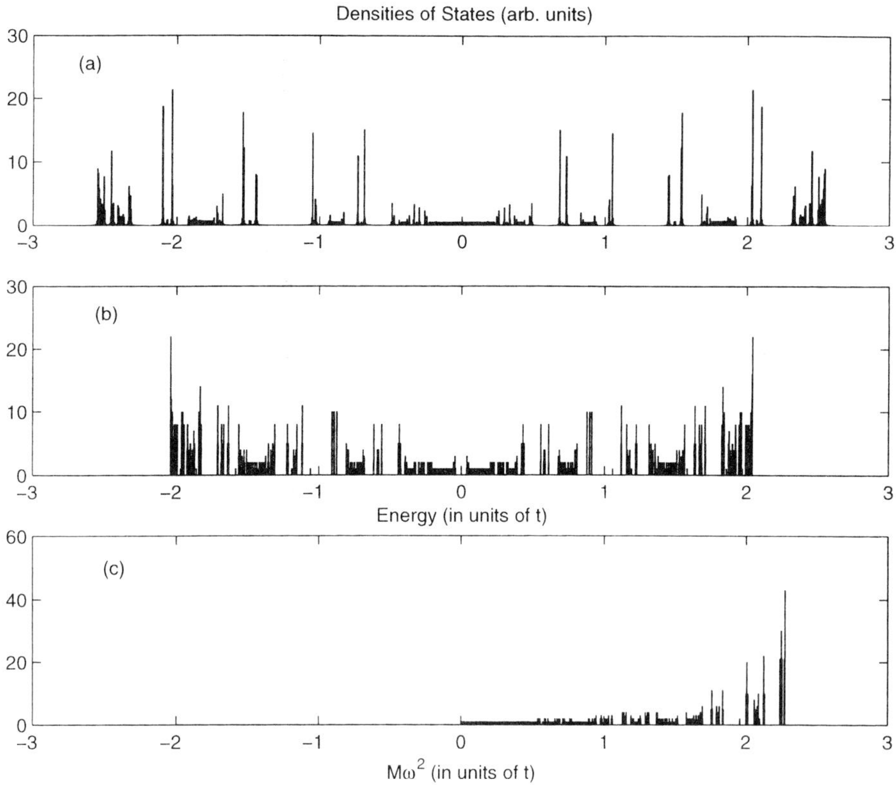

Figure 5. Densities of States for the Thue-Morse chain, calculated following the renormalization procedure for: (a) The bond problem with $t = 1$ and $t_c = 1.5$. (b) The site problem with $\Sigma_0 = -\Sigma_0^c = 0.5$. (c) The bond phonon problem with $t - t_c = 0.5$.

each sub-band), and dimers (with two split states). Therefore, the fundamental unit is a pair of unlike atoms, which could be thought to be attached to the rest of the lattice by an effective bond $K(E)$, which takes into account all the signals coming from the outside world. It is clear that this K should be a function of the energy and its exact form should be investigated in detail. In any case, the complete spectrum can be reduced to the secular equation

$$\Delta = \begin{vmatrix} E + \Sigma - K & 1 \\ 1 & E - \Sigma - K \end{vmatrix} = 0, \tag{8}$$

where all the energies are in units of t, and whose implicit solution is $E = K \pm \sqrt{\Sigma^2 + 1}$. Independently of the functional relation $K(E)$, the magnitude $|K| \leq 1$, and attains its limit values at the band edges, which are the totally symmetric or antisymmetric coherent states. Therefore, we know that the overall symmetry of the lattice allows the states at which $K = \pm 1$ and that these solutions correspond to the true band edges. In Fig. 6 we show a comparison between the wider bands of the Thue-Morse chain and the perfect diatomic 1d-crystal.

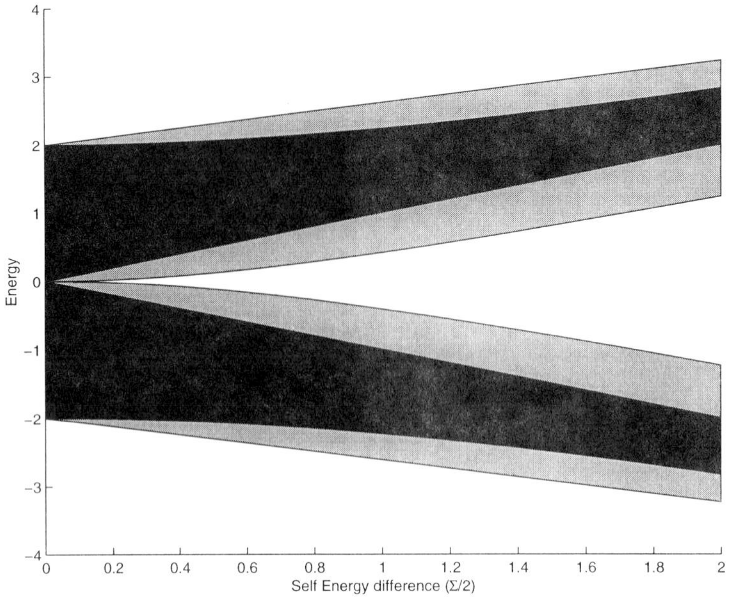

Figure 6. Comparison of the allowed bands between the Thue-Morse chain (outer limits) and the diatomic crystal (darker inner regions) as a function of site energy difference.

It is also possible to find *all* eigenstates exactly, if one examines the properties of the secular matrices for each generation. For example, consider a chain of generation $n = 1$ (two atoms). The secular equation to solve is

$$\Delta_1 = \begin{vmatrix} E+\Sigma & 1 \\ 1 & E-\Sigma \end{vmatrix} = 0.$$

For the next generation, we renormalize the central sites, as before, and the secular equation is

$$\Delta_2 = \begin{vmatrix} E-\Sigma-\dfrac{1}{E+\Sigma} & 1 \\ 1 & E-\Sigma-\dfrac{1}{E+\Sigma} \end{vmatrix} = {\Delta_1}^2 - (E+\Sigma)^2 = 0,$$

and the corresponding equation for the complement.

In general, one finds that

$$\Delta_n = \Delta_{n-2}\left[(\Delta_{n-2})\left(\Delta_{n-1} + \Delta^c_{n-1} - \Delta_{n-2}(\Delta_{n-2}-1)\right) + 1\right],$$

for $n \geq 3$. This result shows that one can find all the solutions for a chain of generation n as $\Delta_1 = \Delta_2 = \ldots = \Delta_{n-1} = 0$, and $\Delta_n = 1$. The last equation gives the extreme eigenvalues. Furthermore, if ϵ is a solution, then $-\epsilon$ also is a solution. This symmetry of the spectrum around $E = 0$ is a property of bipartite lattices, that is retained in the Thue-Morse chain because if one renormalizes every other atom in the sequence, one obtains exactly the same sequence, as one could easily verify in Fig. 1.

This exact solution is closely connected to the one found by Ghosh and Karmakar,[5] which is based on the trace map of the transfer matrices.[10] In all one-dimensional systems a useful way of solving the problem, either numerically or analytically, is the so called transfer matrix method. This method provides a solid basis to investigate the spectrum and the localization properties of the eigenfunctions in a deeper way. The 2×2 transfer matrix is defined by writing Shrödinger's equation as

$$\begin{pmatrix} \Psi_{n+1} \\ \Psi_n \end{pmatrix} = M_n \begin{pmatrix} \Psi_n \\ \Psi_{n-1} \end{pmatrix}.$$

If $\alpha_n = \text{Tr } M_n/2$, then

$$\alpha_n = 4^{n-2}\alpha_{n-2}^2\alpha_{n-3}^2 ... \alpha_2^2\alpha_1^2(\alpha_2 - 1) + 1.$$

with $\alpha_1 = (E^2 - \Sigma^2 - 2)/2$, and $\alpha_2 = 2\alpha_1^2 - 2\Sigma^2 - 1$.

Periodic boundary conditions imply that $\alpha_n = 1$. Therefore, the solutions are found as $\alpha_1^2 = \alpha_2^2 = \ldots = \alpha_{n-2}^2 = 0$ and $\alpha_2 - 1 = 0$. The last equation gives again the band edges, and the other equations give all the doubly degenerate eigenvalues. Ghosh and Karmakar also show that *all states are truly extended.* This is a consequence of the overall mirror symmetries of the Thue-Morse chain, that really distinguishes it from the 1d-quasicrystals, which lack this sort of symmetry. It is easy to see that for a block of n sites (a finite Thue-Morse chain), one can write the transfer matrix as a linear combination of Pauli matrices.

For n odd:

$$\begin{aligned} T_n &= \alpha_n I + \beta_n \sigma_x + \gamma_n \sigma_y + \delta_n \sigma_z , \\ T_n^c &= \alpha_n I - \beta_n \sigma_x + \gamma_n \sigma_y + \delta_n \sigma_z , \end{aligned}$$

and for n even:

$$\begin{aligned} T_n &= \alpha_n I + \gamma_n \sigma_y + \delta_n \sigma_z , \\ T_n^c &= \alpha_n I + \gamma_n' \sigma_y + \delta_n' \sigma_z . \end{aligned}$$

Introducing the eigenvalue conditions is equivalent to setting the values $\beta_n = 0$ and $\gamma_1^2 + \delta_1^2 = 1$. This implies that $[T_{n-1}, T_{n-1}^c] = 0$, and therefore all wave functions are extended: For each eigenvalue there is a "unit cell" that maps the Thue-Morse chain into a periodic lattice. Is is also shown that the Landauer resistivity is zero for all band states. The band edges are special in the sense that the Landauer resistivity scales as the square of the chain length, but this is due to the fixed boundary conditions at infinity.

VI. Conclusions

We have presented various methods to investigate theoretically the spectra of excitations of the Thue-Morse sequence. The structure factor is singularly continuous, as in the case of quasicrystals. The spectrum is found to be formed of extended states only, and many exact results are found. We can conclude that this type of disorder is unable to localize states in infinite chains. Moreover, in quasicrystals, due to the entangled (fractal) disposition of bands and gaps, any localized defects will produce localization in all the spectrum, and this is the reason why in finite Fibonacci chains one finds critical and localized states. In the Thue-Morse chain this picture is also present, although critical states are much harder to find in sequences with defects, since the states near the center of the spectrum remain fully extended.

In quasicrystals there are peculiar mistakes named phasons, which change the thermodynamic properties of the lattice. It would be very interesting to produce phason-like disorder in the Thue-Morse sequence, and this study is currently being carried out.

Acknowledgments

This work was supported by UNAM through grant DGAPA-UNAM projects IN104698 and a scholarship for MN from DGEP-UNAM.

References

1. Chumin Wang and R. A. Barrio, *Phys. Rev. Lett.* **61**, 191 (1988).
2. G. G. Naumis and J. L. Aragón, *Phys. Rev. B* **54**, 15079 (1996).
3. Z. Cheng, R. S. Savit, and R. Merlin, *Phys. Rev. B* **37**, 4375 (1988).
4. F. Axel and J. Perrière, *J. Stat. Phys.* **57**, 1013 (1989).
5. A. Ghosh and S. N. Karmakar, *Physica A* **274**, 555 (1999).
6. A. Prouhet, *C.R. Acad. Sci., Paris Sèr I* **33**, 31 (1851).
7. A. Thue, *Norske vid. Selsk. Skr. I. Math. Nat. Kl. Christiania* **7**, 1 (1906).
8. M. Morse, *Trans. Am. Math. Soc.* **22**, 84 (1921).
9. R. A. Barrio and Chumin Wang, *Proc. Third Int. Conf. on Quasicrystals and Incommensurate Phases in Condensed Matter*, edited by M. José Yacamán, D. Romeu. V. Castaño, and A. Gómez (World Scientific, Singapore, (1989) p. 448.
10. M. Kohmoto, B. Sutherland, and C. Tang, *Phys. Rev. B* **35**, 1020 (1987).

One-Dimensional Adsorbate Systems: Electronic, Dynamic, and Kinetic Features

W. Widdra and D. Menzel

Physik-Department E20
Technische Universität München
D-85747 Garching b. München
GERMANY

Abstract

Anisotropic lateral interactions between adsorbate species on single crystal surfaces can lead to one-dimensional (1D) behavior of various physical properties not only on anisotropic substrates, but even on low index, high symmetry surfaces, for which symmetry breaking by the adsorbate-adsorbate interactions can occur. Examples for both cases are demonstrated here for three different physical properties of adsorbate layers: the electronic adsorbate band structure, the surface phonon bands, and the desorption kinetics. As examples of 1D electronic band structures we use the systems Xe on H-modified Pt(110), and C_2H_4 on single-domain Si(100)-(2×1) surfaces; the reasons and some consequences of their behavior will be discussed, partly also in comparison to very similar 2D systems. 1D-dispersion of extrinsic surface phonon bands is found in (2×1)-row structures of adsorbed oxygen atoms as well as of (O+NO) coadsorbate layers on the hexagonal Ru(001) surface; comparison with dilute [(2×2)-O] and with dense adsorbate [(1×1)-O] and coadsorbate [(2×2)-(NO+3O)] systems on the same surface clearly shows the effect of dimensionality on the vibrational coupling of adsorbates. Finally, the direct influence of reduced dimensionality of an adsorbate system on surface kinetics will be demonstrated for the case of Xe on flat and stepped Pt(111) surfaces. The distinct modification of desorption kinetics observed here can be well understood in terms of the statistical thermodynamics of the adlayer, in particular the critical temperatures of the adsorbate condensate phases and its dimensionality dependence. In all cases the lateral adsorbate-adsorbate coupling—though via quite different coupling mechanisms in the various cases—underlies the found behavior.

Physics of Low Dimensional Systems
Edited by J. L. Morán-López, Kluwer Academic/Plenum Publishers, New York 2001

I. Introduction

Dimensionality is an important feature of physical systems, and systems with decreased dimensionality have received considerable interest in recent years. For instance, in semiconductor physics 2D electron gas layers, 1D quantum wires, and 0D quantum dots are being investigated very actively. The influence of dimensionality in phase transitions is another high interest field.

Here, we want to discuss the causes and the influences of dimensionality on atomic and molecular adsorbates: on their electronic and vibrational bandstructures, and on their kinetic behavior. It will become obvious that one-dimensional behavior can, as may be expected, be induced by an anisotropic surface structure which enforces strongly anisotropic adsorbate-adsorbate coupling; but that even on highly symmetric surfaces symmetry breaking to one-dimensional behavior can be induced by intrinsically highly anisotropic adsorbate-adsorbate coupling. The aim in these investigations is, of course, to understand the underlying physical mechanisms of interaction which lead to the observed behavior, and which can be very different in the different cases. We will also draw additional conclusions about the systems from these results and their detailed understanding.

II. Adsorbate Electronic Bandstructures

Two-dimensional adsorbate bandstructures have been investigated since many years.[1] One-dimensional surface states have been reported before on metal surfaces close to the Fermi level.[2] We are not aware of reports about one-dimensionality of adsorbate valence states prior to our own work. The two cases we show here concern rare gas adsorption on a one-dimensional structured transition metal surface, and a molecular adsorbate on a semiconductor surface. These two cases are very different in terms of the adsorbate bonding: mainly Van der Waals for the former and strongly covalent for the latter. The reason for one-dimensionality is similar in the two cases. It is clearly preformed by the substrate structure in the former. In the latter, there is an additional influence of the anisotropy of the adsorbate. The reason for band formation has similarities as well as dissimilarities in the two cases: while bands are formed from adsorbate orbitals by their overlap on neighboring adparticles (which is limited in extent by Pauli repulsion), the driving force for this overlap is attractive Van der Waals interaction for Xe, and the existence of preformed sites—the surface Si dimers—with covalent bonding for adsorbed ethylene on Si(100)-(2×1).

The method of choice to investigate the dispersion behavior of such adsorbate valence states is angle-resolved photoelectron spectroscopy (ARUPS). It has been preferably carried out using synchrotron radiation, because then the photon energy and angles can be chosen such that final states with $k_{\|}$ vectors over the entire surface Brillouin zone can be obtained with good accuracy and large cross sections. We use for our measurements a multichannel angle-resolving toroidal electron spectrometer developed here,[3] which is particularly suited for this purpose because we can measure simultaneously not only all polar angles, but the azimuthal angles of the light vector and of the detector orientation can be varied independently.

II.1 Xe on Pt(110)-(1×2)-H.

The Pt(110) surface reconstructs in the missing-row fashion, leading to rows of close-packed Pt surface atoms separated by furrows that are two lattice spacings deep. On

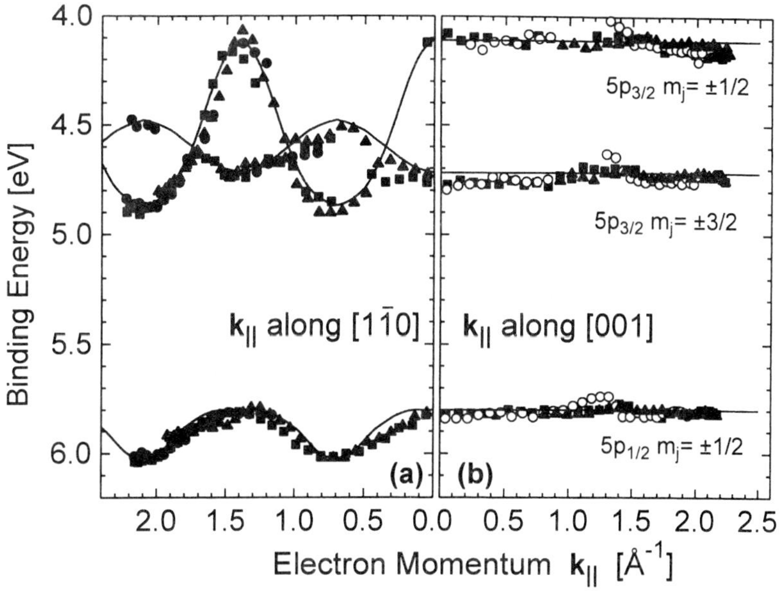

Figure 1. Electronic band structure of chains of Xe atoms on a H-modified Pt(110)-(1 × 2) surface, for momenta $\hbar k_{\|}$ along and perpendicular to the adsorbate chains. The solid lines are the theoretical results for the isolated adsorbate chain (Ref. 6).

such a surface Xe is adsorbed in two different species, likely to be bonded in the furrows and on the Pt rows, respectively. They have quite similar adsorption energies and can therefore not be prepared separately. On such a surface complicated dispersion behavior of the $5p$ levels of the two types of Xe is observed.[4] However, if a certain amount of hydrogen atoms is preadsorbed on the Pt(110) surface, the two types of Xe adatoms are influenced in such a way that now the more strongly bound species can be prepared in pure form.[5] We have reasons to believe that these Xe atoms sit in the troughs, but this has not yet been proven beyond doubt.

Figure 1 shows the band structure obtained for densely packed Xe-rows (only stronger bound Xe species) on the hydrogen-modified Pt(110)-(1 × 2) surface (Xe coverage 0.45 relative to Pt surface atoms, atoms equally spaced), in the two high-symmetry directions parallel to the surface. It is immediately obvious that no dispersion exists perpendicular to the Pt surface rows (which run in the [001] direction), while considerable dispersion (up to 800 meV, for the $5p_{3/2}$ state) results parallel to the Pt (and Xe) rows. Its analysis reveals a periodicity corresponding to 1.4 Å^{-1}, equivalent to a distance of 4.4 Å in real space, in good agreement with LEED results. Annealing the layer carefully leads to partial desorption. At a stage where about half of the Xe atoms are retained (overall coverage 0.23), short one-dimensional chains and free regions coexist, and within the chains the Xe-Xe distance relaxes to 4.52 Å, as shown by LEED. In this state the band width decreases (maximum bandwidth 600 meV, again for the $5p_{3/2}$ state). Finally, when the Xe atoms are spaced so far that the overlap decreases essentially to zero—which situation can be produced by adsorbing them at the steps of a Pt(997) crystal[5]—no dispersion is found, as expected: the system behaves zero-dimensionally. The band width as a function of Xe-Xe distance can be represented by an exponential relation, as expected for the overlap of wave functions.

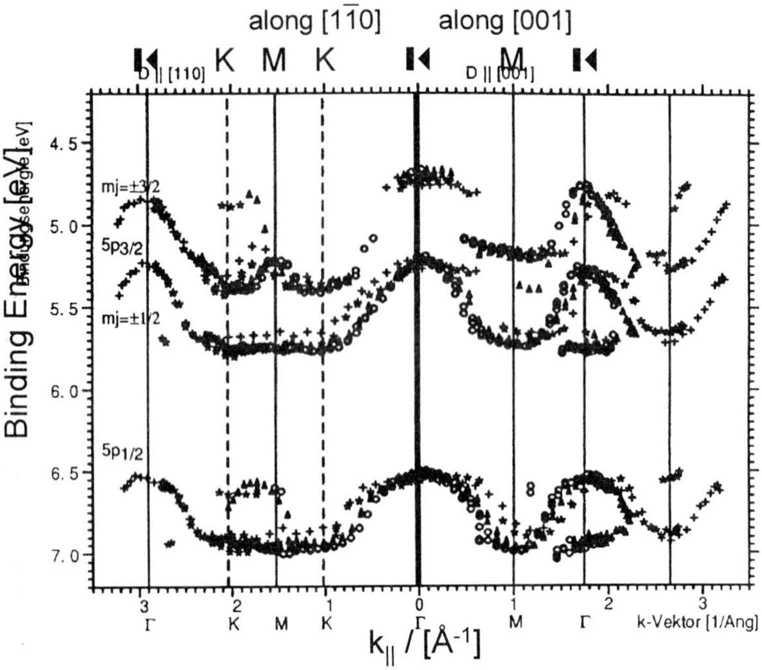

Figure 2. Electronic band structure of a hexagonal Xe layer on a H-modified Ni(110)-(1 × 2) surface, for momenta $\hbar k_{\|}$ along and perpendicular to the two high symmetry directions.[5]

The given assignment to the spin-orbit sublevels of Xe $5p$—the $5p_{3/2}$ level split into $m_j = \pm 1/2$ and $\pm 3/2$, with the former having lower binding energy at Γ—has been derived from measurements at appropriately chosen geometries and corroborated by comparison with fully relativistic first principles calculations for *isolated* Xe chains[6] at the found distance. The results are contained in the figure and are seen to be able to reproduce the observed dispersion—including the crossing of sublevels—very well. The reasons for the directions of dispersion can be understood in detail from the wave functions involved.

A very important result of this analysis is the energetic reordering of the $5p$ sublevels in this one-dimensional coupled state, compared to a two-dimensional layer. While in a 2D hexagonally close-packed layer the $5p_{3/2}$ ($m_j = \pm 3/2$) state is that of lowest binding energy, in the one-dimensional chain the order is reversed. In earlier work these splittings had been attributed to crystal field effects and/or to differences in final state screening;[7] it is now clear that lateral interactions are the reason. The interesting point with regard to dimensionality is that in the 2D system the main axis of the electric field gradient (EFG), due to the charge distribution of the neighboring atoms, is parallel to the surface *normal*; while the charge distribution due to the co-adatoms lies *in* the surface plane. This then leads to the observed quadrupole splitting of the magnetic sublevels with the given energetic ordering. In the 1D chain the main axis of the EFG is *in* the chain axis, and the charge distribution has the same axis. This combination leads to the observed reversal of the ordering of sublevels which at the same time is a corroboration of the proposed mechanisms of coupling. Indeed, when a Ni(110) surface is covered with Xe to saturation, leading to a 2D hexagonal Xe layer (Fig. 2), the measured band structure shows the expected ordering of sublevels

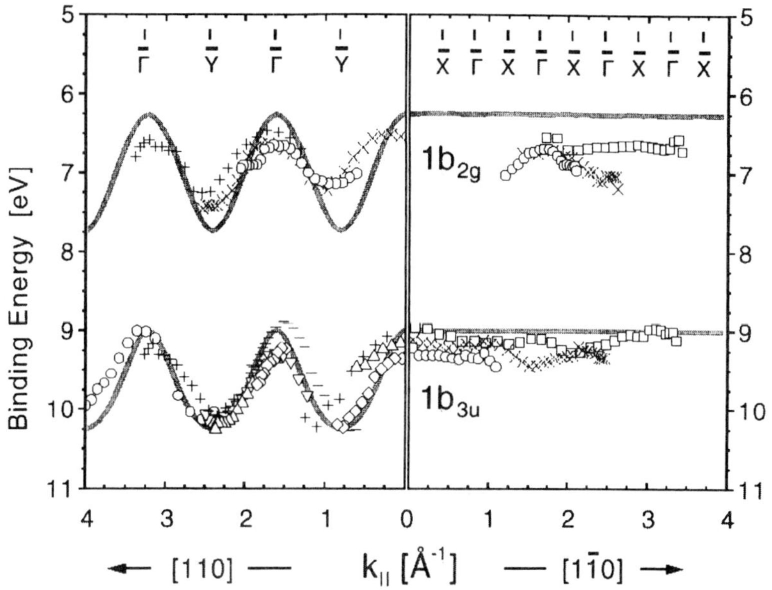

Figure 3. Electronic band structure of ethylene molecules adsorbed on a (single-domain) Si(100)-(2 × 1) surface, along and perpendicular to the Si-Si dimer rows. The full lines are the theoretical results for the (unrotated) adsorbate structure with C_{2V} symmetry.[8]

without any crossings. In fact all dispersions are essentially parallel. For Xe saturation on the Pt(110)-(1 × 2)-H surface an intermediate situation is encountered which can be interpreted on the basis of a coupling between two different one-dimensional Xe chains. All these features can be fully understood from the effects of dimensionality on coupling.

II.2. C_2H_4 on Si(100), and Other Similar Systems

The adsorption of unsaturated hydrocarbons on Si and Ge surfaces is strong and shows high site specificity. For C_2H_4 on Si(100) the behavior of the Si surface states and of the ethylene-derived valence states as observed in angle-resolved photoemission, and the comparison with first principles density functional calculations (LDA with gradient corrections)[8,9] shows that the ethylene molecule is bound by di-σ bonds to the dangling bonds of the surface dimers and saturates them. Recently the corresponding geometry has also been found by photoelectron diffraction.[10] A surprising result of the ARUPS measurements was that the symmetry of the adsorbate complex, as derived using dipole selection rules, turned out to be reduced to C_2 from the expected C_{2v}. This, as well as the reasons for the 1D behavior observed, became understandable by a detailed analysis.

Figure 3 shows the measured band structure of C_2H_4 on single-domain Si(100)-(2 × 1) for two occupied C_{2v}-derived states. The 1D behavior is again very striking. The dispersion along the dimer rows (left side) stems from the overlap of the corresponding wave functions which is determined by the site selectivity of the covalent adsorbate-substrate bonds. The figure also contains the results of the LDA calculations assuming the full C_{2v} symmetry of the adsorbed C_2H_4 molecules. It is seen that, while they agree with the main characteristics of the measurements, the dispersion in particular of the $1b_{2g}$ derived valence state is calculated too high. Further optimization of the calculations explained both this discrepancy and the lower symmetry of the real situation. Very good agreement was reached by rotating the molecules around their centers such that the overlap of the $1b_{2g}$-derived states is decreased. This rotation leads at the same time to the symmetry reduction seen in polarization-dependent ARUPS. Thus the 1D behavior is forced upon the system by the strong site-specific bonds to the dimers; the latter's distances are such that overlap of the most extended wave functions of occupied valence states is forced which leads to Pauli repulsion. The system optimizes the total energy by the rotation of the molecules which decreases the dispersion again. On Ge(100) and on Si/Ge(100) surfaces smaller bandwidths are found which fits well the expectation of the wider lattice spacings and the decreased bonding strength.[11] It is seen that also in this case the detailed results and their consistent interpretation lead to a major step forward in the understanding of the adsorbate bond and the adsorbate-adsorbate interactions.

III. Adsorbate Phonon Bands

Adsorption converts the free translations and rotations of molecules into frustrated movements on the surface, in addition to modification of the internal vibrations. All the resulting normal modes can couple to each other which leads to (extrinsic) surface phonon bands. Several mechanisms contribute to this vibrational coupling which leads to frequency shifts; they can be classified into static and dynamic interactions. For the former a further distinction of field-induced (*e.g.* by static electric dipole fields) and "chemical" shifts (caused by the modification of the electronic and mechanical properties of the adsorbate by its surroundings) is often made, although these effects are interconnected so that such a separation is in most cases not really possible. As to dynamical interactions, the vibrations of neighboring adsorbates can couple to each other either through the substrate, or via the coupling of the dynamical electric dipole moments of an adparticle with the electric dipole fields of the neighbors. (The first type of coupling is important for low frequency adsorbate modes in the range of the substrate phonons, while the second one predominates for high frequency, strongly infrared-active internal modes.) For periodically arranged adsorbates the formation of a phonon band structure is expected which again can have different dimensionality. In the present context the possibility of 1D behavior is of interest. The method of choice for detailed measurement of (intrinsic and extrinsic) surface vibrations and their dispersion is high resolution electron energy loss spectroscopy (HREELS).[12] We use a spectrometer with a resolution of vibrational losses better than 1 meV which can be used at any momentum transfer, so that surface phonon band dispersions can be determined.

We have observed 1D adsorbate phonon bands for adsorbate and coadsorbate systems consisting of one-dimensional adsorbate rows. Interestingly these rows are formed on a high symmetry, hexagonally close-packed metal surface, the Ru(001) surface. If half a monolayer of oxygen atoms is adsorbed on such a surface, close-packed rows of oxygen atoms along one of the main axes of the surface are formed, separated from each other by a factor $\sqrt{3}$ of the lattice constant: a (2×1)-O structure is formed[13] in which all O sit on three-fold *hcp* sites. This breaking of the symmetry of the close-packed surface by the adsorbed O atoms indicates a strongly anisotropic lateral interaction of the O atoms. An approximately 1D dispersion is found for the perpendicular vibration of the O atoms.[14] If such a surface is exposed to NO at low temperatures, the rows of vacant *hcp* sites are covered by NO molecules bound upright

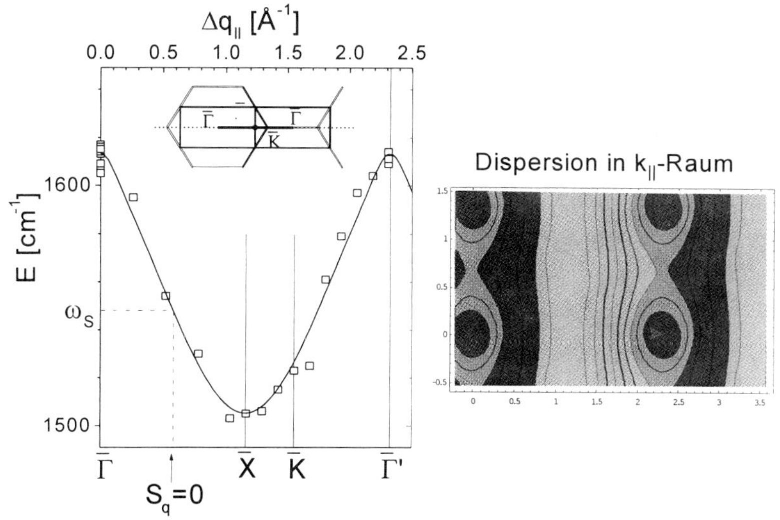

Figure 4. Dispersion of the internal N-O stretching vibration in the quasi one-dimensional Ru(001)-(2×1)- (NO+O)-phase along the NO chains (left side). The line shows the theoretical result for dipole-dipole coupling with best-fit parameters. On the right, the dispersion in the two-dimensional surface Brillouin zone is given schematically in a contour plot. The arrow gives the direction along the NO chains, Γ × Γ. From Ref. 14.

with the N end down:[15] a (2 × 1)-(O+NO) structure with alternating rows of O atoms and NO molecules, all in the same (*hcp*) sites, results. Obviously the modification of the surface by the O-rows is strong enough to make the very close approach of the NO molecules (singly spaced along the rows) possible (pure NO saturates the surface at 0.75 monolayers). We have investigated the phonon bands of this system. The interesting band dispersion is that of the internal stretch vibration of the NO molecules. As shown in Fig. 4, it distinctly disperses down from the maximum frequency at the center of the surface Brillouin zone, as expected for an optical mode.[14] Perpendicular to it much less dispersion is observed (by about a factor 3). The reason for the incomplete 1D behavior is the non-zero interaction between *different* NO rows due to the long-range character of the dipole-dipole interaction.

For a 2D periodic structure containing the same adsorbates, 2D dispersion is observed. For instance, the (2 × 2)-(3O+NO) layer on the same surface consists of a hexagonal arrangement of O atoms with a (2×2) arrangement of vacancies, into which NO molecules can be adsorbed; all adparticles sit again in *hcp* sites.[16] The dispersion of the NO stretch vibration is now the same in the two main axes, and because of the larger NO-NO spacing the band width is smaller. In the full (1 × 1)-O layer and the (2 × 2)-O layer, 2D dispersions of the perpendicular vibration of the O atoms against the substrate are observed,[14,17] which show the same characteristics but—due to the density difference—are different in magnitude by a factor of 8. All these cases can be quantitatively understood in terms of dynamical dipole-dipole coupling with coupling parameters which can be consistently derived from the experiments[17] and can be used to extract the frequency of the isolated particle on the surface, the 'singleton'. The different dimensionality of the adsorbate phonon bands leads to different properties of the corresponding two-phonon bands. In particular, the modified density of states of the two-phonon band will lead to substantially changed interaction with the overtone which is important to understand the bond anharmonicity of adsorbates.[18]

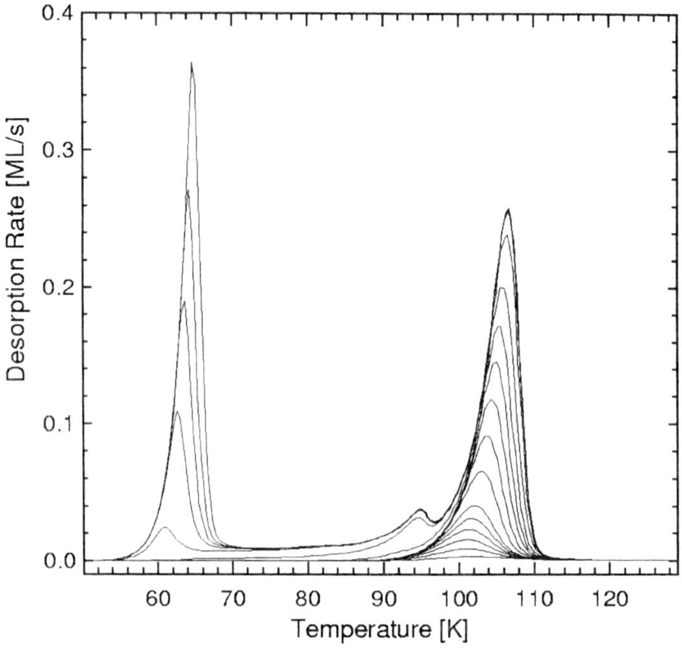

Figure 5. Temperature Programmed Desorption spectra for different Xe coverages on Pt(111) in the range up to 1.8 monolayers. The first monolayer desborbs between 85 and 120 K; the peak around 65 K is due to the second monolayer. The heating rate was 2.0 K/s. From Ref. 21.

IV. Dimensionality and Surface Kinetics

Lateral interactions among adsorbates are the cause of the effects described so far; both, for the formation of the observed structures as well as for their properties. Lateral interactions also have a strong influence on the statistical thermodynamics of adsorbate layers, which in turn influences the kinetics of desorption. In fact, if the dynamical part of the desorption rate—which is given by the sticking coefficient—possesses only small variations with coverage and temperature, and quasi-equilibrium persists. Then the kinetics of desorption is dominated by the thermodynamical properties of the layer.[19] Since a change of dimensionality influences the statistics, this will also affect desorption kinetics. In the following we show an example of such an effect.

For detailed kinetic analysis of thermal desorption, high quality data are required as can be obtained using high resolution temperature programmed desorption (HR-TPD) as developed here.[20]) We have used this methodology to investigate Xe adsorption on various Pt surfaces. Xe adsorption on a flat Pt(111) surface leads, as on many close-packed transition metal surfaces, to an ordered $(\sqrt{3} \times \sqrt{3})$-R30 structure.[21] Its thermal desorption leads to traces (Fig. 5) which are characterized by common leading edges over a wide range of coverages. This zeroth order behavior is due to the existence of a two-dimensional equilibrium between the 2D condensate and 2D gas during desorption. We conclude that the critical temperature of the ordered phase lies at about 120 K, *i.e.* above the desorption temperature. Detailed thermodynamic analysis using kinetic lattice gas theories[22] yields consistent data for the lateral interactions (attractive on near-next neighbor sites because of Van-der-Waals attraction,

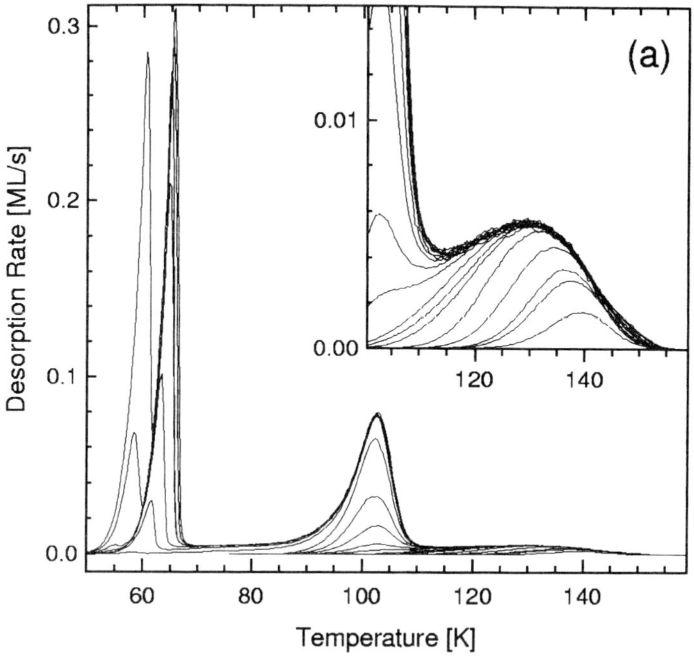

Figure 6. Temperature Programmed Desorption spectra for different Xe coverages on Pt(997) in the coverage range of 2.8 monolayers. The peak between 85 and 110 K is due to desorption from the first monolayer on the terraces, that between 110 and 150 K (see also insert with an amplified scale) to desorption from the steps. The peaks between 55 and 65 K stem from the second and third layers. The heating rate was 1.0 K/s. From Ref. 21

reduced by repulsive dipole-dipole interactions; repulsive on next-neighbor sites because of additional Pauli repulsion) and allows to reproduce the TPD traces. If the same experiment is carried out on the Pt(997) surface which consists of (111) terraces 7 lattice spacings wide separated by monoatomic steps, then no zeroth order behavior is found for the Xe adsorbed on the terraces (Fig. 6, low temperature peak). Indeed, thermodynamic evaluation leads to virtually identical interaction parameters but shows that now the critical temperature is *below* the desorption temperature, so that more complicated kinetics result.

This decrease in the critical temperature—leading to the observed drastic change of desorption kinetics—is due to the fact that now in one direction a finite (low) number of interacting adsorbate atoms exist. Calculations with a suitable lattice gas model show that this finite size effect, the breakdown of 2D condensate-gas equilibrium below the desorption temperature with consequent change of desorption kinetics from zeroth to first order, is expected for terraces below about 9 atom rows wide. Already for 11 row wide terraces, persistent phase equilibrium and zeroth order kinetics are retained.

A second, more strongly bound Xe adsorbate state (Fig. 6) exists at the steps which however stays much farther apart (minimum next-neighbor distance along the steps 5.5 A) because of the very high dipole moment of this species and concomitant strong repulsion. We have discussed in Sec. II.1. above that this leads to electronically 0D behavior, closing the circle of our report on dimensionality and lateral interactions in adsorbate layers.

V. Summary and Conclusions.

We have shown in this survey that anisotropic lateral interactions of adsorbates can lead to reduced dimensionality of the adsorbate properties. These anisotropies can stem from the anisotropic structure of the surface "template", or can arise by intrinsically anisotropic interactions which can break the symmetry of the surface. The resulting 1D behavior is clearest for the electronic band structure of adsorbate valence bands; this is understandable from the rapid (*i.e.* exponential) decrease of the lateral overlap of adsorbate orbitals with increasing distance, as has been demonstrated for two very different types of adsorbate systems. Xenon chains on Pt(110) can be produced with variable spacings and concomitant band width; the resulting inter-adsorbate distances stem from the counteracting forces of Van der Waals attraction, and dipole-dipole plus Pauli repulsion. Compared to situations leading to 2D layers, the 1D chains possess different energetic ordering and dispersion of the spin-orbit split $5p$ bands of Xe which are understandable in terms of the relative orientations of the axes of electric field gradient and charge distribution; this clearly demonstrates the qualitative influence of reduction of dimensionality. For covalently bonded hydrocarbons—example ethylene on Si(100)—1D behavior of the electronic band structure results from arrangement of the surface Si dimers. Their distance is smaller than that which would allow easy relative arrangement of neighboring molecules, so that Pauli repulsion leads to energetic optimization by rotation of the molecules.

For phonon bands derived from inner vibrations of adsorbate molecules, the fall-off with distance of the relevant lateral interaction, dynamic dipole coupling, is much slower. Therefore even for preformed chains of molecules (such as NO coadsorbed with O atoms on Ru(001) with twice the distance between chains than that within the close-packed chains) some dispersion is observed perpendicular to the chains. Still the dispersion along the chains is much stronger, and the behavior can be understood quantitatively in terms of the mentioned interaction mechanism. Again a comparison with similar 2D systems is very illuminating for the analysis of interaction, and allows the separation of static and dynamic contributions.

Finally, the reduction of dimensionality by going from "infinitely" extended terraces to narrower ones has a clear effect on the statistical thermodynamics of the layer which expresses itself in the critical order-disorder temperatures of 2D condensates. Since thermodynamics dominates desorption kinetics for unstructured sticking coefficients, the latter behaves very differently for the two cases. For Xe on Pt(111) and on narrow stripes of the same surface [Pt(997)] the desorption traces indicate zeroth order desorption for the former and more complicated (close to first order) behavior for the latter case, even though the interaction parameters extracted from the results are very similar in both cases. This means that for the extended system, 2D equilibrium between 2D gas and condensate persists up to desorption, while for the latter disorder happens before desorption; demonstrating a finite size effect in one dimension for desorption.

These examples show that anisotropic interactions leading to reduced dimensionality in adsorbate layers have qualitative effects for electronic, vibrational, and thermodynamic/kinetic properties. They can all be fully understood by suitable theories and used to extract the details of interaction mechanisms and parameters.

Acknowledgments

We sincerely thank the many people involved in this work, both in obtaining, evaluating, and interpreting data, and in developing and applying theoretical concepts and calculations which were an absolute prerequisite for the obtained understanding. Among the first group we particularly want to mention the graduate students A. Fink T. Moritz, and P. Trischberger; in the second our theorist collegues J. Henk (Duisburg/Uppsala), U. Birkenheuer and N. Rösch (TU München), and H. J. Kreuzer (Dalhousie university). It was always a great pleasure to work with them. The help of the BESSY staff was mandatory for the success of the synchrotron work.

These projects have been supported financially by the Deutsche Forschungsgemeinschaft in Sonderforschungsbereich 338, by the German Ministry of Education and Research under grants 05625WOA and 05SF8WOA and by the funds of the chemical industry.

References

1. N. V. Richardson and A. M. Bradshaw, in: Electron Spectroscopy-Theory, Techniques and Applications, Vol. 4, edited by C. R. Brundle and A. P. Baker, (Academic Press, New York, 1981); E. W. Plummer and W. Eberhardt, *Advances Chem. Phys.* **49**, 533 (1982); H.-P. Steinrück, *Vacuum* **45**, 715 (1994).
2. R. Courths, B. Cord, H. Wern, H. Saalfeld, and S. Hüfner, *Solid State Commun.* **63**, 619 (1987); U. Bischler and E. Bertel, *Phys. Rev. Lett.* **71**, 2296 (1993); A. Biedermann, O. Genser, W. Hebenstreit, M. Schmid, J. Redinger, R. Podloucky, and P. Varga, *Phys. Rev. Lett.* **76**, 4179 (1996).
3. H. A. Engelhardt, W. Bäck, D. Menzel, and H. Liebl, *Rev. Sci. Instrum.* **52**, 835 (1981); H. A. Engelhardt, A. Zartner, and D. Menzel, *Rev. Sci. Instrum.* **52**, 1161 (1981).
4. M. Weinelt, P. Trischberger, W. Widdra, K. Eberle, P. Zebisch, S. Gokhale, D. Menzel, J. Henk, R. Feder, H. Dröge, and H. P. Steinrück, *Phys. Rev. B* **52**, R17048 (1995).
5. P. Trischberger, H. Dröge, S. Gokhale, J. Henk, H.-P. Steinrück, W. Widdra, and D. Menzel, *Surf. Sci.* **377-379**, 155 (1997); P. Trischberger, Ph. D. Thesis, TU München (1998).
6. W. Widdra, P. Trischberger, and J. Henk, *Phys. Rev. B* **60**, R5161 (1999).
7. K. Horn, M. Scheffler, and A. M. Bradshaw, *Phys. Rev. Lett.* **41**, 822 (1978); B. J. Waclawski and J. F. Herbst, *Phys. Rev. Lett.* **35**, 1594 (1975); J. A. D. Matthew and M. G. Devey, *Solid State Phys.* **9**, L413 (1976); P. R. Antoniewicz, *Phys. Rev. Lett.* **38** (1977) 374; M. Scheffler, K. Horn, A. M. Bradshaw, and K. Kambe, *Surf. Sci.* **80**, 69 (1979); J. Henk, and R. Feder, *J. Phys. Cond. Matter* **6**, 1913 (1994).
8. W. Widdra, A. Fink, S. Gokhale, P. Trischberger, U. Gutdeutsch, U. Birkenheuer, N. Rösch, and D. Menzel, *Phys. Rev. Lett.* **80**, 4269 (1998).
9. U. Birkenheuer, U. Gutdeutsch, N. Rösch, A. Fink, S. Gokhale, P. Trischberger, D. Menzel, and W. Widdra, *J. Chem. Phys.* **108**, 9868 (1998).
10. P. Baumgärtel, R. Lindsay, O. Schaff, T. Giessel, R. Terborg, A. M. Bradshaw, D. P. Woodruff, M. Carbone, R. Zanoni, and M. N. Piancastelli, *J. Chem. Phys.*, (in press).
11. A. Fink, R. Huber, T. Moritz, D. Menzel, and W. Widdra, to be published.

12. H. Ibach and D. Mills, *Electron Energy Loss Spectroscopy and Surface Vibrations* (Academic Press, New York 1982).
13. H. Pfnür, G. Held, M. Lindroos, and D. Menzel, *Surf. Sci.* **220**, 43 (1989).
14. T. Moritz and W. Widdra, in preparation.
15. M. Stichler and D. Menzel, *Surf. Sci.* **419**, 272 (1999).
16. K. L. Kostov, D. Menzel, and W. Widdra, *Phys. Rev. B* **61**, (2000).
17. T. Moritz, D. Menzel, and W. Widdra, *Surf. Sci.* **427-428**, 64 (1999).
18. T. Moritz and W. Widdra, in preparation.
19. See papers of W. Brenig and D. Menzel, in "Kinetics of Interface Reactions", edited by H.J. Kreuzer and M. Grunze, *Springer Series in Surf. Sci.* **8**, (Springer, Berlin, 1987).
20. H. Schlichting and D. Menzel, *Rev. Sci. Instrum.* **64**, 2013 (1993); *Surf. Sci.* **272**, 27 (1992); **285**, 209 (1993).
21. W. Widdra, P. Trischberger, W. Frieβ, D. Menzel, S. H. Payne, and H. J. Kreuzer, *Phys. Rev. B* **57**, 4111 (1998).
22. H. J. Kreuzer and S. H. Payne, *Surf. Sci.* **188**, 235 (1988); **205**, 153 (1988).

Surface States on Clean and Adsorbate-Covered Metal Surfaces

J. Osterwalder,[1] T. Greber,[1] J. Kröger,[1] J. Wider,[1] H.-J. Neff,[1] F. Baumberger,[1] M. Hoesch,[1,2] W. Auwärter,[1] R. Fasel,[3,*] and P. Aebi[3]

[1] *Physik-Institut*
Universität Zürich
Winterthurerstrasse 190
CH-8057 Zürich
SWITZERLAND

[2] *Swiss Light Source Project*
Paul-Scherrer-Institut
CH-5232 Villigen-PSI
SWITZERLAND

[3] *Institut de Physique*
Universite de Fribourg
CH-1700 Fribourg
SWITZERLAND

Abstract

We review our research activities that cover the preparation and characterization of clean metal surfaces and monolayer films in view of using them as a convenient laboratory for studying and manipulating electronic properties in two or less dimensions. In particular, the interactions of surface states with adsorbed molecular and atomic layers, as well as with periodic arrangements of steps as conveniently found on vicinal surfaces, are studied. Photoemission intensity mapping is used in order to map out Fermi surface contours and band dispersions of these systems. The following examples will be discussed: Na adsorption on Al(001) occurs in two different structures, both with a c(2×2) symmetry, depending on the temperature. The Al(001) surface state reveals its different contributions to the chemical bonding in the two situations. On hydrogen-saturated Mo(110) a previously observed phonon anomaly can be related to the occurrence of nested features in the surface state Fermi surface with the proper nesting vectors. On vicinal Cu(111) surfaces the interaction of the Shockley surface state, forming a two-dimensional electron gas, with clean and decorated steps is monitored.

Physics of Low Dimensional Systems
Edited by J. L. Morán-López, Kluwer Academic/Plenum Publishers, New York 2001

I. Introduction

Electronic surface states arise as a consequence of the broken translational symmetry along the surface normal. Periodic boundary conditions for the wave functions inside the solid no longer apply and new solutions to the Schrödinger equation exist typically inside the band gaps of the surface projected three-dimensional band structure.[1,2] A particularly simple and illuminating description for the formation of such surface states is the phase accumulation model that has been brought forward by Smith *et al.*[3] In this model, electrons are trapped by the surface potential step, which evolves into the image potential at farther distances, and on the bulk side by a band gap which exists in the solid for particular momentum components $\vec{k}_{\parallel}$ parallel to the surface. A whole series of states can form for each value of $\vec{k}_{\parallel}$, some of which may be occupied and others not.

On semiconductor surfaces it has long been clear that surface and interface states are crucial for device performance. For metal surfaces it is less clear how surface states are related to their physical and chemical properties. In recent years some research has been devoted to this question. The work by the Bertel group[4] has linked the absence of occupied surface states on transition metal surfaces to the occurrence of non-activated dissociative hydrogen chemisorption. If the surface state is partially occupied, the additional electronic charge density far outside the surface pushes the repulsive part of the molecule-surface interaction potential away from the surface and an activation barrier for dissociative chemisorption results. The same group has also shown that occupied surface states may modify the potential energy landscape for the surface diffusion of atoms, notably by reducing the Schwoebel-Ehrlich barrier across surface steps.[5]

In this brief review we discuss two cases where surface states can be connected directly to adsorbate bonding (Sec. III) and to phonon instabilities (Sec. IV). In a last example we show how surface states can represent a two-dimensional electron gas (2DEG) that interacts with lateral structures of nanoscale dimensions, which are readily available in the form of regular arrays of steps on vicinal surfaces (Sec. V). Through angle-resolved photoemission experiments a very detailed picture can be obtained of its energy and momentum distribution and how it reacts to surface modifications such as induced by the presence of adsorbates or surface steps.

II. Experimental

In photoemission at ultraviolet energies the electron momentum component $\vec{k}_{\parallel}$ is conserved.[6] For surface states, where $\vec{k}_{\parallel}$ represents the full set of spacial quantum numbers, the complete energy dispersion relation can thus be measured by taking photoemission spectra for a large set of emission angles. From the binding energy E_B of a photoelectron line, the polar emission angle θ and the exciting photon energy $h\nu$ one calculates the parallel momentum as

$$|\vec{k}_{\parallel}| = \frac{1}{\hbar}\sqrt{2m(h\nu - \Phi - E_B)}\,\sin\theta. \tag{1}$$

Here, Φ is the work function of the surface under study. Combined with the azimuthal angle ϕ of the photoemission measurement, which is measured from some high-symmetry direction of the surface, this defines the exact position of $\vec{k}_{\parallel}$ with respect to the surface Brillouin zone, and $E_B(\vec{k}_{\parallel})$ can readily be plotted.

In our groups we typically produce two types of photoemission data sets: (*i*) Dispersion plots are generated by taking full photoemission spectra for a continuous series of polar or azimuthal angles.[7] If the measured intensities are plotted as gray scale values in a binding energy *versus* angle diagram, bands dispersing along this particular $\vec{k}$-space section become apparent (see Figs. 1 and 3). Note that $\vec{k}_{\|}$ depends both on the angles and on the binding energy (Eq. 1). (*ii*) Constant energy maps or, more specifically, Fermi surface maps result from taking photoemission intensities at just one binding energy and scanning both emission angles over most of the hemisphere above the surface.[8,9] If the Fermi energy E_F has been selected for this procedure, the resulting two-dimensional gray scale plot shows Fermi surface contours (see Fig. 2).

In our apparatus, which is described extensively elsewhere,[10,11] this angular scanning is done sequentially by means of a computer-controlled high-precision sample goniometer. Dispersion plots typically contain 50 angular settings, Fermi surface maps about 4000. Measuring times range from 30 minutes up to a few hours, depending on the sampling density, the selected angle and energy resolution and the photoelectric cross sections. For the data presented in Figs. 1–3, the angular resolution is of the order of $\pm$ 1° and the energy resolution is better than 40 meV. He I radiation was used for excitation providing a photon energy of 21.21 eV.

Sample preparation procedures have been described in the original papers: Fasel *et al.*[12] for Na/Al(001), Kröger *et al.*[13] for H/Mo(110) and Baumberger *et al.*[14] for the vicinal Cu(111) surfaces.

III. Surface States and Adsorbate Chemical Bonding: Na/Al(001)

On the clean Al(001) surface a surface state has been identified which is centered at the $\overline{\Gamma}$ point of the surface Brillouin zone ($\vec{k}_{\|} = 0$),[15,16] *i.e.* the bottom of its band appears at normal emission and is located at a binding energy of $E_0 = 2.76$ eV. The electrons in this state propagate nearl freely, with an effective mass of 1.18 m_e (see Fig. 1a). From the measured Fermi wave vector $k_F = 0.60(2\pi/a_0)$, which is isotropic to a good approximation, the filling of the first Brillouin zone can readily by calculated ($a_0 = 4.05$ Å is the bulk lattice constant of Al). A band filling of 1.13 electrons per surface Al atom is deduced from this data. Other bands that are visible in Fig. 1(a) arise from bulk bands which we will not discuss in this paper. The surface state character of the former band is supported by its partially falling into a band gap of the projected Al(001) bulk band structure,[17] by synchrotron radiation excited spectra showing that the position of this band is independent of the momentum component perpendicular to the surface,[18] and last but not least by the sensitivity of this band to adsorbates as we shall see just now.

The interest in the Na/Al(001) adsorbate system arises from the fact that at half a monolayer coverage a well defined c(2×2) overlayer structure forms and quite surprisingly that the structural and electronic properties of this phase depend on the preparation and measuring temperature. Structural work by low-energy electron diffraction (LEED) and supported by density functional theory (DFT) total-energy calculations discovered that a metastable low-temperature (LT) structure forms at $T \approx 120$ K.[19] In this case, Na atoms occupy fourfold hollow sites on top of a complete Al(001) surface layer. If this structure is annealed at room temperature (RT), or if the preparation is done at RT, a quite different structure forms where every other Al atom

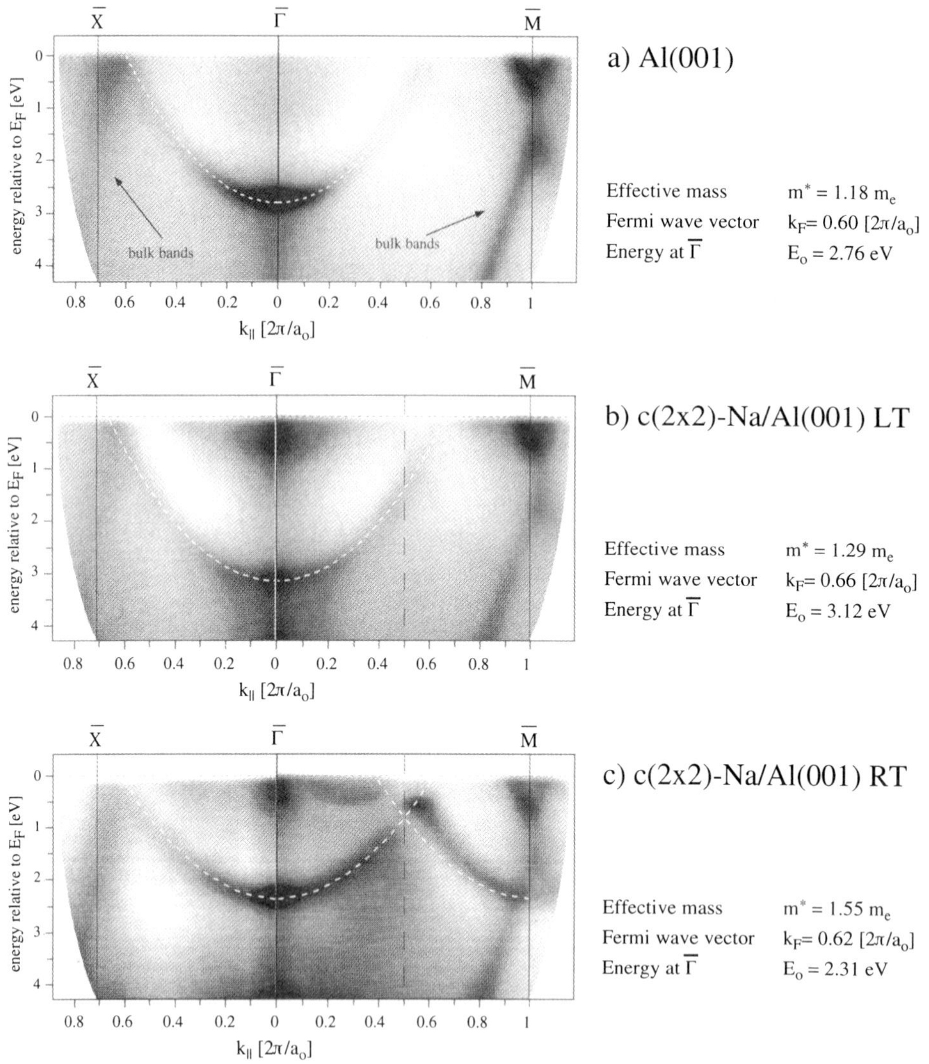

Figure 1. Dispersion plots from (a) clean Al(001); (b) the LT c(2x2)-Na/Al(001) phase; and (c) the RT c(2x2)-Na/Al(001) phase, all measured with He I excitation ($h\nu = 21.21$ eV). These plots represent series of energy spectra taken on a dense grid of polar angles along the $\overline{\Gamma M}$ and $\overline{\Gamma X}$ directions of the Al(001) surface Brillouin zone. The representation is in a linear grey scale with low intensities in white and high intensities in black. The white dashed parabolas centered around the $\overline{\Gamma}$ points represent the fits to the surface state bands. Effective mass, Fermi level crossing and band energy obtained from these fits are indicated at the right-hand side of each plot (data taken from Ref. 12).

is removed from the surface layer, diffusing away to some distant step, and replaced by a Na atom. A corrugated Na-Al alloy layer with Na atoms sitting *ca.* 1.2 Å above the deficient Al layer results.[19,20]

The dispersion plots shown in Figs. 1(b) and 1(c) demonstrate that the Al(001) surface state discriminates strongly the different atomic environments found in the two cases. For the LT-structure (Fig. 1b) the band bottom moves to higher binding energy ($E_0 = 3.12$ eV), and the filling of the surface state increases to 1.37 electrons per Al surface atom. In the RT case (Fig. 1c) the state is found shifted in the opposite way to a maximum binding energy of $E_0 = 2.31$ eV. A further striking difference exists: Along the $\overline{\Gamma M}$ direction the state does not disperse all the way to E_F but is reflected at near half distance which is where the zone boundary of the c(2×2) adsorbate structure is crossed.

These careful measurements have raised a number of questions which could be answered quite consistently by means of recent DFT calculations.[21] The picture that arises is as follows: The Na half-monolayer forms a free-electron-like state with a maximum binding energy lower than that of the Al surface state. On the LT surface, the two states interact and form a simple bonding-antibonding pair of states. The bonding state which is mostly of Al(001) surface state character, is pushed to higher binding energy where we observe it in Fig. 1(b), while the Na-type state is pushed closer to E_F where the data show a broad emission intensity with a parabolic boundary. From the change in k_F of the Al(001) state we deduce a charge transfer of 0.25 electrons from Na to each Al surface atom, while some further charge transfer must exist to Al bulk-like states below the surface, because a filling of 0.5 electrons in the Na-derived state is not compatible with the observed Na band (there are two surface Al atoms for every Na atom in this structure).

On the RT surface the chemical bonding picture should be rather similar. However, the reference for the bonding state is here not the surface state of clean Al(001). Remember that we have on this surface a c(2×2) vacancy structure of Al atoms in the first layer. The calculations show that for this case a similar parabolic surface state arises, but with a much higher binding energy. The reference state is thus much closer to the Na-derived state and pushes it almost above the Fermi level. By the same means the bonding state, being pushed down, now comes to lie still above the position of the pristine surface state with the complete Al surface layer. Another strong indication for the Al vacancy layer origin is that the bonding state is now strongly scattered at the c(2×2) Brillouin zone boundary, which is not the case for the LT structure. The data indicate even the formation of an energy gap at the zone boundary, leading to a neck in the Fermi surface of this state along the $\overline{\Gamma M}$ direction. In order to use the same electron counting arguments for the charge transfer, we now need to compare the new Fermi surface volume to the volume of the c(2×2) Brillouin zone. We cannot give precise numbers for this case, because we see that the Fermi surface now deviates considerably from a circle and we have not measured its complete contours. However, from the observed value of k_F along the $\overline{\Gamma X}$ direction we must expect the filling to be close to 2 electrons per surface Al atom, and the bonding is thus more ionic in this case.

Summarizing this section, we have obtained a very detailed picture of the chemical bonding for this system, and we find that this strongly occupied surface state plays a major role in bond formation with the adsorbate.

IV. Surface States and Surface Phonons: H/Mo(110)

In low-dimensional systems the electron-phonon coupling can lead to dramatic effects in the lattice dynamics, inducing strong phonon anomalies and in some cases leading to surface reconstructions.[22] Prominent examples for such anomalies are found on the (110) surfaces of W and Mo. Upon H adsorption, both surfaces show characteristic reductions of phonon frequencies at few and very specific phonon wave vectors. This effect is seen clearly in high-resolution electron energy loss spectra (HREELS)[23] and much more pronounced in helium atom scattering spectra,[24] where it is enhanced by the direct coupling of the He inelastig scattering channel to the electronic degrees of freedom of the surface.

Recent DFT calculations[25] support the scenario of a giant Kohn anomaly: The dynamic screening of a specific surface phonon mode of wave vector $\vec{Q}_c$ is dramatically enhanced if the Fermi surface exhibits strong nesting features, *i.e.* extended parallel sets of contours that can be connected with each other by a single nesting vector $\vec{Q}_c$. Such nesting features provide a large phase space for low-energy excitations that involve the formation of standing electrons waves of the same periodicity as the specific surface phonon mode.

While the DFT calculations indicate such nesting features on both H/W(110) and H/Mo(110) systems, the experimental situation is less clear. For H/W(110) a recent photoemission study supports this scenario,[26] but the measured Fermi surface contours are complicated by strong spin-orbit splitting which was not included in the DFT calculations. The critical nesting vector was found to connect Fermi surface sheets of different total angular momentum $\vec{J}$. For H/Mo(110) an earlier photoemission study was not able to confirm the nesting properties of the Fermi surface.[27]

We have repeated this experiment on clean and H-saturated Mo(110),[13] as is summarized in Fig. 2. The quarter-pie shaped plots in Figs. 2(a) and 2(b) present the high-resolution photoemission measurements of Fermi surface contours within a part of the surface Brillouin zone for the clean Mo(110) surface (a) and for the H-saturated surface (b). By their very sensitivity to H adsorption some of the contours identify themselves as surface state bands, leading to assignments that are in agreement with an earlier study for clean Mo(110).[28] The most prominent H-induced change is the splitting off of a contour both on the right and left hand side of a strong bulk-related feature which is not affected itself by H adsorption. These split-off contours form, over a range of the order of 0.4 Å^{-1}, straight sections parallel to the $\overline{\Gamma N}$ direction and are thus strongly nested. The corresponding nesting vector measured from these data amounts to 0.85 Å^{-1}. This is in excellent agreement with the nesting vector of 0.86 Å^{-1} found in the DFT calculation[25] and is fully consistent with the critical phonon wave vector of 0.90 Å^{-1} along the $\overline{\Gamma H}$ direction where the anomaly occurs.[23,24] We observe similar nesting also perpendicular to the $\overline{\Gamma S}$ direction (not shown) where the comparison of Q_c yields values of 1.19 Å^{-1} (photoemission[13]), 1.23 Å^{-1} (DFT[25]) and 1.22 Å^{-1}.[23,24] These results provide therefore strong evidence that the mechanism of a giant Kohn anomaly applies for the occurrence of the phonon anomalies on the H-saturated Mo(110) surface.

Reduced dimensionality is an important prerequisite for the scenario of a giant Kohn anomaly. The Fermi surface shows a considerable degree of nesting already on the clean Mo(110) surface, where the Fermi surface contour of the surface state follows closely that of a related bulk band such that our photoemission measurements cannot separate the two at the chosen photon energy. The DFT calculations[25] show that the surface state is here actually slightly outside the projected band gap and should more correctly be described as a surface resonance, making it three- rather than two-dimensional. As Hulpke has pointed out,[22] the effect of H adsorption in inducing the giant Kohn anomaly is twofold: it maintains or even enhances the nesting characteristics of this state *and* it moves this state far into the projected band gap, making it truely two-dimensional.

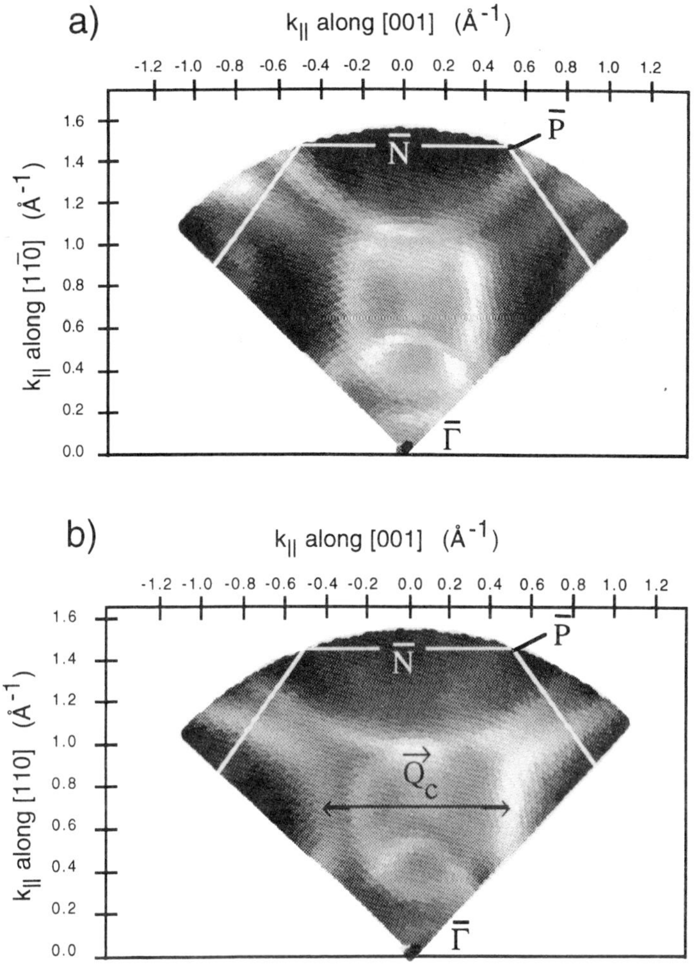

Figure 2. Sector of a high-resolution Fermi surface map of the clean (a) and the hydrogen saturated (b) Mo(110) surface. The photoemission intensity ($h\nu = 21.21$ eV) at the Fermi level is plotted in a logarithmic grey scale as a function of $\vec{k}_{||}$. In both data sets the surface Brillouin zone is indicated as well as high-symmetry points. In b) the nesting vector $\vec{Q}_c$ connecting extended parallel contours is indicated (data taken from Ref. 13).

V. The Interaction of Surface States with Nanostructures: Vicinal Cu(111) Surfaces

The two-dimensional electron gas (2DEG) as realized in semiconductor heterostructures has paved the way to an abundance of exciting physics in mesoscopic dimensions.[29] In these systems the experimental challenge lies in the creation of thin enough structures and well defined interfaces, while the technology for lateral structuring is widely available and is being pushed to smaller and smaller structures. With Fermi wave lengths of the order of 40 nm and phase coherence lengths larger than 10 μm at low temperature, devices can now be brought into the quantum regime.

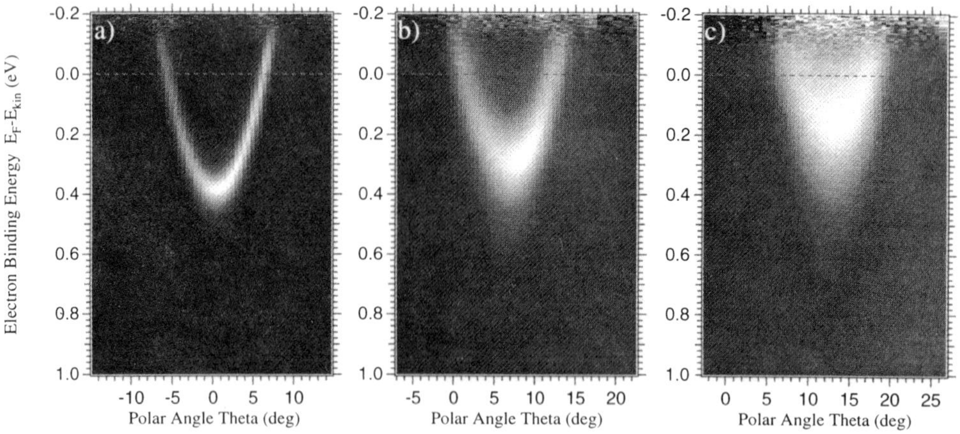

Figure 3. Dispersion plots from (a) Cu(111), (b) Cu(332) and (c) Cu(221) measured with monochromatized He I excitation ($h\nu = 21.21$ eV). The polar angle is scanned along the $[\bar{1}\bar{1}2]$, $[\bar{1}\bar{1}3]$ and $[\bar{1}\bar{1}4]$ directions, respectively, for the three surfaces, *i.e.* along the azimuth of the miscut angle. Photoemission intensities are plotted in linear grey scale after normalizing the data by an experimentally determined Fermi-Dirac distribution function in order to follow the band dispersion beyond the Fermi level. (data taken from Ref. 14).

The Shockley surface states on the (111) surfaces of the noble metals[3] are representants of a quite different type of 2DEG. While these electron gases are intrinsic to these surfaces and can be easily prepared, their small Fermi wave lengths (*ca.* 3 nm on Cu(111)) and phase coherence lengths (*ca.* 10 nm at RT) make the preparation of lateral structures that produce quantum effects much more demanding. Maybe the most astounding work in this field is still the demonstration of a quantum corral by Crommie *et al.*[30] where a circle of Fe atoms was prepared at cryogenic temperatures by manipulating single atoms with a scanning tunnelling microscope (STM) tip on a Cu(111) surface. The surface state electron density was then visualized by means of the same low-temperature STM, and it showed the standing wave pattern characteristic for a 2DEG in a circular quantum corral.

A different approach for the generation of lateral nanostructures of the appropriate sizes on these noble metal surfaces is the preparation of vicinal surfaces. We have done a comparative study of the surface state 2DEG on Cu(111), Cu(332) and Cu(221).[14] The two vicinal surfaces have miscut angles of 10° and 15.8° and mean (111) terrace lengths of 12 Å and 7.7 Å, respectively, bounded by 111-type step facets. The aim of this study was to show how the periodic step array influences the energy dispersion and the Fermi surface of the 2DEG.

The dispersion plot measured for Cu(111) (Fig. 3a) reproduces the perfectly parabolic dispersion of the surface state found in earlier studies.[32,33] The parabola is centered at the $\bar{\Gamma}$ point of the surface Brillouin zone and the band bottom is measured at $E_0^{(111)} = 385(5)$ meV. On the vicinal surfaces we also find an essentially parabolic dispersion (Figs. 3b and 3c) with three significant differences with respect to the (111) surface:

(*i*) The apices appear shifted away from the surface normal in the direction of the surface miscut angle. However, the measured apex angles are 6.9° and 12.8° and thus significantly different from the respective miscut angles. When the apex angles are translated into $k_{\|}$ values, one finds that they are located exactly on the boundaries of the new surface Brillouin zones introduced by the periodic step array. We can thus conclude that the 2DEG does not live on the individual (111) terraces but propagates along the macroscopic surface and feels the steps as a one-dimensional periodic potential, in agreement with earlier studies.[34–36] Very recently, this behaviour has been shown to switch over to (111) propagation at terrace lengths longer than 17 Å.[37]

(*ii*) The maximum binding energies move to lower values with decreasing terrace length. For 12 Å a value $E_0^{(332)} = 293(10)$ meV is found, for 7.7 A the value is $E_0^{(221)} = 150(10)$ meV. This behaviour can be well explained within a one-dimensional Kronig-Penney model[1] where the steps are modelled as a periodic sequence of potential barriers. For Cu(332) potential barriers of 1 A width and 1.3 eV height lead to a dispersion curve that fits very well the one measured perpendicular to the steps. For Cu(221) the same procedure yields a barrier height of 2.05 eV. From this analysis further properties of the 2DEG within these step arrays can be deduced, such as transmission coefficients and transmission phase shifts.[14,34] The application of the one-dimensional Kronig-Penney model is further justified by the finding that the dispersion along the steps shows an effective mass which is essentially unaffected by the presence of the steps.

(*iii*) The surface state photoemission line widths increase dramatically with decreasing terrace length. This effect has different origins, the relative importance of which is not fully understood. It is known from STM analyses that the step arrays are not fully periodic and that the terrace lengths on one particular surface show a distribution centered about the mean value.[35] The 2DEG appears to have a phase coherence length of the order of 2 nm,[37] *i.e.* its dispersion relation samples two or maybe three terraces at a time. Consequently, the terrace length distribution will be mapped onto the measured dispersion plot, thus increasing the photoemission line widths. A further effect that occurs with increasing miscut angle is that the wave vector component $\vec{k}_{\perp}$ perpendicular to the surface, for which the surface state is seen at the given photon energy, moves gradually out of the L-point band gap of Cu(111). The coupling to underlying bulk states is thus strongly enhanced and the residence time of the electrons in the surface state is reduced.

All of these observations are also reflected in the Fermi surfaces of these three Cu surfaces (not shown). On Cu(111) one finds a perfect circle centered at the $\overline{\Gamma}$ point. On Cu(332) and Cu(221) the Fermi surfaces are slightly elliptic due to the partial confinement of the electrons perpendicular to the steps and the concommittant increase of the effective mass. The Fermi surface centers move according to the apex positions observed in the dispersion plots, *i.e.* they are located right on the surface Brillouin zone boundaries. The shrinking volume inside the Fermi surface tells us that the surface state occupation decreases with increasing miscut angle, which is again due to the confinement. The resulting occupation numbers are $\langle n \rangle^{111} = 0.043(2)$ electrons per surface Cu atom, $\langle n \rangle^{332} = 0.031(3)$ and $\langle n \rangle^{221} = 0.021(4)$. In comparison to Al(001) these are highly diluted 2DEG systems, while the electron densities are still 2–3 orders of magnitude higher than in the systems based on semiconductor heterostructures.

Upon closer inspection of the Cu(332) Fermi surface map (not shown) one finds, in addition to the elliptical 2DEG Fermi surface, weak contours that represent straight lines perpendicular to the step direction. Dispersion plots exploring the $\vec{k}_{\|}$ egion around these contours reveal an additional state that disperses freely along the steps but is localized completely in the direction perpendicular to the steps.[38] The occurrence of such a one-dimensional state—or rather a surface state resonance—could be rationalized within a modified Kronig-Penney model in which the periodic step sequence is perturbed at places which then show an energy-dependent locally enhanced surface state occupation number.

In further exploring how the properties of this 2DEG can be tuned we are trying to modify the step potentials by the adsorption of molecules. Preliminary experiments indicate that for low temperature adsorption of CO at coverages where we expect a preferential adsorption at the steps, the step potentials probed by the 2DEG are vastly influenced by the presence of the molecules.[14] On the flat Cu(111) surface CO adsorption pushes the surface state up towards lower binding energies. On the stepped surfaces, however, the surface state binding energy increases strongly. Within the framework of the Kronig-Penney model discussed above this suggests that the adsorption of CO to the steps compensates the repulsive potential barriers, or even transforms the step barriers into shallow attractive potential wells.

VI. Conclusions

In this review we have discussed surface states on metal surfaces in various roles, and how these roles are reflected in their energy and momentum distribution. The surface state on Al(001) is occupied with more than one electron per surface Al atom. It is therefore not surprising that this state contributes strongly to the chemical bonding of adsorbates, as was exemplified by the Na adsorbate system. Another case of a strongly occupied surface state is found on the Mo(110) surface. Upon H-saturation of the surface, the Fermi surface of this *d*-type state shows pronounced nesting features with two nesting vectors that coincide with phonon wave vectors where strong anomalies occur. This finding supports the scenario of a giant Kohn anomaly. On Cu(111) the Shockley surface state is only very weakly occupied, with only about 1/20 of an electron per surface Cu atom. It forms a highly interesting 2DEG which can be used as a sensitive probe of nanoscopic potential landscapes as has been demonstrated by the experiments on vicinal Cu surfaces.

Acknowledgments

We thank W. Deichmann for his assistance with the experiments. This work was supported by the Swiss National Science Foundation.

References

* Present address: EMPA Dübendorf, Überlandstr. 129, CH-8600 Dübendorf, Switzerland.

1. S. G. Davison and M. Steslicka, *Basic Theory of Surface States*, (Clarendon Press, Oxford, 1992).
2. A. Zangwill, *Physics at Surfaces* (Cambridge University Press, 1988).
3. N. V. Smith, *Phys. Rev. B* **32**, 3549 (1985).
4. E. Bertel, P. Sandl, K. D. Rendulic, and M. Beutl, *Ber. Bunsenges. Phys. Chem.* **100**, 114 (1996).
5. N. Memmel and E. Bertel, *Phys. Rev. Lett.* **75**, 485 (1995).
6. S. Hüfner, *Photoelectron Spectroscopy* (Springer, Berlin, 1995).

7. T. J. Kreutz, T. Greber, P. Aebi, and J. Osterwalder, *Phys. Rev. B* **58**, 1300 (1998).
8. P. Aebi, J. Osterwalder, R. Fasel, D. Naumovic, and L. Schlapbach, *Surf. Sci.* **307-309**, 917 (1994).
9. J. Osterwalder, *Surf. Rev. Lett.* **4**, 391 (1997).
10. J. Osterwalder, T. Greber, A. Stuck, and L. Schlapbach, *Phys. Rev. B* **44**, 13764 (1991).
11. T. Greber, O. Raetzo, T. J. Kreutz, P. Schwaller, W. Deichmann, E. Wetli, and J. Osterwalder, *Rev. Sci. Instrum.* **68**, 4549 (1997).
12. R. Fasel, P. Aebi, R. G. Agostino, L. Schlapbach, and J. Osterwalder, *Phys. Rev. B* **54**, 5893 (1996).
13. J. Kröger, T. Greber, and J. Osterwalder, *Phys. Rev. B* **61**, 14146 (2000).
14. F. Baumberger, Diploma Thesis, University of Zürich, 1998.
15. P. O. Gartland, and B. J. Slagsvold, *Solid State Commun.* **25**, 489 (1978).
16. G. V. Hansson and S. A. Flodstrom, *Phys. Rev. B* **18**, 1562 (1978).
17. M. Heinrichsmeier, A. Fleszar, and A. G. Eguiluz, *Surf. Sci.* **285**, 129 (1993).
18. H. J. Levinson, F. Greuter, and E. W. Plummer, *Phys. Rev. B* **27**, 727 (1983).
19. W. Berndt, D. Weick, C. Stampfl, A. M. Bradshaw, and M. Scheffler, *Surf. Sci.* **330**, 182 (1995).
20. R. Fasel, P. Aebi, J. Osterwalder, L. Schlapbach, R. G. Agostino, and G. Chiarello, *Phys. Rev. B* **50**, 14516 (1994).
21. C. Stampfl, K. Kambe, R. Fasel, P. Aebi, and M. Scheffler, *Phys. Rev. B* **57**, 15251 (1998).
22. E. Hulpke, in *Electronic Surface and Interface States on Metallic Systems*, edited by E. Bertel and M. Donath, (World Scientific, Singapore, 1995), p. 91.
23. J. Kröger, S. Lehwald, and H. Ibach, *Phys. Rev. B* **55**, 10895 (1997).
24. E. Hulpke and J. Lüdecke, *Surf. Sci.* **287/288** (1993).
25. B. Kohler, P. Ruggerone, S. Wilke, and M. Scheffler, *Phys. Rev. Lett.* **74**, 1387 (1995).
26. E. Rotenberg and S. D. Kevan, *Phys. Rev. Lett.* **80**, 2905 (1998).
27. R. H. Gaylord, K. H. Jeong, and S. D. Kevan, *Phys. Rev. Lett.* **62**, 2036 (1989).
28. K. Jeong, R. H. Gaylord, and S. D. Kevan, *Phys. Rev. B* **39**, 2973 (1989).
29. T. Ando, Y. Arakawa, K. Furuya, S. Komiyama, and H. Nakashima, eds., *Mesoscopic Physics and Electronics* (Springer, Berlin, 1998).
30. M. F. Crommie, C. P. Lutz, and D. M. Eigler, *Science* **262**, 218 (1993).
31. F. Baumberger, T. Greber, and J. Osterwalder, in preparation.
32. P. O. Gartland and B. J. Slagsvold, *Phys. Rev. B* **12**, 4047 (1975).
33. S. D. Kevan and R. H. Gaylord, *Phys. Rev. B* **36**, 5809 (1987).
34. O. Sanchez, J. M. Garcia, P. Segovia, J. Alvarez, A. L. Vazquez de Parga, J. E. Ortega, M. Prietsch, and R. Miranda, *Phys. Rev. B* **52**, 7894 (1995).
35. J. M. Garcia, O. Sanchez, P. Segovia, J. E. Ortega, J. Alvarez, A. L. Vazquez de Parga, and R. Miranda, *Appl. Phys. A* **61**, 609 (1995).
36. X. Y. Wang, X. J. Shen, and R. M. Osgood, Jr., *Phys. Rev. B* **56**, 7665 (1997).
37. J. E. Ortega, S. Speller, A. R. Bachmann, A. Mascaraque, E. G. Michel, A. Närmann, A. Mugarza, A. Rubio, and F. J. Himpsel, *Phys. Rev. Lett.*, submitted (2000).
38. F. Baumberger, T. Greber, and J. Osterwalder, *Phys. Rev. Lett.*, submitted (2000).

"Pentepistemology" of Biological Structures and Quasicrystals

B. Bolliger,[1] M. Erbudak,[1] A. Hensch,[1] A. R. Kortan,[2]
E. Ott,[3] and D. D. Vvedensky[4]

[1] *Laboratorium für Festkörperphysik*
Eidgenössische Technische Hochschule Zürich
8093 Zürich
SWITZERLAND

[2] *Bell Laboratories, Lucent Technologies*
Murray Hill
New Jersey 07974
U.S.A.

[3] *Kantonsschule Dübendorf*
8600 Dübendorf
SWITZERLAND

[4] *The Blackett Laboratory*
Imperial College
London SW7 2BZ
UNITED KINGDOM

Abstract

There are mathematical progressions which describe the construction of the animal and plant structures. In common with these structures, quasicrystals possess symmetry elements which are not compatible with the translational symmetry of ordinary crystals. We present an algorithm that determines the atomic arrangement within a particular quasicrystal and demonstrate the appearance of the Fibonacci sequence in the natural development of all of these structures.

I. Introduction

Symmetry and geometry are fundamental human concerns, as evidenced by their presence in the artifacts of virtually all cultures. Symmetric objects are aesthetically appealing to the human mind and, in fact, the Greek word *symmetros* was meant originally to convey the notion of "well-proportioned" or "harmonious." This fascination with symmetry first found its rational expression around 400 B.C. in the Platonic solids and continues to this day unabated in many branches of science.

In physics, symmetry is of particular importance because it leads to transformations which leave the laws of nature invariant. The arrangement of atoms in molecules and crystalline solids also obey laws of symmetry due to the quantum mechanical nature of matter. A relatively recent advance which highlights this is the discovery of *quasicrystals*, which have orientational symmetry but no translational symmetry and are thus sometimes called *aperiodic*. This results in complex atomic arrangements called *tilings*,[1] which are analogous to the tessellations of Escher[2] and Penrose.[3]

Symmetry and geometry also find expression in the life sciences. Haeckel[4] devoted much of his life to studying the skeletons of unicellular *foraminifers* and, as a result, discovered mathematical laws behind some of nature's most amazing forms of art. Merz,[5] using elements from nature and combining these with technological materials, has constructed biomorphic artifacts and has reproduced organic architecture similar to certain *mollusks* and *crustaceans* on the basis of numerical progressions such as the celebrated work Leonardo Fibonacci.[6]

In this paper, we describe the application of mathematical progressions, especially Fibonacci sequences, to account for a number of structures found in nature. The diverse nature of these structures reinforces the observation by Goethe[7] that "mathematical and geometrical laws form the basis of nature's construction plan".

II. Living Nature

Two almost universal features of nature are axial symmetry and the spiral. Both exist in all sizes of biological structures, from the molecule to the fine structure of certain organs, and even to the macroscopic appearance of an entire organism. The macromolecules of genes (*deoxyribo nuclei acid*) exhibit this construction principle through the double helix. The globular proteins of the *microtubuli*—the motoric organells in each cell—are organized in a spiral, as well as the cellulose *fibrilles* in the secondary cell wall of plant cells. Other examples are the cellular and tissue structures of chloroplasts of the filamentous *algae spirogyra* and the bundles of *tracheids*, *i.e.*, the conducting elements of wood in trees with spiral growth. However, it has not yet been possible to show a causal connection between the chirality of macroscopic living structures and the asymmetry of molecules.

Descartes, the famous mathematician and philosopher of the 17th century, was fascinated by the chambered nautilus (*nautilus macromphalus*).[8–10] Today, nautilus is known as a four-gilled *cephalopod* of the tropical western Pacific, leading its still mysterious life in great ocean depths in a shell which has not changed since the triassic period.

II.1. Spirals and Fibonacci Sequences

Spirals occur in a variety of structures. One important type of spiral, the *logarithmic* spiral with equal angles, forms the basis of exponentially growing processes of mollusks

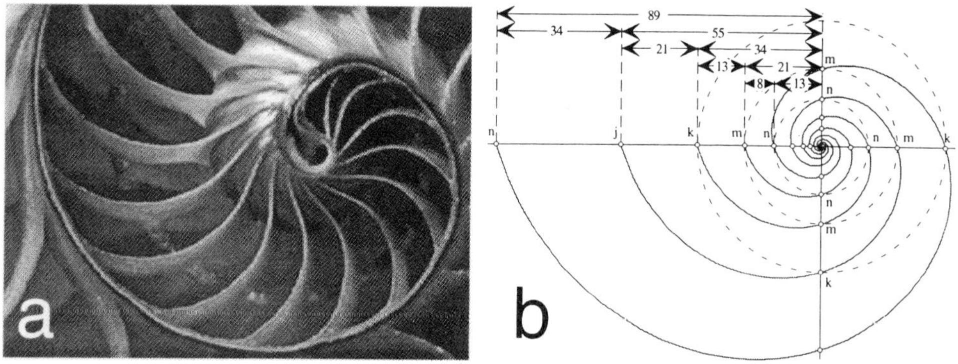

Figure 1. (a) A cross section through the shell of the *nautilus macromphalus*. The siphonal canal between each chamber is for pressure regulation. Nautilus can adjust the gas pressure individually in each compartment and modify the buoyancy force to balance itself.[9] (b) Four equiangular spirals. Each spiral cuts the radii vectors at a constant angle of 73°.

like *architectonica nobilis*, *haliotis asinina*, and *penion dilatus*. The same rule applies in all these cases: the sequential growth rates measured with the neighboring radii all including the same angle are elements of the *Fibonacci progression*,[6] *viz.*,

$$F_n = F_{n-1} + F_{n-2}.$$

Figure 1(a) shows a cut through the nautilus shell. Nautilus constructs this spiral shell by adding ever bigger living chambers while maintaining the same form during the growth process. The soft parts of the shell grow isometrically outward, protected inside the last chamber, thereby depositing new shell material. Thus, the shell grows not only bigger and wider, but also separates the old parts with geometrically similar gas chambers linked with a siphonal canal. Therefore, the regular spiral structure of the nautilus shell is given by the constant form of the chambers winding themselves around the unoccupied nucleus. Descartes discovered that any tangent with its proper radius always includes the same angle, *i.e.*, the angles formed from tangents to the spiral's radii are congruent; hence, the expression *equiangular* spiral.[6,8,9] The radius increases at a geometric rate, so any radius is cut by the spiral into sections that form a geometric progression. Thus, all chambers of the nautilus between the same angles have the same form, *i.e.*, they are geometrically isometric. Each time part of a living chamber is divided off, a new compartment develops which is 6.3% bigger than the previous one.[9] This means that the spiral completes one whole loop after 18 chambers, thereby increasing its volume threefold. Additionally, its wall thickness grows proportionally with the radius, so that the durability of the entire structure is constantly assured during growth.

Figure 1(b) presents four equiangular (logarithmic) spirals. Every other spiral is drawn after rotating the preceding one by 90°. The spirals cut the radius vectors under 73°. Note that the spiral transforms to a circle in the limiting case if the angle of intersection is 90°. In the figure, three such circles passing through the same set of letters are drawn in dashed lines. The intersections of the spirals with the coordinate axes are labelled with small letters. The Fibonacci progression can be identified in the growth rate of the radii, which is evident from the numbers indicated in the figure.

For increasing values of n, the ratio between two consecutive elements of the Fibonacci progression approaches the *Golden Section*, G. In a regular pentagon, the

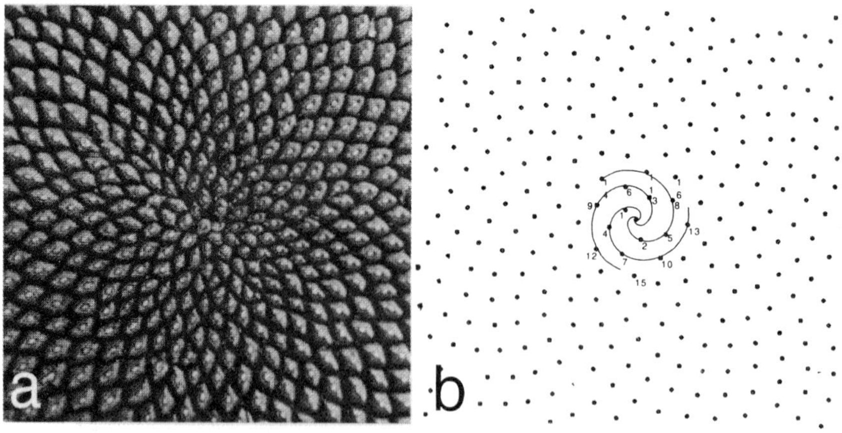

Figure 2. (a) A close-up view of the spiral structure of a dried sun flower. The seeds are arranged on spirals winding in both directions for optimum packing. (b) An idealized, computer-generated spiraling pattern.

horizontal diagonal cuts its height in G. The diagonals also subdivide between themselves in this relation, G and $G-1$, respectively, known by the musicians as the major sixth. G is $(\sqrt{5}+1)/2 \approx 1.618$, $G-1 = g \approx 0.618$, and $g = G-1 = G^{-1}$. From Fig. 1(b), it is evident that the nautilus spiral increases in size by a factor of G every turn.

II.2. Optimum Packing of Natural Objects

Hundreds of small fertile tubular flowers are pressed closely together in the inflorescence of a sun flower and other members of the *compositae* family. We can observe their spiral structure more directly once the seeds have formed. Figure 2(a) shows a dried *helianthus annuus* (sun flower) with its seeds forming spiral shapes winding outward to the left and to the right. The number of spiral arms are members of the Fibonacci sequence. Nature chooses this construction principle because it guarantees an optimal packing of the seeds so that, no matter how large the seed head is, they are uniformly packed; all seeds being the same size, no crowding in the center occurs nor are they too sparse at the edges.

No matter how big the sun flower grows and how large the seed head is, the seeds will always be packed uniformly and a single fixed angle produces the optimal design. Thus, once a seed is positioned on a seed head, the seed continues out in a straight line pushed out by the new seeds, but retaining the original angle on the seed head, *i.e.*, a single fixed angle of rotation produces uniform packing.[11] This turning angle is G cells per turn and g turns per new cell. We note that rotations by $G \pm n$, for integer n, are equivalent, *i.e.*, rotations by G, g, and $g-1$ are equivalent. Note that $1-g = g^2 = G^{-2}$. The seeds or petals are most commonly arranged by $g^2 \times 2\pi \approx 137.5°$. The best packing arrangement renders the best possible occupation of the flower-head area such that the tubular flowers receive the best exposure to the external environment, *e.g.*, sunshine, rain, or insects attracted for pollination. The positions of the individual seeds on the flower head are drawn in Fig. 2(b). Several sets of spirals can be drawn through these points. The simple group of set-3 spirals shown in Fig. 2(b) are obtained by connecting the points 1,4,7, ..., 2,5,8, ..., and 3,6,9, ... The larger groups of spirals are those of set-5, set-8, set-13, etc. The Fibonacci numbers appear in the spiral construction because they form the best integer approximations to G. If the ratio F_n/F_{n+1} is used for the construction of the spiral, so every F_n turns will produce exactly F_{n+1} seeds, *i.e.*, there will be F_{n+1} arms. As long as the ratio is an irrational number, as in this case, the seed head is efficiently covered with seeds. The ratio of Fibonacci numbers, however, is the most efficient for occupying the seed head.

III. Man-Made Objects

Some of the most common patterns of living nature are the regular hexagons found in the honeycombs of bees and wasps and on the surface of the compound eyes of insects composed of *ommatides*. The *geodesic dome*, invented by Fuller is one of the most obvious examples of man-made art and architecture.[12] It encloses space by an almost spherical surface composed of equilateral triangles. Thus, the dome is an ingenious connection between a polyhedron and architecture. This construction can also be realized by starting with an icosahedron and repeatedly truncating its protruding vertices to eventually approach a sphere. The resulting shape is not limited to modern architecture: a chemical polyhedron, *buckminsterfullerene* (C_{60}), has the shape of an icosahedron truncated just once. This molecule with its cage form occurs in nature, can be synthesized to macroscopic crystals, and doped with impurities to obtain tailored superconducting properties.

IV. Quasicrystals

There are several other examples of molecular structures assembled in clusters. Bergman clusters[13] or Mackay icosahedra[14] have attracted particular interest after the discovery of "*quasicrystals.*" In an ideal quasicrystal, long-range orientational order exists, but no structural unit is repeated, so a quasicrystal may be regarded as a crystal with a single unit cell of effectively infinite extent. In the quasicrystal we discuss here, with the bulk composition $Al_{70}Pd_{20}Mn_{10}$, the structure is favored both by the presence of the Al metal, which forms stable compounds with open structures, and Mn, which has a tendency to change its valence to easily form nonconducting salts.[15] Palladium is an almost inert filler because it has no tendency to form salts or oxides nor to show covalent-like bonding with Al or Mn. We have found that with controlled changes in composition, AlPdMn quasicrystals with pentagonal or decagonal local structures can be stabilized.[16]

IV.1. Secondary-Electron Imaging

Our model of the atomic structure of $Al_{70}Pd_{20}Mn_{10}$ is based on secondary-electron imaging (SEI), which provides real-space information about the atomic symmetry near surfaces.[17] SEI involves the excitation of a surface with a primary beam of 2000 eV electrons and the imaging of the quasielastically backscattered secondary electrons on a hemispherical screen. At energies above a few hundred eV, electron propagation is dominated by elastic forward scattering, so the scattered electrons are channeled along chains of atoms. Thus, the intensity of backscattered electrons is enhanced along these directions so, with each atom effectively acting as a source of incoherent secondary electrons, the pattern produced on the screen may be interpreted as a *central*

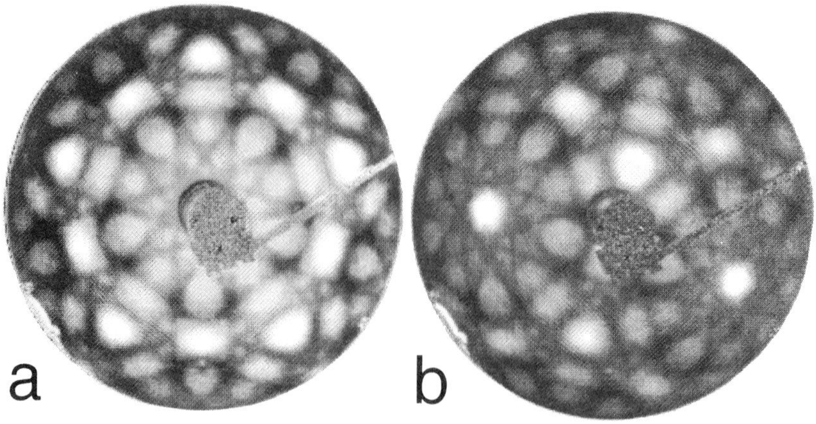

Figure 3. (a) A secondary-electron pattern obtained from the pentagonal surface of an icosahedral $Al_{70}Pd_{20}Mn_{10}$ quasicrystal. The center of the pattern corresponds to electron emission along the surface normal, which coincides with the fivefold-symmetry axis of the sample. (b) A pattern obtained after the sample has been rotated by approximately 30°.

projection of the atomic arrangement near the surface. Patterns can be recorded while the sample is rotated, providing projections along different directions and enabling efficient identification of unknown structure.[16]

IV.2. Secondary-Electron Patterns of Quasicrystals

Figure 3(a) shows the secondary-electron pattern obtained from the pentagonal surface of our quasicrystalline $Al_{70}Pd_{20}Mn_{10}$ sample. Figure 3(b) displays a pattern from the same sample but rotated such that the fivefold-symmetry axis, which was coincident with the display axis in Fig. 3(a), is now on the left-hand side of the pattern.

The most striking feature in the patterns is the pentagonal symmetry. Adjacent to each side of the prominent pentagon are bright patches arranged in five groups of equilateral triangles, the corners of which are the twofold-symmetry axes. The centers of the triangles represent five threefold-symmetry axes. In Fig. 3(b), a twofold-symmetry axis appears behind the electron gun. This bright patch is located in the middle of two fivefold-symmetry axes, both of which are clearly discernible on the left- and right-hand side of the pattern. The bright patches above and below the shadow of the electron gun are each a threefold-symmetry axis. On the extension of these two patches, a third fivefold-symmetry axis can be recognized at the upper part of the pattern near the rim. This distribution of triangles and pentagons form an *icosidodecahedron* and, therefore, these observations are consistent with the local icosahedral symmetry of $Al_{70}Pd_{20}Mn_{10}$.

Besides the point-group symmetry, there is another important feature of the SEI pattern, namely, the groups of bright patches in the pattern which lie within bands like those known as Kikuchi bands. These bands are confined between pairs of dark lines. In the case of single crystals, these phenomena have their origin in electrons diffracted at low-index crystallographic planes.[18] In the present case, however, the observation of Kikuchi bands directly proves the presence of well-defined *planes* on which the atoms of the quasicrystalline solid are placed. Therefore, the SEI patterns shown in Figs. 3(a) and 3(b) carry information which has local (short-range) as well as global (long-range) character.

IV.3. Fibonacci Sequences and Quasicrystals

Based on these observations, we have devised an algorithm for constructing the quasicrystalline solid. We start with an atom at the origin and draw planes along the twofold-, threefold-, and fivefold-symmetry directions of the icosahedron. We then assemble parallel planes with spacings obeying the Fibonacci sequence and place atoms at the intersections of these planes. Within these planes, atoms have an aperiodic distribution and there are only two layer spacings, S and $L = G \times S$, which follow the Fibonacci sequence $LSLLSLSL\ldots$ within the quasicrystalline solid. If we choose $L = 0.67\,\text{Å}$ and $S = 0.41\,\text{Å}$ along the fivefold-symmetry directions, there are only two nearest neighbor distances, 2.34 and 2.54 Å, so the solid has an atomic density of 0.049 Å^{-3}, which is close to that of bulk Al. This model is consistent with the patterns in Fig. 3.[19]

V. Conclusions

A feature common to all the spirals in inflorescences of compositae and in shells of mollusks is that they result from growth processes which could proceed indefinitely without changing their form. The organism does not seek beauty using the Golden Section and it does not count out the Fibonacci progression. Harmony and mathematics are natural outcomes of a simple growing system in relation to its three-dimensional surroundings. It seems that the only genetic basis is simply a certain scope for form modification, so that the economical handling of information, energy, and material is assured.

For packing identical spherical (circular in two dimensions) objects, hexagonal symmetry is the most efficient. There are two crystal structures, face-centered cubic and hexagonal closed packed, that fall into this category. For naturally grown objects, like flower seeds, the best arrangement of packing, while minimizing wasted space, is apparently the spiral. Similarly, quasicrystals arrange in icosahedral symmetry which includes fivefold-rotation axes, and corresponds to its optimal design.

Acknowledgments

The authors appreciate several enlightening discussions with G. Guekos. Financial support was provided for this work by the Eidgenössische Technische Hochschule Zürich and the Schweizerischer Nationalfonds.

References

1. C. Janot, *Quasicrystals, A Primer* (Clarendon Press, Oxford, 1992).
2. J. L. Teeters, *Mathematics Teacher* **67**, 307 (1974).
3. R. Penrose, "Escher and the Visual Representation of Mathematical Ideas", in *M. C. Escher: Art and Science*, edited by H. S. M. Coxeter (North Holland, New York, 1986) p. 143.
4. E. Haeckel, *Kunstformen der Natur* (Prestel Verlag, Munich, 1997).
5. D. Eccher, *Mario Merz* (Hopeful Monster, Torino, 1995).
6. R. Knott, http://www.mcs.surrey.ac.uk/Personal/R.Knott/Fibonacci/fib.html.

7. J. W. von Goethe, "Über die Spiraltendenz der Vegetation", in *Schriften zur Biologie* (Langen Müller, Wien, 1982) p. 330.
8. E. T. Bell, "Gentleman, Soldier, and Mathematician: Descartes", in *Men of Mathematics: The Lives and Achievements of the Great Mathematicians from Zeno to Poincaré* (Simon and Schuster, New York, 1986) p. 35.
9. T. A. McMahon and J. T. Bonner, *On Size and Life* (Scientific American Books, New York, 1983).
10. J. D. Currey, "Problems of Scaling in the Skeleton", in *Scale Effects in Animal Locomotion*, edited by T. J. Pedley (Academic Press, New York, 1977) p. 153.
11. S. Douady and Y. Couder, *J. Theor. Biol.* **178**, 255 (1996).
12. http://www.teleport.com/~pdx4d/domehist.html.
13 G. Bergman, J. L. T. Waugh, and L. Pauling, *Acta Cryst.* **10**, 254 (1954).
14. A. L. Mackay, *Acta Cryst.* **15**, 916 (1962).
15. D. G. Pettifor, *Bonding and Structure of Molecules and Solids* (Clarendon Press, Oxford, 1996).
16. B. Bolliger, M. Erbudak, D. D. Vvedensky, and A. R. Kortan, *Phys. Rev. Lett.* **82**, 763 (1999).
17. M. Erbudak, M. Hochstrasser, E. Wetli, and M. Zurkirch, *Surf. Rev. Lett.* **4**, 179 (1997).
18. K. Shinohara, *Phys. Rev.* **47**, 730 (1935).
19. B. Bolliger, M. Erbudak, D. D. Vvedensky, and A. R. Kortan, *Cz. J. Physics* **49**, 1531 (1999).

Structural and Magnetic Properties of Co-Cu Film Systems

A. R. Bachmann,[1] S. Speller,[1] J. Manske,[1] M. Schleberger,[1]
A. Närmann,[2] and W. Heiland[1]

[1] *Fachbereich Physik,*
Universität Osnabrück,
D–49069 Osnabrück,
GERMANY

[2] *Physikalisches Institut,*
Technische Universität Clausthal,
D-38678 Clausthal-Zellerfeld
GERMANY

Abstract

Thin films of Co on Cu are studied with respect to structural and magnetic properties by means of STM, MOKE and spin sensitive electron capture from surfaces (ECS). Sub-monolayer coverages of Co have been deposited on vicinal Cu(111) surfaces with steps oriented along $\langle 1\bar{1}0 \rangle$. STM topographies revealed that on vicinal surfaces regular arrays of steps with $\{100\}$ as well as with $\{111\}$ minifacets can be prepared. As Co is deposited it aggregates on both types of surfaces at the steps. On the surfaces with the $\{100\}$ facets the step array is rearranged into a configuration where double steps dominate. Along the $\{100\}$ facets of the double steps the Co exists as one-dimensional structures. Magnetic properties, *e.g.* hysteresis loops, are measured for different Co thicknesses above one monolayer on low-indexed Cu(111). The difference between the ECS and the MOKE hysteresis loops, respectively, afford insights into relations between bulk and surface magnetic properties.

I. Introduction

On surfaces a wide range of nanosized low dimensional structures can be realized. Especially structures of non-ferromagnetic combined with ferromagnetic materials are interesting for magnetoelectronics. Even more challenging is the production of one-dimensional structures which might exhibit totally different magnetic properties.[1] Vicinal surfaces often exhibit quite regular step arrangements and preferred aggregation

Physics of Low Dimensional Systems
Edited by J. L. Morán-López, Kluwer Academic/Plenum Publishers, New York 2001

Figure 1. Sphere models of the {100} and {111} step facets

of ferromagnetic materials at steps has been observed for the systems Co/Cu(111),[2] Cr/Cu(111),[3] and Ni/Ag(111).[4] With sub-monolayer coverages of Co on Cu(111) approximately 50 Å wide decoration bands are formed along step edges.[2] Thicker films of Co on Cu(111) have been studied as well. For Co films between 1.5 and 15 monolayers a strong tendency for island formation is observed. The structure changes from face-centered-cubic to hexagonal at a film thickness of approximately 5-6 monolayers.[5] Layer-by-layer growth may be achieved at very high deposition rates, as shown by Zheng *et al.*[6] or by the use of Pb as a surfactant.[7]

In this study we used samples with narrow terraces, below 50 Å and coverages below 0.1 monolayer. With scanning tunneling microscopy (STM) we observe distinct morphologies upon Co deposition on vicinal Cu(111) surfaces with different step facet but identical step (edge) orientation. There are indications for one-dimensional alloying of Co in Cu steps with {100} minifacets. Additionally, we have investigated the magnetic properties of thicker Co films on Cu(111) with ECS and MOKE.

II. Experiment

For the STM measurements we use an Omicron STM I setup, equipped with Auger electron spectroscopy (AES), low energy electron diffraction (LEED), and reflection high energy diffraction (RHEED). Cu(111) disks have been cut off by sparc erosion from a monocrystalline Cu rod after alignment by Laue diffraction. The slices have been cut into two pieces along $\pm[\bar{1}10]$ and wedged using a polishing machine. The surfaces of the samples have been polished mechanically and electrochemically. The surfaces are vicinal to (111) and the 5° miscut is about the $[\bar{1}\bar{1}2]$ direction. Therefore, the steps run along the dense $\pm[\bar{1}10]$ directions and exhibit either exclusively {100}-type or {111}-type minifacets at the step edges (Fig. 1) depending on whether the miscut is clockwise or counterclockwise. The atomic spacing along the steps is $a_{\|} =$ 2.55 Å and the spacing of the rows in the terrace perpendicular to the steps is $a_{\perp} =$ 2.21 Å. The average terrace width of the 5° sample pair is 23.9 Å. Both samples of the pair were mounted together on the same holder. Between the back of the samples and the holder a roof was inserted in order to compensate for the wedge shape of the samples. With this mounting completely equal preparation conditions on both samples were achieved. Sputter-anneal cycles have been applied till a nice beam splitting in LEED patterns was observed as well as AES and STM cleanliness. Deposition of Co was done at room temperature using a thermal evaporator. STM images have been taken at room temperature using tungsten tips. The step type has been identified by Laue diffraction, LEED and atomically resolved STM images.

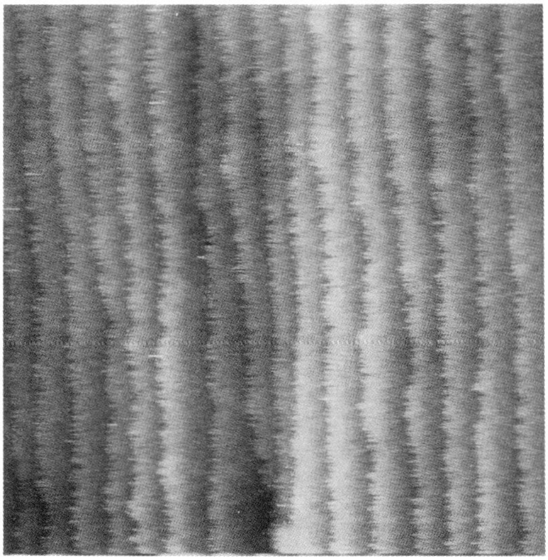

Figure 2. Cu(111) surface with 5° miscut and {100} steps (380 Å, −2.3 V, 0.7 nA).

For the magnetic measurements, a low-indexed Cu(111) substrate was used. It was cut and prepared in the same way as the vicinal Cu(111) samples. We applied two methods, the magneto-optical Kerr effect (MOKE) and electron capture spectroscopy (ECS). The former is a standard tool and will not be described here (see *e.g.* Ref. 8, the latter is somewhat uncommon and will be briefly described in the following (for details see *e.g.* Ref. 9). The scattering of low energy He^+-ions yields mostly neutralized ground state He atoms. But a fraction is neutralized into excited states. The subsequent deexcitation occurs via light emission. The electron spin is conserved during the neutralization process. Because of dipole selection rules, the information about the spin orientation is eventually transferred to the emitted photons. If the electrons at the surface have a netto spin polarization (*e.g.* at a magnetic surface) their spin has a preferred orientation. The magnetic sublevels of the He atom are occupied accordingly and the emitted light is circularly polarized. The sense of rotation and the degree of polarization are directly related to the spin polarization. Combining ion scattering and optical spectroscopy, we are thus able to obtain information about the surface magnetization. Recently, we could show that ECS is extremely surface sensitive and that it is possible to measure hysteresis loops.[10]

III. Results and Discussion

In Fig. 2 we show the typical STM topography of the vicinal surface as prepared prior to deposition. The surfaces with the {100} steps and the {111} steps appear qualitatively equal (except for atomic resolution). The steps arrange themselves in a regular array. Occasionally, impurities are visible which pin steps (see *e.g.* at the bottom center of Fig. 2). Apart from that no multiple steps are present. In STM images step edges appear frizzy, indicating kink mobility that is undersampled in time by the STM. Such a high mobility at ambient temperature is typical for materials with a relatively low melting point. Directly at the pin points the frizziness is absent. We found that the {111} steps display stronger frizziness and wider terrace width distributions compared to {100} steps and inferred different elastic step-step-interactions.[11]

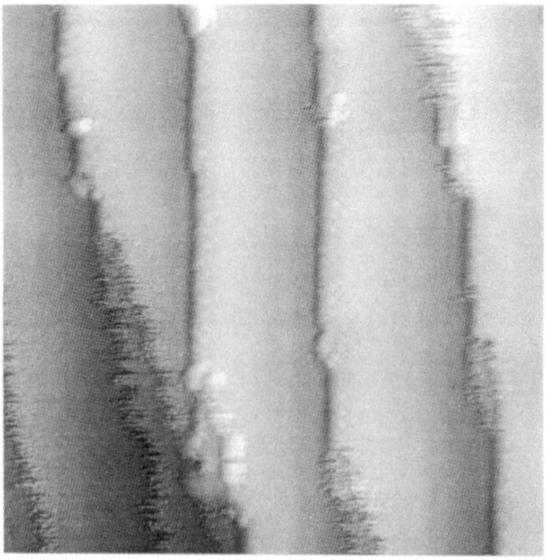

Figure 3. Cu(111) surface with 5° miscut, {100} steps and low Co coverage. The 'beady' step edges contain Co and have double height (185 Å, −0.3 V, 0.7 nA).

After small amounts of Co have been deposited onto the vicinal surfaces the Co aggregates at both types of steps (Figs. 3 and 4). In this coverage and terrace width regime islands are almost never located on terraces. There are striking differences in the morphology of the two surfaces with respect to the step type. The most remarkable effect of the Co is that on the surfaces with the {100} steps the step array is rearranged into a configuration where double steps prevail (Fig. 3). These rearrangements afford considerable mass transport in the substrate surface. These double steps do not display frizziness. Since according to AES no third element is present in the surface region we must attribute this to incorporated Co. The stronger binding immobilizes larger agglomerates of Co. Thus kink diffusion becomes slow compared to usual STM sampling rates of some kHz when Co is involved. In smaller STM frames the edges of the double steps appear corrugated (Fig. 3). The amplitude of this corrugation is 0.3 Å and its periodicity is 5.1 A. This period is compatible with $2 \cdot a_{\|}$, the double atomic spacing of Cu(111) along $\langle 1\bar{1}0 \rangle$. Occasionally, single steps with fringes are found. The topography of these steps and the clean steps is the same and we conclude that they do not contain Co. We did not find any steps that are both single and without fringes nor did we find any frizzled steps of double height. The corrugation period and the lack of mobility suggest an arrangement of alternating Co and Cu atoms at the double step edges. The STM topographies do not allow to deduce the complete atomic arrangement of the Co atoms within the {100} step in detail. The analysis of our data yields that the non-frizzy, double steps are always very accurately aligned with $\langle 1\bar{1}0 \rangle$ whereas monoatomic, frizzy steps can be misoriented by up to 30°. Misalignments are rarely observed with the clean steps of the uncovered surface. Thus, the misalignments occurring with Co coverage must be a consequence of the step pairing. It is instructive to look more closely at the transition zones between the double steps and single step pairs. Four of such transitions are visible in Fig. 3. This kind of transition resembles a zipper. The closed part of the zipper is represented by the double step which is always straight. The open part of the zipper is represented by two frizzy simple steps. This special configuration suggests that the formation process of the double steps works

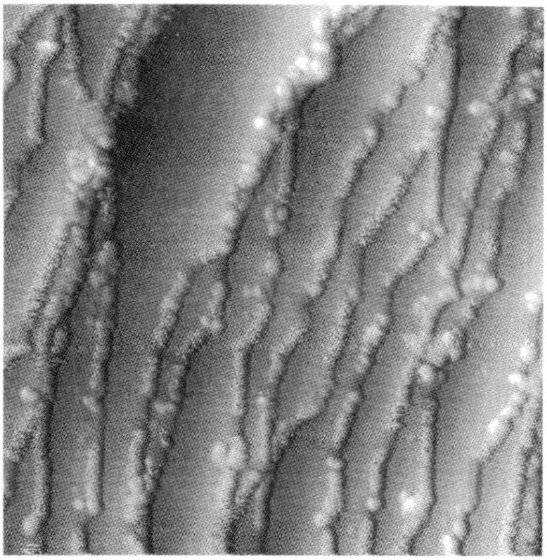

Figure 4. Cu(111) surface with 5° miscut, {111} steps and low Co coverage. At the step edges there are islands and noisy bands present (380 Å, −2.2 V, 0.5 nA).

zipper-like as well. It is likely that Co atoms diffuse towards such transition points, are bound there and the two Cu steps arrange themselves in line with the 'closed' part of the zipper. In this way, the zipper can close more and more till the step misalignments in the open part are maximal.

In the topography of the surfaces with the steps of the {111} type two major changes can be observed after the deposition (Fig. 4). First, islands of approximately 8 Å in diameter and of monoatomic height with respect to the higher terrace are present at steps, and secondly the regularity of the step array is reduced compared to the clean state (see Fig. 4). But unlike the steps with {100} facets these steps are single steps and the mass transport is moderate. A new type of frizziness is observed along the step edges in between the islands (Fig. 4). The difference compared to the usual frizziness of clean step edges is that here the height is affected as well in an approximately 6 Å wide region. Thus, this new kind of frizzes extends parallel and perpendicular to the surface. The mean level the height fluctuates about is approximately 0.4 Å above the upper terrace. Since the width is very similar to the width of the islands we suspect behind these noisy bands small mobile Co agglomerates or an electronic effect of Cu kinks diffusing in front of embedded Co structures. As the coverage is increased this effect disappears.

The formation of double steps requires a high mobility on the surface and is actuated by the presence of Co atoms on the surface. Without Co there was no evidence for step doubling in our data. Step doubling represents a mechanism to reduce step-step-repulsion since in most cases this depends on the distance x as $1/x^2$. Molecular dynamics studies of clean vicinal Cu(111) surfaces yielded similar surface energies for single, double and triple steps when the step minifacet is {100}.[12] With {111}-type steps surfaces with single steps turned out to be more stable than with multiple steps. This result is applied to clean steps. On the other hand, double steps are common in chemically ordered alloys when stacking of layers with alternating composition is present such as $Pt_3Sn(100)$[13] and $Cu_3Pd(110)$.[14]

Since double steps have double{100} minifacets their formation might be related with the situation on Cu(100) where site exchanges are prevalent. It is helpful to compare with the Co structures found on flat Cu{100} and Cu{111}. Because of the larger surface energy of Co it is unfavorable that the Co stays exposed to the vacuum. With both substrate orientations a certain degree of encapsulation or embedding of the Co by Cu was observed. On Cu(111) a partial capping or replacement of the Co islands by substrate material was found.[2,5,15] On Cu(100) there are indirect indications of strong intermixing after annealing.[16,17] In STM topographies, the apparent height of small Co islands on Cu(100) depended on the gap voltage which is rather unusual for metallic systems.[17] This finding points towards a change in the electronic structure, possibly a consequence of surface intermixing or even alloying. According to an STM and a density-functional theory study Co occupies substitutional sites in the Cu(100) substrate.[18] The exchanged Cu atoms form larger islands on the surface and become decorated by Co. These results are helpful to interpret our findings. The islands on the vicinal Cu(111) surface with the {111} step facets are not surprising, since they are similar to the islands at the steps of the flat surfaces. Considering, that for the capping of the islands substrate material is removed, also the decrease of the step regularity of the vicinal surface must be expected when Co islands are present. On Cu(111) substrate damages have been observed in form of circular vacancy islands.[2,5,15] At ambient temperatures this so called etching is not saturating within reasonable times scales.[19] At the Co islands the steps are pinned and thus the etching must be stronger at uncovered step regions further away from islands. The step etching is therefore highly non-uniform. The reduction of the step regularity observed on the vicinal surface with {111} step facets and Co islands is in line with these results. On the surface with the {100} steps site exchanges are likely to occur when Co atoms reach the {100} minifacets. Since the minifacets exist only along the step lines one-dimensional Co structures result. Thereby, substrate atoms are released which are supposed to attach to the steps of the open part of the zipper.

We did not investigate films in the sub-monolayer range by ECS since the Co-Cu(111) system is too volatile. The additional energy of the impinging ion beam is already sufficient to give rise to a substantial intermixing of the surface layers. A decrease of the Co/Cu ratio in the topmost layer on a time scale of hours has already been observed by low-energy-ion scattering for Co/Cu(111).[20] Thus, very thin films of Co on low-indexed Cu(111) are unstable and Co is removed efficiently out of the top layer by the ion beam. This was further substantiated from our Auger-measurements as well as from our ECS experiments. We estimate the life-time of one monolayer Co on top of a Cu(111)surface under ion bombardment to be around the order of minutes. For MOKE measurements, the signal stemming from a sub-monolayer coverage is too small to be quantitatively analyzed with the present setup. Therefore we investigate the magnetic properties of Co films between one and twenty monolayers. In this range the films are sufficiently thick so that we are able to record hysteresis loops with ECS without destroying the film under investigation.

Figure 5 shows the magnetic ECS signal from a clean Cu(111) surface. The circular polarization $-S/I$ is stable around 40% independent of the magnetizing current. This clearly shows that at the Cu(111) surface no spin-polarization exists. That changes drastically if we deposit 6 monolayers of Co on top of the Cu(111) surface. As shown in Fig. 6, the surface now responds to the magnetizing field. A hysteresis is observed that is clear evidence for the ferromagnetic order of the topmost layer. For comparison we have included the hysteresis that we obtained from the same system be means of MOKE. Neither MOKE nor ECS data yield quantitative information about the

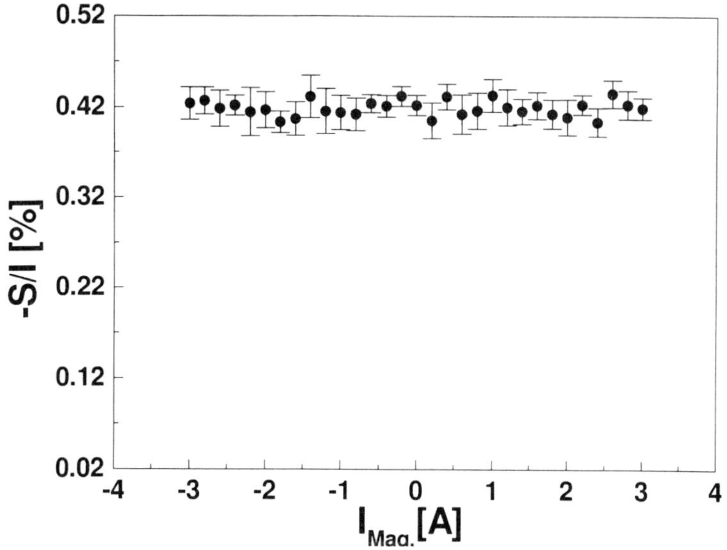

Figure 5. Circular polarization from a clean Cu(111) sample as a function of the applied magnetic field. Primary energy $E_P = 10$ keV, angle of incidence $\Psi = 3°$.

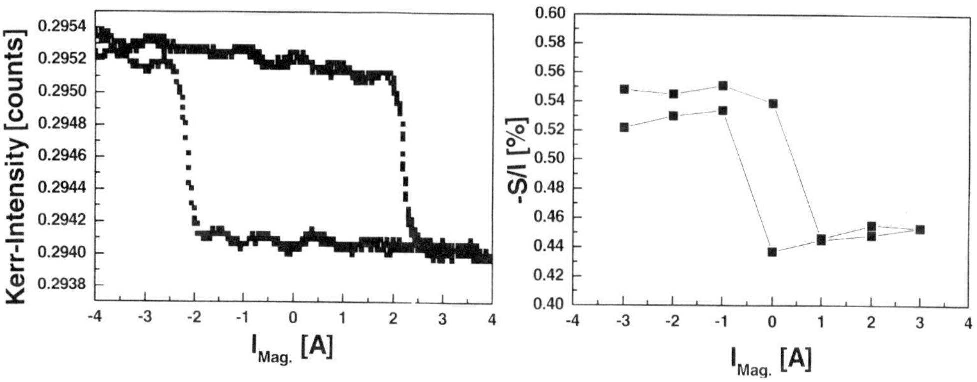

Figure 6. Hystereses of 6 monolayers Co/Cu(111) measured with MOKE (right panel) and with ECS (left panel). The oscillations seen in the MOKE hysteresis are due to our experimental setup.

magnetic moment but the difference in the coercivity of the hystereses is obvious from Fig. 6. This is clear evidence that the magnetic behaviour at the very surface of the system is intrinsically different from the bulk. From several experiments with different Co film thicknesses we conclude that the magnetization signal at first increases linearly with the film thickness and then saturates at a thickness of $\simeq$ 6–8 monolayers. This could be due to a change of the geometrical structure[5,21] and/or a change of the magnetic structure (in-plane to off-plane[22]) of the topmost layer. To establish the precise dependence of the magnetization signal as a function of the Co film thickness and of the film structure more data is required which will be acquired in future experiments.

Acknowledgments

We thank Aitor Murgaza and Enrique Ortega for their contribution to this work. This work is supported by the Deutsche Forschungsgemeinschaft.

References

1. F. J. Himpsel, J. E. Ortega, G. J. Mankey, and R. F. Willis, *Adv. in Phys.* **47**, 511 (1998).
2. J. de la Figuera, J. E. Prieto, C. Ocal, and R. Miranda, *Surf. Sci.* **307-309**, 538 (1994).
3. S. Speller *et al*, unpublished.
4. S. Morin, A. Lachenwitzer, O. M. Magnussen, and R. J. Behm, *Phys. Rev. Lett.* **83**, 5066 (1999).
5. J. de la Figuera, J. E. Prieto, G. Koska, S. Müller, C. Ocal, R. Miranda, and K. Heinz, *Surf. Sci.* **349**, L139 (1996).
6. M. Zheng, J. Shen, C. Mohan, P. Ohresser, J. Barthel, and J. Kirschner, *Appl. Phys. Lett.* **74**, 425 (1999).
7. W. Kuch, A. Dittshar, M.-T. Lin, M. Salvietti, M. Zharnikov, C. M. Schneider, J. Kirschner, J. Camarero, J. J. de Miguel, and R. Miranda, *J. Magn. Magn. Mat.* **170**, L13 (1997).
8. S. Bader, *J. Magn. Magn. Mat.* **100**, 400 (1991).
9. A. Närmann, M. Dirska, J. Manske, and M. Schleberger, *Surf. Sci.* **398**, 84 (1998), and references therein.
10. M. Schleberger, M. Dirska, J. Manske, and A. Närmann, *Appl. Phys. Lett.* **71**, 3156 (1997).
11. A. R. Bachmann, A. Mugarza, J. E. Ortega, A. Närmann, and S. Speller, *Phys. Rev. B*, submitted.
12. P. Hecquet and B. Salanon, *Surf. Sci.* **366**, 415 (1996).
13. M. Hoheisel, J. Kuntze, S. Speller, A. Postnikov, W. Heiland, I. Spolveri, and U. Bardi, *Phys. Rev. B* **60**, 2033 (1999).
14. L. Barbier, B. Salanon, and A. Loiseau, *Phys. Rev. B* **50**, 4929 (1994).
15. M. O. Pedersen, I. A. Bönicke, E. Lægsgaard, I. Stensgaard, A. Ruban, J. K. Nørskov, and F. Besenbacher, *Surf. Sci.* **387**, 86 (1997).
16. U. Ramsperger, A. Vaterlaus, P. Pfäffli, U. Maier, and D. Pescia, *Phys. Rev. B* **53**, 8001 (1996).
17. J. Fassbender, R. Allenspach, and U. Dürig, *Surf. Sci.* **383**, L742 (1997).
18. F. Nouvertné, U. May, M. Bamming, A. Rampe, U. Korte, G. Güntherodt, R. Pentcheva, and M. Scheffler, *Phys. Rev. B* **60**, 14382 (1999).
19. S. Speller, S. Degroote, J. Dekoster, G. Langouche, J. E. Ortega, and A. Närmann, *Surf. Sci. Lett.* **405**, 542 (1998).
20. A. Rabe, N. Memmel, A. Steltenpohl, and T. Fauster, *Phys. Rev. Lett.* **73**, 2448 (1994).
21. S. Müller, G. Kostka, T. Schäfer, J. de la Figuera, J. E. Prieto, C. Ocal, R. Miranda, K. Heinz, and K. Müller, *Surf. Sci.* **352,** (1996).
22. W. Kuch, A. Dittschar, M. Salvietti, M.-T. Lin, M. Zharnikov, C. M. Schneider, J. Camarero, J. J. de Miguel, R. Miranda, and J. Kirschner, *Phys. Rev. B* **57**, 5340 (1998).

The Role of Interfaces in Magnetic and Electron Transport Properties of Au/Fe/Cu/Fe/GaAs(001) and Fe/MgO/Fe-Whisker(001) Systems

T. L. Monchesky,[1] A. Enders,[1] R. Urban,[1] J. F. Cochran,[1] B. Heinrich,[1] W. Wulfhekel,[2] M. Klaua,[2] F. Zavaliche,[2] and J. Kirschner[2]

[1] *Physics Department,*
Simon Fraser University,
Burnaby, BC, V5A 1S6
CANADA

[2] *Max-Planck-Institut für Mikrostrukturphysik,*
Weinberg 2,
D-06120 Halle/Saale
GERMANY

Abstract

Smooth and As free Fe layers were grown directly on semi-insulating GaAs(001) templates with a pseudo 4×6 surface reconstruction. The STM studies revealed that the pseudo 4×6 reconstruction consists of 1×6 and 4×2 domains. The relative areas of the individual domains depend sensitively on the substrate temperature and that explains the large variations observed in the strength of the in plane uniaxial anisotropy. The temperature dependence of the magneto conductance and sheet resistance measured from 4 K to 300 K was modeled by the Boltzman's equation. Spin asymmetry at the metallic interfaces is $|T^{\uparrow} - T^{\downarrow}| = 0.34$ and $(T^{\uparrow} + T^{\downarrow})/2 = 0.79$. Single crystal epitaxial Magneto-Tunneling Junctions (MTJ) were grown using Fe whisker substrates, MgO oxide spacer and Au/Ni/Fe(001) top electrodes. The RHEED and LEED patterns show that the deposition of MgO proceeds pseudomorphically in a nearly layer by layer mode up to 6 atomic layers. The Fe whisker/MgO/Au/Ni/Fe(001) tunneling

Physics of Low Dimensional Systems
Edited by J. L. Morán-López, Kluwer Academic/Plenum Publishers, New York 2001

junctions were investigated using an Atomic Force Microscope (AFM) with a conducting tip. The AFM images showed low tunneling currents with an *I-V* characteristics corresponding to a tunneling barrier of 3.6 eV. This barrier height is in perfect agreement with the potential barrier in the crystalline MgO layer. AFM scans revealed some localized spikes in the tunneling current. It seems that the spikes originated in the oxide barrier. The total current over the scanned area is dominated by low current areas. The presence of spikes indicates that an AFM system using a conducting tip can be used to study the local tunneling *I-V* characteristics with nanometer lateral resolution.

I. Giant Magnetoresistance in Au/Fe/Cu/Fe/GaAs(001) Epitaxial Structures

An understanding of the mechanism responsible for giant magnetoresistance (GMR) remains a controversial issue due to the lack of high quality samples. Although detailed studies of the current in plane (CIP) magnetoresistance of samples prepared by sputtering have been made, poor interface quality makes quantitative comparison to theoretical calculation difficult. The work presented in this article examines the magnetoresistance of high quality epitaxial trilayer structures characterized by scanning tunneling microscopy (STM).

The lattice constant of Fe(100) is well matched to GaAs(100), only 1.4% larger than half the lattice constant of GaAs(100), but the interdiffusion of As and Ga into deposited Fe layers presents a challenge for producing high quality Fe films. Previous studies have shown that the amount of interaction between Fe and GaAs can be reduced by growing Fe on GaAs(001) at 50°C.[1] Room temperature growth of Fe on (4×6) GaAs(100)[2] prevents the formation of magnetic dead layers which is observed for growth at elevated temperature.[1,3] *Ex situ* magnetometry measurements show that the onset of ferromagnetism of (4×6) Fe/GaAs(001) occurs at the first monolayer.[2] A magnetic moment per atom of 2.1 μ_B at the Fe/GaAs(100) interface is inferred from a 7 monolayer (ML) film. Even though the interface alloying is suppressed by deposition of Fe on room temperature (4×6) GaAs(001), As continues to segregate on the top surface of Fe. For some special cases surfactants have the advantageous effect of smoothing a growing surface,[4] this is not true for As/Fe(100). It will be demonstrated that by sputtering the As away from the surface of Fe deposited on GaAs(100), smooth Fe layers may be grown with larger atomic terraces, suitable for high quality GMR structures.

I.1. Growth of Au/Fe/Cu/Fe/GaAs(001)

The GaAs surface was prepared by inserting an epi-ready wafer in UHV without prior treatment and annealing to roughly 500°C to desorb the carbon from the substrate. The oxide was removed by 500 eV Ar^+ sputtering at an angle of 75° with respect to the surface normal. The sputtering was performed at room temperature under Auger observation until the contaminants were removed. The sample temperature was gradually raised under RHEED observation until a well ordered (4×6) reconstruction was obtained at a temperature of roughly 600°C, see Refs. 5 and 6 for further details. STM studies presented in this paper, show that the GaAs surface has a pseudo (4×6) reconstruction, which is a combination of (4×2) domains and (1×6) domains, (see Fig. 1). Terraces are typically 400 Å wide.

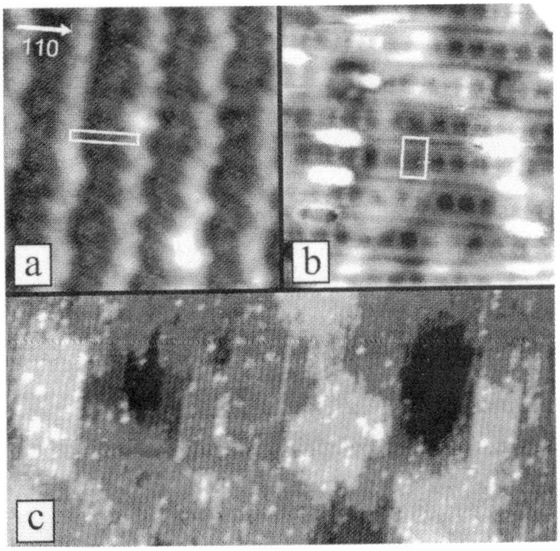

Figure 1. Filled state STM images of pseudo (4×6)GaAs(001) with 0.3 nA tunneling current and -1.80 V bias voltage. (a) 100×100 Å^2 image of a (1×6) domain with a vertical RMS roughness $\sigma = 0.7$ Å. The size of the (1×6) unit cell is shown by the white box. The crystal orientation labeled in the top right hand corner is the same for all three images. (b) 100×100 Å^2 image of a (4×2) domain where $\sigma = 0.3$ Å. A white box shows the size of the (4×2) unit cell. (c) 770×2000 Å image of pseudo (4×6) GaAs(001), $\sigma = 2.1$ Å.

The GaAs was cooled to room temperature before the deposition of ultrathin films. Fe was deposited from a thermal effusion cell at a rate of roughly 1 Å/min as determined by a quartz crystal monitor and RHEED intensity oscillations. STM images of a GaAs terrace covered by 20 ML of Fe indicated that the RMS roughness, $\sigma = 1.9$ Å, and the mean seperation between island centers $L = 45$ Å, giving a characteristic angle of surface roughness of $4\sigma/L = 10°$. X-ray photoelectron spectroscopy indicated that As acts like a surfactant, floating on the top of the Fe surface. In order to improve the surface quality, the As was removed from the surface of a 20 ML thick Fe film by sputtering with 500 eV Ar^+ ions at an angle of 75° with respect to the surface normal until the As $2p_{3/2}$ line disappeared. In the process of cleaning the surface, 2.2 ML of Fe was removed as determined from the decrease in the intensity of the Fe $2p_{3/2}$ peak. From the Ga $2p_{3/2}$ line, it was found that roughly 0.2–0.3 ML of Ga was released into the Fe layer as a result of cascade mixing from the sputtering.

Once a continuous Fe film was deposited, a bulk diffusion activation energy prevented any further intermixing for annealing temperatures less than 200°C. In order to repair the damage created by sputtering, additional Fe layers were further deposited at a temperature of 200°C. The RHEED intensity increased by a factor of three with increased Fe thickness and strong RHEED intensity oscillations were recovered. Narrow RHEED streaks indicated an increase in the terrace size. The advantage of As removal was supported by the STM data: the characteristic roughness angle as determined from STM decreased monotonically from 8.7° to 3.0° with increasing substrate temperature T_{sub} from 20°C to 170°C.

The deposition of Cu on Fe/GaAs(001) was complicated by the high mobility of Cu ad-atoms which created the tendency to form nanocrystallites. However, by cooling to temperatures below 190 K, the Cu mobility is reduced to the point where the Cu film grows in a quasi layer by layer mode as indicated by the presence of RHEED oscillations which persist up to 13 ML of *bcc* Cu. After roughly 10 ML of growth, a typical lattice reconstruction was visible in the RHEED patterns.[7] The Cu surface in 13 ML Cu/28 ML Fe/GaAs(001) sample was imaged with STM showing that the reconstruction consisted of 0.3 A high corrugations along either the [010] or [100] directions, separated by 12.1 ± 1.8 A. The average separation between Cu islands was 100 A and the RMS roughness $\sigma = 1.5$ A.

Ferromagnetic resonance (FMR) studies have shown that the As-free Fe layers grown on GaAs(001) have their magnetic properties well described by interface and bulk contributions.[5] MOKE and magnetoresistance measurements were performed on 20 ML Au / 10 ML Fe / X Cu / 28 ML Fe / GaAs(001) trilayers, where X (X= 9, 11, 13.2) is the number of Cu atomic layers. The samples were patterned *ex situ* in order to measure MOKE and magnetoresistance in a 4-probe geometry simultaneously. The coupling in samples with 9 ML and 11 ML Cu spacers was ferromagnetic. Only the samples with a 13.2 ML Cu spacer showed non-collinear coupling for small applied magnetic fields, and were used for GMR studies.

I.2. Magnetoresistance in Au/Fe/Cu/Fe/GaAs(001)

The resistance was measured over an area of 1.2×0.5 mm^2. In parallel configuration of magnetic moments the resistance measured at 300 K was 25.83 Ω for current perpendicular to the applied magnetic field and dropped to 16.35 Ω at 4 K. A typical magnetoresistance measurement is shown in Fig. 2(a), where regions of anti-parallel, non-collinear and parallel configurations of the magnetic moments of the two films are clearly visible. The GMR ratio increased from 2.0% to 5.5% as the temperature was dropped from room temperature to 4 K. A weak temperature dependence of the resistance suggested that the electron transport is strongly affected by diffuse scattering at both the outer interfaces.

Solutions to the Boltzmann equation enable a simple comparison of the measured magneto-conductance of thin films with theoretical predictions based on bulk conductivities and bulk electron mean free paths; for example see Ref. 8. A solution which takes only interface scattering into account was considered, based on the fact that in Fe the ratio of mean free paths for majority to minority electrons is roughly 0.7,[9] while the spin-asymmetry scattering is expected to be approximately 10 for Fe/Cu.[10] Furthermore, the simulation for the case of a 10 ML thick top Fe layer shows that the bulk spin-dependent scattering contribution to the magneto-conductance is negligible. The electron mean free paths used in the simulations were determined from temperature dependent bulk resistivity measurements. Using a free electron model, the temperature dependent conductivity $\sigma(T)$ can be related to the temperature dependent mean free path $\lambda(T)$, through the simple formula $\sigma(T) = (2me^2/\ 3\pi^2\hbar^3)E_F\lambda(T)$,[9] where m is the electron mass, e is the electron charge, and E_F is the Fermi energy. The mean free paths at room temperature and resistance ratios for Fe, Au and Cu are 9.3 nm(430), 32 nm(100) and 42 nm(840) respectively where the number in brackets represents the resistance ratio $R(293\ \text{K})/R(10\ \text{K})$.[11] The reflection and transmission coefficients were treated as fitting parameters.

Assuming complete diffuse scattering at the outer interfaces leads to the temperature dependence of the change in magneto-conductance, ΔG, which is much smaller than that observed. ΔG is defined in the usual way as the difference in sheet conductance between parallel and anti-parallel orientations of the magnetic moments in the two Fe layers. Some degree of specular reflection at the interfaces was required

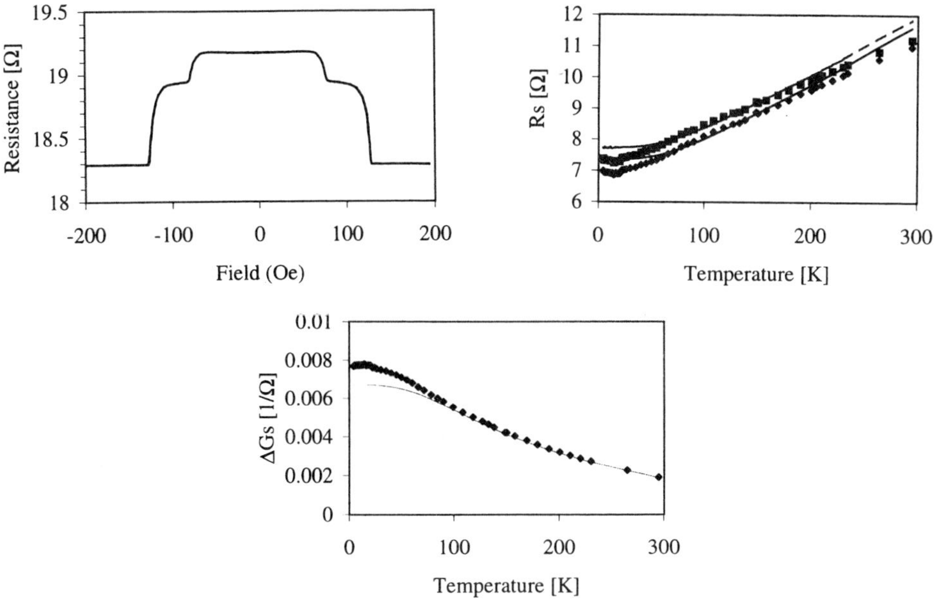

Figure 2. Magneto-transport measurements of 20 ML Au/10 ML Fe/13.2 ML Cu/28 ML Fe/GaAs(001). The integers represent the number of atomic layers. (a) The magnetoresistance at 83 K for the current perpendicular to the applied field direction. The hysteresis was removed by an 100 Oe transverse ac field for each applied dc field as described in Ref. 1. (b) The temperature dependence of the calculated sheet resistance at saturation and zero applied field respectively (solid line, dashed line). (c) The temperature dependence of the change in sheet magneto-conductance ΔG, where the solid line is the calculated ΔG given by the parameters explained in the text.

to bring the modeled resistance into agreement with the data, see Fig. 2(b). Pippard's formula was used to decide which interfaces contribute to a partial specular scattering. The effective angle of interface facets is given by $4\pi\sigma/L$, where L is the average terrace size.[12] From the STM data, this parameter was found to be 0.002 at the Fe/GaAs interface, and ranged between 0.1–0.5 at the metal-metal and the outer Au interfaces.[6] Based on STM measurements, partial specular scattering was confined only to the Fe/GaAs interface. The fit of the temperature dependence of ΔG could be further improved by reducing the Fe defect scattering mean free path to 25 nm, $1/l_{\rm Fe}^{\rm film} = 1/25{\rm nm} + 1/l_{\rm Fe}^{\rm bulk}$, yielding a resistance ratio of 3.6. This additional scattering may be due to the presence of a small concentration of Ga in the thick Fe film, as measured by XPS, see Sec. I.1.

The magneto-conductance and the sheet resistance were fitted simultaneously using the following assumptions: (1) the reflection coefficients, R, at the Fe/GaAs interface were spin independent, (2) the Fe/Cu and Fe/Au interfaces had the same spin-dependent transmission coefficients $T^{\uparrow}, T^{\downarrow}$ with $R = 0$, (3) the scattering at the outer Au interface was purely diffuse. The best fits were obtained for $|T^{\uparrow} - T^{\downarrow}| = 0.34$, $(T^{\uparrow} + T^{\downarrow})/2 = 0.79$, and $R_{\rm Fe/GaAs} = 0.45$. Using these values, good overall fits are obtained above 80 K for the magneto-conductance ΔG, see Fig. 2. Although the minority and majority electron-transmission coefficients cannot be separated in this kind of experiments, the difference and average transmission coefficients indicate that one spin-channel has high transmitivity of 0.96, while the second channel has a transmitivity of 0.62. For low temperatures, less than 80 K, fits of resistance and ΔG could be improved by a small increase in the transmission coefficients. At helium temperatures one spin-channel becomes nearly transparent, with transmission coefficient equal to 1, and the other channel equal to 0.65. It appears that one does not have to involve an appreciable contribution from spin-channel mixing due to the presence of thermally excited spin-waves. At this point it is interesting to compare the measured transmission coefficients with first-principle calculations by Stiles.[13] The calculations were done for Au/Fe which is isoelectronic to Cu/Fe and also represents one of our interfaces. The calculated $|T^{\uparrow} - T^{\downarrow}|$, which is the quantity determining the ΔG, equals 0.44 in good agreement with our experiment. The calculated $(T^{\uparrow} + T^{\downarrow})/2 = 0.37$ is somewhat lower than that obtained from the fits. This would imply that the ratio of the spin-dependent transmission coefficients, $\alpha = T^{\downarrow}/T^{\uparrow}$, is lower than that expected from first principle calculations. The experimentally determined $\alpha_{\text{exp}} = 0.65$ is higher that than from calculations $\alpha_{\text{calc}} = 0.27$. This difference can be attributed to our strong assumption of no specular scattering at the Fe/Cu and Fe/Au interfaces.

A full film thickness dependence of the GMR is currently being performed to extract the mean-free-paths, and the transmission and reflection coefficients. The preliminary *in-situ* 4-probe resistance measurements on Fe/GaAs(001) with variable Fe thickness indicates that the room temperature mean free path of electrons is 72 Å, and the reflection coefficient at the Fe/GaAs interface is $R_{\text{Fe/GaAs}} = 0.41$. These results are in accord with our interpretation of magnetoresistance data. The observed mean free path is much longer than the average mean free path of 18 Å measured for sputtered Fe films[9] and shows the high structural quality of MBE grown samples.

A full thickness dependence of the GMR is currently being performed to extract the mean-free path, and the transmission and reflection coefficients. These parameters are deduced from fits to the Fe and Au thickness dependence of the sheet resistance of Fe- and Au/Fe-layers on GaAs(001) measured *in-situ* with a four probe technique. Preliminary results show that the extracted values for the mean free paths, transmission and reflection coefficients depend strongly on the underlying model. Fits based on the semiclassical Boltzmann equation and the assumption of a free electron gas turned out to be too simple for Fe. More sophisticated fits which take the full band structure of the investigated systems into account are currently being performed in collaboration with W. Butler and X. Zhang, Oak Ridge National Laboratory. These more realistic calculations indicate that the mean free path is strongly spin dependent with a ratio $\lambda^{\uparrow}/\lambda^{\downarrow} \sim 1/10$.

II. Electron Tunneling in Au/Ni/Fe/MgO/Fe-Whisker(001) Structures

The electron tunneling studies in structures with two metallic ferromagnetic (FM) films separated by an insulator spacer is currently a hot topic in the study of magnetic nanostructures. Magnetic Tunnel Junctions (MTJ) play a significant role in the development of non-volatile Random Access Memories (MRAM). MTJ used so far are mostly based on amorphous Al_2O_3 oxides.[14] The electron tunneling in such systems is complex, it consists of a random hopping between oxide resonant states which are created by defects. The lateral component of the electron k-vector, $k_{\parallel}$, is randomized by scattering in the barrier resulting in non-coherent tunneling. The tunneling current is not only determined by the intrinsic properties but is significantly affected by the lattice disorder in the barrier. The total randomization of $k_{\parallel}$ results in averaging of the junction magnetoresistance (JMR) over the Fermi surface of the *s*-*p* valence band electrons.[14] Theoretical calculations based on a full spin dependent electron band formalism with the tunneling transport calculated by the Landauer formalism is valid in the limit of complete coherent tunneling. These calculations provide challenging predictions. The tunneling resistance in Fe/MgO/Fe(001) is governed by electrons having a transversal $k_{\parallel} \approx 0$. The tunneling conductance for the parallel oriention of the magnetic moments is estimated to be 100 times larger than that for the antiparallel orientation of the magnetic moments (see Ref. 15). To our knowledge no significant experimental effort was performed to test the wealth of these first priciple calculations. The coherent tunneling can be achieved only in nearly perfect oxide layers. The MTJ studies using MgO have not been successful so far. One is able to grow crystalline MgO layers on Fe,[16] but crystalline tunneling Fe/MgO/Fe/MgO(001) structures[17] have been affected by the poor growth of Fe on the MgO(001) substrates. Pinholes resulted in an ohmic contact. This poor behaviour is connected with the rough initial Fe layer grown on a MgO(001) wafer. We have extensively employed Fe whiskers in many magnetic studies requiring high quality samples. Fe whiskers are prepared by chemical vapor transport at atmospheric pressure using the chemical reduction of $FeCl_2$ by H_2 gas. Fe whiskers can be prepared as long rectangular bars with {001} facets. They are usually several (3–10) mm long and 100–300 micrometers wide. Their surfaces can be prepared having nearly perfect {001} atomic planes having a width of atomic terraces in excess of 3 μm. The Fe whisker templates allow one to grow ultrathin film structures with abrupt and crystalline interfaces. The success of Fe whiskers in the exchange coupling studies of Fe/Cr/Fe-whisker(001) structures[18] led us to employ Fe whiskers in the growth of crystalline MTJ. We have prepared perfect tunneling junctions using MgO(001) crystalline films as tunneling barrier.[19]

II.1. Growth of MTJ

We have grown single crystal epitaxial Magneto Tunneling Junctions (MTJ) using Fe whisker substrates, MgO(001) oxide spacers and Au/Fe(001) and Au/Ni/Fe(001) top electrodes. Samples were prepared by MBE with a growth rate of 1 ML/min. The MgO layers were deposited by an electron beam evaporator using a pure bulk MgO target. Reflection High Energy Electron Diffraction (RHEED) and Low Energy Electron Diffraction (LEED) patterns showed that the MgO films are crystalline with Fe[100] parallel to MgO[110]. The MgO layers grow pseudomorphically up to 6 ML. This results in a 3.5 percents compression of MgO. The growth shows well defined RHEED intensity oscillations and the RHEED diffraction spots are similar to those of Fe whiskers indicating that the growth of MgO proceeds in a layer by layer mode. MgO layers equal and thicker than 7 ML exhibit sharp satellite features in the LEED diffraction patterns due to the onset of misfit dislocations. The strain is gradually relieved between MgO and Fe by a network of edge misfit dislocations. The distance on the LEED screen between the satellite spots and the central diffraction spot scales inversely with the LEED primary beam energy indicating that the satellite diffraction spots arise from a tilted MgO surface. The top Fe, Ni and Au(001) layers were deposited at RT using thermal deposition with a growth rate of 1 ML/minute. The growth of Fe on MgO was significantly more rough than that of MgO layer on Fe. At RT no RHEED oscillations were observed. After deposition of Fe, a Ni layer was deposited at RT with a thickness of 12 ML. The Ni grows for the first 3ML in *bcc* form, after that it develops a typical RHEED superlattice structure which is indicative of

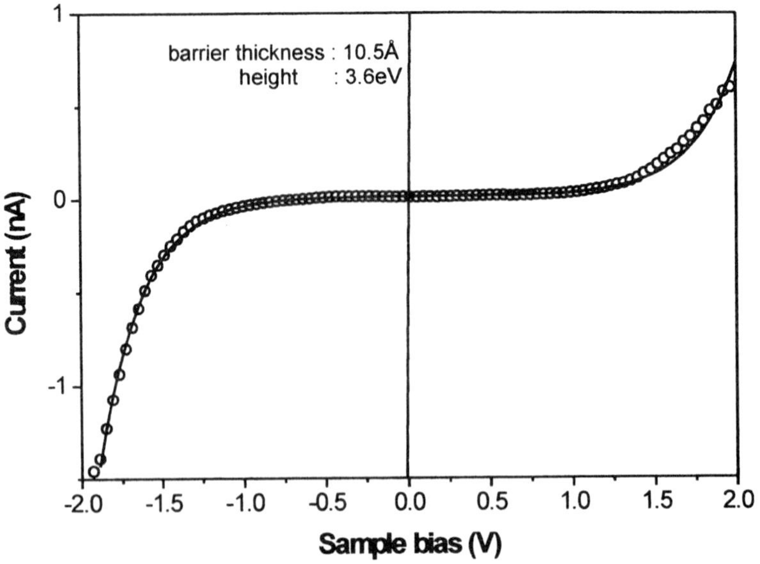

Figure 3. *I-V* characteristics of the tunneling junction 24Au/12Ni/9Fe/5MgO/Fe-whisker(001) measured by an AFM with a conducting tip, see the text for further details. The integers represent the number of atomic layers.

the transition from *bcc* to *fcc* Ni stable lattice. The Ni layer remains crystalline. The reconstructed Ni layer on Fe is known to create a large coercive field in a Ni/Fe(001) sandwich. The use of Ni was motivated by our desire to create a parallel and antiparallel configuration between the magnetic moments of the Fe whisker and the top metallic Ni/Fe(001) electrode. 20 ML of Au(001) was deposited on the top of the Ni layer for protection. Au(001) grows epitaxially with a typical (5×1) surface reconstruction. The crystallinity was maintained through the whole growth of this multilayer structure. The total peak to peak roughness of the top metallic electrode was measured by AFM and was found to be 1.7 nm.

II.2. Tunneling Studies

The tunneling studies were carried out using a 5 ML thick MgO(001) layer for which no misfit dislocations were present. A tunneling barrier of $V_0 = 3.6$ eV was determined by UPS using the upper edge of the MgO Oxygen $2p$ valence band, the Fermi surface of Fe and the band gap in MgO of 7.8 eV.[20] After the preparation, samples were transferred into an AFM/STM system, and were degassed for 24 hours to remove the wetting layer of water. In order to achieve reproducible and stable contacts between the conducting AFM tip and the sample, additional 4 ML of Au were on the sample deposited inside the UHV system. The tungsten tip of the AFM was cleaned by sputtering and the contact resistance was reduced by depositing 4 ML of Au on the tip. During the measurement the conducting tip was in dynamic contact mode touching the upper Au/Ni/Fe(001) electrode. We were able to observe the rows of the (5×1) surface reconstruction of Au(001) and the atomic steps with a resolution of 2 nm indicating that the contact area can be as small as 2×2 nm^2. The *I-V* characteristics were measured between -2 to $+2$ V, see Fig. 3. The *I-V* curve is somewhat asymmetric and

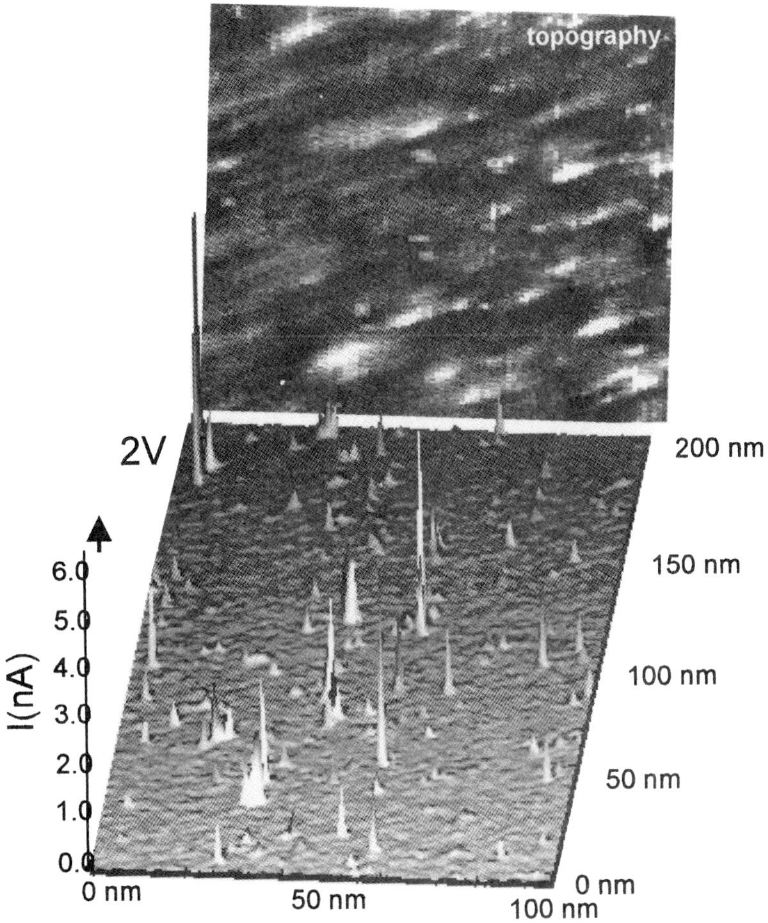

Figure 4. Topography and tunneling current images. The tunneling current images were taken at a sample bias of 2 V. The current spikes and the local topography show no correlations.

it was fitted using hot electron tunneling through a square barrier (see *e.g.* Bardou,[20] and DaCosta[21]) where the effective barrier $V_{eff} = V_0 - V$. The fits were carried out for the positive (from −0.8 to 2 V) and negative (from −2 to 0.8 V) parts of the I-V curve. Both fits led to the same values for the MgO spacer thickness of 10.2 ± 1.8 Å and the tunneling barrier of 3.7 ± 0.1 eV. This is in very good agreement with the oxide film thickness of 10.5 Å measured by the RHEED intensity oscillations and the barrier heigh determined by UPS. The asymmetry in I-V characteristics is caused by the difference in the intensity parameters indicating that the electron density of states at the MgO/Fe whisker and Fe/MgO film are not equal. Assuming that the Fermi level of metallic electrodes are pulled up by the applied voltage, the Simmons' rule becomes applicable.[22] In this case the effective barrier $V_{eff} = V_0 - V/2$. Fits resulted in barrier height $V_0 = (1.8 \pm 0.1)$ V and the MgO spacer thickness (13.3 ± 0.3) Å. Agreement with the measured film thickness and barrier height is not good. The 5 ML thick MgO(001) layer was easily able to withstand 2 V. The MgO layer breaks down at an applied voltage of 10 V; after the dielectric breakdown the I-V characteristics are

Ohmic, see the vertical line in Fig. 4. No pinholes were observed prior to the high applied voltage. The tunneling current was imaged by scanning the AFM tip over the film surface. Most of the scanned areas show low tunneling currents of the order of a fraction of nA at 2 V as expected for the MgO tunneling barrier. However, there are spikes in the current up to several nA in strength, see Fig. 4. The width of the spikes varies from 10 nm to the contact resolution of 2 nm. The position of spikes showed no obvious correlation to the surface topography indicating that the sharp increases in the tunneling current were caused by localized defects in the MgO layer. The *I-V* characteristics of the spikes are very asymmetric and the tunneling barrier is significantly decreased. The decrease in the tunneling barrier indicates the presence of dopants in the MgO band gap, and the highly asymmetric *I-V* characteristics show that the distribution of dopants is asymmetric around the metal/oxide interfaces. The tunneling current recovers its low value corresponding to a proper tunneling barrier once the tip is moved away from the spike. The fact that one is able to see the individual spikes in tunneling current is very surprizing. The AFM tip touches the upper conducting electrode, the sheet resistance of the metallic upper electrode is by many orders of magnitude smaller than that of the MgO layer. In the limit of a simple electric continuum, the upper electrode should reach the potential level of the tunneling tip and consequently the lateral contrast on a nanoscopic scale should be absent. The observed high lateral resolution in tunneling indicates that a simple electric continuum picture does not apply in our geometry. The electrons are emitted from the tip in a direction nearly perpendicular to the film surface, and traverse the upper metallic electrode with a negligible scattering in the transversal component. This supports interpretation of *I-V* curves using the hot electron model (see above). The realization of such tunneling in our samples is possible because the electron mean free path in the upper electrode is much larger than the film thickness and the lateral dimensions of the contact area of the AFM tip. Butler *et al.*[15] have shown that mostly electrons with a vanishing parallel momentum contribute to the tunneling in the high quality crystalline MgO barrier. In the present version of the AFM system we were not able to apply an external field to investigate the MR effect. The AFM system is being modified to carry out the MR studies in the near future.

References

1. A. Filipe and A. Schuhl, *J. Appl. Phys.* **81**, 4359 (1997).
2. M. Zolfl, M. Brockmann, M. Köhler, S. Kreuzer, T. Schweinbck, S. Miethaner, F. Bensch, and G. Bayreuther, *J. Magn. Magn. Mat.* **175**, 16 (1997).
3. J. Krebs, B. Jonker, and G. Prinz, *J. Appl. Phys.* **61**, 2596 (1987).
4. W. F. Egelhoff, P. J. Chen, C. J. Powell, M. D. Stiles, R. D. McMichael, J. H. Judy, K. Takano, and A. E. Berkowitz, *J. Appl. Phys.* **82**, 6142 (1997).
5. T. L. Monchesky, B. Heinrich, R. Urban, K. Myrtle, M. Klaua, and J. Kirschner, *Phys. Rev. B* **60**, 10243 (1999).
6. T. L. Monchesky, R. Urban, B. Heinrich, M. Klaua, and J. Kirschner, *J. Appl. Phys.* **87**, (2000), to be published.
7. B. Heinrich and J. Cochran, *Adv. Phys.* **42**, 523 (1993).
8. H. J. Swaggten, M. M. H. Willekens, and W. J. M. D. Jonge, *Frontiers in Magnetism of Reduced Dimension Systems*, edited by V. G. Bar'yakhtar, P. E. Wiegen, and N. A. Lesnik, Vol. 49 of NATO-ASI Series B3: High Techology (Kluwer Academic Publishers, Dordrecht, 1998).

9. B. A. Gurney, V. S. Speriosu, J.-P. Norzires, H. Lefakis, D. R. Wilhoit, and O. U. Need, *Phys. Rev.* **24**, 4023 (1993).
10. I. Meertig, *Rep. Prog. Phys.* **62**, 237 (1999).
11. CRC *Handbook of Chemistry and Physics*, edited by D. R. Lide, (CRC Press LLC, 80th Edition, Boca Raton, 1999).
12. J. M. Ziman, *Electrons and Phonons* (Oxford University Press, Oxford, 1960).
13. M. Stiles, *J. Appl. Phys.* **78**, 5805 (1996).
14. J. Moodera and G. Mathon, *J. Mag. Magn. Mat.* **200**, 6605 (1999).
15. W. Butler, J. MacClaren, and X.-G. Zhang, private communication, to be published.
16. M. Dynna, J. Vassent, A. Marty, and B. Gilles, *J. Appl. Phys.* **80**, 2650 (1996).
17. D. Keavney, *Phys. Rev. Lett.* **71**, 1641 (1993).
18. B. Heinrich, J. Codmam, T. Monchesky, and R. Urban, *Phys. Rev. B* **59**, 14520 (1999).
19. R. Whited, C. Flaten, and W. Walker, *Solid State Commun.* **13**, 1903 (1973).
20. F. Bardou, *Europhys. Lett.* **39**, 239 (1997).
21. V. Dacosta, F. Bardou, C. Beal, Y. Henry, J.P. Bucher, and K. Ounadjela, *J. Appl. Phys.* **83**, 6703 (1998).
22. J.G. Simmons, *J. Appl. Phys.* **34**, 1793 (1963).

Heat-Induced Effective Exchange: A New Coupling Mechanism in Magnetic Multilayers

Michael Hunziker and Martin Landolt

Laboratorium für Festkörperphysik
ETH Zürich
CH–8093 Zürich
SWITZERLAND

Abstract

The effective exchange interaction between two ferromagnetic layers separated by a semiconductor spacer layer is found to be reversibly heat induced. First we present experimental data on Fe/a-ZnSe/Fe trilayers. We report the thickness and temperature dependences of the exchange energy. In order to model the observations we then study a coupling which is mediated via impurity states situated at the interfaces of the layers. Weakly bound impurity states may interfere across the spacer layer. The interaction of these "molecular orbitals" depends on the spin configuration of the electrons. Thermal repopulation of the impurity levels yields a positive temperature coefficient of the coupling. The results of the calculations are found to reproduce well the experimental observations.

I. Introduction

Two ferromagnetic Fe layers separated by a nonmagnetic, amorphous semiconductor spacer-layer are exchange coupled. An outstanding property of this effective exchange interaction is that its temperature coefficient can be positive. The coupling strength reversibly increases with rising temperature until it saturates. This property, which we label "heat induced exchange", is of basic interest.

Heat-induced effective exchange coupling has been found in trilayers with Si,[1] Ge,[2] and ZnSe[3] spacers. In Sec. II of this paper we present experimental data measured on Fe/ZnSe/Fe trilayers. We report the thickness and temperature dependences of the exchange energy. In Sec. III, we then introduce a new model of effective exchange interaction in magnetic multilayers, which we call the "molecular-orbital model". We find that the model, even with rough approximations, describes the experimental data amazingly well.

II. Magnetic Measurements on Trilayers with ZnSe Spacers

The experiments on Fe/ZnSe/Fe trilayers, the sample preparation as well as the magnetic measurements, are performed in an UHV chamber with a base pressure of 5×10^{-11} mbar. A sputter cleaned and annealed Cu(100) crystal is the substrate. We first evaporate at 90°C a 70 Å thick Co film which serves as a magnetic driver. A 6 Å Fe layer, which is strongly coupled to the Co film, makes sure that the interface conditions are comparable to the previous experiments with Si and Ge spacers. Then the sample is cooled down to 20 K for the preparation of the ZnSe spacer, which is evaporated from powder in a W crucible.[4,5] The sample is completed by a 15 Å thick Fe layer. The cleanliness of the sample is checked by standard Auger Electron Spectroscopy (AES). Within the resolving power of AES we do not find any interdiffusion to occur. We note that evaporation at low temperatures is strictly necessary for the heat-induced coupling to occur. Prior to the measurements of the coupling strength the sample must be briefly annealed at 150 K presumably to form the appropriate interfaces.

We use surface magnetometry by Spin Polarized Secondary Electron Emission (SPSEE) to determine the thickness dependence of the effective exchange coupling and to address the importance of a possible 90° component. A 1–5 keV primary electron beam produces a cascade of secondary electrons on the sample surface. A subsequent spin analysis of the emitted secondary electrons with reference to the two in plane quantization axes is carried out in a 100 keV Mott detector. The spin polarization P, defined as $P = (N\uparrow - N\downarrow)/(N\uparrow + N\downarrow)$, is proportional to the magnetization of the sample at the surface.[6,7] $N\uparrow$ and $N\downarrow$ are the number of electrons with spin parallel and antiparallel to the chosen quantization axis, respectively. The high surface sensitivity allows to directly probe the magnetization of the outermost layer. In our coupling experiments, we monitor the response of an exchange coupled surface layer with respect to the magnetization of a bottom layer on the magnetic driver and in this way study the exchange coupling across a particular spacer material between surface layer and bottom layer. In order to determine coupling strength as a function of temperature we use *in situ* magnetometry by the Magneto-Optical Kerr Effect (MOKE). As opposed to the SPSEE measurements, MOKE is less surface sensitive and therefore a signal originating from top layer, bottom layer, and magnetic driver is detected.

First, we address the thickness dependence of the exchange coupling across amorphous ZnSe. To do so we prepare a sample with a wedge-shaped spacer layer. Then we magnetize the Co magnetic driver by an external magnet field pulse and perform SPSEE measurements of the polarization P at remanence along the wedge. The signal P originates from the top layer only. It therefore reflects the direction of the top layer magnetization with respect to the one of the magnetic driver. As a result the dependence of the two in-plane components of P at remanence on the spacer thickness is shown in Fig. 1. Data measured on the same sample at 40 K and at 150 K are presented. The P signal yields that for different spacer thicknesses different types of exchange coupling do occur. Below 14 Å we find that the top layer magnetization is parallel to the magnetic driver magnetization and saturated, which stands for

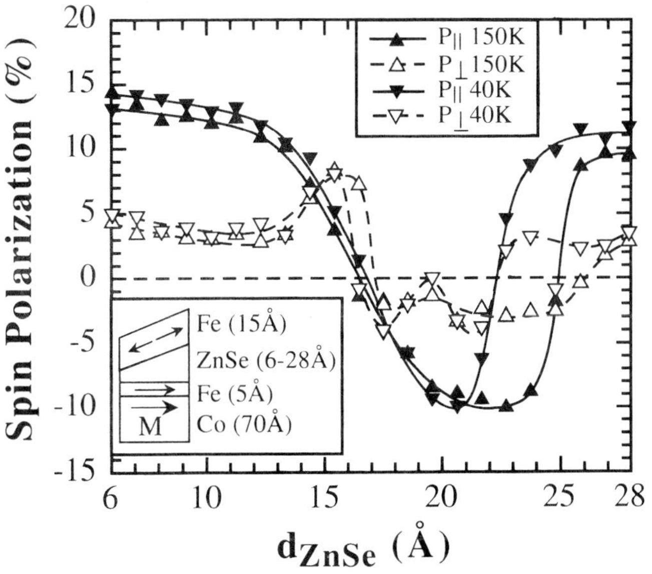

Figure 1. Spin polarization P of secondary electrons at remanence of the top Fe layer of an Fe/a-ZnSe(wedge)/Fe sample deposited on a 70 Å Co/Cu(100) substrate, *versus* ZnSe spacer thickness. In-plane components parallel and perpendicular to the magnetizing field are shown. The change between the behavior at 40 K and at 150 K is fully reversible.

strong ferromagnetic coupling. Then, between 14 Å and 17 Å, a strong perpendicular component indicates that in this intermediate range 90°-coupling prevails. In a spacer thickness range between 18 Å and 22 Å the top layer magnetization points in the negative direction with respect to the magnetic driver magnetization and the external field pulse direction. This unambiguously demonstrates the occurrence of antiferromagnetic coupling. Most striking, between 23 Å and 25 Å, the coupling is antiferromagnetic at 150 K, whereas at 40 K we find a magnetization pointing in the positive direction. This corresponds to a reversible transition from ferromagnetic to antiferromagentic exchange coupling upon heating. Above 25 Å the Fe top layer magnetization is parallel to the driver magnetization indicating either ferromagnetic or very weak coupling with a coupling strength that is not sufficient to overcome the top layer coercivity.

The positive temperature coefficient of the coupling strength certainly is the most outstanding aspect of exchange coupling in multilayers with semiconducting spacers. The observed broadening of the antiferromagnetic region with increasing temperature shown in Fig. 1 indeed points at an exciting temperature dependence. In the following, we investigate the temperature dependence of the coupling strength by measurements of the compensation field H_{comp} using MOKE. H_{comp} is the external field necessary to cancel out the ferromagnetic or antiferromagnetic interlayer exchange coupling. It is strictly proportional to the coupling strength. An example of a MOKE measurement on an antiferromagnetically coupled Fe/ZnSe/Fe/Co/Cu(100) sample is shown in Fig. 2. Since for most of the samples the coercivity of the Co/Cu(100) magnetic driver is higher than the ferromagnetic or antiferromagnetic coupling strength, we use the following procedure for the determination of H_{comp}: We apply a magnetic field pulse in one direction in order to define the magnetization of the Fe/Co bottom layer. Then a hysteresis loop of the top layer is measured, during which the applied field

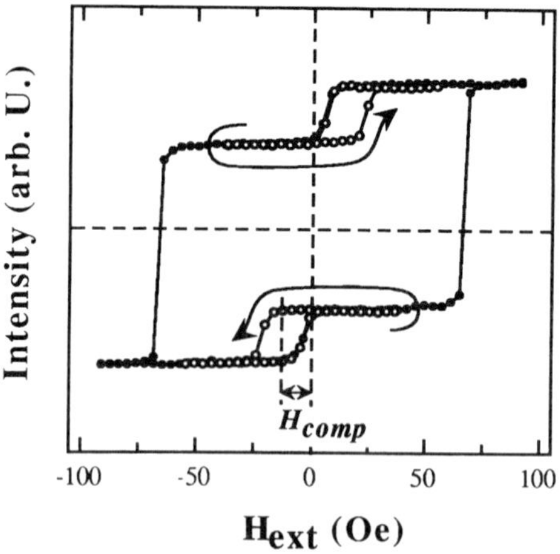

Figure 2. MOKE signal of a 15 Å Fe/25 Å ZnSe/6 Å Fe/70 Å Co/Cu(100). Two minor loops originating from the top layer only are shown by open circles; the dots represent a complete hysteresis loop.

does not exceed the coercivity of the bottom layer. A *minor loop* results, as depicted by the circles in Fig. 2. Next the bottom layer magnetization is reversed and again a top layer hysteresis loop is measured. The shift between the centers—the remanent states—of the two loops along the external field axis then reveals the compensation field $2H_{\text{comp}}$ and hence the strength of the antiferromagnetic or ferromagnetic coupling. The major steps in the MOKE signal in Fig. 2 (dots) stem from the reversal of the Fe/Co substrate magnetization. With this scheme we identify heat induced antiferromagnetic interaction and a sign change from ferromagnetic to antiferromagnetic coupling upon heating, depending on the spacer thickness. We strongly emphasize that the temperature dependences to be discussed below and hence the term 'heat induced' always refer to *reversible* effects.

The determination of the effective coupling strength from a compensation field is straightforward. In an external magnetic field the top layer magnetostatic energy HVM_S competes the coupling energy JA, if the magnetization of the Fe/Co bottom layer is fixed. M_S, V, and A are saturation magnetization, volume and area of the Fe top layer, respectively. H_{comp} by definition is the field at which the two energies are equal. Thus,

$$J = t_{\text{Fe}} M_S H_{\text{comp}}. \tag{1}$$

Measurements of the coupling strength J as a function of temperature for different samples with a variety of spacer thicknesses are compiled in Fig. 3. We find that the data exhibit a considerable thickness dependence concerning the absolute coupling strength whereas, however, the temperature dependence looks remarkably uniform for all antiferromagnetically coupled samples. In all cases, the positive temperature coefficient is most evident at low temperatures, and the coupling strength reaches saturation at $T \approx 100$ K. We note that the thickness dependence of the coupling strength at 40 K is not completely consistent with the measurement shown in Fig. 1. We attribute this to the fact that the data presented in Fig. 3 have been recorded later in the course of the experiments. A slight shift to higher spacer thicknesses for which the coupling occurs is observed if the ZnSe evaporators have been used for a long time.

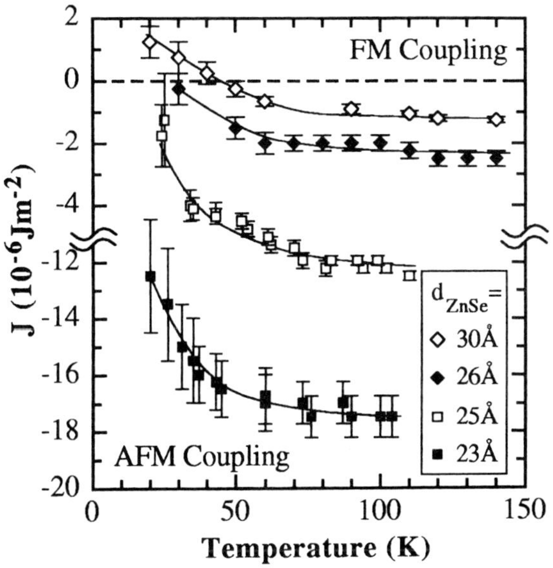

Figure 3. Coupling strength J *versus* temperature for a variety of spacer thicknesses. In the temperature range chosen, all temperature dependences are fully reversible. We observe that the coupling of samples with larger spacer thicknesses undergoes a sign change.

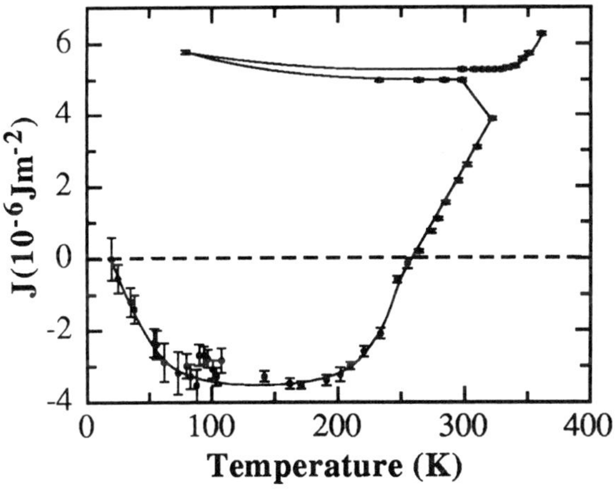

Figure 4. Coupling strength J *versus* temperature of a multilayer with spacer thickness t_{ZnSe} = 30 Å. Upon heating to above 200 K we observe an irreversible transition tc ferromagnetic coupling with a weak temperature dependence.

Next, we address the irreversible part of the temperature dependence of J. In Fig. 4 the effect of heating an antiferromagnetically coupled sample to higher temperatures than 200 K is presented. We find that upon heating beyond 200 K the antiferromagnetic coupling strength reduces and the coupling becomes ferromagnetic. This transition is irreversible. After the transition the coupling is always ferromagnetic and almost independent of temperature, as shown in Fig. 4.

In this section we present evidence for heat-induced, antiferromagnetic exchange interaction in Fe/a-ZnSe/Fe trilayers. We find that completion of the sample below 150 K is essential and that annealing above 200 K irreversibly removes the antiferromagnetic coupling. We conclude that during this irreversible transition defect states in the semiconducting spacer material are healed out. This leads us to the assumption that the existence of localized electron states which are situated near or at the interfaces is the key to the understanding of the coupling mechanism. Based on this we have devised a new model, the molecular-orbital model, which is presented in the following section, to describe the experimental findings reported above.

III. Molecular-Orbital Model

The molecular-orbital model relies on weakly bound impurity states which are situated at the interfaces between the ferromagnets and the spacer and interact across the spacer layer. It is entirely different from earlier approaches by other authors which have used k-dependent tunneling,[8] direct tunneling,[9] or excitons in a direct band gap.[10]

Let us consider an amorphous semiconductor layer of thickness d embedded between two ferromagnetic metal layers of thicknesses much larger than d. The semiconductor material is characterized by a dielectric constant ε and an effective mass m^*. Most importantly, as mentioned, localized, weakly bound electron states shall be present in the semiconducting material near the interfaces. These shallow donor states are described as ground states of hydrogen like impurities, the central charge of which is screened by the dielectricity ε of the semiconductor. Furthermore, the weakly bound electrons are magnetically coupled to the nearby ferromagnet at the corresponding interface. This coupling is described in mean-field approximation by a strong exchange field, the Weiss field, to which these electrons are exposed. Finally, the impurity states interfere across the spacer layer and in this way mediate the effective exchange coupling between the ferromagnets. The density of the impurity states per interface area, however, shall be low enough so that the localized electrons do not interact along the interfaces. A possible charge transfer between the metal layers and the semiconductor gives rise to electrical dipoles across the interfaces. Since d is small the effect of these dipoles can be approximated by a shift in energy between the band structures of the metal layers and the semiconductor.

In order to calculate the electronic states of the impurities we first neglect the exchange field. The familiar ground state then is a $1s$-like orbital,

$$\phi(\varepsilon, m^*) \sim \left(\frac{m^*}{m\varepsilon\, a_0}\right)^{3/2} \exp\left\{\frac{-m^*\, r}{m\varepsilon\, a_0}\right\}, \tag{2}$$

where a_0 is the Bohr radius. The corresponding energy is

$$E_S = \frac{m^*}{m}\frac{1}{\varepsilon^2}\, E_0, \tag{3}$$

where E_0 is the Rydberg energy.

The introduction of a screening $\varepsilon > 1$ and a reduced mass $m^* < m$ obviously enlarges the spatial extension of the impurity states, which is of decisive importance. We note that the variation of both parameters qualitatively has the same effect. Therefore, for the sake of simplicity, we choose $m^* = m$ for the remainder of the calculations and only retain ε as a free parameter.

Because of their large spatial extensions the impurity orbitals from the "right" and the "left" interface may overlap across the semiconductor to form giant molecular orbitals. The Pauli principle requires that the molecular-orbital wave function with parallel spins has a lower probability amplitude within the spacer layer than the one with antiparallel spins. This causes a difference between the energies of the two spin configurations which eventually drives the effective exchange coupling. Therefore, in order to obtain the coupling strength, we have to determine these energies. As an approximation we combine via a Slater determinant a "left" and a "right" undisturbed impurity wave-function to form a molecular orbital. We use this particular approximation because a singlet-state configuration of the two electrons does not exist in the presence of the Weiss fields. The electrons must remain in defined orbitals with defined spin states. For the first step of the calculation we neglect the magnetic energies of the Weiss fields. The hamiltonian then consists of the single electron energies $H_{S,L}$ and $H_{S,R}$, where L and R denote "left" and "right", and the Coulomb interactions across the spacer layer. The electrons are labeled 1 and 2, and $r_{j,i}$ stands for the distance $|r_j - r_i|$ with $j = R, L$ and $i = 1, 2$.

$$H = H_{S,L} + H_{S,R} + \frac{e^2}{\varepsilon}\left(-\frac{1}{r_{L,2}} - \frac{1}{r_{R,1}} + \frac{1}{r_{1,2}} + \frac{1}{d}\right). \tag{4}$$

The wave functions are Slater determinants as mentioned, where ϕ denotes the orbital and χ the spin functions.

$$\Psi_{\uparrow\downarrow}(1,2) = \frac{1}{\sqrt{2}}[\phi_L(1)\chi_\uparrow(1)\phi_R(2)\chi_\downarrow(2) - \phi_L(2)\chi_\uparrow(2)\phi_R(1)\chi_\downarrow(1)], \tag{5}$$

$$\Psi_{\uparrow\uparrow}(1,2) = \frac{1}{\sqrt{2(1-S^2)}}\chi_\uparrow(1)\chi_\uparrow(2)\left[\phi_L(1)\phi_R(2) - \phi_L(2)\phi_R(1)\right]. \tag{6}$$

S is the overlap integral $\int \phi_L(r)\phi_R(r)\, d^3r$. If we assume the single electron energies to be the same at both sites, the average energies per electron become

$$E_{\uparrow\downarrow} = E_S(\varepsilon) + \frac{1}{2}\left[\frac{e^2}{\varepsilon\, d} + V_c(d,\varepsilon)\right], \tag{7}$$

$$E_{\uparrow\uparrow} = E_S(\varepsilon) + \frac{1}{2}\left[\frac{e^2}{\varepsilon\, d} + \frac{V_c(d,\varepsilon) - V_{ex}(d,\varepsilon)}{1 - S^2(d,\varepsilon)}\right], \tag{8}$$

V_c and V_{ex} are the Coulomb and exchange energies, respectively.

$$V_c = \iint \phi_L^2(r_1)\phi_R^2(r_2)\frac{e^2}{\varepsilon}\left(-\frac{1}{r_{L,2}} - \frac{1}{r_{R,1}} + \frac{1}{r_{1,2}}\right) d^3r_1 d^3r_2, \tag{9}$$

$$V_{ex} = \iint \phi_L(r_1)\phi_R(r_2)\frac{e^2}{\varepsilon}\left(-\frac{1}{r_{L,2}} - \frac{1}{r_{R,1}} + \frac{1}{r_{1,2}}\right) \phi_L(r_2)\phi_R(r_1)\, d^3r_1\, d^3r_2. \tag{10}$$

The orbitals $\phi_{L(R)}$ are inserted from Eq. (2).

Next we briefly discuss the coupling between the donors and the nearby ferromagnet. This coupling is approximated by a Weiss field $\mathbf{H_w}$. The magnetic energy of an electron with magnetic moment $\mathbf{m}$ in this field is given by $E_m = -\mathbf{m}\mathbf{H_w}$. In the ground state $\mathbf{m}$ is parallel to the field. A magnetic moment standing antiparallel to $\mathbf{H_w}$ thus gains an energy of $2mH_w$ relative to the ground state. A Weiss field is approximately $H_w \sim 10^7$ Oe which yields $2mH_w$ to be of the order of 1 eV. At ambient temperatures this energy is huge compared to kT. Therefore the levels with one or both moments antiparallel to the corresponding Weiss field can be ignored

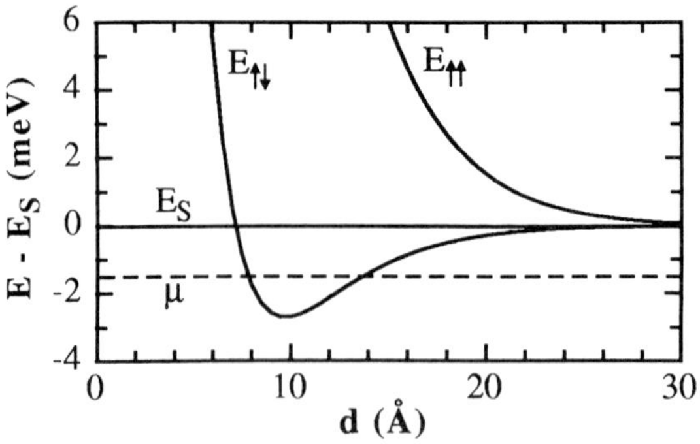

Figure 5. Calculated electron energies *versus* spacer thickness d for $\varepsilon = 10$. E_S is the energy of a noninteracting impurity state, set to zero. $E_{\uparrow\uparrow}$ and $E_{\uparrow\downarrow}$ are the energies of interacting impurity states for the magnetizations of the ferromagnets parallel and antiparallel to each other, respectively. μ is the chemical potential, treated as a free parameter.

for thermodynamical considerations. It is sufficient to retain the ground states, *i.e.* the states with $\mathbf{m}$ parallel to $\mathbf{H_w}$. Thus, for antiparallel and parallel alignment of the magnetizations of the two ferromagnets, the energies of the interfering electrons are described by the Eqs. (7) and (8), respectively. Using the integrals by Sugiura,[11] we numerically evaluate these two equations with the free parameter $\varepsilon = 10$, which roughly corresponds to the value of ZnSe. The results are shown in Fig. 5.

V_c is always larger than V_{ex} and therefore the energy of the antiparallel spin configuration is always lower than the one of the parallel spins. If only these states are occupied antiferromagnetic coupling exists but no temperature dependence is found.

The temperature dependence of the effective coupling strength J is a consequence of thermal repopulation of the states depicted in Fig. 5. In order to maintain overall charge neutrality we must allow for transfer of electrons from or to the metals, which serve as particle reservoirs with chemical potential μ. We then need to consider the thermodynamic potential $\Omega = \sum_i n_i E_i - \mu \sum_i n_i$, where E_i are the one particle energies and n_i their respective occupations. The energy position of μ within the semiconductor is determined by the details of the electronic structures of the semiconductor and of the metal layers. As an approximation we set the energy difference $E_S - \mu =$ const. throughout the semiconductor layer. μ is shown in Fig. 5 as a dashed line, and we treat the quantity $E_S - \mu$ as the second free parameter of the calculation. The occupations are governed by the Fermi distribution $f(E_i) = (1 + \exp\{(E_i - \mu)/kT\})^{-1}$ since the interfering electrons still are fermions. Considering that the molecular levels $E_{\uparrow\downarrow}$ and $E_{\uparrow\uparrow}$ are doubly occupied and the levels E_S are the same impurity orbitals but singly occupied, we obtain the thermodynamical potentials $\Omega_{\uparrow\downarrow}$ and $\Omega_{\uparrow\uparrow}$ for antiparallel and parallel alignment of the ferromagnetic layers, respectively,

$$\Omega_{\uparrow\downarrow(\uparrow\uparrow)} = AN\{2f(E_{\uparrow\downarrow(\uparrow\uparrow)})[E_{\uparrow\downarrow(\uparrow\uparrow)} - \mu] + [1 - f(E_{\uparrow\downarrow(\uparrow\uparrow)})]f(E_S)[E_S - \mu]\}, \quad (11)$$

where N is the number of impurities at one interface per interface area A. N is treated as the third free parameter of the calculation.

The effective coupling strength J is identified with the difference of the thermodynamic potentials per unit area $J = (\Omega_{\uparrow\downarrow} - \Omega_{\uparrow\uparrow})/A$.[12] This yields

$$J = N\{2f(E_{\uparrow\downarrow})[E_{\uparrow\downarrow} - \mu] - 2f(E_{\uparrow\uparrow})[E_{\uparrow\uparrow} - \mu] + [f(E_{\uparrow\uparrow}) - f(E_{\uparrow\downarrow})]f(E_S)[E_S - \mu]\}. \quad (12)$$

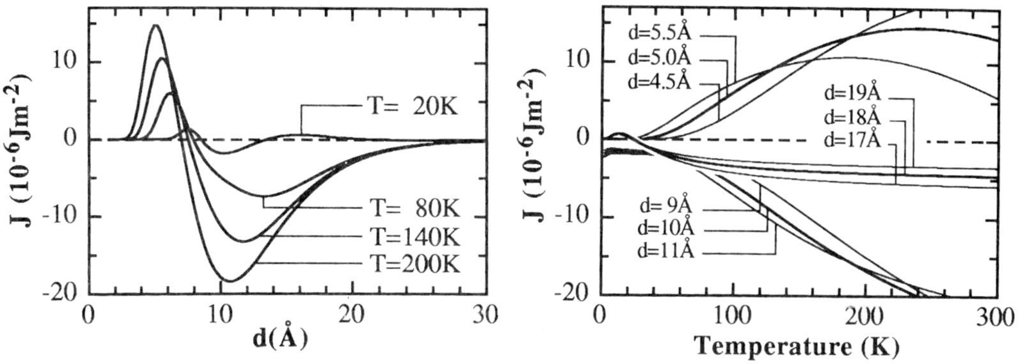

Figure 6. Calculated effective coupling strength J *versus* spacer thickness d (left panel), and *vs.* temperature (right panel). The free parameters are set $N = 1 \times 10^{12}$ cm^{-2}, $\varepsilon = 10$, and $E_S - \mu = 2$ meV.

Numerical results of J are presented in Fig. 6 with appropriate choices of the free parameters, namely $\varepsilon = 10$, $E_S - \mu = 2$ meV and $N = 1 \times 10^{12}$ cm^{-2}. We find that the results well reproduce the experimental observations shown in Figs. 1 and 3.

Of particular interest is the role of the three free parameters of the calculation. The easiest to understand is the number of impurities per interface area N. The coupling strength J is proportional to N, and there is no qualitative change of J upon variation of N. The chosen value of $N = 1 \times 10^{12}$ cm^{-2} is realistic for this type of interfaces. It is supported by preliminary transport measurements. The dielectric constant ε and the position of the chemical potential μ, on the other hand, influence the thickness and temperature dependences of J. We wish to emphasize, however, that the gross features of J persist upon considerable variation of ε and μ. The chosen value of $\varepsilon = 10$ lies within the range of the dielectric constants of the semiconductors under investigation. The energetic position of the chemical potential in the semiconductor, on the other hand, is hard to estimate because it depends on particularities of the electronic structures of the materials. It is likely to be near the impurity levels and thus we choose $E_S - \mu = 2$ meV as a physically meaningful value.

In conclusion, we propose a molecular-orbital model of heat-induced exchange coupling between ferromagnetic layers separated by an amorphous-semiconductor spacer layer. The mechanism is based on the assumption of localized, weakly bound electron states to exist at the semiconductor-metal interfaces. They are descibed as ground states of hydrogen like impurities which are screened by the dielectricity of the semiconductor. They can form molecular orbitals across the spacer layer and thus mediate an exchange interaction. The present calculation, even with rough approximations, yields antiferromagnetic coupling across amorphous semiconductors with a positive temperature coefficient. With a physically meaningful choice of the three free parameters of the calculation we find that the antiferromagnetic coupling occurs in a range of spacer thicknesses which well corresponds to the experimental observations. At low temperatures this range is quite narrow and broadens towards larger spacer thicknesses with rising temperature. At very small spacer thicknesses the calculation yields no coupling, at variance with the experimental result. This may be explained by pinholes which are likely to occur at very small spacer thicknesses. The calculated temperature dependences exhibit a positive temperature coefficient or heat-induced behavior, which is the key feature of exchange coupling across amorphous semiconductors. In particular, one finds good agreement to the experimental data for larger spacer thicknesses. The magnitude of the coupling strength is well reproduced, the calculated and experimental results turn out to be within the same order of magnitude.

We are confident that the present molecular-orbital model for heat-induced effective exchange, which we have introduced in the simplest possible approximation, will stimulate further theoretical and experimental work on exchange coupling across semiconductors in magnetic multilayers. As a particular challenge we propose to study the effect of an applied voltage across the spacer on the exchange coupling. This is of interest since non-dissipative steering of exchange interaction with an applied voltage has a great technological potential.

Acknowledgments

Happy birthday Hans-Christoph Siegmann! You have been a brillant teacher and good friend over years, we thank you for that and are looking forward to many more years of collaboration to come! It also is a pleasure to thank P. Walser for his important contributions and K. Brunner for expert technical assistance. With great appreciation we acknowledge many helpful comments by W. Baltensperger and J. W. Blatter. Financial support by the Schweizerischer Nationalfonds is gratefully acknowledged.

References

1. S. Toscano, B. Briner, H. Hopster, and M. Landolt, *J. Magn. Magn. Mater.* **114,** L6 (1992).
2. P. Walser, M. Schleberger, P. Fuchs, and M. Landolt, *Phys. Rev. Lett.* **80,** 2217 (1998).
3. P. Walser, M. Hunziker, T. Speck, and M. Landolt, *Phys. Rev. B* **60,** 4082 (1999).
4. K. Ohkawa, H. Takeishi, S. Hayashi, S. Yoshi, A. Tsujimura, T. Tarasawa, and T. Mitsuyu, *Phys. Stat. Sol. B* **187,** 291 (1995).
5. P. Goldfinger and M. Jeunehomme, *Trans. Faraday Soc.* **59,** 2851 (1963).
6. M. Landolt, *Appl. Phys. A* **41,** 83 (1986).
7. H. C. Siegmann, *J. Phys.: Condens. Matter* **4,** 8395 (1992).
8. J. C. Slonczewski, *Phys. Rev. B* **39,** 6995 (1989).
9. P. Bruno, *Phys. Rev. B* **49**, 13231 (1994).
10. C. A. R. Sá de Melo, *Phys. Rev. B* **51,** 8922 (1995).
11. Y. Sugiura, *Z. Phys.* **45,** 484 (1927).
12. D. M. Edwards, J. Mathon, R. B. Muniz, and M. S. Phan, *Phys. Rev. Lett.* **67,** 493 (1991).

Exchange Bias Theory: The Role of Interface Structure and of Domains in the Ferromagnet

Miguel Kiwi, José Mejía-López,
Ruben D. Portugal,* and Ricardo Ramírez

Facultad de Física
Pontificia Universidad Católica
UMR46 Casilla 306
6904411 Santiago
CHILE

Abstract

Exchange bias (EB) is a shift of the hysteresis loop from its normal position, symmetric around $H = 0$, to $H_E \neq 0$. It occurs when ferromagnetic (F) materials are in close contact with a variety of antiferromagnets (AF). Here we propose a model that describes the EB phenomenon for a magnetically compensated interface. Based on extensive experimental evidence, especially the memory effect, and ample computer simulations, we suggest that a canted spin configuration in the AF interface freezes into a metastable (spin glass like) state, close to the Néel temperature. As a consequence a spring-like magnet or incomplete domain wall (IDW) structure develops in the F slab. The EB energy is reversibly stored in this IDW. The results we extract, both analytically and through simulations, are qualitatively and quantitatively compatible with the available experimental information.

Exchange anisotropy (EB) was discovered more than 40 years ago by Meiklejohn and Bean,[1] but a full understanding of the phenomenon has not yet emerged,[2] in spite of the revived interest and the technological applications that EB has. Here we propose a model for the compensated (110) antiferromagnetic (AF) interface, which is compatible with the data on the most extensively investigated and best characterized EB systems: Fe/FeF_2 and Fe/MnF_2. Because of the outstanding characterization of the interfaces, its rather simple crystal and interface structure, its large H_E value and the extensive experimental information available[2] we choose, as our prototype system, a 13 nm Fe film deposited on the (110) *compensated* crystal face[3] of FeF_2 and MnF_2.

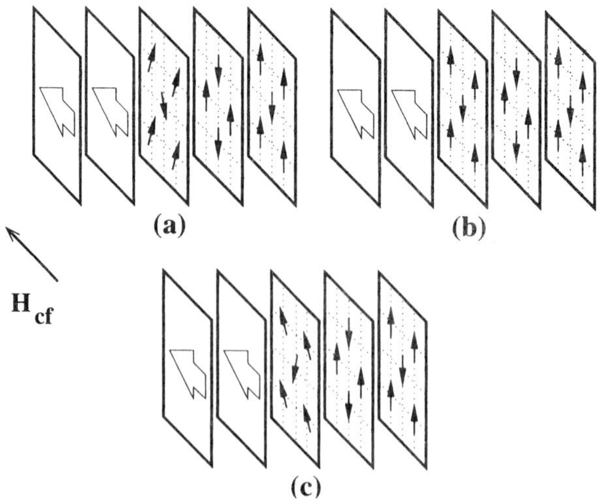

Figure 1. Spin configuration of the AF interface monolayer and the two F and AF monolayers closest to the interface, for low (a), medium (b) and large (c) cooling field H_{cf}. The canting angle θ_c is measured relative to $\vec{H}_{cf}$, applied parallel to the $(\bar{1}10)$ AF crystal direction.

First we implemented extensive numerical simulations (in particular, we used simulated annealing and Camley's method[4]) for a unit interface magnetic cell, which in the direction orthogonal to the interface extends for up to 65 F, and many AF, monolayers. Next we developed an analytic model whose main features are *i*) that EB is an interface governed phenomenon; *ii*) that a spin glass type magnetic structure freezes in, close to the Néel temperature T_N, at the AF interface layer; and *iii*) that the energy is mainly stored in an incomplete domain wall (IDW) within the ferromagnetic (F) slab.

A theoretical model was put forward as early as 1962 Meiklejohn.[5] It assumes a perfect, flat, *uncompensated* AF interface; however, it yields EB fields H_E two orders of magnitude larger than observed. In addition, it fails to explain biasing for a fully compensated AF interface. Later on, a full description of an AF-DW model, for a F/AF system, was proposed by Néel.[6] Néel's model incorporates several hypothesis: weak anisotropy, F interface coupling, an uncompensated AF interface layer and a *continuum* approximation, which requires considerable thickness of the bilayer slabs to be valid. These assumptions are not fulfilled by the most extensively investigated EB systems.

What makes this problem a hard nut to crack are the two very different characteristic length scales it involves: the F and AF domain wall widths, d_w^{F} and d_w^{AF}, respectively. While $d_w^{\mathrm{F}} \sim 100$ nm, d_w^{AF} amounts to just a few monolayers due to the large FeF_2 and MnF_2 anisotropies, which casts doubts on the validity of the continuum approximation (a table with the angular variation as a function of layer index can be found in a paper by the present authors in Ref. 7). This led Koon[8] to explore the compensated case by micromagnetic simulation, to conclude that the bulk AF magnetization is perpendicular[9] to the F, while the AF interface layer, when cooled in low field, adopts a canted (often incorrectly called spin-flop) configuration, as illustrated, for low $|\vec{H}_{cf}|$, in Fig. 1(a). However, the model by Koon, as well as Ref. 10, coauthored by our guest of honor Prof. Hans Christopf Siegmann, and the one by Malozemoff,[11] do not fully explain important EB features. Several additional models have recently been put forward,[12,13] but limitations of space do not allow to discuss them in detail here. However, all of them share a different, but crucial, assumption on the interface structure; ours will be specified below.

A relevant feature, which emerged from our simulations, is the incompatibility of EB (*i.e.* $H_E \neq 0$) with two symmetric mirror image magnetic configurations in the $H \to \pm\infty$ limit; the latter leads to two M *vs.* H plots shifted along the H-axis by the coercive field, but to a centered $H_E = 0$ magnetization loop. By the same token, the presence of canted spins at the AF interface layer is no guarantee for $H_E \neq 0$ in the absence of a symmetry breaking mechanism. Certainly there are several possible symmetry breaking mechanisms: *i*) it is quite likely that the canted spin configuration corresponds to a metastable, spin glass like, minimum; *ii*) a fraction of the canted spins may pin to interface irregularities like twin borders, lattice mismatch defects between the deposited films and the substrate, impurities and/or dislocations. Actually, we performed simulations with a fraction of AF interface magnetic moments clamped, obtaining the experimental values[2] of H_E by adjusting the fraction of clamped moments. In addition, in our simulations we confirmed that while a long range IDW develops in the F, when the applied field points opposite to the cooling field H_{cf}, the interface layer AF spins deviate significantly relative to the AF bulk, but only a tiny deviation develops in the second AF layer and no appreciable canting in the third, which is consistent with results by Camley *et al.*[14] (see table in Ref. 7). This leads to the crucial conclusion that the relevant region for EB encompasses all the F, but at most only two AF monolayers.

Therefore, it is essential to understand in detail what happens to the AF interface layer during the cooling process under the external field H_{cf}. This layer is exposed to a significantly different environment to all the others since, in addition to H_{cf}, it is subject to the strong exchange coupling of the ordered F slab ($T < T_C$). We suggest that this interface AF monolayer reconstructs into an almost rigid canted magnetic structure close to the Néel temperature T_N, which fully freezes when the AF bulk orders. Moreover, it remains frozen, in a metastable state, during the cycling of the external magnetic field, when performed for $|H| < H_{\mathrm{cf}}$. This constitutes *our* assumption on the interface structure.

Several recent experiments lend compelling support to this picture. Among them the most compelling one is a strong memory effect at the F/AF interface, observed by Chien *et al.*,[15] that persists well above T_N and up to fields two orders of magnitude larger than the coercive field. These experiments imply that the memory can only be stored at the interface, since neither the F nor the AF above T_N can clench this information. Since $t < d_w^{\mathrm{F}}$, where t is the thickness of the F slab, a discrete treatment is in order. Analytically

$$\mathcal{H} = \mathcal{H}_{\mathrm{AF}} + \mathcal{H}_{\mathrm{F/AF}} + \mathcal{H}_{\mathrm{F}} \,, \tag{1}$$

where $\mathcal{H}_{\mathrm{AF}}$, $\mathcal{H}_{\mathrm{F/AF}}$ and $\mathcal{H}_{\mathrm{F}}$ describe the AF substrate, interface coupling and the F slab, respectively. Using this model, which we described in detail in Ref. 16, we obtain an expression for the energy per unit interface area ϵ which reads

$$\epsilon = -h\sum_{k=1}^{N}\cos\theta_k - \sum_{k=1}^{N-1}\cos\left(\theta_{k+1}-\theta_k\right) - \kappa\cos\theta_1 - D\sum_{k=1}^{N}\cos^2\theta_k . \tag{2}$$

Above k is the F layer index, and we define the dimensionless applied field $h = \mu_B g H/2J_{\mathrm{F}} < 10^{-3}$, $D = K_{\mathrm{F}}/2J_{\mathrm{F}} < 10^{-5}$ and the effective interface coupling $\kappa = -(|J_{\mathrm{F/AF}}|/J_{\mathrm{F}})\cos\theta_c$. θ_c is the AF interface canting angle, which depends on the AF parameters (J_{AF} and K_{AF}) and on H_{cf}. We differentiate ϵ respect to θ_j, equate to zero

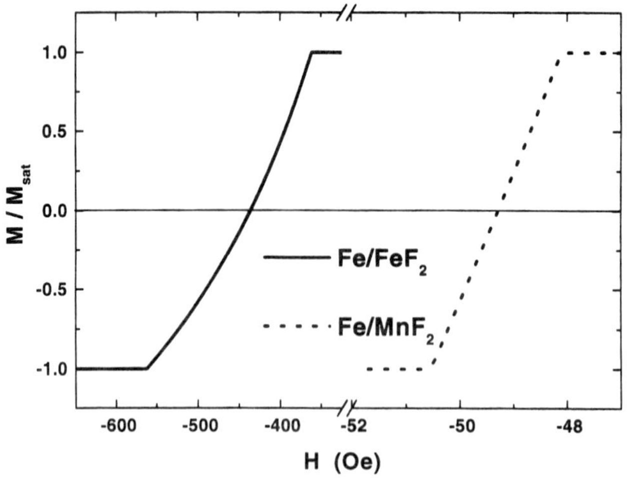

Figure 2. Magnetization M *versus* applied field H for Fe/FeF$_2$ and Fe/MnF$_2$. The values of $J_{F/AF}$ are -1.2 and -0.35 meV for FeF$_2$ and MnF$_2$, respectively.

to minimize the energy and thus obtain the set of nonlinear equations to be solved for $\{\theta_k\}$

$$h\sin\theta_j - (1-\delta_{j,N})\sin(\theta_{j+1}-\theta_j) + \delta_{j,1}\kappa\sin\theta_1 + (1-\delta_{j,1})\sin(\theta_j-\theta_{j-1}) + 2D\sin\theta_j\cos\theta_j = 0, \qquad (3)$$

where $\delta_{i,j}$ is the Kronecker symbol. Below we apply this model to two well documented systems: Fe/FeF$_2$ and Fe/MnF$_2$. They have in common a very small AF-DW width, (of the order of monolayers) and a well characterized, controlled and simple F/AF interface structure.[2] Our computations are carried out for the *compensated* (110) AF interface. A crucial feature to be stressed is the fact that, consistent with our assumption on the freezing of the AF interface layer, in these systems only this layer canting angle differs significantly from the F and AF magnetic bulk. Results for the magnetization vector angle θ_k, of the k-th layer relative to the cooling field direction, can be found in Ref. 7. The condition required for negative EB ($H_E < 0$) is that the minimum energy configuration corresponds to a net interface magnetization component opposite to H_{cf}, or equivalently $\theta_c = \theta_{k=-1} > 90°$, as illustrated in Fig. 1(c). This provides the symmetry breaking necessary to generate EB.

We now turn to the most relevant test of the model outlined above: the M *versus* H plot, which is illustrated in Fig. 2. The lines represent the results of the numerical solution of Eq. (3), while the discrete points were obtained by a micromagnetic numerical calculation (in particular, we used simulated annealing). The values we used were $H_{cf} = 2000$ Oe and $J_{AF} = -1.2$ meV and $K_{AF} = 2.5$ meV/spin,[14] for the FeF$_2$ exchange and anisotropy parameters, respectively. For iron the parameters are $J_F = 16$ meV and $K_F = 5 \times 10^{-4}$ meV/spin.[17] While the solution of Eq. (3) is readily evaluated the simulation is quite involved and requires lengthy computation. In fact, each discrete point corresponds to the lowest energy value obtained after exploring 40 different random initial configurations (in addition to the preceding one) for each value of the applied field, in order to be reasonably confident that the computed minimum is global rather than local. In addition, the agreement of the values achieved by simulation and the ones obtained analytically, reinforces the reliability of our results.

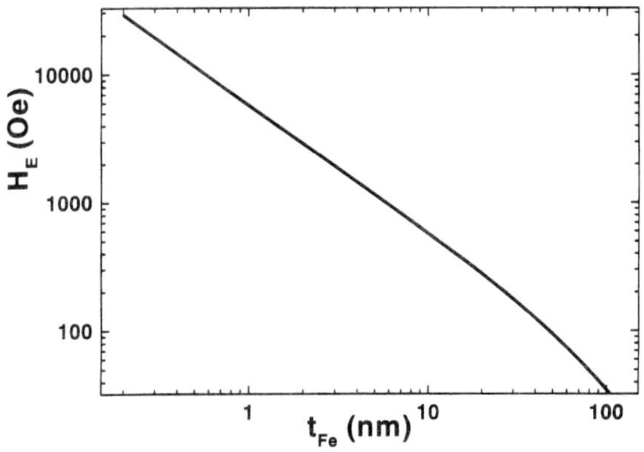

Figure 3. H_E dependence on Fe layer thickness t_F.

Several features of these results are important to notice: *i)* the strong dependence of H_E on the value of $J_{F/AF}$, the only adjustable parameter of our theory; *ii)* the fact that for $J_{F/AF} = J_{AF} = -1.2$ meV and $t_F = 13$ nm we obtain $H_E = -436$ Oe, quite close to the experimentally observed[3] value $H_E \approx -500$ Oe. Moreover, for an Fe/FeF_2 slab of $t_F = 6.7$ nm we obtain $H_E = -870$ Oe. For $t_F = 130$ nm we obtain by extrapolation of $H_E \propto t_F^{-1}$ the value of -56 Oe, while a full calculation with $t_F = 550$ monolayers yields -50 Oe. Both are in excellent agreement with experiment,[3] which yields $H_E \approx -700$ and -46 Oe, for $t_F = 6.7$ and 130 nm, respectively; *iii)* the asymmetric shape of the *M versus H* plot respect to H_E, which is due to the formation of an IDW in the F, close to the interface; *iv)* the full reversibility, for decreasing and increasing applied field H, of both our simulation and numerical results for M *versus* H; *v)* the excellent agreement between the two independent approaches we implemented, *i.e.* the simulation and the numerical solution of Eq. (3).

In Fig. 3 we plot the thickness dependence of H_E *versus* the thickness t_F of the magnetic slab. In agreement with experiment, for values of $t_F < d_w^F$ the values of $H_E \propto t_F^{-1}$, while for $t_F > d_w^F$ the continuum approximation becomes valid and a crossover to $H_E \propto t_F^{-1/2}$ is observed. This constitutes an important success of our model.

In Fig. 4 we display the effect of the cooling field H_{cf} on H_E. An excellent quantitative agreement for Fe/MnF_2 is obtained, while the results for Fe/FeF_2 are less satisfactory.

In conclusion, we have put forward a model that provides an adequate description of the EB phenomenon for a compensated AF interface, on the basis of a theory with only one adjustable parameter: $J_{F/AF}$, the F/AF interface exchange coupling. The crucial elements of the model are the freezing of the canted spin configuration of the AF interface and the formation of an IDW magnetic structure in the F slab. Our results are in qualitative and quantitative agreement with experiment.

Acknowledgments

We gratefully acknowledge enlightening discussions with Prof. I. K. Schuller and Drs. M. Fitzsimmons and T. Schulthess. Partially supported by the *Fondo Nacional de Investigaciones Científicas y Tecnológicas* (FONDECYT, Chile) under grants # 8990005, 1980811 and 2980013.

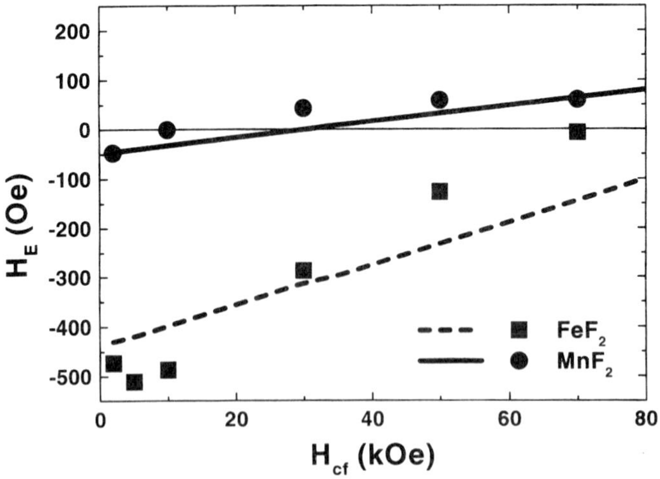

Figure 4. H_E *versus* cooling field H_{cf} for Fe/FeF_2 and Fe/MnF_2.

References

* PEDECIBA-Física Graduate Fellowship, Universidad de la República, Uruguay.

1. W. P. Meiklejohn and C. P. Bean, *Phys. Rev.* **102**, 1413 (1956); **105**, 904 (1957).
2. J. Nogués and I. K. Schuller, *J. Mag. Magm. Mat.* **192**, 203 (1999).
3. J. Nogués, D. Lederman, T. J. Moran, I. K. Schuller, and K. V. Rao, *Appl. Phys. Lett.* **68**, 3186 (1996).
4. R. E. Camley, *Phys. Rev. B* **35**, 3608 (1987); R. E. Camley and D. R. Tilley, *Phys. Rev. B* **37**, 3413 (1988).
5. W. P. Meiklejohn, *J. Appl. Phys.* **33**, 1328 (1962).
6. L. Néel, *Ann. Phys. (Paris)* **2**, 61 (1967).
7. M. Kiwi, J. Mejía-López, R. D. Portugal, and R. Ramírez, *Appl. Phys. Lett.* **75**, 3995 (1999).
8. N. C. Koon, *Phys. Rev. Lett.* **78**, 4865 (1997).
9. T. J. Moran, Ph.D. thesis, University of California, San Diego, 1995 (unpublished); T. J. Moran, J. Nogués, D. Lederman, and I. K. Schuller, *Appl. Phys. Lett.* **72**, 617 (1998).
10. C. Mauri, H. C. Siegmann, P. S. Bagus, and E. Kay, *J. Appl. Phys.* **62**, 3047 (1987).
11. A. P. Malozemoff, *Phys. Rev. B* **35**, 3679 (1987); *Phys. Rev. B* **37**, 7673 (1988); and *J. Appl. Phys.* **63**, 3874 (1988).
12. T. C. Schulthess and W. H. Butler, *Phys. Rev. Lett.* **81**, 4516 (1998).
13. M. D. Stiles and R. D. McMichael, *Phys. Rev. B* **59**, 3722 (1999).
14. A. S. Carriço, R. E. Camley, and R. L. Stamps, *Phys. Rev. B* **50**, 13453 (1994).
15. T. Ambrose, R. L. Sommer, and C. L. Chien, *Phys. Rev. B* **56**, 83 (1997); S. M. Chou, Kai Liu, and C. L. Chien, *Phys. Rev. B* **58**, R14 717 (1998).
16. M. Kiwi, J. Mejía-López, R. D. Portugal, and R. Ramírez, *Europhys. Lett.* **48**, 573 (1999).
17. C. Kittel, *Introduction to Solid State Physics*, sixth edition, (Wiley, New York, 1986), p. 423.

The Magnetic Order of Cr in Fe/Cr/Fe(001) Trilayers

D. T. Pierce, J. Unguris, R. J. Celotta, and M. D. Stiles

Electron Physics Group
National Institute of Standards and Technology
Gaithersburg, MD 20899-8412
USA

Abstract

The temperature dependence of the short period oscillatory coupling in Fe/Cr/Fe(001) whisker trilayers, when analyzed in light of recent theory, provides strong evidence that incommensurate spin-density-wave antiferromagnetic order is induced in the Cr over a wide range of thickness and up to temperatures at least 1.8 times the Néel temperature of bulk Cr. Because the portion of the Cr Fermi surface connected with the short-period interlayer coupling in paramagnetic Cr does not exist in antiferromagnetic Cr, the conventional quantum well model of interlayer coupling is not applicable. Rather, it is necessary to use a model for the coupling that includes the electron-electron interactions that lead to the antiferromagnetic order.

Measurements of Fe/Cr multilayers have been the source of numerous important discoveries in thin film magnetism. The fundamental phenomena of exhange coupling in transition metal multilayers,[1] giant magnetoresistance,[2,3] long-[4] and short-period[5,6] oscillations in the exchange coupling with varying Cr spacer thickness, and the imaging[7] of 90 degree biquadratic coupling were all first reported for Fe/Cr multilayers. There have been extensive studies of Fe/Cr multilayers ranging from trilayers to superlattices with many bilayer repeats. The apparent discrepancies between results reported for different systems can be understood in a unified picture that accounts for the sensitivity of the Cr magnetic order and the interlayer exchange coupling to a variety of structural details.[8]

Directly measuring the Cr magnetic order by neutron scattering requires superlattices with both enough layers and large enough area to provide sufficient sensitivity.

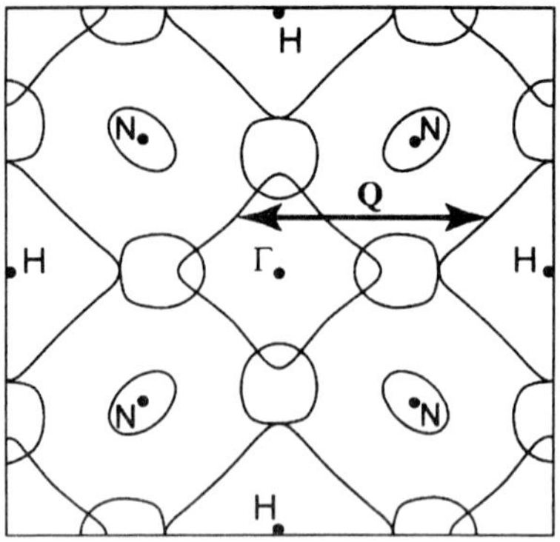

Figure 1. A slice through the paramagnetic Cr Fermi surface for a (001) interface. The wavevector **Q** connects parallel "nested" regions of the Fermi surface.

Depending on temperature and Cr spacer layer thickness, paramagnetic, and commensurate and incommensurate spin-density-wave antiferromagnetic phases are observed. With Fe whisker substrates, it is possible to grow Fe/Cr/Fe trilayers with interfaces that are much closer to ideal than those in superlattices. Unlike the superlattices, Fe/Cr/Fe trilayers are not amenable to neutron scattering measurements. The issue of whether the Cr spacer layer is paramagnetic or antiferromagnetic in the temperature and thickness ranges of interest must be inferred examining the behavior of the magnetic coupling.

The focus of this paper is to elucidate the magnetic state of Cr in optimally prepared Fe/Cr/Fe(001) whisker trilayers and the implications for the models which describe the interlayer exchange coupling. The *conventional* model for interlayer exchange coupling through a paramagnetic metal spacer layer, such as a noble metal, is the quantum well picture.[9–11] The strength of the coupling depends on the reflection amplitudes of electrons at the interfaces between the paramagnetic spacer layer and the ferromagnetic layers as well as on the spacer Fermi surface geometry. The periods of the coupling are determined from the critical spanning vectors of the spacer layer Fermi surface. Thickness fluctuations of the spacer layer average the coupling, and short-period oscillations of the coupling are not observed if the thickness fluctuations are too large. An extrinsic biquadratic coupling may also be induced by thickness fluctuations.[12]

The paramagnetic Cr Fermi surface for an (001) interface is shown in Fig. 1. In Cr, there are substantial regions of the Fermi surface that are parallel, or nested, and separated by the nesting wavevector of magnitude $Q = (2\pi/a)(1-\delta)$ where the lattice constant is twice the layer spacing. The nesting wave vector is slightly incommensurate with the lattice wavevector $(2\pi/a)$ where δ is the incommensurability parameter. In the conventional model for interlayer exchange coupling, the nesting wavevector **Q** of paramagnetic Cr is the extremal spanning vector of the Fermi surface that leads to the observed short-period oscillatory exchange coupling which is hence slightly incommensurate with the *bcc* lattice. The long-period oscillatory coupling has been attributed to the extremal spanning vectors at the N-centered ellipses.[13]

There is also an enhancement of the magnetic susceptibility at **Q** that leads to a transition from paramagnetic to antiferromagnetic Cr below the Néel temperature, T_N. When Cr orders antiferromagnetically, a small gap opens at the Fermi level and that part of the nested Fermi surface connected by **Q** in Fig. 1 disappears.[14] This has been known for some time and is manifested, for example, as a resistivity anomaly at T_N in bulk Cr.[14] Recent photoemission measurements that map the energy bands show explicitly that a gap opens at the Fermi level.[15] Because that part of the Fermi surface connected with the short-period oscillations in the conventional coupling model no longer exists in antiferromagnetic Cr, it is necessary to go beyond the conventional model, which treats the spacer material in an itinerant electron picture, and include a treatment of the electron-electron interactions of the type that stabilize the incommensurate order in Cr. Thus, the appropriate model for the short-period oscillatory exchange coupling in Fe/Cr/Fe depends on whether the Cr spacer layer is paramagnetic or antiferromagnetic. In contrast, the N-centered ellipses do not differ significantly between paramagnetic and antiferromagnetic Cr and the conventional model is presumed adequate for either state.

Several descriptions of the antiferromagnetic order of Cr in Fe/Cr/Fe trilayers have been advanced. First-principle calculations show that short-period oscillatory magnetic coupling occurs for both paramagnetic and antiferromagnetic Cr spacers.[16–18] Recent model calculations of Shi and Fishman[19] are particularly useful in their explanations of the temperature dependence of the Cr magnetic order. We will analyze our measurements of the temperature dependence of the coupling in light of the model calculations of Shi and Fishman[19] and show that one is led to the conclusion that in optimally grown Fe/Cr/Fe trilayers on Fe(001) single crystal whiskers, the Cr is antiferromagnetic over a wide range of temperature and thickness.

The model calculations of Shi and Fishman[19] consider the competition between *a)* the interface magnetic coupling, which maximizes the amplitude of the Cr spin-density-wave at the boundaries (*i.e.* an antinode at the Fe/Cr interface), and *b)* the intrinsic antiferromagnetism of the Cr spacer, which favors bulk values of the spin-density-wave amplitude and wave vector. The magnitude of the spin-density-wave ordering wavevector $Q' = (2\pi/a)(1 - \delta')$ is always closer to commensuration than the nesting wave vector $\mathbf{Q}$, that is, $0 \leq \delta' < \delta$.[19] Two examples of transverse spin-density-waves are shown in Fig. 2. A commensurate spin-density-wave, *i.e.*, $\delta' = 0$, is shown in Fig. 2(a). An incommensurate spin-density-wave is shown in Fig. 2(b); the nodes are separated by $1/\delta'$ ML.

We begin by reviewing the room temperature measurements[5,20] of the interlayer coupling of trilayers where the Cr is grown as a very shallow wedge on a near perfect Fe single crystal whisker substrate as shown in Fig. 3. The oscillations in the exchange coupling with varying Cr thickness are readily observed in a Scanning Electron Microscopy with Polarization Analysis (SEMPA) image of the top Fe layer magnetization. The SEMPA image is formed by measuring the spin polarization of the secondary electrons as the SEM beam is rastered across the sample surface. Such an image is shown in Fig. 4(b). A line scan through Fig. 4(b) gives the measured polarization profile shown in Fig. 4(c) of the Fe overlayer. The Fe overlayer is seen to be coupled ferromagnetically to the Fe whisker substrate for the first four layers, and then the coupling begins to oscillate. It changes from ferromagnetic coupling [white regions in Fig. 4(b)] to antiferromagnetic (overlayer magnetization antiparallel to the whisker magnetization) and back as the Cr increases in thickness by two additional layers. The change in direction of the coupling continues with each additional Cr layer up to 24 layers. The coupling is ferromagnetic for both 24 and 25 layers of Cr and only for 26 layers does it change to antiferromagnetic. This phase slip in the coupling is repeated each 20 layers. In the conventional model of the exchange coupling, the phase slip is a beating phenomenon resulting from the slight difference between the extremal spanning vector of the Fermi surface $\mathbf{Q}$ and the lattice wavevector.

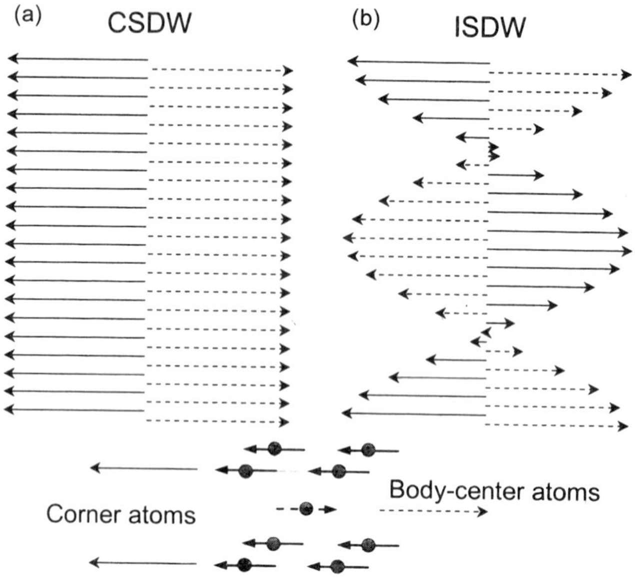

Figure 2. The solid arrows and dashed arrows represent the Cr moments on corner atom and body-center atom sites respectively. (a) Commensurate antiferromagnetic order. (b) One period of an incommensurate spin-density-wave showing two nodes.

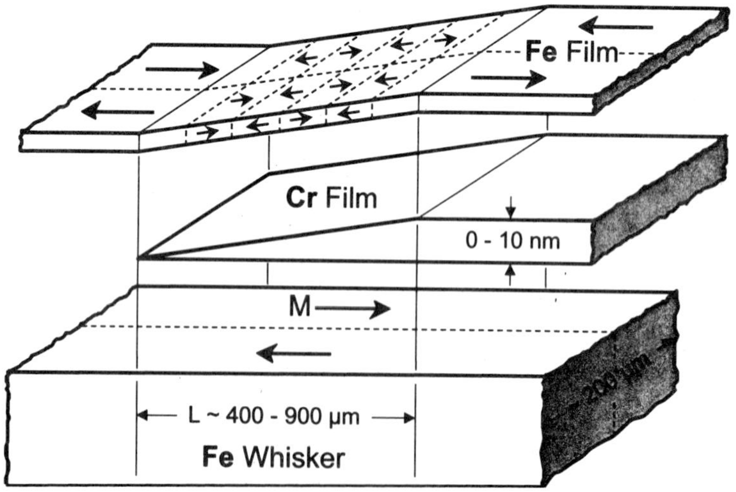

Figure 3. A schematic view of the wedge trilayer sample structure showing the Fe(001) single crystal whisker substrate, the evaporated Cr wedge, and the Fe overlayer. The vertical scale is expanded many times.

If the Cr is antiferromagnetic,[17–19] the phase slip occurs at Cr thicknesses where an additional node is introduced in the spin-density-wave. In this picture, the data of Fig. 4 results from changes in the Cr spacer as shown in Fig 4(a). There is commensurate spin-density-wave order through 24 layers. At 25 layers, a node pops in, and then there is incommensurate spin-density-wave order. With increasing Cr thickness, the incommensurability δ' decreases, that is the spin-density-wave period increases, so as

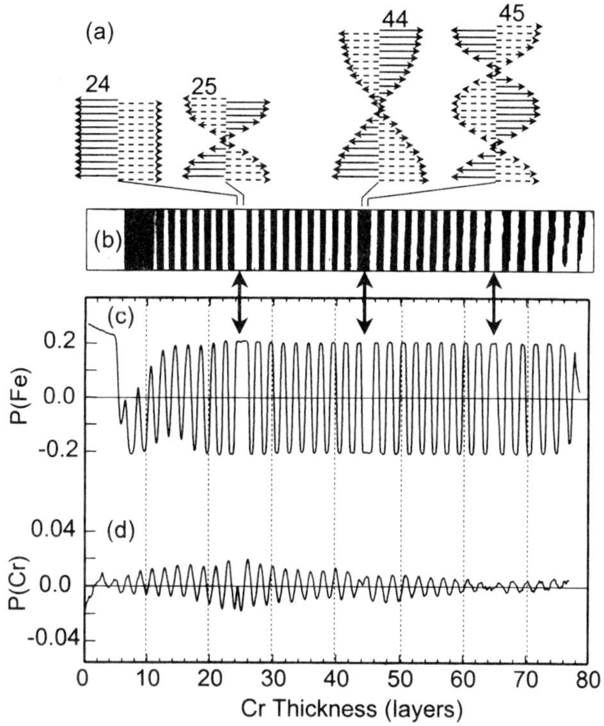

Figure 4. (a) Presumed spin-density-wave structures at each side of the first two phase slips.[17–19] (b) SEMPA image of the component of magnetization, M_x, in the Fe overlayer along the Fe whisker. The arrows mark the Cr spacer layer thicknesses where phase slips in the short-period oscillations of the magnetization occur. (c) a line scan through (b) showing the measured spin polarization profile of the Fe overlayer. (d) The spin polarization of the Cr layer P(Cr), before depositing the Fe overlayer, after subtracting the contribution from the whisker.

to maintain maximum spin-density-wave amplitude at the Fe interface up through a thickness of 44 layers. At 45 layers, another node is formed in the spin-density-wave and the distance between nodes contracts toward bulk values. Thus, the period of the spin-density-wave is not rigid in these Cr films, but adjusts to match the boundary conditions. Although the internal structure of the Cr illustrated in Fig. 4(a) cannot be measured directly for these Fe/Cr/Fe trilayers, this picture is consistent with our results. Exactly what occurs in the first few Cr layers at the interface with the Fe whisker is not known. There is evidence of interfacial alloying at the temperatures for optimal Cr growth.[21,22] Calculations[23] show that 25% of the first Cr layer intermixing with the Fe is sufficient to cause the observed π phase change in the antiferromagnetic stacking of Cr. Note that the polarization profile P(Cr) of the Cr wedge shown in Fig. 4(d), measured before adding the top Fe overlayer, also exhibits short-period oscillations with phase slips every 20 layers.[20] Apparently, one Fe interface is sufficient to induce spin-density-wave antiferromagnetic order in the Cr.

Whereas the room temperature coupling measurements of Fig. 4 could be explained by the conventional model of interlayer exchange coupling, measurements of the position of the first phase slip at different temperatures provides definitive evi-

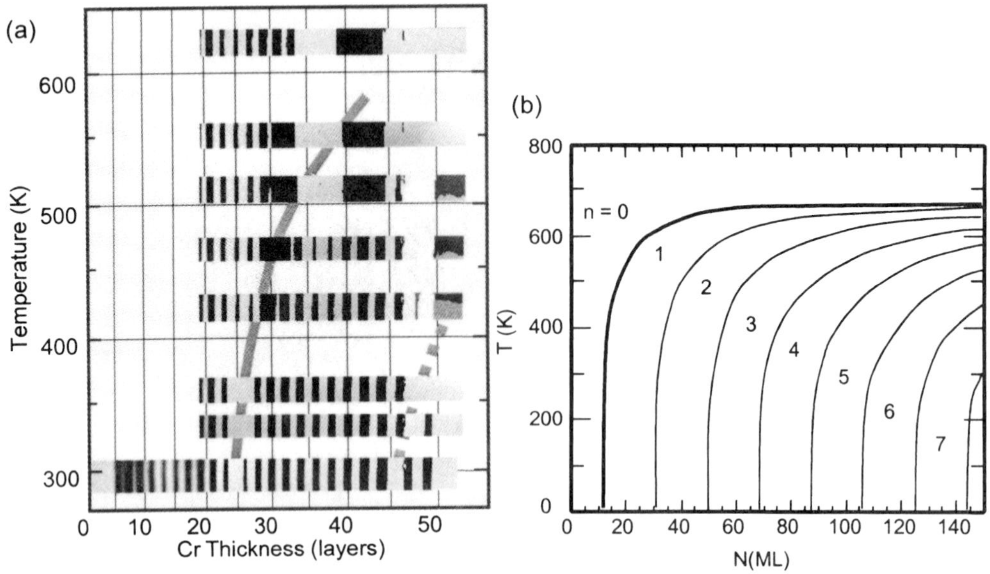

Figure 5. (a) SEMPA images of the coupling in Fe/Cr/Fe(001) at different temperatures. The phase slips measured on the bare Cr are shown by the solid gray line; the dashed line is the estimated position of the next phase slip. Note that, where visible, the short-period oscillations in the SEMPA images of Fe/Cr/Fe have opposite direction above and below these lines. (b) Calculated[19] curve showing the transition (heavy line) from commensurate order ($n = 0$) to an incommensurate spin-density-wave with one node ($n = 1$), as a function of temperature and Cr thickness. The lighter curves show the transitions between incommensurate spin-density-waves with different number of nodes.

dence for antiferromagnetic order in the Fe/Cr/Fe trilayers. The change in the position of the phase slip with temperature is most easily seen in measurements of the bare Cr wedge on the Fe(001) whisker. The displacement by 14 layers of the first phase slip from its position at 24–25 layers at 310 K to 38–39 layers at 550 K is shown by the solid curved line in Fig. 5(a). SEMPA images in the same temperature range[24] are also shown in Fig. 5(a). The SEMPA images show that the long-period coupling can obscure the short-period coupling at higher temperatures. The magnetization direction of the short-period oscillations is reversed where it can be seen above and below the phase slip line, for example at Cr thicknesses of 25 to 30 layers. The dashed line is the solid curve displaced by 20 layers to indicate where the next phase slip occurs. The temperature dependent measurements on bare Cr and on the trilayers were completely reversible.

This large temperature dependence of the position of the phase slips cannot be explained in the conventional coupling model because the temperature dependent changes in the Cr paramagnetic Fermi surface are much too small to cause a sufficient change in the nesting wavevector and hence in the incommensurability.[19] Rather, a model is required, such as that of Shi and Fishman,[19] that takes into account the electron-electron interactions in the Cr that stabilize the antiferromagnetic state. In this picture, the solid line in Fig. 5(a) marks the transition from a commensurate spin-density-wave with no nodes ($n = 0$) where the Cr is thinner, to an incommensurate spin-density-wave with one node. The dashed line then corresponds to the $n = 1$ to 2 incommensurate to incommensurate transition. The calculated thickness and temperature dependence of these transitions is shown in Fig. 5(b). The model gets the node-to-node distance correct assuming a 0.6% decrease in the lattice constant of Cr grown on Fe, but the phase slips occur at smaller distances than found experimentally. The overall trend of the change in the calculated phase slip position with temperature agrees well with that found experimentally in the range of overlap. The strong temperature dependence found experimentally, and the agreement of the temperature variation calculated by Shi and Fishman, provides solid evidence that the Cr in Fe/Cr/Fe(001) trilayers is in an antiferromagnetic state. It is quite striking that incommensurate spin-density-wave antiferromagnetic order is induced in the Cr layers up to at least 1.8 times the bulk Néel temperature. This antiferromagnetic order that exists well above T_N of bulk Cr is due to the proximity of the Fe and the strong interfacial coupling of the Cr and Fe.

In summary, the strong temperature dependence of the thickness of the Cr film at which the phase slips occur in the short-period oscillatory coupling in Fe/Cr/Fe(001) trilayers is evidence for spin-density-wave order in the Cr. The results are in qualitative agreement with the temperature dependence predicted by Shi and Fishman[19] and consistent with a picture of magnetic order in the Cr consisting of a commensurate spin-density-wave up to the first phase slip and incommensurate spin-density-wave order beyond that. For the nearly ideal interfaces of these structures grown on Fe whiskers, the Fe-Cr interfacial coupling induces antiferromagnetic order in the Cr to temperatures well above the bulk Néel temperature. The short-period oscillatory coupling cannot be explained in the conventional quantum well picture; instead, it is closely tied to the spin-density-wave order of the Cr.

Acknowledgments

We are grateful to R. S. Fishman, J. A. Stroscio, and A. Davies for helpful discussions. This work was supported in part by the office of Naval Research.

References

1. P. Grünberg, R. Schreiber, Y. Pang, M.B. Brodsky, and H. Sowers, *Phys. Rev. Lett.* **57**, 2442 (1986).
2. M. N. Baibich, J. M. Broto, A. Fert, F. Nguyen Van Dau, F. Petroff, P. Etienne, G. Creuzet, A. Friederich, and J. Chazelas, *Phys. Rev. Lett.* **61**, 2472 (1988).
3. G. Binasch, P. Grünberg, F. Saurenbach, and W. Zinn, *Phys. Rev. B* **39**, 4828 (1989).
4. S. S. P. Parkin, N. More, and K.P. Roche, *Phys. Rev. Lett.* **64**, 2304 (1990).
5. J. Unguris, R. J. Celotta, and D. T. Pierce, *Phys. Rev. Lett.* **67**, 140 (1991).
6. S. T. Purcell, W. Folkerts, M. T. Johnson, N. W. E. McGee, K. Jager, J. Vaan de Stegge, W. B. Zeper, W. Hoving, and P. Grünberg, *Phys. Rev. Lett.* **67**, 903 (1991).
7. M. Rührig, R. Schäfer, A. Hubert, R. Mosler, J. A. Wolf, S. Demokritov, and P. Grünberg, *Phys. Stat. Sol. A* **125**, 635 (1991).
8. D. T. Pierce, J. Unguris, R. J. Celotta, and M. D. Stiles, *J. Magn. Magn. Mater.* **200**, 290 (1999).
9. D. M. Edwards, J. Mathon, R. B. Muniz, and M. S. Phan, *Phys. Rev. Lett.* **67**, 493 (1991).

10. M. C. Stiles, *Phys. Rev. B* **48**, 7238 (1993).
11. P. Bruno, *J. Magn. Magn. Mater.* **121**, 248 (1993); *Phys. Rev. B* **52**, 411 (1995).
12. J. C. Slonczewski, *Phys. Rev. Lett.* **67**, 3172 (1991).
13. M. D. Stiles, *J. Magn. Magn. Mater.* **200**, 322 (1999).
14. E. Fawcett, *Rev. Mod. Phys.* **60**, 209 (1988).
15. J. Schäfer, E. Rotenberg, G. Meigs, S. D. Kevan, P. Blaha, and S. Hüfner, *Phys. Rev. Lett.* **83**, 2069 (1999).
16. M. van Schilfgaarde and F. Herman, *Phys. Rev. Lett.* **71**, 1923 (1993).
17. K. Hirai, *Phys. Rev. B* **59**, R6612 (1999).
18. A. M. N. Niklasson, B. Johansson, and L. Nordström, *Phys. Rev. Lett.* **82**, 4544 (1999).
19. Z. P. Shi and R. S. Fishman, *Phys. Rev. Lett.* **78**, 1351 (1997); R. S. Fishman and Z. P. Shi, *J. Phys.: Condens. Matter* **10**, L277 (1998).
20. J. Unguris, R. J. Celotta, and D. T. Pierce, *Phys. Rev. Lett.* **69**, 1125 (1992).
21. A. Davies, J. A. Stroscio, D. T. Pierce, and R. J. Celotta, *Phys. Rev. Lett.* **76**, 4175 (1996).
22. B. Heinrich, J. F. Cochran, T. Monchesky, and R. Urban, *Phys. Rev. B* **59**, 14520 (1999).
23. M. Freyss, D. Stoeffler, and H. Dreyssé, *Phys. Rev. B* **56**, 6047 (1997).
24. J. Unguris, R. J. Celotta, D. A. Tulchinsky, and D. T. Pierce, *J. Magn. Magn. Mater.* **198-199**, 396 (1999).

Magnetic Domain Imaging of Thin Metallic Layers Using PEEM

G. Schönhense

Institut für Physik
Johannes Gutenberg-Universität Mainz
Staudinger Weg 7
55099 Mainz
GERMANY

Abstract

Photoemission electron microscopy (PEEM) in combination with resonant excitation by circularly polarized soft X-rays has proven to be a powerful analytical tool for the study of magnetic microstructures and multilayers. In this type of electron microscope the lateral intensity distribution of the emitted low-energy secondary or photoelectrons is imaged by an electron-optical system. Owing to its fast parallel image acquisition, its wide zoom range allowing fields of view from almost 1 mm down to a few μm combined with a high base-resolution of the order of 20 nm, the method offers a unique access to many aspects in surface and thin film magnetism on the mesoscopic length scale. Magnetic contrast is achieved by the magnetic circular or linear dichroism. Sum-rule analysis allows to extract quantitative information about the spin and orbital magnetic moments of $3d$-transition metals. Two other modes of magnetic imaging via PEEM work with simple UV light sources and are therefore highly attractive for standard laboratory applications. The magnetic stray-field close to the sample surface leads to a Lorentz-type contrast. A third type of contrast arises as a consequence of a small rotation of the displacement vector inside a magnetic material, a phenomenon which is also responsible for the well-known magneto-optical Kerr-effect. Examples and typical applications of magnetic imaging using PEEM are discussed.

I. Introduction

The world-wide efforts in the study of surface and 2D magnetism[1] with a large potential of applications for novel magnetic devices[2] like the various elements in the new field of spin electronics are a considerable challenge for magnetic imaging techniques. Often, the properties of interest can be tailored only in ultra-thin magnetic films or multilayers which emphasizes the need of a highly surface sensitive magnetic imaging

Physics of Low Dimensional Systems
Edited by J. L. Morán-López, Kluwer Academic/Plenum Publishers, New York 2001

technique. On the other hand devices like the M-RAM require patterned structures with the size of each element tending to the minimum limit which is technically possible. The next generation of such devices is characterized by a size in the 100 nm region. Hence, an analysis of the micromagnetic behavior of such structures is no longer possible by the "classical" approach of magneto-optical Kerr microscopy working with visible or UV-light. Consequently, new analysis techniques need to be developed for the study of such objects.

In recent years, it has been demonstrated that the magnetic imaging techniques which are based on the high-brilliance X-ray beams from electron storage rings are a highly promising tool for both magnetic spectroscopy and microscopy. These techniques utilize the *X-ray magnetic circular dichroism* (XMCD) or *X-ray magnetic linear dichroism* (XMLD) at selected absorption edges. Hence, the analysis is both element specific and magnetically sensitive. A sum-rule analysis of the magnetic dichroism spectra allows a quantitative determination of the spin and orbital components of the magnetic moments. A lot of experimental and theoretical work has been performed for the $L_{2,3}$-edges of the $3d$-transition metals. The study of these edges requires tunable soft X-ray radiation in the photon-energy range up to about 1000 eV. Today, a number of storage rings all over the world offer beamlines with very high brilliance in this energy range.

There are basically three different approaches of magnetic imaging and local magnetic spectroscopy exploiting XMCD or XMLD. *Transmission X-ray microscopy* (TXM)[3] and *scanning transmission X-ray microscopy* (STXM)[4] can be used for the study of transparent magnetic samples. Usually these are thin-film samples grown on transparent substrates. Since the magnetic signal scales with the X-ray absorption, these techniques are sensitive to the bulk properties of the sample. *Photoemission electron microscopy* (PEEM)[5] is a surface-sensitive technique owing to the escape depth of the electrons used for image formation. Thus the information obtained by X-ray microscopy and PEEM are complementary. A certain advantage of the PEEM technique is that the samples do not need to be transparent, *i.e.* whole devices can be directly studied. On the other hand, TXM and STXM are possible in the presence of external magnetic fields. XMCD is sensitive to ferromagnetism whereas XMLD can reveal the domain structure of antiferromagnetic materials.[6]

Magnetic imaging using a PEEM is not necessarily connected with the exploitation of XMCD or XMLD. Recently, the occurrence of a *magnetic linear dichroism in threshold* PEEM has been found,[7] *i.e.* for photon energies in the UV-range around 5 eV. In analogy to the XMLD we introduce the acronym UMLD for this mode of PEEM operation. UMLD-PEEM is highly attractive for standard laboratory applications since it does not require circularly polarized X-rays. Another magnetic contrast mechanism exists which has a different nature as these dichroic effects. It arises due to the electromagnetic forces experienced by the low-energy electrons when leaving the sample which lead to a *Lorentz-type contrast*. It thus bears much resemblance to the type I contrast in a scanning electron microscope. This way of magnetic imaging via PEEM is actually the oldest method and was discovered and investigated as early as 1957.[8]

It is the purpose of the present article to explain the different modes of magnetic imaging using a PEEM and to illustrate its performance by typical examples of magnetic thin films. We will focus on those aspects and special applications which are hardly possible or even impossible by means of other surface-sensitive magnetic imaging techniques like spin-polarized SEM (SEMPA) or magnetic force microscopy (MFM). Finally, an outlook on some ongoing future developments will be given.

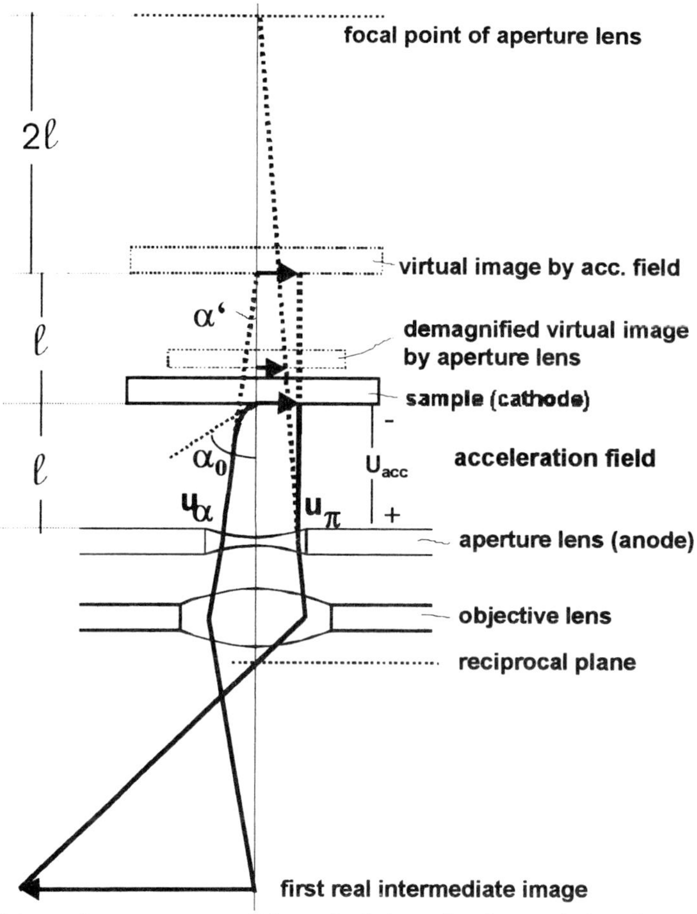

Figure 1. Schematic arrangement of the cathode lens of a photoemission electron microscope with the two principal rays. For illustrative purposes the transverse scale of the two rays shown has been strongly expanded. Virtual images and corresponding rays are denoted by dashed lines. For simplicity the objective lens is shown as a thin lens.

II. Types of Magnetic Contrast in Photoemission Microscopy

II.1. Image Formation and Instrumental Aberrations

The key feature of a high-resolution PEEM is the cathode lens, an electron-optical immersion lens using a strong electric field between sample surface (cathode) and an extractor electrode (anode). Figure 1 shows a schematic diagram of the basic electron optical elements together with the fundamental rays. The electric field given by the acceleration voltage U_{extr} and the gap distance ℓ accelerates the electrons being photoemitted from the sample surface towards the extractor electrode and the objective lens. Let us consider the two fundamental rays defining the arrow (object) on the sample surface, *i.e.* the ray u_α starting under an arbitrary angle α_0 with the starting energy E_0 on the optical axis and the ray u_π starting off-center parallel to the optical axis. The u_α-ray follows a parabola-shaped trajectory, whereas the u_π-ray remains parallel to the optical axis until both rays reach the extractor electrode.

When forming the tangent to the u_α-ray in the plane of the extractor electrode we see that the acceleration field yields a virtual image of the sample in a distance of approximately ℓ behind the sample surface.[10] Since u_π stays parallel to the optical axis, the magnification of this virtual image is $M = 1$. The bore in the extractor electrode acts as a thin diverging electron lens, the so-called aperture lens. In the geometry shown, such an aperture lens terminating an accelerating field has a focal distance of 4ℓ.[9] The focal length of the aperture lens gives rise to a second virtual image in a distance of $\ell/3$ behind the real sample. This image is demagnified by $M = 2/3$. Finally, the objective lens forms a real intermediate image with a typical magnification of $M = 10$–40. This image can be further magnified by projector lenses (not shown). A key feature of the PEEM is its capability of high resolution. In the following we will summarize the leading aberration terms of the electron optics. The aberrations of the acceleration field have been derived by Bauer[11] and Rempfer *et al.*[12]

The *transversal spherical aberration* of a homogeneous acceleration field is given by

$$\delta_s^{\mathrm{acc}} = 2\ell \frac{E_0}{eU_{\mathrm{extr}}}(\sin\alpha_0 - 0.5\sin 2\alpha_0). \tag{1}$$

Here, E_0 and α_0 are the starting energy and starting angle on the object surface, respectively. We have used Snell's law which connects α_0 and α', the latter being the apparent starting angle on the virtual image produced by the acceleration field (see Fig. 1)

$$\alpha' = \left(\frac{E_0}{eU_{\mathrm{extr}}}\right)^{1/2} \sin\alpha_0, \tag{2}$$

which holds under the condition $E_0 \ll eU_{\mathrm{extr}}$ (valid for threshold or secondary-electron PEEM). For small starting angles α_0 the trigonometric functions can be expanded yielding the well-known equation of third-order theory

$$\delta_s^{\mathrm{acc}} = \ell \left(\frac{eU_{\mathrm{extr}}}{E_0}\right)^{1/2} \alpha'^3 = C_s^{\mathrm{acc}} \alpha'^3. \tag{3}$$

We see that in this approximation the term for the spherical aberration takes the form well-known for electron-optical lenses, whereas the correct function given in Eq. (1) clearly deviates from the behavior valid for electron lenses.

The lateral chromatic aberration of the acceleration field is given by

$$\delta_s^{\mathrm{acc}} = \ell \left(\frac{E_0}{eU_{\mathrm{extr}}}\right)^{1/2} \left[\left(\frac{E_0 + \Delta E/2}{eU_{\mathrm{extr}}}\right)^{1/2} - \left(\frac{E_0 - \Delta E/2}{eU_{\mathrm{extr}}}\right)^{1/2}\right] \sin 2\alpha_0, \tag{4}$$

ΔE is the energy width of the initial electron distribution centered around E_0. If $\Delta E \sim 0.5 E_0$, a condition which is approximately fulfilled in PEEM both in the UV- and X-ray-range, we obtain

$$\delta_c^{\mathrm{acc}} = -0.5\ell \frac{\Delta E}{eU_{\mathrm{extr}}} \sin 2\alpha_0. \tag{5}$$

For small α_0 we can again use the lowest order of the expansion series yielding simply α' for the angular dependence. If, in addition, $\Delta E \ll E_0$ we can expand the square root functions in Eq. (4), yielding

$$\delta_c^{\mathrm{acc}} = -\ell \left(\frac{eU_{\mathrm{extr}}}{E_0}\right)^{1/2} \frac{\Delta E}{eU_{\mathrm{extr}}} \alpha' = C_c^{\mathrm{acc}} \frac{\Delta E}{eU_{\mathrm{extr}}} \alpha'. \tag{6}$$

Similar to the case of the spherical aberration the term of the chromatic aberration takes the form which is well-known for electron lenses in this approximation. Note that all equations refer to sample coordinates. The condition of small starting angles is fulfilled in high-resolution PEEM. It should be mentioned, however, that the condition $\Delta E \ll E_0$ can be approximately valid in LEEM but is usually not fulfilled in PEEM. The spherical and chromatic aberration coefficients are equal in this approximation.[11]

If we take into account that the circle of least confusion is located at a considerable distance before the paraxial focus (Gauss plane), the aberrations reduce to $\delta^{\mathrm{acc}}_{s,\min} = 0.25\,\delta^{\mathrm{acc}}_{s}$ and $\delta^{\mathrm{acc}}_{c,\min} = 0.5\,\delta^{\mathrm{acc}}_{c}$ if the electron lenses are focused to this plane of least confusion.

Numerically, we obtain $C^{\mathrm{acc}}_c = C^{\mathrm{acc}}_s = 245$ mm for typical conditions ($E_0 = 1$ eV, $U_{\mathrm{extr}} = 15$ kV, $\ell = 2$ mm). Actually, these aberrations of the acceleration field are the major contribution of the total aberration of the PEEM instrument. The aberrations of the aperture lens are very small and can be neglected.[9] As a rough guide for the spherical aberration of the objective lens we can use tabulated values for three-aperture-lenses.[13] For a simple einzel-lens at a magnification of $M = 10$ we find $C^{\mathrm{Ob}}_s \sim 26$ mm, *i.e.* about an order of magnitude smaller than the value for the acceleration field. For C^{Ob}_c the situation is similar.

In principle, the spherical and chromatic aberrations can be corrected by multipole elements[14] or electron mirrors,[12,15] both ways being very demanding. The instruments in use employ only a contrast aperture in a reciprocal plane in order to reduce the angle α' (see Fig. 1).

For small diameters of the contrast aperture the diffraction contribution δ_d[9] must be taken into account

$$\delta_d = \frac{0.61\lambda}{\alpha_{\max}} = \frac{0.73\,\mathrm{nm}}{\sqrt{E_0}\sin\alpha_0}. \tag{7}$$

Where λ is the wavelength of the electrons, $\alpha_{\max}$ the maximum angle of the u_α-ray passing the contrast aperture and E_0 the starting energy in eV. For the second term we have made use of the Helmholtz-Lagrange condition. The total resolution is given by $\delta_{\mathrm{tot}} = (\delta_s^2 + \delta_c^2 + \delta_d^2)^{1/2}$.

Figure 2 shows the joint action of the different aberration contributions of the acceleration field and the diffraction term calculated as a function of the emission angle α_0. For threshold photoemission, *i.e.* at photon energies in the UV-range around $h\nu = 5$ eV, the energy distribution of the photoelectrons emitted from the sample is typically $E \sim 0.1$–0.5 eV. For excitation in the X-ray range, however, the distribution of secondary electrons is a few eV wide. We assumed an acceleration voltage of 15 kV, a gap of $\ell = 2$ mm and for *UV-excitation* an electron emission energy distribution centered around 0.3 eV with a width of $\Delta E = \pm\,0.2$ eV (full curve) and for *X-ray excitation* a distribution centered at 2 eV with $\Delta E = \pm\,1.5$ eV (dashed curve). The analytical functions according to Eqs. (1), (4), and (7) were used. The figure clearly reveals that the best resolution is in the range below 10 nm and the chromatic aberration of the acceleration field is indeed the leading term.

For the conditions of threshold photoemission ΔE can be small so that the spherical aberration becomes significant. In contrast, in PEEM in the X-ray range an energy width of 3 eV and more strongly rises the dominant term δ^{acc}_c. A restriction of the beam pencil angle α' by means of a small contrast aperture in a reciprocal plane leads to a reduction of the effective energy width ΔE of the electrons being transmitted through the optical system. However, the intensity drops proportional to α'^2 so that the increase of resolution is paid by a considerable loss in intensity.

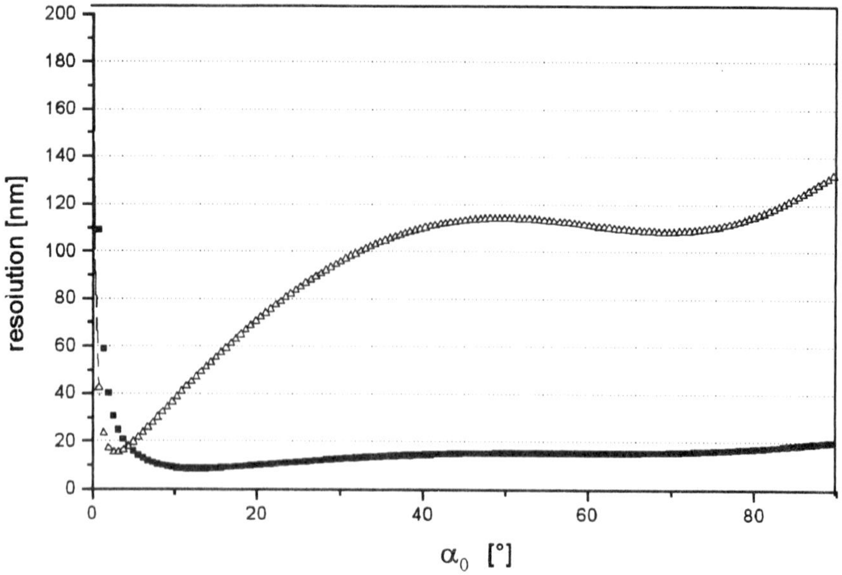

Figure 2. Calculated leading aberration terms of the cathode lens shown in Fig.1. The curves represent the joint action of the spherical and chromatic aberrations of the acceleration field as well as the diffraction term. The full and open symbols correspond to typical parameter sets for threshold (UV)- and X-ray excitation, respectively.

II.2. Domain Imaging Exploiting Magnetic Circular or Linear Dichroism

In the following we will briefly recall the different mechanisms leading to a magnetic contrast in PEEM, Fig. 3 schematically illustrates the four categories. Cases (a) to (c) are connected with magnetic dichroism effects, case (d) is a consequence of the Lorentz force. It has been shown by Schütz *et al.*[16] by the example of the Fe K-edges that absorption of circularly polarized X-rays depends on the magnetization state of the sample. For a fixed photon helicity (left or right circularly polarized) a characteristic change of the absorption spectra has been observed when the magnetization vector **M** was switched from parallel to antiparallel orientation with respect to the direction of photon incidence **q**, or, more precisely, photon spin s_γ. This XMCD can be considered as the high-energy analog of the magneto-optical Kerr-effect. Both are based on the simultaneous action of spin-orbit coupling and exchange interaction in the electronic states being involved in the optical excitation. The fact that XMCD arises at the characteristic X-ray absorption edges provides an outstanding advantage for the element-selective investigation of magnetic phenomena.

The XMCD-signal occurs in the X-ray absorption and electron yield signal, the latter containing more surface specific information on magnetism. This effect is well-suited for a combination with the PEEM-technique for imaging of magnetic domains as has first been shown by Stöhr *et al.*[5] The XMCD in the initial absorption signal is transferred to the emitted electrons in a two-step-process. First, the optical excitation creates a core hole, in the example shown below in the $2p$-shell. Owing to optical spin orientation by circularly polarized light the excited electrons are spin polarized.[17,18] Just above the absorption edge, the final state of the initial photo-excitation lies in the

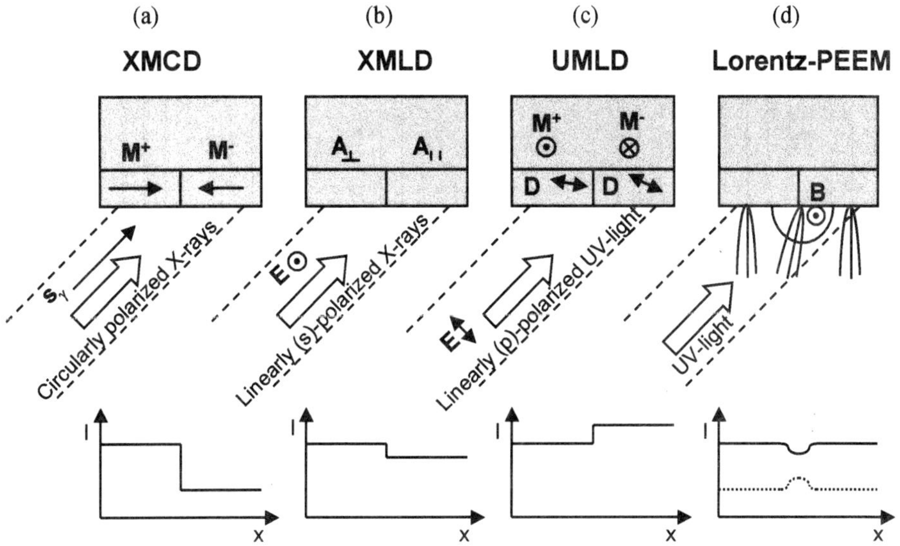

Figure 3. Magnetic contrast mechanisms observed for a thin magnetic film in a photoemission microscope. XMCD and UMLD are sensitive to the local orientation of the magnetization **M**, XMLD arises for antiferromagnets characterized by the alignment **A** and Lorentz-type contrast is a result of a local stray magnetic field **B** close to the sample surface. The plots of I *vs.* x schematically denote the corresponding intensity behaviours as observed in a line scan.

region of the unoccupied d-bands above the Fermi energy E_F. Since there is a higher unoccupied density of states in the minority spin channel, primary electrons with this spin orientation are favored, whereas majority spins find only a smaller part of unoccupied band structure. This is the origin of the XMCD asymmetry in the initial photoabsorption step. The resulting different absorption cross section for opposite magnetization directions is equivalent to a different probability of the creation of a core hole. The spin quantization axis for the density of states is defined by $-\mathbf{M}$, the optical spin orientation is aligned along the photon spin s_γ. Consequently, the projection of s_γ onto $\mathbf{M}$ is a measure of the observed XMCD signal. As illustrated in Fig. 3(a) this results in a magnetic domain contrast.

In a secondary step, the core hole decays with a final lifetime either through fluorescence or via an Auger process. In the latter case the magnetic dichroism in the absorption channel is directly transferred to the Auger electron yield. Since in an Auger transition various angular momenta are involved, there could be quantitative changes of the XMCD signal when being transferred to the Auger channel.

On their way to the sample surface the characteristic Auger electrons experience inelastic scattering events and thus produce a cascade of secondary electrons. In a good approximation, the intensity of the secondary electrons is proportional to the number of initially excited Auger electrons. In this way, the XMCD-signal being created in the initial step of the photoexcitation is transferred via the intermediate step of the Auger electron emission finally to the low-energy secondary electrons. Except for the persisting XMCD asymmetry, these electrons carry no direct information about the specific electronic transition in the sample. However, element selectivity is ensured by the *initial* excitation at a characteristic absorption edge.

From near-edge X-ray absorption fine structure (NEXAFS) spectroscopy[19] it is well known that the symmetry and orientation (or alignment) of electronic states can be probed using linearly polarized X-rays. In case of a spin alignment as present in antiferromagnets the anisotropic interaction between the electric vector **E** and the axis defined by the antiferromagnetic order within a domain can give rise to a XMLD contrast in PEEM. Recent experiments[6] on NiO have shown that the phenomenon occurs for s-polarized light and perpendicularly arranged alignment directions $\mathbf{A}_{\|}$ and $\mathbf{A}_{\perp}$ as indicated in Fig. 3(b).

The third and most recent achievement of magnetic imaging using PEEM has been discovered by Marx *et al.*[7] It resembles the XMLD but it arises in threshold emission, *i.e.* in the UV-range. Consequently, we term it UMLD. This effect arises for p-polarized light with **E** being perpendicular to **M**, *i.e.* in the geometry of the transversal magneto-optical Kerr-effect, see Fig. 3(c). The UMLD can be quantitatively understood in the framework of the Kerr-effect being based on a dependence of the dielectric tensor on the magnetization of the sample. The Kerr-effect causes a small rotation of the electric vector upon reflection on the sample surface as well as a magnetization-dependent intensity modulation of the reflected light beam in certain geometries. Optical Kerr-microscopy is up to now the most common technique for magnetic domain imaging.[20] It employs light in the visible or near UV-range, its lateral resolution is diffraction limited. By using an immersion objective lens and an optimized dielectric coating of the magnetic sample surface a resolution of about 300 nm can be achieved in favorable cases. In contrast, photoemission microscopy is characterized by a base resolution in the region of 20 nm.

Since the XMCD and XMLD effects occur at characteristic absorption edges this approach requires a tuneable light source in the soft X-ray range. In threshold photoemission using a UV-lamp the photon energy is only slightly larger than the workfunction. Consequently only electrons in a narrow energy window of typically less than 1 eV can contribute to the image. In this case the amount of inelastically scattered secondary electrons is small and the entire spectrum is dominated by direct photoelectrons. At such low energies, the escaping electrons have only a small wave-vector component $\mathbf{k}_{\|}$ to the surface, *i.e.*, their total momenta are close to the normal-emission condition. In a magnetised sample the dielectric response on the external **E** vector is characterized by the refraction at the solid-vacuum interface leading to macroscopic rotation of the displacement vector **D** inside the material. In addition, the Lorentz force acts on the quasi-free metal electrons leading to a small additional rotation (the Kerr rotation) of **D** as schematically illustrated in Fig. 3(c). This rotation is the origin of the magneto-optical Kerr-effect[20] and it also gives rise to the UMLD. In the limit of $\mathbf{k}_{\|} = 0$ the projection of **D** onto the surface normal **n** is a measure of the transition amplitude of the emitted electron signal.[21] In this simple picture of quasi-free electrons, *i.e.* neglecting all band-structure effects, the emission intensity I, along the direction **n** is proportional to $(\mathbf{D}\cdot\,\mathbf{n})^2$.

In Fig. 3(c) we have assumed a magnetization vector perpendicular to the drawing plane and pointing upwards in the left-hand domain ($\mathbf{M}^+$) and downwards in the right-hand domain ($\mathbf{M}^-$). The resulting Kerr rotation of **D** is thus counterclockwise or clockwise, respectively. In the case of the clockwise rotation the projection of **D** onto the surface normal **n** is *larger*, thus giving rise to a *higher* photoelectron intensity normal to the surface than in the opposite case. The corresponding domain (right) will therefore appear *brighter* than the other one. The orientation of the magnetization vector on the right-hand side leads to an enhanced photoelectron signal and also to a *larger reflected intensity* as compared to the left-hand side in the optical Kerr-signal.

More details on this analogy and a quantitative estimation of the expected size of the UMLD-effect taking into account metal optics are discussed by Marx *et al.*[7]

This simple picture of quasi-free electrons being essentially based on the Drude-theory extended for the presence of a sample magnetization is particularly useful for polycrystalline materials, where no **k**-resolution is given. For the case of single-crystal surfaces the microscope detects electrons around a well-defined direction in **k**-space. In this case, a band structure calculation is mandatory. Henk and Feder[22] have performed *ab-initio* photoemission calculations for the Ni(110) surface at $h\nu = 5.1$ eV photon energy. In a group-theory based analytical approach they treat the photoexcitation in a fully relativistic electron system. This treatment indeed predicts the existence of a sizeable UMLD in particular for *sp*-polarized light. The model calculation used the formalism which had previously been successfully applied to the analysis of magnetic linear and circular dichroism in Ni.[23] It accounts for the optical response of the surface on the basis of Fresnel's equations.

II.3. Magnetic Contrast Due to the Lorentz Force

In this mode of magnetic imaging using a PEEM the magnetic contrast occurs due to a magnetic stray field in the surface region of the sample. Like the UMLD this method is not element specific but it has the important practical advantage that no special light source like a synchrotron is required. It can be performed with simple laboratory UV-lamps.

The Lorentz-force acting on a moving charge in a magnetic field causes a deflection of the trajectories. The resulting contrast in the electron microscope is therefore called Lorentz contrast. It is well-known in transmission and scanning electron microscopy (TEM and SEM) and is utilized for magnetic imaging, for details see, *e.g.*, Ref. 24. In Lorentz-microscopy using a TEM the contrast occurs due to the deflection of the electron beam usually inside a magnetic material. In a SEM two different types of magnetic contrast are possible. The so-called type I contrast arises in the secondary electron image and results from the action of an external magnetic stray field. The type II contrast occurs around certain impact angles inside of the material in the case when the elastically backscattered electrons are detected.

The origin of the Lorentz-contrast in PEEM resembles the type I contrast in a SEM. If the stray magnetic field has a component parallel to the sample surface, the Lorentz force causes a deflection of those trajectories which experience the action of the stray field. If we assume a local stray field **B** perpendicular to the drawing plane as indicated in Fig. 3(d), the trajectories are bent either to the left or to the right, depending on the local field direction. As discussed in Sec. II.1, the contrast aperture in the backfocal plane of the objective lens selects the solid angle interval of the electron emission distribution contributing to the image. If the contrast aperture is placed off-center, a magnetic contrast results and parts of the sample with a certain stray field direction will appear either brighter or darker than the field-free parts, depending on the aperture position.

It can be shown[25] that a weak magnetic stray-field contrast persists even when no aperture is used. This appearance of a Lorentz-force-based contrast has been discovered as early as 1957 by Spivak *et al.*[8] Later Mundschau *et al.*[26] and Marx *et al.*[27] exploited the Lorentz-type contrast for magnetic imaging. A highly interesting aspect of Lorentz-PEEM is the possibility of a quantitative determination of local magnetic microfields above surfaces, nanostructures or miniaturized building elements like write heads, sensors or magneto-electronic devices.

III. Applications for Thin Magnetic Films

III.1. Domain Imaging and Sum-Rule Analysis of Co-Films Via X-ray Magnetic Circular Dichroism (XMCD-PEEM)

Owing to their magnetic properties, Co films or alloys are a promising candidate for an application in spin valves or tunnel junctions. The micromagnetic behavior depends on the film thickness, therefore high-resolution imaging techniques are mandatory for the characterization of such devices. Magnetic force microscopy (MFM)[28] revealed for sputtered polycrystalline Co films up to a thickness of 30 nm Neél walls, above 110 nm Bloch walls and in the intermediate range the so-called "cross-tie" walls, being characterized by a complex magnetic interaction due to the simultaneous appearance of Bloch-and Neél-like walls. In the region of cross-tie domain-walls the local magnetization generally has an out-of-plane component giving rise to external stray magnetic fields. In view of the various contrast mechanisms discussed in Sec. II it seemed worthwhile to study Co films in this intermediate thickness range using PEEM.

The polycrystalline film has been grown by UHV-evaporation on a Si wafer (12×12 mm^2) being covered by its natural oxide (growth conditions: rate about 2.3 nm/min, room temperature, pressure during evaporation 5×10^{-9} mbar). The wafer surface showed a miscut of about 3.75° with respect to the (111) surface normal. The thickness of the film discussed below was determined *in-situ* by a quartz balance to be 71 nm and confirmed *ex-situ* by X-ray diffraction yielding 73 nm. X-ray diffraction also revealed that the film exhibited a weak texture with c-axis orientation perpendicular to the sample surface. *In-situ* detection of hysteresis loops yielded a coercive force of the in-plane magnetization of $\mu_0 H_c = 5$ mT and a saturation field of $\mu_0 H_s = 10$ mT. The thin-film sample was mounted on a manipulator allowing xyz-shift and azimuthal rotation of the sample. The sample was located in the center of a soft-magnetic yoke thus allowing the observation of magnetization processes.[29] In order to reduce the noise and optimize the dynamic range, the imaging system did not use an electron intensifier (multichannel plate). Instead, the electron-optical image was directly viewed by a fluorescent screen followed by a slow-scan CCD camera. The measurements have been performed at the PM 3 monochromator at BESSY I in Berlin. The out-of-plane angle of the monochromator was set to 0.3 mrad, corresponding to a degree of circular polarization of about $P_{\rm circ} = 70$–80%.[30]

The XMCD contrast in the images was obtained as follows. First the exact positions of the maxima for the Co L_3- and L_2-edges have been determined via the absorption spectra (see below). Then two images have been taken at these photon energies with the same helicity of the radiation. The magnetic contrast is strongly enhanced by subtraction of the two images with the intensity differences being taken into account by multiplying the L_2 image with the corresponding factor (pixel by pixel). This subtraction removes practically all contrast contributions other than magnetic. All images shown are oriented so that the photon impact direction (projection of photon spin s_γ) points from top to bottom.

Figure 4 shows a series of images taken for different azimuthal sample orientations. Before, the sample has been macroscopically demagnetized by an external magnetic AC-field (50 Hertz) with decreasing amplitude. This procedure results in a random domain distribution. The image reveals that the local magnetization exhibits components oriented *perpendicular* to the external field $\mathbf{H}_{\rm ext}$. The magnetization direction could be easily determined by exploiting the $\cos\phi$ dependence of the XMCD contrast.

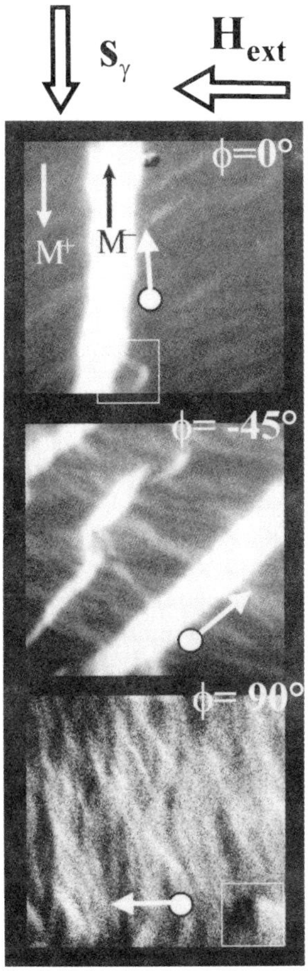

Figure 4. Domain structure of a polycrystalline Co film observed by XMCD-PEEM utilizing circularly polarized synchrotron radiation. The film was grown on a Si-wafer with a miscut of 3.75°. The three panels correspond to different azimuthal orientations $\phi = 0°, -45°$, and $90°$. Note the variation of the magnetic contrast. The dot with arrow serves for lateral and angular orientation, the white squares mark a characteristic feature. Field of view $150 \times 150 \mu m^2$ (from Marx *et al.*[29])

The maximum black and white contrast corresponds to $\phi = 0°$, whereas the orientation $\phi = 90°$ can be very precisely derived from the vanishing contrast of the bright domain (note that the center of rotation did not exactly coincide with the center of the images). The reason for this unexpected behavior of $\mathbf{M} \perp \mathbf{H}_{ext}$ is that the miscut of the silicon wafer of 3.75° induces a uniaxial anisotropy in the cobalt film. A similar behavior has been observed by Berger *et al.*[31]

In order to explore the reason of the magnetic anisotropy the growth of the Co layer has been controlled by atomic force microscopy (AFM). The result is shown in Fig. 5. Indeed, the miscut of the substrate induces an anisotropic growth of the Co film. The atomic steps or step bunches of the Si-substrate give rise to oblate Co islands with elliptical shape which in turn results in a contribution of the shape anisotropy. In the geometry shown in Fig. 4 the uniaxial anisotropy along the long axis of the Co islands is oriented perpendicular to $\mathbf{H}_{ext}$.

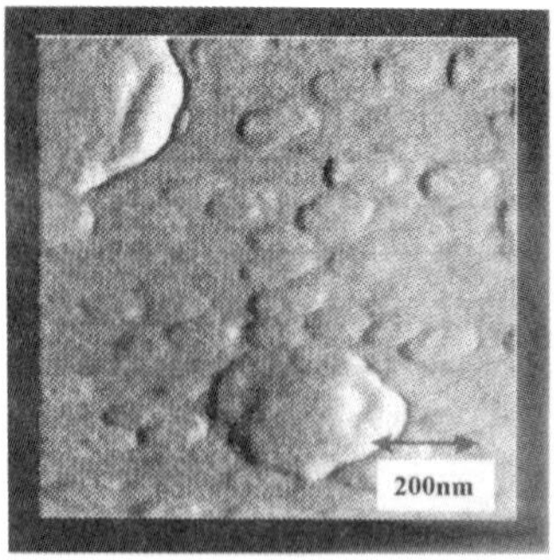

Figure 5. Structural image of the Co film taken by atomic force microscopy. Field of view 800×800 nm^2. (courtesy A. Dion, Mainz).

In addition to the large bright and dark domains a fine magnetic ripple is observed in the images of Fig. 4, which is characteristic for cobalt films. The ripple is largely masked by the strong contrast of the striped domain in the $\phi = 0°$ orientation but it becomes visible in the $-45°$ and especially in the 90° orientation (bottom panel) where the striped domain is invisible due to $\cos\phi = 0$. The ripple is connected with the microcrystalline structure of the layer. The ripple pattern is stabilized by symmetric Neél walls which strongly interact. The contrast of the magnetization ripple is indicative of the behavior of Neél walls. Owing to the strong mutual interaction the Neél walls can even cross the Bloch walls as can be seen in the center image where the faint ripple pattern penetrates from dark regions into the bright regions. For orientation a dot with arrow marks identical spots on the images. Depending on the magnetization procedure also cross-tie walls have been observed.[20] The cross-tie wall consists of a Bloch wall and perpendicularly oriented shorter Neél walls. This type of complex magnetization structure is energetically favored in a certain film-thickness range.

The white squares in Fig. 4 denote a particularly interesting oval feature consisting essentially of two oppositely magnetized domains visible in the bottom image (bright and dark corresponds to **M** pointing upwards and downwards, respectively). In the top image (rotated by 90°) only a bright rim appears which obviously represents a closure domain. Detection of such subtle features like a magnetic ripple crossing the main domains or small closure domains with **M** perpendicular to the large domains necessarily requires the possibility of azimuthal sample rotation.

The escape depth of the electrons is of the order of 5 nm.[32] This means that the magnetic signal is averaged over many atomic layers, weighted by the exponential decay function. This information depth can result in a deterioration of magnetic resolution in cases of complex wall structures in the surface. In addition, stray magnetic fields in particular due to out-of-plane components of the magnetization vector can set a practical limit to the obtainable lateral magnetic resolution.[27] The stray fields themselves give rise to an additional Lorentz-type contrast.

For a *sum-rule* analysis the XMCD-spectra in the photoyield across the $L_{2,3}$ white lines are exploited. The scans of photon energy have been facilitated by a synchronization of the monochromator control and the image acquisition software. The energy resolution in the spectra is about 0.5 eV at a typical exposure time per energy of 1 s. The photon energy has been varied between typically 770 and 830 eV thus covering the range of the $L_{2,3}$ spin-orbit doublet.

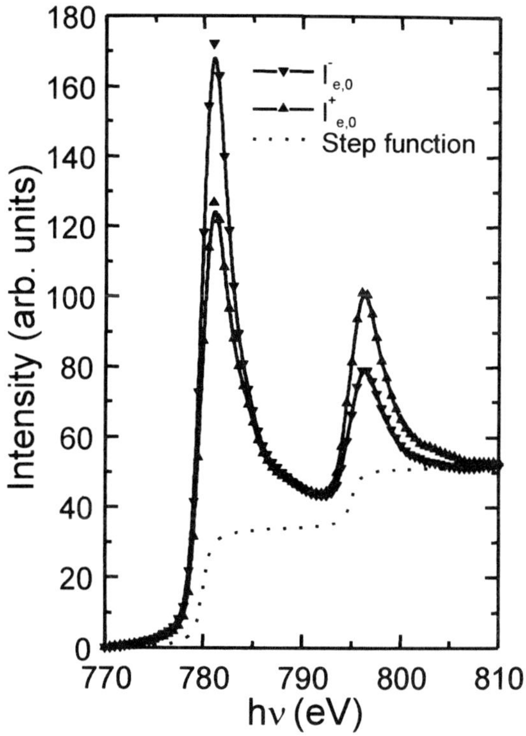

Figure 6. Photoabsorption spectra of the $L_{2,3}$ absorption edge of a Co film taken with circularly polarized X-rays. The two spectra corresponding to opposite sample magnetization show a XMCD asymmetry (from Marx *et al.*[29]).

The spectra have been accumulated as follows. First, the sample was magnetized along the $\mathbf{M}^+$ direction using the external magnetic field. The monochromator was swept between $h\nu = 770$ eV and 830 eV with a step width of 0.5 eV; after each step a full image was taken. Subsequently, the sample magnetization was reversed ($\mathbf{M}^-$ direction) and a second sequence of images was taken in the same way. After inspection of the images a uniformly magnetized region has been defined via the camera software. In this way saturation magnetization could be chosen although microscopically the sample sometimes still exhibited small domains of opposite magnetization direction, *i.e.* the film was not in a single-domain state. This possibility to select regions with full magnetization is a general advantage of spatially-resolving techniques such as PEEM. Finally, the intensity signal of all pixels within the region has been integrated thus giving a reliable intensity value. Every intensity point shown in the spectra thus corresponds to the integral of a spot of about 40 μm$\times$40 μm on the sample surface.

Figure 6 shows a typical example for spectra across the Co $L_{2,3}$-doublett taken for both magnetization directions (intensity curves I^+ and I^-). The dichroic signal, *i.e.* the intensity differences in the two peaks is very pronounced. The magnetic dichroism signal of the two so-called "white lines" exhibits different sign as known from earlier experiments.[5] This fact can be understood from the angular-momentum dependence of the initial photoexcitation step (optical spin orientation with circularly polarized light).[17,18] The good signal-to-noise ratio of larger than 200, facilitates a quantitative evaluation of the spectra and a derivation of the magnetic moments. We apply the formalism as described in Refs. 33 and 34. It has been demonstrated that the sum-rule analysis is well applicable for $3d$-transition metals.[35,36]

The expectation values $\langle L_z \rangle$ and $\langle S_z \rangle$ correspond to the z-components of the magnetic moments as $\mu L = -\mu_B \langle Lz \rangle$ (orbital moment) and $\mu_s = -2\mu_B \langle S_z \rangle$ (spin moment). The sum rules relate these expectation values to the signal strengths of the spectra in Fig. 6 as follows:

$$\frac{\mu_L}{\mu_B} = -\langle L_z \rangle = \frac{2n_h}{P_{\mathrm{circ}} \cos 15^\circ} \frac{A_3 + A_2}{A_{\mathrm{ISO}}}, \tag{8}$$

$$\frac{\mu_S}{\mu_B} = -2\langle S_z \rangle = \frac{3n_h}{P_{\mathrm{circ}} \cos 15^\circ} \frac{A_3 - 2A_2}{A_{\mathrm{ISO}}}. \tag{9}$$

Here n_h is the number of d-holes per Co-atom, P_{circ} is the circular polarization of the exciting radiation and the angular term takes into account the photon impact direction. The signal strengths in the spectrum are represented by integrals over the two white lines after subtraction of the background (indicated by the step function in Fig. 6) which is due to the excitation from $2p_{1/2}$ and $2p_{3/2} \to 4s$. This is necessary because these electrons carry no information about the magnetic properties of the system. Since the density of states of the $4s$-electrons is very flat and unstructured, the contribution of these electrons to the spectrum can be well approximated by a step function. The integrals A_3 and A_2 correspond to the differences of the background-corrected I^+ and I^- curves in Fig. 6 integrated over the L_3 and L_2 line, respectively. The white-line intensity A_{ISO} is obtained by integration of the sum of the corrected I^+ and I^- curves over both lines. For the discussion of various correction terms we refer to the literature.[33–38]

For the analysis of the effective magnetic moments we first derive the *ratio between spin and orbital moments*. According to Eqs. (8) and (9) this ratio is independent of the number of d-holes and the degree of photon polarization. We obtain $\mu_L/\mu_S = 0.113$. If we further use the literature value for the circular polarization,[30] for the given setting $P_{\mathrm{circ}} \cos(15^\circ) = 0.76$, we can derive the *magnetic moments per d-hole*: $\mu_L = 0.035\,\mu_B/d$-hole and $\mu_S = 0.61\,\mu_B/d$-hole.

Finally, for the determination of the *absolute moments* the number n_h has to be taken from the literature. We assume that our polycrystalline cobalt film has the same number as *hcp*-cobalt, *i.e.* $n_h = 2.43$. Entering this value into Eqs. (8) and (9) we obtain for the orbital and spin moment $\mu_L = 0.084\,\mu_B$ and $\mu_S = 1.48\,\mu_B$. This results in a total magnetic moment of the polycrystalline cobalt film of $\mu_{\mathrm{tot}} = \mu_L + \mu_S = 1.56\,\mu_B$.

These values must be considered as *effective* moments, because they are averaged over the faint ripple structure mentioned above. This will generally lead to a reduction of the apparent moments. The absolute values have a relatively large error. For the number of holes n_h we find in the literature for Co 2.43 and 2.8.[36] Also, for the mean free path of the Auger electrons λ_e different values are given. The results of Marx *et al.*[29] compare reasonably well with the data reported in the literature.[37–42]

III.2. Element-Selective Magnetic Imaging of Exchange-Coupled Fe-Cr-Co-Film: Investigation of "Buried Layers"

The potential to view "buried" layers through top layers of different constituents using the element-resolved imaging technique is one of the striking advantages of magnetic XMCD-PEEM. Except for the trivial case of a non-magnetic top layer none of the various other imaging techniques is capable to reveal the magnetic structure of a

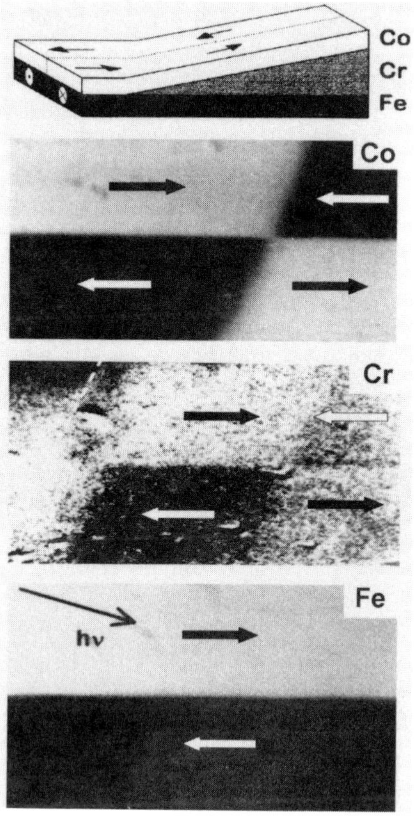

Figure 7. Magnetic exchange coupling between Fe and Co across a Cr-wedge observed using XMCD-PEEM. The film was prepared at 500 K substrate temperature. On an Fe(100) whisker a Cr-wedge was grown and covered with 5 ML Co. The thickness of the Cr-wedge increases from left to right from 0-3 ML as indicated in the top panel. For technical reasons the orientation of the Cr-wedge is tilted with respect to the Fe-whisker, see dashed line. The element-resolved domain images for Co, Cr and Fe (top, centre, bottom, respectively) have been taken at the corresponding $L3$ absorption edges. The arrows denote the local magnetization direction. Note the change of the magnetic coupling between Fe and Co at a Cr thickness of $\sim$ 2 ML and the magnetic contrast visible in the Cr-image. (from Kuch *et al.*[43])

buried layer. TXM could work in principle if the layered structure is sufficiently thin and the X-ray optics allows tunability over a photon energy range comprising the relevant absorption edges. In the following we will present a typical example in order to illustrate the performance of the PEEM technique.

Kuch *et al.*[43] have investigated the sandwich structure of a Co *film being separated from an Fe substrate by a Cr-wedge* for deposition at 500 K. Figure 7 shows a result obtained by the XMCD-PEEM technique. The sample was prepared in the following way: The Cr-wedge with increasing thickness from 0–3 monolayers (ML) was grown on a Fe(100)-whisker (single crystal) surface. Finally the whole structure was covered by a 5 ML Co-film. The film structure is schematically illustrated in the top panel. This sandwich system was imaged element selectively at the L_3-edges of Co, Cr and Fe.

The typical domain structure of the Fe-whisker, two regions of opposite magnetization, is visible in the bottom panel. Bright areas correspond to a magnetization direction essentially opposite to the direction of photon incidence (slightly tilted, see $h\nu$-arrow) and dark areas to magnetization along the photon incidence. This simple domain configuration often encountered in Fe whiskers is very convenient for imaging magnetic coupling phenomena in wedge-shaped overlayers.[44–46] The magnetic structure of Fe is clearly visible through 3 ML Cr plus 5 ML Co. The Cr-selective image (center) shows the onset (dashed line) and increase of the Cr-wedge. The onset is tilted for technical reasons.

The Co image (top) reveals that up to a certain thickness of the Cr-wedge the Co magnetization is *parallel* to the Fe magnetization. For a Cr thickness above about 2 ML the magnetic contrast in the Co image is reversed with respect to the Fe image. At these Cr thicknesses, Co obviously displays an *antiferromagnetic* coupling to the Fe substrate. Close to the transition from ferromagnetic to antiferromagnetic coupling, a small region with an intermediate gray level is observed. This could be due to a simultaneous presence of ferromagnetic and antiferromagnetic coupling in that region, or to a biquadratic coupling behavior, which would result in a 90° rotation of the Co magnetization.[47]

The element selectivity of the method allows to study the residual ferromagnetic ordering of the Cr-wedge induced by the adjacent Fe and Co magnetic layers. The center image of Fig. 7 shows the residual magnetization of Cr with the photon wavelength being tuned to the Cr L_3-edge. There is indeed an net ferromagnetic moment also in the Cr layer. The Cr magnetization obviously follows that of the Co top layer. A darker gray level occurs in the lower half of the image in the center region and in the upper half on the right-hand side. Only at the lowest Cr coverages below about 0.3 ML, just right of the broken line, the Cr magnetization appears to be opposite to that of Fe and Co. The comparison with measurements for films prepared at other temperatures,[48] exhibiting ferromagnetic coupling independent of the Cr thickness, suggests that in such exchange-coupled systems the preparation conditions play a crucial role. From these results it becomes clear that elemental selectivity is an absolutely necessary prerequisite for studying the coupling behavior of sandwich-like structures on a ferromagnetic surface. These images were taken at the ESRF, Grenoble, with total exposure times of about 5 minutes for each helicity, thanks to the high brightness of the undulator beamline ID 12 B.[49]

III.3. Detection of Very Weak Magnetic Signals from Submonolayer Cr-Films

Owing to the element selectivity the XMCD-PEEM method is characterized by a very high magnetic sensitivity. The asymmetry at the Cr L_3-edge in Fig. 7 is only 0.3–0.5% in contrast to about 20% at the Fe or Co L_3-edges. This indicates that the total moment of the Cr layer is much lower than the Fe or Co moments. Furthermore, the asymmetry visible in the Cr image is weaker in the region of antiferromagnetic coupling (right hand side of the image). This can be explained by the fact that the contributions at the Fe-Cr and Cr-Co interfaces tend to cancel out in the antiferromagnetic coupling region, whereas they add up in the region of ferromagnetic coupling. Kuch *et al.*[43] draw further conclusions on the relative coupling strengths and the effect of a possible interface roughness.

The magnetic coupling of a *sub-monolayer Cr-wedge to a Fe(100) single-crystal surface* has been investigated by Schneider *et al.*[48,50] This experiment also gives an in-

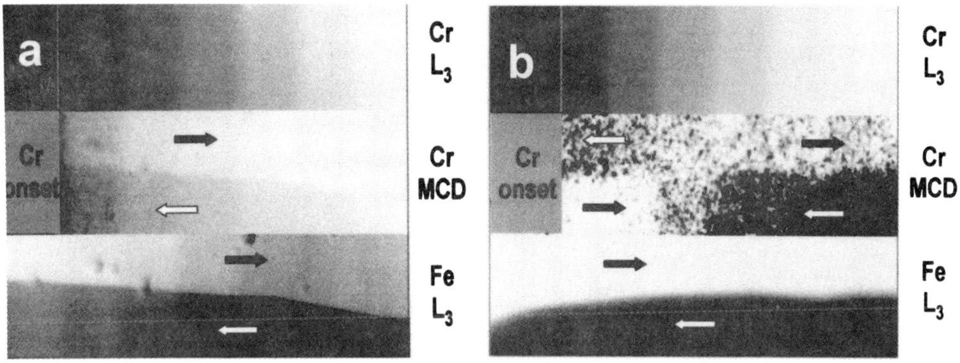

Figure 8. XMCD-PEEM study of a stepped sub-monolayer structure (0.1–0.6 ML) of Cr on Fe(100) deposited at elevated temperature (a) and at room temperature (b). The chemical and magnetic contrast of Cr is depicted in the top and centre panels, respectively. The domain structure of the Fe-whisker is given in the bottom panels. Horizontal field of view 80 μm. (from Schneider *et al.* Ref. 48 and Ref. 50 for (a) and (b), respectively).

formation about the sensitivity of the PEEM method. Results for two different preparations are shown in Fig. 8. The Cr-films were grown as discontinuous wedges forming a sequence of equidistant terrasses of 10 μm width per terrasse. The Cr-coverage was increased stepwise by 0.1 ML increment, the chemical contrast taken at the Cr L_3-edge clearly reflects this stepwise increase of coverage (top-panels). The images of the dichroism asymmetry in the Cr L_3-signal (center panels) reveal a magnetic contrast of Cr, which must be related to the domain pattern of the Fe substrate (bottom panels). In Fig. 8(a) (film preparation at elevated temperatures) the domain structure of the Cr-film reflects a ferromagnetic coupling of the Cr atoms to the Fe substrate all over the coverage range.

The maximum magnetic contrast in the Cr image is only about 0.8% XMCD asymmetry. Despite this fact it was possible to discern the magnetic contrast already at the smallest film thickness of 0.1 ML. For preparation at room temperature (Fig. 8b), Schneider *et al.*[50] have found a different behavior with the Cr-Fe coupling character changing from ferromagnetic to antiferromagnetic in this submonolayer coverage range (see arrows). Again, the behavior of the exchange coupling critically depends on the preparation conditions. The reasons may be found in the formation of a Cr-Fe interfacial alloy or in two-dimensional cluster formation depending on growth temperature and/or substrate-surface conditions.

Concluding this section about X-ray excited PEEM it is worthwhile to comment on the sensitivity and spatial resolution of the method. The sensitivity allowing to image the magnetic structure of a tenth of a monolayer material on a bulk ferromagnet (and even less at future high-brilliance beamlines) is outstanding and cannot be reached by any other means. MFM, SEMPA, Lorentz- and magneto-optical Kerr-microscopy would detect *averaged* signals with the main contributions resulting from the substrate. In Sec. II.1 we have seen that the chromatic aberration of the electron optics is the limiting factor for the resolution. Due to this effect and the relatively large fields of view the resolution of the images shown above is of the order of 200–400 nm. As discussed in II.1 the effective energy width can be reduced by decreasing the diameter of the contrast aperture provided the photon intensity is sufficiently high (see also paper by Sinkovic in Ref. 6).

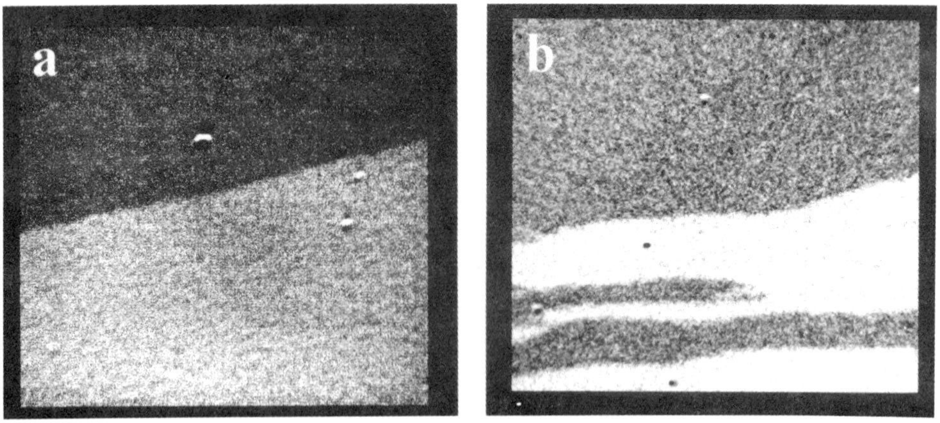

Figure 9. Magnetic domain contrast for a thin Fe film observed in threshold (UV)-photoemission microscopy (UMLD-PEEM) with p-polarized (a) and unpolarized light (b). The images were taken for different areas. This experiment requires only a simple UV-source for excitation. Field of view $100 \times 100\ \mu m^2$. (from Marx *et al.*[7])

III.4. Magnetic Contrast in Fe-Films Due to Magnetic Linear Dichroism in the UV-Range (UMLD-PEEM)

The novel magnetic contrast mechanism in threshold photoemission UMLD has been investigated very recently by Marx *et al.*[7] The experimental geometry corresponds to a typical arrangement for detecting the transversal Kerr-effect with an angle of photon incidence of 75° with respect to the surface normal (*cf.* Fig. 3c). The photon beam from a Xe-Hg high-pressure arc lamp ($h\nu \leq 5.2$ eV) passed through a linear polarizer (Glan-Thompson prism) and was focussed onto the sample surface. By rotating the prism the linear polarization could be changed continuously between s- and p-polarized.

A polycrystalline iron film with a thickness of 100 nm has been deposited by means of UHV-evaporation on a Si wafer with native oxide layer. The sample holder allowed the *in-situ* application of an external magnetic field. The coercive field of the iron film was determined to be about $\mu_0 H = 4$ mT. The workfunction of the film was about $\Phi = 4.84$ eV, so that the energy width of the photoelectron distribution was only $\Delta E \leq 0.4$ eV keeping the chromatic aberration small (*cf.* Sec. II.1).

In threshold photoemission a "background image" can be taken for a sample in saturation magnetization (or in a completely demagnetized state), a technique which is also used in Kerr-microscopy.[20] In the resulting difference image, other contrast contributions such as workfunction contrast, topographical contrast, impurities etc. are largely eliminated. Figure 9 shows two examples of the magnetic contrast observed [for p-polarized (a) and unpolarized light (b)].[7] In this case a background image of the sample being in a single-domain state has been subtracted. The asymmetry value extracted from regions of opposite magnetization is $A_{\mathrm{UMLD}} = (0.37 \pm 0.10)\%$. Quantitatively, this asymmetry agrees well with the corresponding value for the magnetooptical Kerr-effect measured for Fe.[51,52]

As compared with the demanding experimental requirements of circularly polarized tunable radiation in the soft X-ray range, the novel approach is extremely simple. It works with a *standard laboratory UV-source*. If we take into account that unpolarized light contains a 50% contribution of p-polarization (which could be enhanced inside the material due to metal optics) the mechanism illustrated in Fig. 3(c) should also exist for unpolarized light. Indeed, Marx *et al.*[7] were able to image the magnetic domains in the polycrystalline iron film using unpolarized light, see Fig. 9(b). In this case the magnetic asymmetry is $A = 0.18\%$, *i.e.* about half of the value obtained for linearly polarized light. The field of view is $150 \times 150\,\mu m^2$ at a resolution of the CCD-camera of 1024×1024 pixel. The total exposure time was 120 s. In the future, the difference image will be obtained by using two photon sources placed symmetrically with respect to the surface normal. Upon switching between the two light sources the magnetic contrast will be reversed. Then, there is no need to change the magnetization of the sample for taking a background image and the magnetic contrast will be enhanced by a factor of two. In principle, a second pair of lamps placed in the azimuth perpendicular to the first one allows to probe the second in-plane component of **M**.

The information contents of UMLD-PEEM and Lorentz-PEEM (Sec. II.2) are complementary to each other because there are important differences. The Lorentz-contrast is based on the existence of an external magnetic stray field close to the sample surface. Therefore, the method is well-suited for perpendicularly magnetized structures and for the observation of domain walls. A uniform in-plane magnetization of a thin film will generally not lead to a contrast. A thin-film structure (like a square) being uniformly magnetized will essentially show up in the regions of the magnetic poles. In this respect the technique bears some resemblance to MFM. Optical Kerr-microscopy as well as its counterpart UMLD in photoemission microscopy yield a true domain contrast. In this case the phenomenon arises inside of the material. A thin-film structure being uniformly magnetized will appear in an unbroken gray level.

Although the first results have been measured for thin Fe films, the phenomenon has the same general physical origin as the magneto-optical Kerr-effect. It will thus not be restricted to a specific class of materials. Due to its potential for high lateral resolution and the efficient parallel image acquisition, the new method is highly attractive for applications. It is a relatively simple laboratory method and does not require special light sources like synchrotron radiation.

IV. Conclusions and Future Developments

The unsolved problems in fundamental research concerning properties of ultrathin ferromagnetic layers and the demands of research and development of novel magnetic devices like the fascinating possibilities opened up by the young field of magneto-electronics are a challenge for magnetic imaging and microspectroscopic techniques. Owing to its spectroscopic capabilities, X-ray excited PEEM provides a promising tool meeting most requirements. Especially in combination with tunable circularly polarized radiation in the soft X-ray range from electron storage rings the method can launch into its full potential. Exploiting X-ray magnetic circular dichroism (XMCD) the micromagnetism and domain structure of thin films and horizontally or vertically patterned structures can be investigated element resolved. This article gave a status report with emphasis of those applications which are hardly possible by alternative techniques.

The performance of XMCD-PEEM has been illustrated for *extremely weak magnetic signals* from submonolayer films (coverage 0.1 to 0.6 ML) of Cr on an Fe(100) single-crystal surface. We know of no other magnetic imaging technique which can view the magnetic structure of a tenth of a monolayer material on a bulk ferromagnet. Another special feature of XMCD-PEEM is its capability to detect the magnetic signal of *buried layers*. As a typical example results for a ternary sandwich system consisting of a Co-layer separated from an Fe-substrate by a Cr wedge have been shown. The domain pattern of Fe was viewed through the Co- and Cr-layers. For such a system, spin-polarized SEM (SEMPA) would observe the magnetization structure of the topmost layer only, magnetic force microscopy and Lorentz-or Kerr-microscopy would detect the magnetic signal averaged over the three layers.

In the *microspectroscopy mode* the magnetic dichroism signal of a selected microspot on the sample can be measured. A *sum-rule analysis* of the XMCD-spectra gives direct access to spin- and orbital magnetic moments for $3d$-transition metals. Given the previous conditions at BESSY I (monochromator PM3 at a bending-magnet beamline) microspot sizes of down to 500×500 nm^2 have been reached.[53] There is potential of considerable improvement at synchrotron sources of the third generation like BESSY II. Since a PEEM acquires the image in a parallel mode, it is principally much faster than all scanning-beam or scanning-probe techniques. Magnetic PEEM also works in the UV-range exploiting either the Lorentz-force or a Kerr-effect-like magneto-optical phenomenon. Since these modes do *not* require synchrotron radiation they are highly attractive for standard laboratory applications.

In the near future we expect a major step towards a *high time resolution* in magnetic imaging. The possibility of "real-time" magnetic imaging arises in the context of the so-called *time-of-flight* (TOF)-PEEM. This novel instrument exploits the time structure of the synchrotron radiation, characterized by photon pulses with a width of several 10 ps and a repetition period of a few 100 ns. In the first experiments[54] we used an ultrafast CCD-camera behind a scintillator screen which set a limit of 1.4 ns to the possible time resolution. More recently, a new approach was explored[55] which is based on a time- and space-resolving electron counting detector. Present time resolution is about 500 ps (with potential for further improvement) using a delayline detector.[56] In the future, magnetic switching phenomena can thus be directly observed in real time with high-spatial resolution.

X-ray photoelectron spectroscopy (XPS) is a very powerful tool for *chemical spectromicroscopy*. It can be implemented into a PEEM by means of a device for energy-selective imaging. The conventional solution is an imaging dispersive energy filter being integrated into the electron-optical column (see, *e.g.* papers by Bauer *et al.* and Tonner *et al.* in Refs. 57 and 58). TOF-PEEM provides a new approach to this XPS-technique by making use of the time-of-flight of the photoelectrons through the electron-optical system. The main advantage of this solution is that it requires only rather simple electron-optics. Since the energy- (*i.e.* time-) dispersion is facilitated in a drift tube, a linear microscope column is retained. The TOF-spectra[55] reveal a remarkably high signal-to-noise ratio due to effective suppression of all unwanted electrons and a good energy resolution which can be driven down to the 100 meV-range. For energy-selective imaging the time-of-flight is set to the desired peak maximum. In addition, TOF-PEEM provides a means for aberration correction of the electron-optical system, a very promising possibility for future instrumental improvements.

The pioneering experiments of Bauer and coworkers[59] have proven the superior performance of *spin-polarized low-energy electron microscopy* (SP-LEEM) for the investigation of magnetic structures on single-crystalline films and surfaces. This method is based on the exchange-scattering asymmetry arising in polarized electron scattering from magnetic surfaces at low kinetic energies. The basic SP-LEEM instrument is a cathode-lens microscope similar to a PEEM with the special feature of a primary polarized electron beam instead of the photon beam. The primary and backscattered

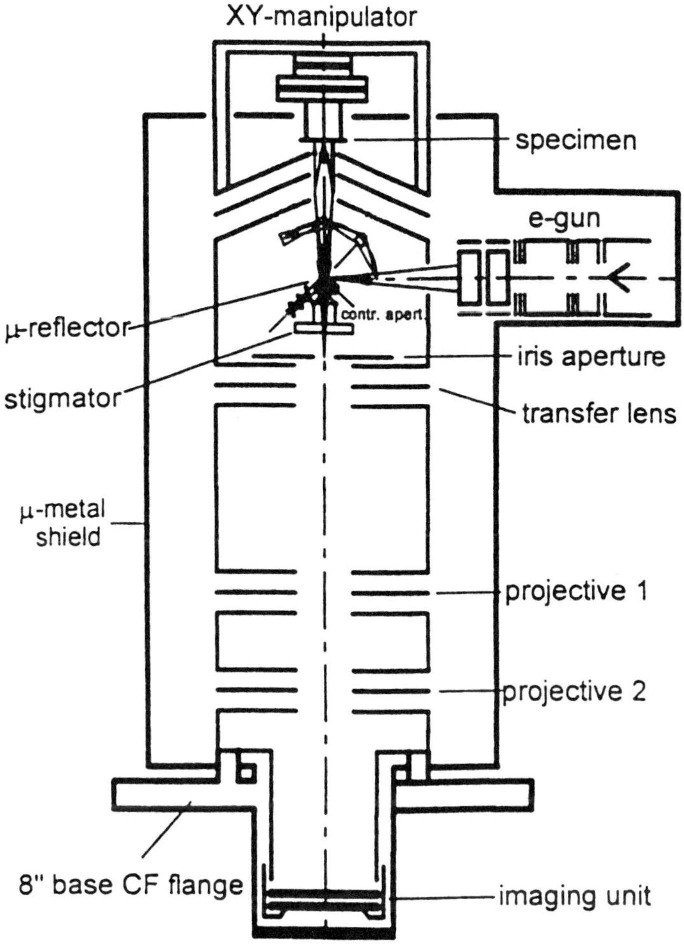

Figure 10. Schematic arrangement of the double-reflection emission electron microscope DREEM. The primary electron beam from the gun is directed along the center of the electron-optical axis of the microscope column by a specular reflection at a W(100)-microcrystal (the 'heart' of the instrument) oriented at 45° with respect to the axis. The microcrystal is positioned very close to the backfocal point of the objective lens, in this case an electrostatic tetrode lens. The electron beam is decelerated to the desired energy by choosing the proper potential of the gun cathode with respect to the sample being kept at ground potential. A relatively large divergence of the illuminating beam in the crossover at the microcrystal allows a parallel illumination of areas up to 500 μm at the sample surface. (from Grzelakowski.[62])

electron beams travel along the same optical axis but with opposite directions. This implies that a beam separator is required, in common LEEM-instruments a magnetic sector field.

Figure 10 shows a novel approach to this technique which employs the specular reflection of the primary beam at a microcrystal surface close to the backfocal point of the objective lens[60] instead of the magnetic beam separator. Since this solution involves two reflections (the first at the microcrystal mirror and the second at the sample itself) and the instrument belongs to the class of emission microscopes, it is termed *double-reflection emission electron microscope* (DREEM). A prototype instrument has been built and tested.[61–63] The concept of DREEM permits the exploitation of many contrast mechanisms and techniques. Scattering energies as low as 2 eV have been achieved. Figure 10 illustrates the simple and extremely compact solution with the small electron gun mounted at right angles to the microscope column. A W(100)-microcrystal serves as electron mirror. Tungsten has been chosen because of its high polarizing power for electrons, which is important for future applications as SP-LEEM. It is known from spin-polarized LEED[64] that degrees of polarization of several 10% can be reached.

Alternatively, a GaAs-source of polarized electrons can replace the standard electron gun. By using a pulsed laser for the excitation of the GaAs,[65] the source can be pulsed with pulse widths of down to 40 ps. This will facilitate time-resolved magnetization studies and real-time observation of magnetic switching phenomena *without the need for synchrotron radiation.* A size-variable and adjustable contrast aperture is located very close to the microcrystal. The present design of microcrystal-reflector and contrast aperture being separately adjustable allows a very easy change between bright- and dark-field imaging by just moving the contrast aperture located below the reflector.

The novel modes of operation mentioned in this outlook will largely expand the field of applications of PEEM in thin-film magnetism. In particular, time-resolved imaging by means of TOF-PEEM (using synchrotron radiation) and SP-DREEM (using a pulsed GaAs-source) have a large future potential. Suitable instruments of both types are under development.

Acknowledgments

The author would like to thank all co-workers being involved in the PEEM development and experiments, in particular to G. K. L. Marx, who has performed the measurements of Secs. III.1 and III.4. Further thanks are due to K. Grzelakowski, M. Escher and M. Merkel (FOCUS GmbH) for fruitful support. The experiments of Secs. III.2 and III.3 have been performed in cooperation with the Max-Planck-Institut für Mikrostrukturphysik, Halle. Sincere thanks are due to C. M. Schneider (now IFW Dresden), R. Frömter, W. Kuch and J. Kirschner. The experiments and the interpretation of the Kerr-effect like contrast benefited a great deal from the cooperation with H. J. Elmers, Mainz. The Fe whisker was kindly provided by B. Heinrich, Simon Fraser University, Canada. Last but not least, I would like to thank H. C. Siegmann for rising my interest in magnetism during a stay in his group in 1984/85. The experiments were funded by BMBF (05 SL8 UM10 and 05 SC8 UMA0), by DFG through Sonderforschungsbereich 262 and by Materialwissenschaftliches Forschungszentrum MWFZ der Universität Mainz.

References

1. H. C. Siegmann, *J. Phys. Condens. Matter* **4**, 8395 (1992).
2. J. B. Kortright, D. D. Awschalom, J. Stöhr, S. D. Bader, Y. U. Idzerda, S. S. P. Parkin, I. K. Schuller, and H. C. Siegmann, *JMMM* **207**, 7 (1999).
3. P. Fischer, G. Schütz, G. Schmahl, P. Guttmann, and D. Raasch, *Z. Phys. B* **101**, 313 (1996); *J. Phys. D* **31**, 649 (1998); G. Schmahl, D. Rudolph, B. Niemann, P. Guttmann, J. Thieme, G. Schneider, *Naturwissenschaften* **83**, 61 (1996).

4. J. Kirz, C. Jacobsen, and, M. Howells, *Rev. Biophysics* **28**, 1 (1995); J. Kirz, H. Rarback, *Rev. Sci. Instrum.* **56**, 1 (1985); T. Warwick, K. Franck, J. B. Kortright, G. Meigs, M. Moronne, S. Myneni, E. Rotenberg, S. Seal, W. F. Steele, H. Ade, A. Garcia, S. Cerasari, J. Denlinger, S. Hayakawa, A. P. Hitchcock, T. Tylizczak, J. Kikuma, E. G. Righter, H.-J. Shin, B. P. Tonner, *Rev. Sci. Instrum.* **69**, 2964 (1998).
5. J. Stöhr, Y. Wu, B. D. Hermsmeir, M. G. Samant, G. R. Harp, s. Koranda, D. Dunham, and B. P. Tonner, *Science* **259**, 658 (1993).
6. D. Spanke,V. Solinus, D. Knabben, F. U. Hillebrecht, F. Ciccacci, L. Gregoratti, and M. Marsi, *Phys Rev. B* **58**, 5201 (1998); F. U. Hillebrecht, H. Ohldag, U. Mick, M. Weiss, and J. Barth, *Science*, submitted; B. Sinkovic, private communication.
7. G. K. L. Marx, H. J. Elmers, and G. Schönhense, *Phys. Rev. Lett.*, june (2000).
8. G. V. Spivak, T. N. Dombrovskaya, and N. N. Sedov, *Doklady Akad. Nauk S. S. S. R.* **113**, 78 (1957).
9. P. W. Hawkes and E. Kasper, *Principles of Electron Optics*, Vols. 1-3, (Academic Press, London, 1996).
10. E. Brüche, *Z. Physik* **86**, 448 (1933); *Kolloid Zeitschr.* **100**, 192 (1942); A. Recknagel, *Z. Physik* **117**, 689 (1941); **120**, 331 (1943).
11. E. Bauer, *Ultramicroscopy* **17**, 51 (1985).
12. O. H. Griffith and G. F. Rempfer, *Advances in Optical and Electron Microscopy* **10**, 269 (1987); G. F. Rempfer, W. P. Skoczylas, O. H. Griffith, *Ultramicroscopy* **36**, 196 (1991).
13. E. Harting and F. H. Read, *Electrostatic Lenses* (Elsevier, Amsterdam, 1976).
14. M. Haider, S. Uhlemann, E. Schwan, H. Rose, B. Kabius, and K. Urban, *Nature* **392**, 768 (1998); M. Haider, H. Rose, S. Uhlemann, E. Schwan, B. Kabius, and K. Urban, *Ultramicroscopy* **75** 53 (1998); H. Rose, *Ultramicroscopy* **78**, 13 (1999).
15. R. Fink, M. R. Weiss, E. Umbach, D. Preikszas, H. Rose, R. Spehr, P. Hartel W. Engel, R. Degenhardt, R. Wichtendahl, H. Kuhlenbeck, W. Erlebach, K. Ihmann, R. Schlögl, H-J. Freund, A. M. Bradshaw, G. Lilienkamp, Th. Schmidt, E. Bauer, and G. Benner, *J. Electron Spectrosc. Relat. Phenom.* **84**, 231 (1997); R. Wichtendahl *et al., Surf. Rev. And Lett.* **5** 1249 (1998).
16. G. Schütz, W. Wagner, W. Wilhelm, P. Kienle, R. Keller, R. Frahm, and G. Materlik, *Phys. Rev. Lett.* **58**, 737 (1987).
17. F. Meier and B. P. Zakharchenya, (Eds.) *Optical Orientation*, (North Holland, Amsterdam, 1984).
18. U. Heinzmann and G. Schönhense, in *Polarized Electrons in Surface Physics*, edited by R. Feder, (World Scientific, Singapore, 1985), Chap. 11.
19. J. Stöhr, "NEXAFS Spectroscopy", Vol 25 of Springer Series in Surface Sciences, (Springer, Heidelberg, 1992).
20. A. Hubert and R. Schäfer, *Magnetic Domains* (Springer, Berlin, 1998).
21. For threshold photoemission, see G. D. Maran, *Phys. Rev. B* **2**, 4334 (1970); J. L. Erskine *et al., Phys. Rev. B* **8**, 1239 (1973); J. K. Sass, *Surface Science* **51**, 199 (1975).
22. J. Henk and R. Feder (1997), private communication (unpublished).
23. J. Henk, S. V. Halilov, T. Scheunemann, and R. Feder, *Phys. Rev. B* **50**, 8130 (1994); T. Scheunemann, S. V. Halilov, J. Henk, and R. Feder, *Solid State Commun.* **91**, 487 (1994).
24. L. Reimer, *Scanning Electron Microscopy* (Springer, Berlin, 1985), Chap. 6.3; L. Reimer, *Transmission Electron Microscopy*, 4th edition, (Springer, Berlin, 1997), Chap. 6.8.

25. S. A. Nepijko, N. N. Sedov, and G. Schönhense, in *Advances in Imaging and Electron Physics*, edited by P. W. Hawkes, (2000) Vol. 113, p. 205.
26. M. Mundschau, J. Romanowicz, J. Y. Wang, D. L. Sun, and H. C. Chen, *J. Vac. Sci. Technol. B* **14**, 3126 (1996); *Surf. Rev. and Lett.* **5**, 1269 (1998).
27. G. K. L. Marx, M. Dunin, V. Przychowski, G. Schönhense, C. M. Schneider, submitted.
28. B. Pfeifer, Dissertation, Wiesner Verlag Augsburg, 1999.
29. G. K. L. Marx, M. J. Klais, P. Haibach, and G. Schönhense, submitted.
30. L. Baumgarten, C. M. Schneider, H. Petersen, F. Schäfers, and J. Kirschner, *Phys. Rev.Lett.* **65**, 492 (1990).
31. A. Berger, U. Linke, and H. P. Oepen, *Phys.Rev. Lett.* **68**, 839 (1992).
32. J. Stöhr, H. A. Padmore, S. Anders, T. Stammler, and M. R. Scheinfein, *Surf. Rev. Lett.* **5**, 1297 (1998).
33. C. T. Chen *et al.*, *Phys. Rev. Lett.* **75**, 152 (1995).
34. J. Stöhr, *J. Electron Spectrosc. Relat. Phenom.* **75**, 253 (1995).
35. J. Vogel and M. Sacchi, *Phys.Rev. B* **49**, 3230 (1994).
36. X. Le Cann, C. Boeglin, B. Carriere, and K. Hricovini, *Phys. Rev. B* **54**, 373 (1996); J. Hunter, D. Arvanitis, and N. Martensson, *Phys. Rev. B* **54**, 11157 (1996).
37. C. T. Chen, F. S., Y. Ma, S. Modesti, *Phys. Rev. B* **42**, 7262 (1990).
38. G. V. D. Laan, *Phys. Rev. Lett.* **66**, 2527 (1991).
39. W. Clemens, Forschungszentrum Jülich, 1994. Jül-2994 (ISSN 0944-2952).
40. H. A. Dürr, G. V. D. Laan, J. Vogel, G. Panaccione, N. B. Brookes, E. Dudzik, and R. McGrath, *Phys. Rev. B* **58**, 11853 (1998).
41. G. Y. Guo, H. E., W. M. Temmermann, P. J. Durham, *Phys. Rev. B* **50**, 3861 (1994).
42. J. Hunter Dunn, D. Arvanitis, N. Martensson, M. Tischer, F. May, M. Russo, and K. Baberschke *J. Phys. Condens. Matter.* **7**, 1111 (1995).
43. W. Kuch, R. Frömter, J. Gilles, D. Hartmann, Ch. Ziethen C. M. Schneider, G. Schönhense, W. Swiech, and J. Kirschner, *Surf. Rev. and Lett.* **5**, 1241 (1998).
44. J. Unguris, R. J. Celotta, and D. T. Pierce, *Phys. Rev. Lett.* **67**, 140 (1991).
45. J. Unguris, R. J. Celotta, and D. T. Pierce, *Phys. Rev. Lett.* **79**, 2734 (1997).
46. C. M. Schneider, K. Meinel, J. Kirschner, M. Neuber, V. Wilde, M. Grunze, K. Holldak, Z. Celinski, and F. J. Baudelet, *Magn. Magn. Mater.* **162**, 7 (1996).
47. M. Rührig, R. Schäfer, A. Hubert, R. Mosler, J. A. Wolf, S. Demokritov, and P. Grünberg, *Phys. Stat. Sol. (a)* **125**, 635 (1991).
48. C. M. Schneider, R. Frömter, Ch. Ziethen, W. Swiech, N. B. Brookes, G. Schönhense, and J. Kirschner, *Mat. Res. Soc. Symp. Proc.* **475**, 381 (1997); C. M. Schneider, *J. Magn. Magn. Mater.* **175**, 160 (1997).
49. P. Elleaume, *J. Synchrotron Rad.* **1**, 19 (1994).
50. C. M. Schneider, R. Frömter, Ch. Ziethen, W. Swiech, N. B. Brookes, G. Schönhense, and J. Kirschner, *Internatl. Colloquium on Magnetic Films and Surfaces (ICMFS)*, Brisbane, Austr., 1997, Conference Digest, p. 132
51. D. B. Dove, *Journal of Applied Physics* **34**, 2067 (1963).
52. T. N. N. Katayama, N. Okusawa, Y. Miyauchi, T. Koide, T. Shidara, Y. Suzuki, and S. Yusas, *J. Magn. Magn. Mater.* **177-181**, 1251 (1998).
53. W. Swiech, G. H. Fecher, Ch. Ziethen, O. Schmidt, G. Schönhense, K. Grzelakowski, C. M. Schneider, R. Frömter, H. P. Oepen, and J. Kirschner, *J. Electron Spectrosc. Relat. Phenom.* **84**, 171 (1997); Ch. Ziethen, O. Schmidt, G. H. Fecher, C. M. Schneider, G. Schönhense, R. Frömter, M. Seider, K. Grzelakowski, M.

Merkel, D. Funnemann, W. Swiech, H. Gundlach, and J. Kirschner, *J. Electron Spectrosc. Relat. Phenom* **88**, 983 (1998).
54. H. Spiecker, O. Schmidt, Ch. Ziethen, D. Menke, U. Kleineberg, R. G. Ahuja, M. Merkel, U. Heinzmann, and G. Schönhense, *Nucl. Instr. and Meth. A* **406**, 499 (1998).
55. A. Oelsner, O. Schmidt, V. Mergel, H. Schmidt-Böcking, and G. Schönhense, 1998 BESSY report p. 495; A. Oelsner, O. Schmidt, M. Schicketanz, V. Mergel, O. Jagutzki, H. Schmidt-Böcking, and G. Schönhense, submitted.
56. R. Mooshammer, M. Unverzagt, W. Schmitt, J. Ullrich, H. Schmidt-Bcking, *Nucl. Inst. Meth. B* **108**, (1996).
57. H. Ade, (Ed.): 1997 "Spectromicroscopy", Special Issue of *J. Electron Spectrosc. Relat. Phenom.* **84** Nos. 1-3
58. E. Bauer (Ed.), "Low Energy Electron Microscopy", Special Issue of *Surf. Rev. and Lett.* **5**, No. 6 (1998).
59. Th. Duden and E. Bauer, Ref. 58, p. 1213.
60. German patent. 195 43 652, USA pat. 5.969.356, Japan pat. pending 0-33/08/755, 781.
61. K. Grzelakowski, *Rev. Sci. Instrum.* **70**, 3346 (1999).
62. K. Grzelakowski, *Surf. Science* **454/456**, (2000), (in print).
63. K. Grzelakowski, J. Settemeyer, M. Escher, M. Merkel, Ch. Ziethen, and G. Schönhense, to be published
64. J. Kirschner, *Polarized Electrons at Surfaces* (Springer Tracts, Berlin 1985).
65. see paper by E. Reichert, These proceedings, p. 453.

Ultra-Thin Magnetic Films with Finite Lateral Size

F. Marty,[1] C. Stamm,[1] U. Maier,[1] U. Ramsperger,[2] and A. Vaterlaus[1]

[1] *Laboratorium für Festkörperphysik*
Mikrostrukturforschung
ETH-Zürich
8093 Zürich
SWITZERLAND

[2] *National Research Institute for Metals*
1-2-1 Sengen
Tsukuba
Ibaraki 305
JAPAN

Abstract

In this paper we review some recent experimental results on the magnetic properties of ultra-thin magnetic dots with μm and sub-μm lateral size grown by the technique of shadow molecular beam epitaxy.

I. Introduction

In the following we introduce a new class of magnetic structures: Two dimensional (2D) magnetic particles. They consist of ultra-thin magnetic films with a finite lateral size. They are grown on top of non-magnetic single crystal surfaces by the technique of shadow MBE (molecular beam epitaxy). In this technique, a mask is placed between the MBE source and the substrate, producing laterally patterned ultra-thin magnetic structures of cobalt and iron with variable size and shape. The magnetic properties of 2D particles differ profoundly from the magnetic properties of their 3D counterpart, which extend to a finite size in all three space coordinates. In particular:

1. In-plane magnetized 2D Co particles grown on Cu(100) are single domain, independent of their lateral size. Moreover, their shape does not affect the magnetization direction, *i.e.* they do not have shape anisotropy. Finally, the mutual

Physics of Low Dimensional Systems
Edited by J. L. Morán-López, Kluwer Academic/Plenum Publishers, New York 2001

magnetostatic interaction of 2D cobalt particles is negligible and the magnetic state of a single particle could be switched without modifying the state of the neighbors.

2. Perpendicularly magnetized ultra-thin Fe particles grown on Cu(100) decay into a stripe domain phase consisting of "up" and "down" magnetization, provided a critical lateral size is exceeded. Below that size they are single domain. Thus, ultrathin magnetic dots are of potential technological interest as elementary small magnets with well defined magnetization state.[1]

The following Secs. II and III address the technical details of producing and characterizing two dimensional particles. The magnetic properties of Co on Cu(001)—magnetized in plane—and of Fe on Cu(001)—magnetized out of plane—are described in Secs. IV and V, respectively. The energies relevant to understand these experimental results are discussed in Sec. VI.

II. Fabrication and Characterization of Two Dimensional Structures

The experimental challenge is to produce small, single crystalline epitaxial magnetic structures with reproducible magnetic properties and with a thickness as small as a few atomic layers. In addition, we require control of their shape and their mutual lateral arrangement. This task is achieved by *in situ* MBE (molecular beam epitaxy) growth through a diaphragm—a technique which we call shadow—MBE.[2,3] The diaphragm contains a sequence of holes with well defined pattern and size. When placed between MBE source and substrate the hole pattern is projected onto the substrate, producing epitaxial films with finite lateral size and well defined geometry. The sharpness of the projected structures is very much dependent on the exact distance between diaphragm and substrate. Thus, the diaphragm is placed on a inertial slider[3,4] allowing accurate positioning as close as possible to the substrate surface. The diaphragm consists of a 1 μm thick titanium foil with micro holes etched through using a commercial focused ion beam (FIB) system. With this combination of MBE and diaphragm technique we are able to fabricate clean, sharp and atomically thin particles, control their number, size and shape at will. Diaphragms can be exchanged *in situ* and are cleaned by standard means of surface physics like argon ion sputtering and mild annealing. This reduces contamination of the substrate surface during growth to a minimum. Evaporation source and diaphragm are typically 300 mm apart. This together with the ability to place the diaphragm extremely close to the sample surface reduces boundary effects which occur due to the non-point like character of the MBE source. The open symbols in Fig. 1(a) represent a cross section of a hole in a titanium foil obtained by SEM (scanning electron microscopy). The width of the hole is 63 nm. The filled symbols show a cross section of a Co dot grown through this 63 nm hole. The Co particle has a width of 160 nm. The difference in diameter between island and hole arises probably from the non vanishing distance D between diaphragm and sample surface. This is confirmed by Monte Carlo simulations of the growth process. The solid line in Fig. 1 is a simulation based on the known diameter of the evaporation source, the source-to-sample distance and the diameter of the hole. The only adjustable parameter used to fit the curve to the data points is the distance between diaphragm and sample surface (D). Taking $D = 20\,\mu$m turns out to give a good fit. This result suggests that a further reduction of D is inevitable for growing structures with a boundary smaller than 50 nm. The calculated width of a dot grown through a 10 nm hole is plotted in

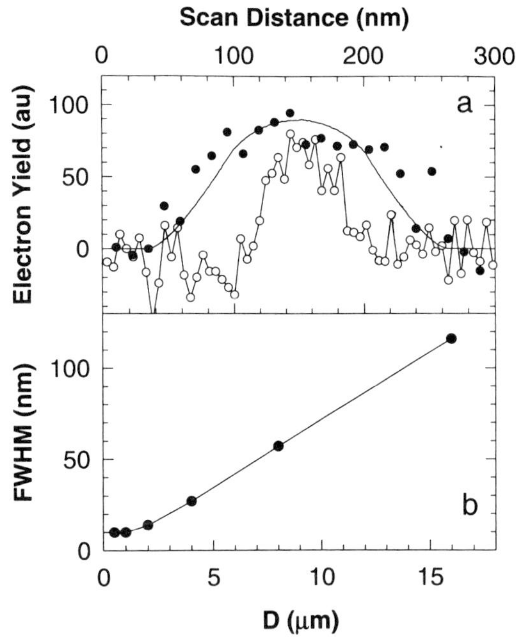

Figure 1. (a) Scanning Electron Microscope cross section through a 63 nm wide hole in a diaphragm (open symbols) and through a Co island grown through this hole (filled symbols). The smooth line is a Monte Carlo simulation of the Co island growth assuming a diaphragm to sample distance of 20 μm. (b) Growth simulation: Full width at half maximum of an island grown through a 10 nm hole as function of the diaphragm to sample distance.

Fig. 1(b) as function of D. The diameter of the dot grown with $D > 3$ μm already exceeds 20 nm. Growth rate and cleanliness of the magnetic layers are checked by Auger electron spectroscopy on layers grown without diaphragm in front of the sample surface. Typically we use a rate between 0.02 AL/min and 1 AL/min and have a pressure of 1×10^{-10} mbar during growth and 6×10^{-11} mbar during the measurement.

III. Magnetic Measurements

One technique used for obtaining images of the surface magnetization is Kerr effect[5,6] with a focused laser beam as light source (SKEM). The spatial resolution of this technique is given by the size of the laser focus and can be estimated from the sharpness of the boundary between magnetic and non magnetic areas. Our current set up has a lateral resolution of 1.3 μm.[3] Notice that Secondary Electron Microscopy with Polarization Analysis (SEMPA) (see below) has a better spatial resolution (typically 20 nm). However, this technique does not allow measurements in an applied magnetic field.

The second technique employed to investigate small magnetic structures is SEMPA. This technique was first introduced by Koike *et al.*[7] in 1984. It uses an electron gun with a beam diameter of typically 10–20 nm to excite secondary electrons out of the top 5 surface layers and a spin sensitive detector for measuring the electron spin polarization of the secondary electrons. Among the possible detection systems,[8-10] we use a modified high voltage Mott detector[11] in combination with a Hitachi 4100S UHV field emission Scanning Electron Microscope (SEM). The best lateral resolution possible is 2 nm. In the geometry used for the SEMPA operation we achieve a resolution of about 10 nm.

For SEMPA measurements the sample is oriented so that its normal makes an angle of 45° with the incoming electron beam. The secondary electrons are collected using a specially designed electron optics. The first lens has a potential of about 500 V with respect to the sample. By means of a 90° deflection, the low energy electrons ($E_{\mathrm{kin}} < 6$ eV) are selected and directed into a medium voltage (60 kV) Mott detector. This detector is equipped with 4 counters and measures one in-plane component and the normal component of the electron spin polarization vector. An external field of up to 0.08 Tesla can be applied in the measuring position, which must be switched off during SEMPA operation. The temperature can be varied between 170 K and 400 K and the sample can be rotated around its surface normal. Apart from the superior resolution with respect to SKEM, the SEMPA is able to measure an absolute magnetic signal—the electron spin polarization (P)—whereas the Kerr effect measures a reflectivity difference between the magnetization at an applied magnetic field $+H$ and an applied field $-H$. Notice that both SKEM and SEMPA detect the magnetization directly, while other known techniques for magnetic microscopy—such as Magnetic Force Microscopy[12]—use the stray field originating from the magnetization for obtaining a magnetic contrast. Below we will argue that the stray field arising from ultrathin magnetic structures is extremely small, so that detection with MFM could be difficult. Magnetic tunneling[13,14] as well as near field optical microscopy[15] are not yet well established in the ultrathin limit.

IV. Magnetic Properties of 2D Co Particles

Ultrathin Co films on Cu(100) with mm lateral size are magnetized along the 110-in-plane directions in virtue of a four-fold magnetic anisotropy.[16-20] The presence of monoatomic steps might introduce a two-fold magnetic anisotropy favoring the step direction as easy magnetization axis.[19] The spatially resolved remanent (*i.e.* at zero applied magnetic field) magnetization M of ultrathin particles (thickness d in the range 2 AL $< d <$ 10 AL) is shown in Fig. 2 for lateral sizes varying from 1 mm down to 100 nm. All particles are ferromagnetic starting from $d \sim 2$ AL. The easy axis of the magnetization is along the same crystallographic direction for all particles, irrespective of their shape and size. Each particle has a roughly square hysteresis loop (Fig. 2d) with a nearly fully magnetized remanent state, and M is homogeneously distributed: we do not observe magnetic domains even when the lateral size is varied over decades. There is a minimum magnetic field H_{rev} required to switch the magnetization of the particles in the opposite direction, see *e.g.* the transition from Fig. 2(f) to Fig. 2(g) and Fig. 2(d). We observe that the single-domain state is maintained up to reverse applied fields as close to H_{rev} as 0.995 H_{rev}. We conclude that even when the dots are brought very close to the state of instability toward magnetization reversal, no static domains are actually able to form within the magnetic dots. A remarkable exception is the mm sized Co-film reported in Fig. 2(b). There, mm-sized domains are produced in the vicinity of H_{rev}. However, over such large scales, the Co-film is bound to meet with the major structural defects provided by the Cu-surface and to develop enough magnetic charge to create domains and pin their walls. Notice that even structures measured as grown, *i.e.* without any applied field being applied prior to the magnetic

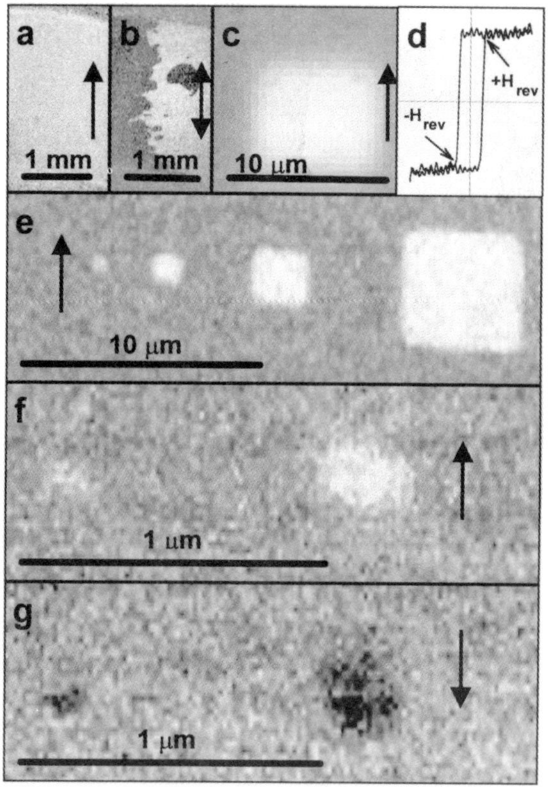

Figure 2. SEMPA and SKEM images of Co particles with various sizes. The magnetization is measured in the remanent state. The gray scale range is proportional to the electron spin polarization or to the Kerr asymmetry. The arrows indicate the direction of the magnetization, which is parallel to a [110] in-plane axis. **a, b**: extended Co layer (d = 4.5 AL, H_{rev} = 3.52 kA/m). The layer is shown as grown (a) and after applying a field amounting to $-0.995\ H_{\mathrm{rev}}$ (b). SKEM image (c) and corresponding hysteresis loop (d) of a 6 AL Co dot (H_{rev} = 0.76 kA/m). (e) SEMPA image of square dots with a size from 0.5 to 4.5 μm ($d = 10\ AL$, H_{rev} = 8 kA/m). (f) SEMPA image of dots with diameter 130 nm (d = 2 AL) and 300 nm (d = 2.5 AL). (g) like (f) but after applying a switching field of 27.2 kA/m (H_{rev} = 11.2 kA/m). The magnetization changed sign (dark corresponds to a negative electron spin polarization).

imaging, are in a single domain state, see Fig. 3. Here the lateral size ranges from 5 μm down to 500 nm. The magnetization of the different dots is pointing in opposite directions, indicating that minute magnetic fields during growth have oriented the magnetization according to their local direction. However, the no-domain rule is still strictly observed in this case as well.

There are exceptions to the no domain rule which are investigated in Fig. 4. Figures 4(a) and 4(b) show a spatially resolved SKEM image of a 5 μm wide Co stripe. The remanent magnetization is shown in Fig. 4(a). There is a section giving no magnetic signal. This apparently "magnetically dead" section becomes "magnetic" when the SKEM image is collected with an applied magnetic field (Fig. 4b). The hysteresis loop collected in the "magnetically dead" region is shifted (see inset). Shifted

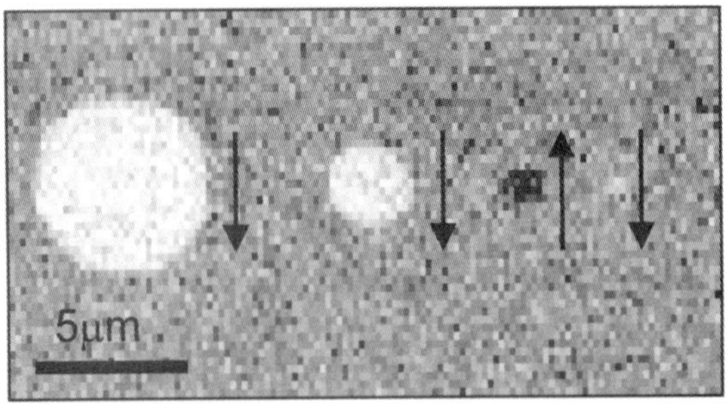

Figure 3. SEMPA image of Co islands in the as grown state. The islands have a size ranging from 5 μm down to 500 nm. Arrows indicate the orientation of the magnetization.

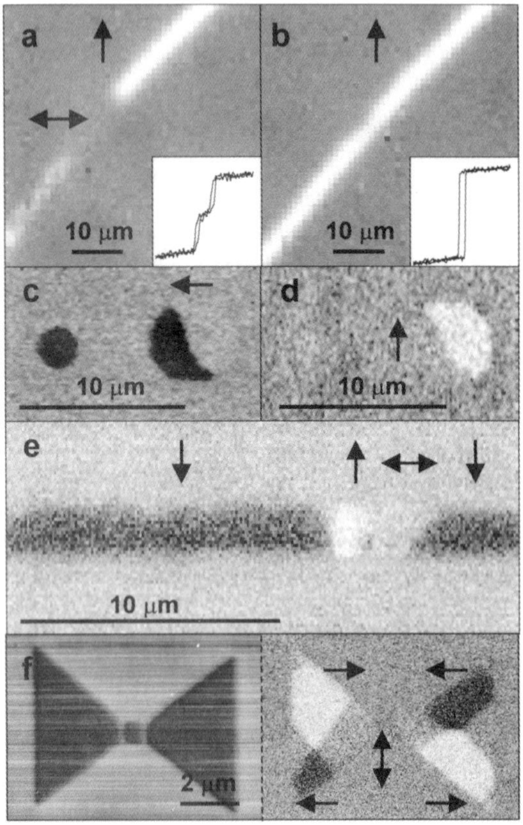

Figure 4. Magnetic domains in 2D particles. (a) and (b) are SKEM images showing a 5 μm wide stripe (d = 2.5 AL). The contrastless part of the stripe in (a) is due to a 90° domain penetrating the stripe. (a) is measured in remanence (b) in an applied field of 0.32 kA/m. The insets show hysteresis curves measured in the contrast-less part [inset in (a)] and the part visible in (a) and (b) [inset in (b)]. (c) and (d) SEMPA images of a 6 AL dot, taken in remanence and showing both in-plane components. (e) section of a 122 μm times 1.9 μm stripe (d = 17 AL) exposed to 82.5% of $H_{\rm rev}$. A white 180° domain is stabilized within the stripe. (f) Topography (left) and magnetization (right) image of a butterfly like structure with a thickness of 80 AL. Domains are visible in the as grown state.

hysteresis loops occur for Co on Cu(001) because of the presence of monoatomic steps inducing locally an uniaxial anisotropy, which is able to rotate the easy axis in the region surrounding the steps.[19] This suggests that the magnetically dead zone is not "dead" but simply magnetized along the in-plane direction not measured in Fig. 4(a). The existence of 90° domain walls is confirmed by the SEMPA images of Figs. 4(c) and 4(d). These pictures show the two in-plane magnetization directions of a small Co particle. The fraction of the particle which is not visible in Fig. 4(c) gives a magnetic signal along the perpendicular in-plane direction (Fig. 4d).

A second possible reason for the appearance of domains is a large film thickness. Fig. 4(e) shows for example a section of a 2 μm wide and 17 AL thick stripe which was exposed to a field being equal to 82.5% of $H_{\rm rev}$. The section of the stripe visible in Fig. 4(e) shows a 180° domain wall (transition between black and white area). Exposing thinner $d < 10$ AL particles and stripes to reverse magnetic fields as large as 99% of $H_{\rm rev}$ did not produce any 180° domains. In thicker structures domains appear in the as grown state as well (see Fig. 4f). The "butterfly" like structure of Fig. 4(f) has a thickness of 80 AL and is imaged in the as-grown state. The figure shows topography (left) and magnetization (right). A complex domain pattern, indicated by arrows, evolves.

To explore the influence of the particle shape on its magnetic state, we have investigated the spatial distribution of the magnetization in atomically thin magnetic stripes. They are the two-dimensional counterpart of (three-dimensional) magnetic needles. The magnetostatic energy stored in needles magnetized along their axis is smaller than the magnetostatic energy of needles magnetized perpendicularly to their axis. This shape anisotropy forces the needles to be magnetized along the axis.[21–24] No shape anisotropy is observed in two-dimensional stripes (see Fig. 5). Figure 5(a) shows an array of 40 μm long and 500 nm wide stripes oriented at a variable angle α with respect to the easy magnetization axis, which is vertical in the figure. By inspection of Fig. 5(a), we recognize that the magnetization is uniform within the stripes and points in the same direction irrespective of α. A detailed analysis of stripe-images is reported in Fig. 5(b) as a plot of the vertical component of the magnetization *versus* α. It confirms that varying α does not produce any measurable deviation from the vertical orientation. This is even true when the long axis of the stripe is exactly orthogonal to the easy magnetization direction, see Fig. 5(c). Evidently, the magnetostatic energy stored in this perpendicular configuration is not enough to demagnetize the stripe, either by domain-formation or magnetization rotation into equivalent in-plane easy-directions, although $H_{\rm rev}$ can be as small as 10 G!

A more elaborate measure of the shape anisotropy in small stripes of Co on Cu(001) is obtained when a Cu surface with steps along the [110] direction is used as substrate. The steps impose an uniaxial anisotropy on the Co layer grown on top. This results in "shifted" hysteresis loops,[19] *i.e.* a hysteresis loop which consists out of two single square loops separated by a characteristic field $2H_S$ (see inset Fig. 4a). For thin layers the shift field H_S is proportional to the uniaxial anisotropy induced by the steps[19,20,25] and is therefore a very sensitive measure of anisotropies. By measuring H_S as function of the layer thickness it was possible to detect morphology induced oscillations of the anisotropies[20] as well as quantum oscillations in a three dimensional electron gas.[26] Growing a stripe parallel to the steps (along [110]) its shape anisotropy represents an effective additional uniaxial anisotropy along [110] and should increase the shift field (H_S). Growing the stripe perpendicularly to the steps should reduce H_S. Thus, we expect the shift fields of identical stripes, which include a variable angle α with the steps, to decrease with increasing α. Figure 6 shows H_S as a function of

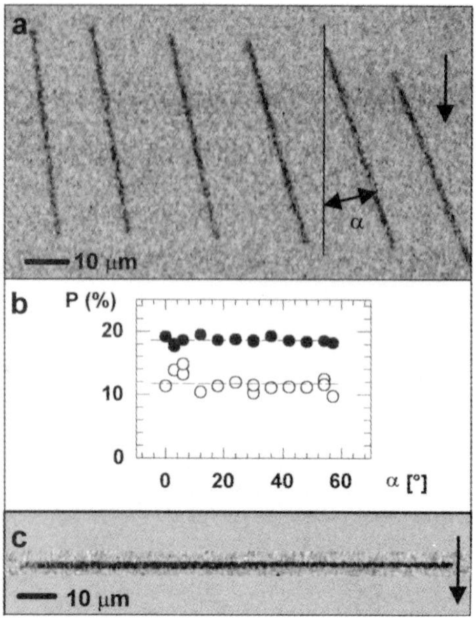

Figure 5. 2D magnetic stripes. (a): SEMPA image showing a set of Co stripes (23 AL thick) with a variable angle α with respect to the vertical easy direction. M was measured in the remanent state along the vertical axis. No spin polarization could be measured along the horizontal axis. (b): The electron spin polarization measured in the remanent state along the vertical easy axis is plotted as function of α for two sets of stripes (40 μm by 0.5 μm), with $d = 3$ AL (open circles) and $d = 6$ AL (filled circles). (c): SEMPA image of a 122 μm by 1.9 μm Co stripe ($d = 10$ AL) magnetized perpendicularly to its long axix.

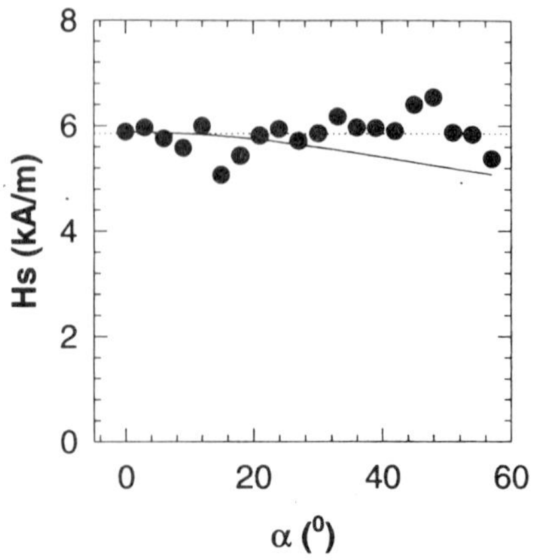

Figure 6. Shift field (H_S) measured on 1 μm wide, 40 μm long and 8 AL thick stripes of Co grown onto a stepped Cu(001) surface. Identical stripes with a variable angle α to the [110] direction were grown. H_S is plotted as function of α. The dotted line represents the average shift field, the solid line is calculated according to Sec. IV.

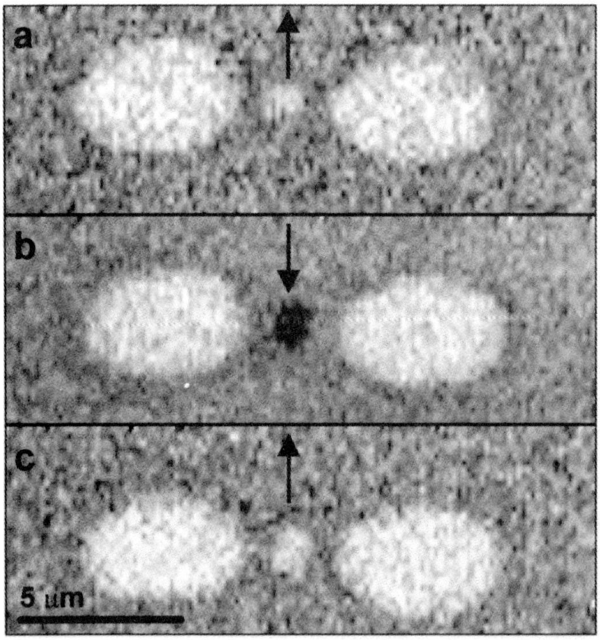

Figure 7. SEMPA images showing a small Co element ($H_{\rm rev}$ = 2.64 kA/m, d = 4 AL) sandwitched between two elliptical particles ($H_{\rm rev}$ = 3.2 kA/m and 2.9 kA/m, d = 4 AL). Before (a) was taken, all particles were magnetized up by a large positive magnetic field and then exposed to a negative magnetic field of −2.48 kA/m. This negative fields did not switch the state of the particles. (b) A magnetic field of $H_{\rm rev-}$ = −2.64 kA/m has reversed M of the smaller Co dot, which now appears dark in the image. (c) A magnetic field of $H_{\rm rev+}$ = 2.64 kA/m has switched M of the smaller Co dot in the positive direction.

α for a set of 1 μm wide, 40 μm long and 8 AL thick stripes. The stripes are grown onto a Cu(001) surface with a miscut of 3° (the steps are running along [110] and the average width of a terrace is 3.5 nm). H_S was determined from hysteresis loops which were measured for each stripe with the help of the SKEM. No systematic variation of H_S with α is found, the experimental uncertainty of ±8 G (±0.64 kA/m) arises from variations in the step density on the substrate surface. The solid line in Fig. 6 is the calculated expected dependence of H_S on α:

$$H_S(\alpha) = H_S^{\rm steps} + H_S^{\rm shape}(2\cos^2\alpha - 1).$$

$H_S^{\rm steps}$ is the anisotropy field due to the step induced uniaxial anisotropy. $H_S^{\rm shape}$ is the shape anisotropy field and can be calculated from $H_S^{\rm shape} = M\ (N_X - N_Y)$, where N_X and N_Y are the demagnetizing coefficients along and perpendicular to the stripe and M is the magnetization. From Refs. 20 and 27 we find that $H_S^{\rm shape}$ = 7 G (0.57 kA/m). The measured and calculated dependence of H_S on α shows that shape anisotropies in 2D stripes are essentially negligible.

Each magnet produces a stray magnetic field that influence the magnetic state of neighboring magnets. How strong is the field arising from a 2D magnetic particle? We test this mutual interaction by growing a small sensor particle between two elliptically shaped islands which act as sources for the magnetic stray field. In Fig. 7(a)

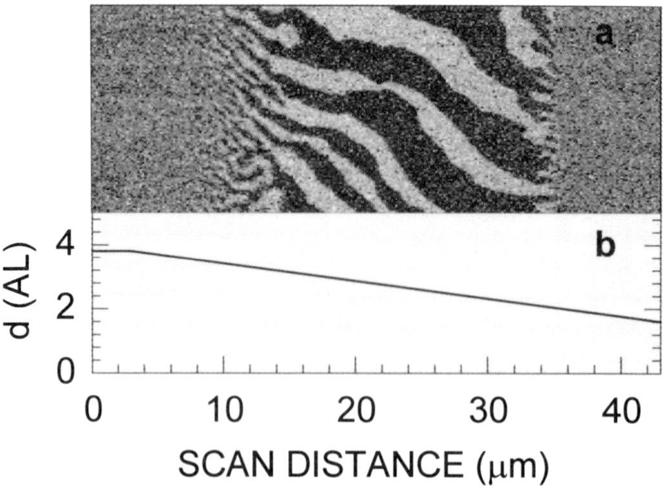

Figure 8. Fe wedge grown on Cu(001). (a) SEMPA image of the perpendicular polarization. The coverage is constant on a line parallel to the y-axis and increases according to (b) along the x-axis. The picture is taken at room temperature. There is no detectable in-plane polarization. Domains are present when ever Fe on Cu(001) is magnetic. The picture corresponds to the as grown state and is taken at room temperature.

all three particles are magnetized along the vertical direction. In this arrangement the test particle is exposed to a bias field pointing opposite to its magnetization. This bias field should help switching the magnetization of the test particle from *up to down* (white to black) and should oppose the switching from black to white. Exactly this is probed in Fig. 7. We calculate the asymmetry in switching field to be about 2 G. Experimentally, we can probe this asymmetry by applying successively higher reversing fields and detecting the state of the particle after the field has been applied. Within the uncertainty of our experiment we are not able to detect a sizeable asymmetry. Thus, within the length scales of the present experiment 2D particles can be regarded as independent fundamental units.

When a magnetic particle is becoming smaller, the energy barrier ΔE separating stable orientations of the magnetization is decreasing and the particle might become unstable towards thermal fluctuations:[28,29] the particle enters the superparamagnetic regime and no net magnetization is recorded above the blocking temperature. Clearly, our ultrathin dots are still sufficiently large to escape the superparamagnetic limit. The size at which the superparamagnetic fluctuations start is still unknown.

V. Perpendicular Magnetized Fe on Cu(001)

Ultrathin films of Fe on Cu(001) grown at room temperature with mm lateral size are magnetized perpendicularly to the film plane in a well defined thickness and temperature range.[30–32] Figure 8 shows a 43 μm wide SEMPA image of a Fe wedge at room temperature. Only the perpendicular component of the electron spin polarization is shown, the in-plane components being zero within experimental accuracy. Between 2 and 3.4 AL the film shows the predicted stripe domain phase.[33] The width of the stripe domains is small close to the onset of magnetic order (right hand side in Fig. 8)

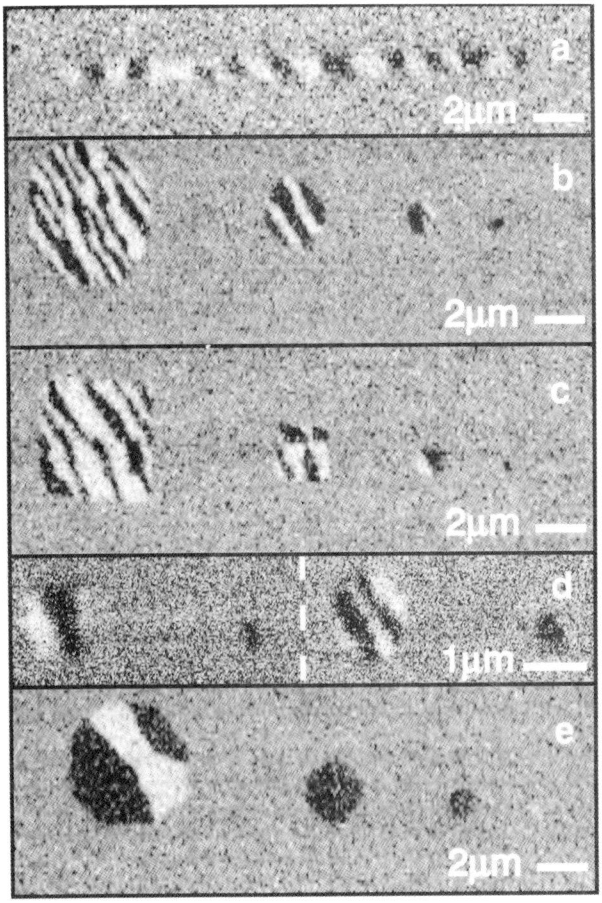

Figure 9. SEMPA images of Fe particles on Cu(001). Stripe (a), circular dot (b) and squares (c) with a thickness of 2.7 AL are shown. (d) Smaller dots. The 500 nm dots appears single domain. (e) Thicker Fe dots (2.8 AL) showing wider stripes. The pictures are taken at room temperature, in remanence and represent the as grown state.

and close to the Curie temperature (left hand side in Fig. 8).[33] The wedge was obtained by placing a wire in front of the substrate at about 7 mm distance from it. This increases boundary effects (see Sec. II) and produces a wedge. The coverage is constant along the vertical axis in the figure and varies as described in Fig. 8(b) along the horizontal axis. Fe particles of various size and shape are shown in Fig. 9. Most islands contain stripe domains irrespective of their size and shape (Figs. 9a, 9b, and 9c), similarly to the extended film. However, a single domain state is observed as soon as the dot size L is smaller than the width of the stripe domains (Figs. 9d and 9e). As the stripe width depends on the thickness of the Fe layer (Fig. 8a), the particle size at which the transition to the single domain state occurs varies with the thickness as well (compare Figs. 9e with 9b, and 9c). Clearly, the observation of domains and of a multi-domain to single-domain transition shows that perpendicularly magnetized ultrathin dots behave very much differently than in-plane magnetized ultrathin dots, for which a no-domain and no-shape-anisotropy rule apply. All these results are in striking contrast to the familiar observation of domains and shape anisotropy in 3D systems.[21–24,34–37] In the following section we will try to give an explanation for these results.

VI. Evaluation of the Relevant Energies and Length Scales

Domains appear whenever the reduction of the magnetostatic energy E_m due to domain formation overcomes the energy needed to introduce a domain wall E_w.[38,39] Shape effects are present as soon as the difference of E_m for different orientations of the magnetization is of the same size as the anisotropy energy E_a and close-packed particles cease to be independent when the stray field is of the same order as the coercive field. Therefore, the key task is the evaluation of the magnetostatic energy E_m

$$E_m = \frac{(g\mu_B)^2}{2} \sum \frac{\vec{S}_i \vec{S}_j}{|\vec{r}_{ij}|^3} - \frac{3(\vec{S}_i \vec{r}_ij)(\vec{S}_j \vec{r}_ij)}{|\vec{r}_{ij}|^5}, \tag{1}$$

where a is the lattice constant, S_j is the spin at the lattice site $\vec{r}_j$ and $\vec{r}_{ij}$ is $(\vec{r}_i - \vec{r}_j)$.

First we calculate the reduction in the magnetostatic energy (ΔE_m) of a square shaped in plane magnetized dot (size L and thickness 1 AL) due to the formation of a domain wall. The wall is running parallel to the edge and the magnetization of the dot and cuts it into two equivalent parts with opposite magnetization. The resulting energy gain per atom (ΔE_m), calculated by performing the dipolar sums Eq. (1) is plotted in Fig. 10(a) (circless) as function of the particle size L. ΔE_m can be fitted with a $1/L$ dependence (solid line through circles). The wall energy (E_w) acts as an energy barrier for the formation of a domain. E_w scales with L so that the wall energy per atom scales with $1/L$. Thus, both energies, ΔE_m as well as E_w have the same scaling behaviour as a function of L. Consequently the balance between ΔE_m and E_w does not change when L is varied: if the dot contains domains, then they will occur at any size L. On the contrary, should the dot be in a single domain state for one size, then the single domain state is energetically preferred at any size L. Evidently, our in-plane dots realize this last option.

The situation is different for a perpendicular magnetized dot (squares in Fig. 10a). Here the reduction in magnetostatic energy per atom due to domain formation scales with $\ln L/L$, see the continuos line through the squares. Thus, the energy balance between ΔE_m and E_w changes as function of L. Domain formation becomes energetically favorable for a particle size exceeding a critical dimension L_c, while the single domain state is energetically preferred below this critical size. The critical size depends on some magnetic parameters and is given by[40]

$$L_c \cong 4w \exp^{E_w/2\Omega},$$

where w is the domain wall width, E_w is the energy of a domain wall per unit length and $\Omega = (g\mu_B s)^2/a^3$ is the dipolar interaction. For perpendicular magnetized Fe grown on Cu(001) w is of the order of 10 nm, 2Ω is of the order of 1 K and a rough estimate for E_w is 5 K. This would lead to an L_c of the order of 1 μm (depending critically on the material parameters). This is in agreement with the observations reported in Sec. V. The magnetostatic energy E_m produces also a shape anisotropy which competes with the magnetocrystalline anisotropy E_a for determining the easy magnetization axis. For a two dimensional stripe with length L and width $L/50$ the difference of the magnetostatic energy δE_m per spin for the state with M perpendicular and M parallel to the long axis of the stripe is plotted in Fig. 10(b) as a function of L. δE_m scales with

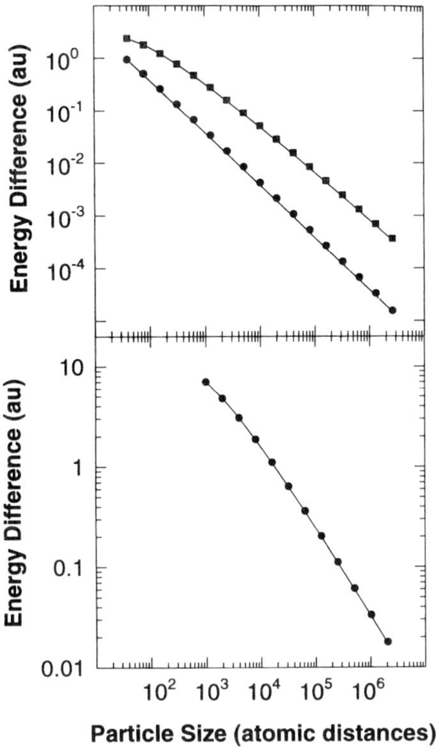

Figure 10. (a) Calculated reduction in magnetostatic energy per atom due to domain formation for an in plane magnetized system (circles) and for a perpendicular magnetized system (squares). The functional dependence in L is $1/L$ and $\ln L/L$ respectively (lines). (b) Difference of the magnetostatic energy (dE_m) of a stripe (aspect ratio 1:50) for magnetization parallel and antiparallel to the long axis of the stripe. The functional dependence of dE_m on the stripe's length L is $\ln L/L$ (solid line).

$\ln L/L$ while E_a (per spin) is constant as function of L. Accordingly, at sufficiently large length scales the shape anisotropy plays a negligible role with respect to the magnetocrystalline anisotropy for determining the easy magnetization axis. This explains the experimental findings of Fig. 5. The mutual interaction of particles with size $O(L)$ and distance $O(L)$ is mediated by the stray field H_m which is given by $\sim MV/L^3$, V being the volume of the magnet. Taking 2D particles of size L and thickness d we obtain $H_m = Md/L$. There is a critical dimension at which the particles cease to be independent, given by Md/H_{rev}, where H_{rev} is the coercive field of the particles. With $M \sim 1$ kG, $d \sim 0.3$ nm, and $H_{\text{rev}} \sim 10$ G we obtain that the critical distance equals a few nm. Thus, the mutual interaction is small until lateral sizes of a few nm are reached.

In summary, 2D in-plane magnetized magnetic structures are single-domain independently of their size, have a negligible mutual interaction and their shape does not influence the magnetic state. Stability against superparamagnetic fluctuations was observed for dots as small as 130 nm. Perpendicular magnetized 2D particles show a multi-domain to single-domain transition for a critical size of the order of micrometers. Both types of 2D particles behave very much differently than their 3D counterparts.

Acknowledgments

We acknowledge some financial support from the Swiss National Science Foundation. The present work was mainly supported by ETH Zürich through a special grant. We thank Danilo Pescia for giving us the possibility to work in his group and for many stimulating discussions and Henning Fuhrmann (ETH Zürich) for producing some of the diaphragms.

References

1. J. Harris and D. Awschalom, *Physics World* **12**, 19 (1999).
2. R. Allenspach, A. Bischof, M. Stampanoni, D. Kerkmann, and D. Pescia, *Appl. Phys. Lett.* **60**, 1908 (1992).
3. A. Vaterlaus, U. Maier, U. Ramsperger, A. Hensch, and D. Pescia, *Rev. Sci. Instrum.* **68**, 2800 (1997).
4. D. W. Pohl, *Rev. Sci. Instrum.* **58**, 54 (1987).
5. C. H. Back, Ch. Würsch, D. Kerkmann, and D. Pescia, Z. *Phys. B* **96**, 1 (1994).
6. C. H. Back, Ch. Würsch, A. Vaterlaus, U. Ramsperger, U. Maier, and D. Pescia, *Nature,* **378**, 597 (1995).
7. K. Koike and K. Hayakawa, *Appl. Phys. Lett.* **45**, 585 (1984); *J. Appl. Phys.* **57**, 4244 (1985).
8. M. R. Scheinfein, J. Unguris, M. H. Kelley, D. T. Pierce, and R. J. Celotta, *Rev. Sci. Instrum.* **61**, 2501 (1990).
9. H. P. Oepen and J. Kirschner, *Scanning Microscopy* **5**, 1 (1991).
10. R. Allenspach, *Phys. World* **7**, 44 (1994).
11. V. N. Petrov, M. Landolt, M. S. Galaktionov, and B. V. Yushenkov, *Rev. Sci. Instrum.* **68**, 4385 (1997).
12. P. Grütter, D. Rugar, H.J. Mamin, *Ultramicroscopy* **47**, 393 (1992).
13. M. Bode, M. Getzlaff, and R. Wiesendanger, *Phys. Rev. Lett.* **81**, 4256 (1998).
14. P. N. First, J. A. Stroscio, D. T. Pierce, R. A. Dragoset, and R. J. Celotta, *J. Vac. Sci. Technol. B* **9**, 531 (1991).
15. C. Durkan, I. V. Shvets, and J. C. Lodder, *Appl. Phys. Lett.* **70**, 1323 (1997).
16. A. K. Schmid and J. Kirschner, *Ultramicroscopy* **42-44**, 483 (1992)
17. U. Ramsperger, A. Vaterlaus, P. Pfäffli, U. Maier, and D. Pescia, *Phys. Rev. B* **53**, 8001 (1996).
18. H. P. Oepen *J. Magn. Magn. Mat.* **93**, 116 (1991).
19. A. Berger, U. Linke, and H. P. Oepen, *Phys. Rev. Lett.* **68**, 839 (1992).
20. W. Weber, C. H. Back, A. Bischof, Ch. Würsch, and R. Allenspach, *Phys. Rev. Lett.* **76**, 1940 (1996).
21. A. D. Kent, S. von Molnár, S. Gider, and D. D. Awschalom, *J. Appl. Phys.* **76**, 6656 (1994).
22. J. F. Smyth, S. Schultz, D. R. Frekin, D. P. Kern, S. A. Rishton, H. Schmid, M. Cali, and T. R. Koehler, *J. Appl. Phys.* **69**, 5262 (1991).
23. M. S. Wei and S. Y. Chou, *J. Appl. Phys* **76**, 6679 (1994).
24. U. Ebels, A. O. Adeyeye, M. Gester, C. Daboo, R. P. Cowburn, and J. A. C. Bland, *J. Appl. Phys.* **81**, 4724 (1997).
25. A. Rettori, L. Trallori, M. G. Pini, C. Stamm, Ch. Würsch, S. Egger, and D. Pescia, *IEEE Transactions on Magnetics* **34**, 1195 (1998).

26. Ch. Würsch, C. Stamm, S. Egger, D. Pescia,W. Baltensperger, and J. S. Helman, *Nature* **389**, 937 (1997).
27. E. C. Stoner, *Phil. Mag.* **36**, 803 (1945).
28. L. Néel, *Rev. Mod. Phys.* **25**, 293 (1953); *Ann. Geophys.* **5**, 99 (1949).
29. W. F. Brown, *Phys. Rev.* **130**, 1677 (1963).
30. J. Thomassen,F. May, B. Feldmann, M. Wuttig, and H. Ibach, *Phys. Rev. Lett.* **69**, 3831 (1992).
31. D. A. Steigerwald, W. F. Egelhoff, *Phys. Rev. Lett.* **60**, 2558 (1988).
32. J. Giergiel, J. Shen, J./ Woltersdorf, A. Kirilyuk, and J. Kirschner, *Phys. Rev. B* **52**, 8528 (1995).
33. R. Czech, J. Villain, *J. Phys. Condens. Matter* **1**, 619 (1989); B. Kaplan and G. A. Gehring, *J. Magn. Magn. Mater. 128*, 111 (1993); Ar. Abanov, V. Kalatsky, V. L. Pokrovsky, and W. M. Saslow, *Phys. Rev. B* **51**, 1023 (1995).
34. M. Hehn, K. Ounadjela, J.-P. Bucher, F. Rousseaux, D. Decanini, B. Bartenlian, and C. Chappert, *Science* **272**, 1782 (1996).
35. E. Gu, E. Ahmad, S. J. Gray, C. Daboo, J. A. C. Bland, L. M. Brown, M. Rührig, A. J. McGibbon, and J. M. Chapman *Phys. Rev. Lett.* **78**, 1158 (1997).
36. R. M. H. New, R.F.W. Pease, and R.L. White, *J. Vac. Sci. Technol. B* **13**, 1089 (1995).
37. C. Beeli, B. Doudin, J.-Ph. Ansermet, and P. A. Stadelmann, *Ultrmicroscopy* **67**, 143 (1997).
38. C. Kittel, *Phys. Rev.* **70**, 965 (1946).
39. W. F. Brown, *J. Appl. Phys. 39*, 993 (1968).
40. P. Politi and M. G. Pini, *Eur. Phys. J. B* **2**, 475 (1998).

Spin-Dependent Transmission and Spin Precession of Electrons Passing Across Ferromagnets

W. Weber, S. Riesen, and D. Oberli

Laboratorium für Festkörperphysik
ETH Zürich
CH–8093 Zürich
SWITZERLAND

Abstract

The transmission of spin-polarized electrons across self-supported ferromagnetic films has been investigated. A larger transmission probability for electrons with their spin polarization $\vec{P}_0$ parallel to the magnetization $\vec{M}$ than for electrons with $\vec{P}_0$ antiparallel to $\vec{M}$ is found. By choosing $\vec{P}_0$ perpendicular to $\vec{M}$, the spin polarization rotates into the direction of $\vec{M}$ and simultaneously also precesses around it. The precession around $\vec{M}$ is the electron analogue to the Faraday rotation observed with polarized light. It is caused by the exchange energy of the ferromagnetic film.

I. Introduction

I.1. Spin Filtering

During the last decade, magnetic nanostructures have become a hot topic in physics. The quest for higher storage density in magnetic recording media and for faster access to magnetically stored information triggered a lot of basic research on the physical phenomena that are connected with magnetic nanostructures.

One example for a discovery that involves both interesting basic physics and powerful applications is the giant magneto resistance (GMR). Here, the resistance in a metallic ferromagnet/non-magnet/ferromagnet trilayer depends upon the relative orientation of the magnetization of the two ferromagnetic layers.[1] Because large effects in small fields are obtained yielding a high signal to noise ratio, GMR is successfully used in applications such as sensitive magnetic field sensors or read heads for hard disks.[2]

Physics of Low Dimensional Systems
Edited by J. L. Morán-López, Kluwer Academic/Plenum Publishers, New York 2001

The essential physical process that is behind GMR is the spin-dependent scattering of conduction electrons in ferromagnetic materials. According to the two-current model,[3] one spin component of the current is more strongly scattered than the other component, if the two ferromagnetic layers are magnetized parallel to one another. In the antiparallel alignment, both spin components are equally strongly scattered. This results in a lower resistance for the parallel than for the antiparallel alignment. The origin of the spin-dependent scattering in transition metals can be understood by considering the fact that the charge carriers, namely the *sp*-electrons, are predominantly scattered into unoccupied *d*-states.[4] This leads to a spin-dependent scattering in ferromagnetic transition metals, because there are more empty minority-spin *d*-states available for scattering than empty majority-spin *d*-states.

In electron spectroscopy, where electrons are excited above the vacuum level, spin-dependent scattering has also been discussed for a while. Probably the first evidence of spin-dependent electron scattering of excited electrons above the vacuum level in ferromagnetic materials was found with spin-polarized photoemission,[5] where an enhanced spin polarization of the photoyield from Ni was observed as compared to the bulk value. Other measurements involving spin-polarized low-energy electron diffraction[6] and spin-resolved electron-energy-loss spectroscopy[7] pointed towards a spin-dependent scattering as well. In addition, the polarization enhancement at low energies in spin-polarized secondary electron spectroscopy[8–10] is understood in terms of spin-dependent scattering. More recent experimental evidence of spin-dependent scattering comes from overlayer experiments with spin-polarized photoemission,[11–13] where a shorter mean free path was found for minority-spin electrons compared to majority-spin electrons. Furthermore, it has been shown[14] by means of spin- and time-resolved two-photon photoemission spectroscopy that the lifetime of excited electrons between the Fermi level and the vacuum level is spin-dependent.

To explain the polarization enhancement at low energies in spin-polarized secondary electron spectroscopy, Glazer and Tosatti proposed a spin-flip excitation across the ferromagnetic Stoner gap.[15] In this process—called a Stoner excitation—the incoming minority-spin electron scatters into an empty minority-spin state, and a majority-spin electron is reemitted with a small energy loss. In total, this process would enhance the spin polarization of the transmitted electron beam. Such Stoner excitations have indeed been detected in spin-polarized scattering experiments.[16,17] However, they do not account for the fact that the polarization enhancement is a very general feature in secondary electron spectroscopy.

A different origin for the spin-dependent transmission observed in photoemission experiments is proposed by Gokhale and Mills.[18] They suggest a spin dependence of the elastic scattering. Whereas a small elastic contribution to the total scattering cannot be excluded from the experiments, these calculations fail to describe why the spin polarization is enhanced in any investigated material, over a wide energy range, and independent of whether the samples are amorphous, polycrystalline or single crystalline.

On the other hand, all the above mentioned experiments can be understood by applying the model proposed by Mott[4] for the conductivity of transition metals to excited electrons, in which the different number of unoccupied *d*-states for the two spin directions is taken into account. In fact, by compiling many attenuation experiments on a number of materials, Siegmann and Schönhense[19] found an empirical rule for the scattering of hot electrons in transition metals, which justifies the application of this model to excited electrons. This rule says that the absorption of electrons well above the *d*-bands is given by a term that accounts for scattering into unoccupied states other than d-states, and a term that is proportional to the number of holes available to one spin state in the d-orbitals. The most interesting consequence of this proportionality is that the absorption becomes spin-dependent for ferromagnetic transition metals, because the number of holes is different for the two spin directions.

I.2. Spin Precession

It has been known since long that there is an analogy between the mathematical description of a polarized light beam and of a non-relativistic spin-polarized electron beam.[20] In both cases a definite state of polarization can be fully described by a wave function

$$\psi = c_1\xi_1 + c_2\xi_2\,,$$

in which ξ_1 and ξ_2 are a complete set of two orthogonal wave functions. In the case of electrons, these two wave functions correspond to two opposite spin orientations, while in the case of photons, they may correspond to a right- and a left-circularly polarized wave.

A different way to characterize the state of polarization of photons or electrons is to give the expectation values of the Pauli matrices $\sigma_i (i = x, y, z)$. One defines the polarization vector

$$\vec{P} = \begin{pmatrix} \langle\sigma_x\rangle_\psi \\ \langle\sigma_y\rangle_\psi \\ \langle\sigma_z\rangle_\psi \end{pmatrix}.$$

Two orthogonal states are then characterized by polarization vectors with opposite direction ($\vec{P}$ and $-\vec{P}$). However, for the interpretation of this polarization vector, one has to distinguish between electrons and photons. In the case of electrons, $\vec{P}$ defines a direction in real space, whereas in the case of photons, $\vec{P}$ is a vector of the 3-dimensional Poincaré representation in an abstract polarization space. Poincaré suggested to map the settings of a complete polarization analyzer for light onto a sphere in order to get a simple expression for the transmitted intensity through the analyzer.

We now consider a linearly polarized light beam which propagates along the z-axis and whose plane of polarization lies in the xz-plane. If it is a pure state, this light beam is described by the wave function

$$\psi_0 = \frac{1}{\sqrt{2}}(\xi_1 + \xi_2) = \frac{1}{\sqrt{2}}\begin{pmatrix} 1 \\ 1 \end{pmatrix}, \tag{1}$$

where $\xi_1 = \begin{pmatrix} 1 \\ 0 \end{pmatrix}$ and $\xi_2 = \begin{pmatrix} 0 \\ 1 \end{pmatrix}$ is a basis describing right- and left-circularly polarized light. Its polarization vector is then

$$\vec{P}_0 = \begin{pmatrix} 1 \\ 0 \\ 0 \end{pmatrix}.$$

Let us further assume this light beam passes through a ferromagnetic material with its magnetization along the propagation direction. Because of the spin-orbit coupling, there will be a different velocity for left- than for right-circularly polarized light propagating along the magnetization direction. Furthermore, the amplitudes of the two wave functions are differently attenuated. This leads to a phase difference ϵ between the two components and to different (real-valued) amplitudes A_1, A_2 and thus to a new wave function of the photons after leaving the ferromagnetic material:

$$\psi = \frac{1}{\sqrt{2}}\left(A_1 e^{-i\epsilon/2}\xi_1 + A_2 e^{+i\epsilon/2}\xi_2\right) = \frac{1}{\sqrt{2}}\begin{pmatrix} A_1 e^{-i\epsilon/2} \\ A_2 e^{+i\epsilon/2} \end{pmatrix}. \quad (2)$$

Hence, its polarization vector becomes:

$$\vec{P} = \begin{pmatrix} A_1 A_2 \cos\epsilon \\ A_1 A_2 \sin\epsilon \\ (A_1^2 - A_2^2)/2 \end{pmatrix}. \quad (3)$$

This corresponds to a precession—called Faraday rotation—of the polarization vector by an angle of ϵ around the magnetization direction and a rotation by an angle ϕ towards the magnetization direction. The angle ϕ is given by $\tan\phi = (A_1^2 - A_2^2)/2A_1A_2$.

If we now consider a totally polarized electron beam with its spin along the x-axis, which passes through a ferromagnetic material magnetized along the z-axis, then this electron beam is described mathematically in exactly the same way as the linearly polarized light beam described above. The incoming wave function is described as a coherent superposition ψ_0 (see Eq. 1) of a majority-spin wave function ξ_1 (spin parallel to the magnetization) and a minority-spin wave function ξ_2 (spin antiparallel to the magnetization). Then, due to a phase difference between the two basis spin functions and a spin-dependent attenuation of the amplitudes, the wave function describing the electron leaving the ferromagnet is again ψ (see Eq. 2). Hence, the polarization vector of the transmitted electron is $\vec{P}$ (see Eq. 3).

In view of this analogy, it seems possible that the magneto-optic phenomena observed with polarized light, such as the Faraday rotation, should also be observed with spin-polarized electrons.

II. Experiment

In order to investigate both the spin filtering and the spin precession of electrons upon transmission across a ferromagnet, a "complete" spin-polarized electron transmission experiment has been set up. It is schematically shown in Fig. 1. A spin-modulated electron source based on a GaAs photocathode produces a transversely spin-polarized free electron beam having a spin polarization $\vec{P}_0$ with $|\vec{P}_0| = 25\%$. By switching from right- to left-circularly polarized light for excitation of the source, we can invert the vector $\vec{P}_0$ of the spin polarization. By applying a combination of electric and magnetic fields to the electron beam, we can also rotate $\vec{P}_0$ into any desired direction in space. We can produce an unpolarized electron beam as well by applying linearly or unpolarized light. The electron beam impinges perpendicular onto a ferromagnetic film of varying thickness sandwiched between Au layers, which serve both as support and protection layers. It is important that spin-orbit coupling cannot produce any spin dependence of the transmission or spin polarization in this geometry. The transmitted electrons are energy analysed by a retarding field and their spin polarization is detected by Mott scattering. Besides the elastic electrons, there is a broad distribution of inelastically scattered electrons as well. However, the elastic electrons can be separated by applying a retarding field.[21]

The trilayer is made in a separate chamber on a substrate consisting of a film of nitrocellulose supported by a Si wafer with a number of 0.5 mm wide apertures. The Au layer of 20 nm thickness is deposited on top of the nitrocellulose by evaporation of Au from a heated crucible. On top of this layer, polycrystalline films of Co or Fe are deposited by electron bombardment. Their thicknesses are measured by a calibrated

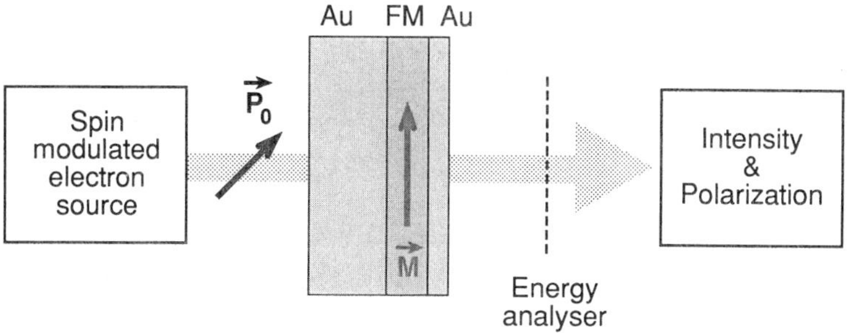

Figure 1. Schematics of the experiment. The experiment consists of a spin-modulated electron source of the GaAs-type with variable spin polarization direction, a self-supported metallic trilayer (Au/ferromagnet/Au), in which the ferromagnetic film (FM) is magnetized remanently in-plane, a retarding field energy analyser, and a detection system in which the intensity and the degree of spin polarization is measured for the electrons transmitted by the trilayer.

quartz microbalance. The ferromagnetic films are capped with a protecting Au layer of 2 nm thickness. The first set of hysteresis loops is measured right after deposition by *in-situ* magneto-optic Kerr effect. The in-plane hysteresis loops are square and exhibit full magnetic remanence. After the magnetic tests are completed, the whole sample is let to air. The nitrocellulose on the apertures is removed in a solution of pentyl acetate. The sample is then introduced through a load-lock system into the chamber with the GaAs electron source where the measurements are done. There, the sample is sputtered to thin the supporting Au layer until low energy electrons are transmitted at an attenuation of 10^{-5}–10^{-6}. The Kerr hysteresis loops taken later show no difference to the loops obtained just after deposition of the samples. In the electron transmission measurements, the ferromagnetic films are remanently magnetized in-plane by applying a magnetic field pulse.

Of crucial importance for the transmission experiments is the preparation of pinhole-free self-supported metal films. This is evident from Fig. 2 where the relative intensity transmitted through a Au/Co/Au trilayer is shown *vs.* the energy E of the incident electron beam. The attenuation increases by 3 orders of magnitude when E increases from 6 eV above E_F to 16 eV. If there is the tiniest hole, the main part of the elastic signal observed at the backside of the trilayer is caused by electrons that have passed through the hole. We suspect that this is the reason why in experiments by Drouhin *et al.*[22] an almost constant attenuation factor of ~ 1–2×10^{-5} has been found over the same energy range. The steep increase of the attenuation with increasing E in our data is in reasonable agreement with the energy dependence of the electron mean free path in Au.[23]

III. Results

III.1. Spin Filtering

In this paragraph we consider the situation where the spin polarization $\vec{P_0}$ of the incident electron beam is chosen parallel (or antiparallel) to the magnetization direction $\vec{M}$. By measuring the transmitted current for both $\vec{P_0}$ parallel (I^+) and antiparal-

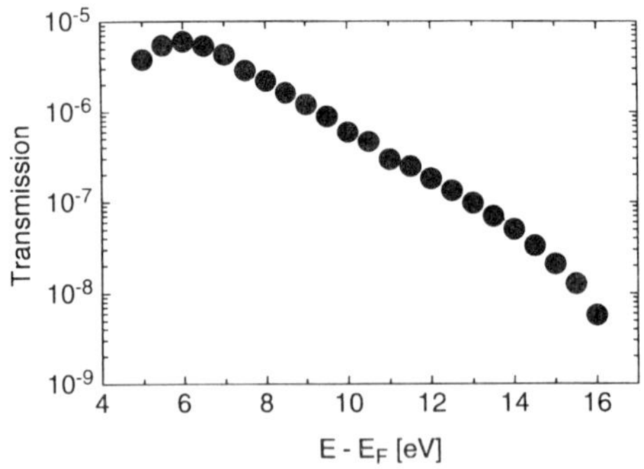

Figure 2. The attenuation of the elastic electrons after penetration of a Au/Co/Au trilayer *vs.* the energy above the Fermi energy E_F. The Co thickness is 2.5 nm.

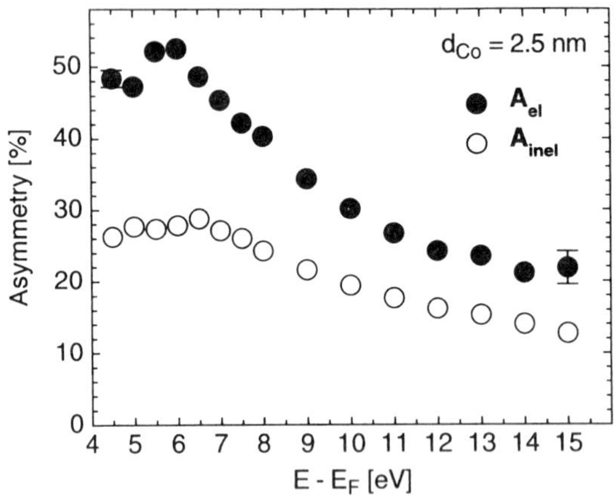

Figure 3. The transmission asymmetry for elastically (A_{el}) and inelastically scattered electrons (A_{inel}) as a function of the primary electron energy is shown. The Co thickness of the trilayer is 2.5 nm.

lel (I^-) to $\vec{M}$ at a given primary electron energy E the transmission asymmetry is defined by:

$$A(E) = \frac{1}{P_0}\frac{I^+ - I^-}{I^+ + I^-}.$$

Depending on the analyzing conditions, one can distinguish between the asymmetry A_{el} of the elastically scattered electrons and the asymmetry A_{inel} of the inelastically scattered electrons. A_{el} is measured when the retarding voltage selects only the elastically scattered electrons, while A_{inel} corresponds to the entire inelastic part of the energy distribution curve.

Figure 3 shows A_{el} and A_{inel} as a function of the primary electron energy. The sample has a Co thickness of 2.5 nm. Over the whole investigated energy range A_{el} is quite large. This finding is explained in terms of the density of states of Co. Because there are much more empty states available for minority-spin electrons to scatter into than for majority-spin electrons, minority-spin electrons have a shorter lifetime[14] and

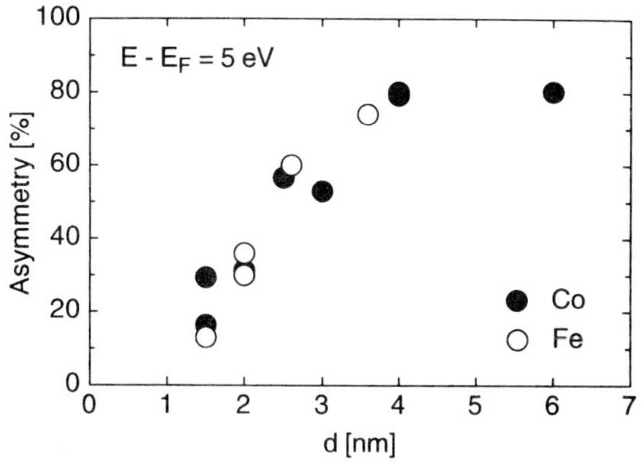

Figure 4. The transmission asymmetry A_{el} of elastically scattered electrons as a function of ferromagnetic film thickness for both Co and Fe. The primary electron energy is 5 eV above E_F.

hence a smaller inelastic mean free path. The decrease in A_{el} with increasing energy is due to the decreasing matrix element for spin-dependent scattering into the Co $3d$-shell.[21]

Furthermore, higher values for A_{el} than for A_{inel} have been observed, which is at variance with earlier investigations by Lassailly *et al.*[24] In order to explain this reduced asymmetry, one has to take into account that the inelastic part of the energy distribution curve does not only consist of inelastically scattered primary electrons, but also of secondary electrons. Though the secondary electrons are spin polarized—as they reflect the bulk magnetization—they have lost the memory of the primary spin polarization. Thus, they reduce the transmission asymmetry.

The thickness dependence of A_{el} for both Co and Fe is shown in Fig. 4. The strong thickness dependence proves that the observed asymmetry is mainly due to bulk scattering. An additional small interface effect, however, cannot be excluded. This is in contrast to the behavior of the conduction electrons at the Fermi level, where spin filtering and hence giant magnetoresistance is believed to be mainly produced by interface scattering.[25]

By measuring P of the transmitted electrons for an unpolarized primary beam, it can be tested how important exchange scattering is in spin-polarized transmission experiments. Exchange scattering removes a minority-spin electron in the primary beam and ejects a majority-spin electron. In the absence of such spin productive scattering events a polarizing spin filter must be equal to an analyzing spin filter, *i.e.* $P = A$. Experimentally, the comparison of P and A shows that the contribution of exchange scattering is below 5% and thus of minor importance in spin-dependent transmission.[21] Thus, a thin ferromagnetic film can be considered as a polarizer foil for spin-polarized electrons in analogy to a polarizer foil for light.

III.2. Spin Precession

As discussed in paragraph Sec. I.2 the interaction of the electron beam with the ferromagnetic film leads to spin-dependent amplitudes A_1, A_2 and a phase shift ϵ between the two basis spin functions. Using the definition of the transmission asymmetry A (see Sec. III.1) and of the intensities I^+ and I^- ($I^+ = A_1^2$, $I^- = A_2^2$) the amplitudes A_1

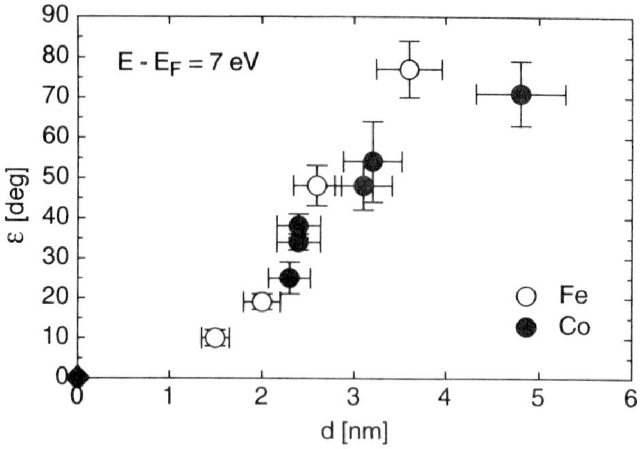

Figure 5. The precession angle ϵ as a function of the ferromagnetic film thickness for both Co and Fe, measured with elastic electrons of energy $(E - E_F) = 7$ eV. The point at zero thickness was measured with a pure Au film of 20 nm thickness.

and A_2 are given by $\sqrt{1+A}$ and $\sqrt{1-A}$, respectively. For a pure spin state, the angle ϕ is thus given by $\phi = \arctan(A/\sqrt{1-A^2})$, which is confirmed by experiments.[21] The angle ϵ, on the other hand, is determined by the energy difference $\Delta E_{\text{ex}} = E_2 - E_1$ between minority- and majority-spin states—the so-called exchange energy—and the time t the electrons spent within the ferromagnet: $\epsilon = \Delta E_{\text{ex}} t/\hbar$. Neglecting quantum resonance effects t is simply given by $t = d/v$, with d the thickness of the ferromagnetic film and v the group velocity of the electrons. The precession angle ϵ is thus given by $\epsilon = \Delta E_{\text{ex}} d/\hbar v$.

The angle ϵ can be estimated by assuming a free electron behaviour, which is reasonable for electrons in the energy range of interest. Then, the group velocity is $v = \sqrt{2E/m_e}$, with m_e the free electron mass and E the energy of the primary electron beam measured with respect to the inner potential of the ferromagnet. One finally obtains:

$$\epsilon = \sqrt{\frac{m_e}{2\hbar^2}} \frac{\Delta E_{\text{ex}}}{\sqrt{E}} d. \tag{4}$$

Assuming $\Delta E_{\text{ex}} = 0.3$ eV for the sp-bands,[26] one obtains $\epsilon \sim 10°$ per 1 nm ferromagnetic film thickness at energies close to the vacuum level.

In fact, the precession of the electron spin polarization could experimentally be found.[21,27] Figure 5 shows the experimental precession angle ϵ for different Co and Fe thicknesses at a primary electron energy $E - E_F = 7$ eV. In the case of Co a reasonable linear fit starting at zero ferromagnetic film thickness is possible, yielding a slope of ≈ 15 degrees per 1 nm of Co film thickness. In the case of Fe one might assume a linear dependence at higher coverages, yielding a slope of ≈ 20 degrees per 1 nm of Fe film thickness. However, too few data at coverages above 2 nm are available at the moment to really confirm the linear thickness dependence. Nevertheless, both sets of data exhibit an increase of ϵ with increasing ferromagnetic film thickness.

The energy dependence of ϵ for both Co and Fe is shown in Fig. 6. In the case of Co the variation of ϵ is weak and the decrease at higher energies can mainly be explained by the $1/\sqrt{E}$ factor in Eq. (4). Thus the exchange energy ΔE_{ex} is quite constant over the investigated energy range. On the other hand, Fe exhibits a much

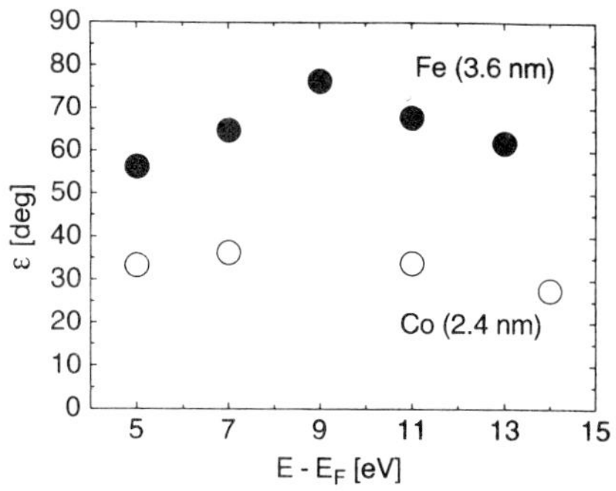

Figure 6. The precession angle ϵ as a function of the electron energy for both Co and Fe. The Co thickness is 2.4 nm, the Fe thickness 3.6 nm.

stronger energy dependence with a maximum at 9 eV above E_F. One is tempted to attribute this maximum of ϵ to a maximum of the exchange energy. However, we have reasons to believe that this non-monotonic behavior is rather the result of a varying group velocity v around 9 eV above E_F. In fact, band structure calculations of Fe,[28] reveal a flattening of bands and thus a decrease of the group velocity around 9 eV above E_F.

The observed precession of the spin polarization vector is in analogy to the Faraday rotation with light passing through a ferromagnetic material. However, the precession angle per unit length observed with electrons is two orders of magnitude larger as compared to the one observed with light. This difference in the strength of the "magneto-optic" phenomena arises because the electron spin couples directly to the magnetization whereas the coupling of the photons to the magnetization must be mediated by the spin-orbit interaction.

We note that the precession around $\vec{M}$ can also be viewed as the Larmor precession of the electron spin around a hypothetical magnetic field, the Weiss field.[29] Such a point of view is justified by the fact that the exchange interaction between the spins in a ferromagnet acts in a way as if there were a magnetic field acting on each spin. In fact, in order to produce the observed precession angles magnetic fields of a few 1000 Tesla are necessary. This is the same order of magnitude as the Weiss field within the classical molecular field theory.

A similar precession of the electron spin polarization has been proposed by Byrne and Farago[30] in electron-atom scattering and is caused by exchange scattering. Depending on the phase shift between the direct and the exchange scattering amplitude the precession angle can assume both negative and positive values and should strongly vary (including sign changes) with electron energy. This is in contrast to our observation of precession angles with constant sign, which are caused by the interference between the two spin wavefunctions of the electron. Moreover, a possible occurrence of such an exchange-induced contribution to the observed precession angle can be excluded. Since the analysing power (*i.e.* asymmetry for polarization measurement on a totally polarized electron beam) equals the polarising power (for the scattering of an unpolarized electron beam) of the self-supported ferromagnetic films (see Sec. III.1) the exchange scattering amplitude is zero and thus any exchange-induced precession of the spin polarization must vanish.

IV. Conclusion

We have shown that with a "complete" spin-polarized transmission experiment new insight into the physics of spin-dependent transport of hot electrons is gained. For $\vec{P}_0$ parallel to $\vec{M}$ the ferromagnetic film acts as a spin filter exhibiting a strong energy dependence of the transmission asymmetry. For $\vec{P}_0$ perpendicular to $\vec{M}$ two kinds of rotations are found. One rotation is caused by the different damping of the spin wave-functions (spin filtering). The precession around $\vec{M}$, on the other hand, is caused by different phase factors having their origin in the exchange energy of the ferromagnetic film. The spin precession represents the electron analogue to the Faraday rotation observed with polarized light.

References

1. M. N. Baibich, J. M. Broto, A. Fert, F. Nguyen van Dau, F. Petroff, P. Etienne, G. Creuzet, A. Friederich, and J. Chazelas, *Phys. Rev. Lett.* **61**, 2472 (1988).
2. S. S. P. Parkin, Z.G. Li and D. J. Smith, *Appl. Phys. Lett.* **58**, 2710 (1991).
3. T. Valet and A. Fert, *Phys. Rev. B* **48**, 7099 (1993).
4. See for example: N. F. Mott and H. Jones, *The Theory of the Properties of Metals and Alloys* (Clarendon Press, Oxford, 1936).
5. A. Bringer, M. Campagna, R. Feder, W. Gudat, E. Kisker, and E. Kuhlmann, *Phys. Rev. Lett.* **42**, 1705 (1979).
6. R. J. Celotta, D. T. Pierce, G. C. Wang, S. D. Bader, and G. P. Felcher, *Phys. Rev. Lett.* **43**, 728 (1979).
7. D. Venus and J. Kirschner, *Phys. Rev. B* **37**, 2199 (1988).
8. J. Unguris, D. T. Pierce, A. Galejs, and R. J. Celotta, *Phys. Rev. Lett.* **49**, 72 (1982).
9. E. Kisker, W. Gudat, and K. Schröder, *Solid State Commun.* **44**, 591 (1982).
10. H. Hopster, R. Raue, E. Kisker, G. Güntherodt, and M. Campagna, *Phys. Rev. Lett.* **50**, 70 (1983).
11. D. P. Pappas, K.-P. Kämper, B. P. Miller, H. Hopster, D. E. Fowler, C. R. Brundle, A. C. Luntz, and Z. X. Shen, *Phys. Rev. Lett.* **66**, 504 (1991).
12. M. Getzlaff, J. Bansmann, and G. Schönhense, *Solid State Commun.* **87**, 467 (1993).
13. E. Vescovo, C. Carbone, U. Alkemper, O. Rader, T. Kachel, W. Gudat, and W. Eberhardt, *Phys. Rev. B* **52**, 13497 (1995).
14. M. Aeschlimann, M. Bauer, S. Pawlik, W. Weber, R. Burgermeister, D. Oberli, and H. C. Siegmann, *Phys. Rev. Lett.* **79**, 5158 (1997).
15. J. Glazer and E. Tosatti, *Solid State Commun.* **52**, 905 (1984).
16. J. Kirschner, D. Rebenstorff, and H. Ibach, *Phys. Rev. Lett.* **53**, 698 (1984).
17. H. Hopster, R. Raue, and R. Clauberg, *Phys. Rev. Lett.* **53**, 695 (1984).
18. M. P. Gokhale and D. L. Mills, *Phys. Rev. Lett.* **66**, 2251 (1991).
19. G. Schönhense and H. C. Siegmann, *Ann. Physik* **2**, 465 (1993).
20. See for example: H. A. Tolhoek, *Rev. Mod. Phys.* **28**, 277 (1956).
21. D. Oberli, R. Burgermeister, S. Riesen, W. Weber, and H. C. Siegmann, *Phys. Rev. Lett.* **81**, 4228 (1998).
22. H.-J. Drouhin, G. Lampel, Y. Lassailly, A. J. van der Sluijs, and C. Marliere, *J. Mag. Mag. Mat.* **151**, 417 (1995).
23. O. Paul, Diss. ETH No. 9210 (1990).

24. Y. Lassailly, H.-J. Drouhin, A. J. van der Sluijs, G. Lampel, and C. Marliere, *Phys. Rev. B* **50**, 13054 (1994).
25. S. S. P. Parkin, *Phys. Rev. Lett.* **71**, 1641 (1993).
26. V. L. Moruzzi, J. F. Janak, and A. R. Williams, *Calculated Electronic Properties of Metals* (Pergamon, New York, 1978)
27. W. Weber, D. Oberli, S. Riesen, and H. C. Siegmann, *New J. Phys.* **1**, 1 (1999).
28. J. Callaway and C. S. Wang, *Phys. Rev. B* **16**, 2095 (1977).
29. P. Weiss, *J. Phys. Radium* **6**, 661 (1907).
30. J. Byrne and P.S. Farago, *J. Phys. B* **4**, 954 (1971).

Theory of Tunneling Magnetoresistance

J. Mathon and A. Umerski

Department of Mathematics,
City University
London EC1V 0HB
UNITED KINGDOM

Abstract

Rigorous theory of the tunneling magnetoresistance (TMR) based on the real-space Kubo formula and fully realistic tight-binding bands fitted to an *ab initio* band structure is decribed. It is applied to calculate the TMR of two Co electrodes separated by a vacuum gap. The calculated TMR ratio reaches $\approx 65\%$ in the tunneling regime but can be as high as 280% in the metallic regime when the vacuum gap is of the order of the Co interatomic distance (abrupt domain wall). It is also shown that the spin polarization P of the tunneling current is negative in the metallic regime but becomes positive $P \approx 35\%$ in the tunneling regime. Using the nonequilibrium Keldysh formalism, the Kubo formula is generalized to a Co junction under a finite bias. The TMR of the Co junction calculated from the Keldysh formula decreases very rapidly with applied bias in good qualitative agreement with experiment. It is also demonstrated that the TMR calculated from the Kubo formula remains nonzero when one of the Co electrodes is covered with a copper layer. It is shown that nonzero TMR is due to quantum well states in the Cu layer which do not participate in transport. Since these only occur in the down-spin channel, their loss from transport creates a spin asymmetry of electrons tunneling from a Cu interlayer, *i.e.* nonzero TMR. Finally, it is shown that diffuse scattering at the ferromagnet/nonmagnet interface may cause quantum well states to evolve into propagating states, in which case the spin asymmetry of the nonmagnetic layer is lost and with it the TMR.

Physics of Low Dimensional Systems
Edited by J. L. Morán-López, Kluwer Academic/Plenum Publishers, New York 2001

The conductance $\Gamma(H_s)$ of a tunnel junction with two ferromagnetic electrodes whose magnetic moments are aligned parallel in an applied saturating field H_s is much higher than its conductance $\Gamma(0)$ in zero field when the moments are antiparallel.[1–3] The effect is called tunneling magnetoresistance (TMR) and the relative change in the resistance of the junction, *i.e.*, the so called 'optimistic' magnetoresistance ratio

$$R_{\mathrm{TMR}} = \frac{\Gamma(0)^{-1} - \Gamma(H_s)^{-1}}{\Gamma(H_s)^{-1}}, \tag{1}$$

can be as high as 60%. The traditional explanation of the TMR effect is based on the assumption that electrons tunneling from a ferromagnet are spin-polarized and their polarization P is given in terms of the spin-dependent density of states D^σ of the ferromagnet by $P = [D^\uparrow(E_F) - D^\downarrow(E_F)]/[D^\uparrow(E_F) + D^\downarrow(E_F)]$. Since the classical theory of tunneling[4] states that the junction conductance is proportional to the product of the densities of states of the left and right electrodes, it is easy to show that the TMR ratio (1) can be written in terms of the spin polarizations P_L, P_R of the left and right electrodes

$$R_{\mathrm{TMR}} = \frac{2P_L P_R}{1 - P_L P_R}. \tag{2}$$

This is the well-known Julliere's formula.[5] Although the Julliere's formula is quite successful in predicting the TMR ratios from the observed values[4] of the spin polarization of electrons tunneling from Fe, Ni and Co into a superconductor, it suffers from several fundamental defects. First of all, it has been known for long time that the polarization of the tunneling current predicted from the total density of states (DOS) of the ferromagnetic electrodes has the wrong sign. The bulk DOS's for Fe, Co, and Ni are reproduced in Fig. 1. One would expect from Fig. 1 that the tunneling current from Fe, Co, and Ni should be dominated by down-spin (minority) electrons since their density of states at E_F is high. In fact, the observed P has just the opposite sign for all three metals.

The second problem is that the simple density of states argument used in the Julliere's formula fails to explain the observed rapid decrease of the TMR ratio with applied bias.

The third problem that has come to light only recently is that the Julliere's formula when applied to a tunneling junction with a thin nonmagnetic metallic interlayer, such as Cu, inserted between one of the ferromagnetic electrodes and the insulating barrier fails to explain the observed[6] nonzero TMR ratio. In fact, since the density of states of the Cu layer adjacent to the barrier is spin independent, $P_{\mathrm{Cu}} = 0$ and, therefore, it follows from Eq. (2) that $R_{\mathrm{TMR}} = 0$, which contradicts the experiment.[6]

The three problems we have identified call into question the validity of the whole classical theory of tunneling based on the density of states of the ferromagnetic electrodes. We shall examine the reasons for the failure of the density of states approach using the rigorous real-space Kubo formula.[7,8] The Kubo formula is exact in the linear response regime (low bias limit). To investigate the bias dependence of the TMR, we shall use the non-equilibrium Keldysh formalism[9] which is a rigorous extension of the Kubo formula beyond the linear-response regime.

To obtain clear-cut answers, we consider tunneling between cobalt electrodes across a vacuum gap. In the case of a junction with a metallic nonmagnetic interlayer, tunneling takes place from a cobalt electrode covered with an overlayer of N atomic planes of copper across a vacuum gap into another cobalt electrode. This is illustrated in Fig. 2(a). Initially, we assume that the electrodes are perfect so that

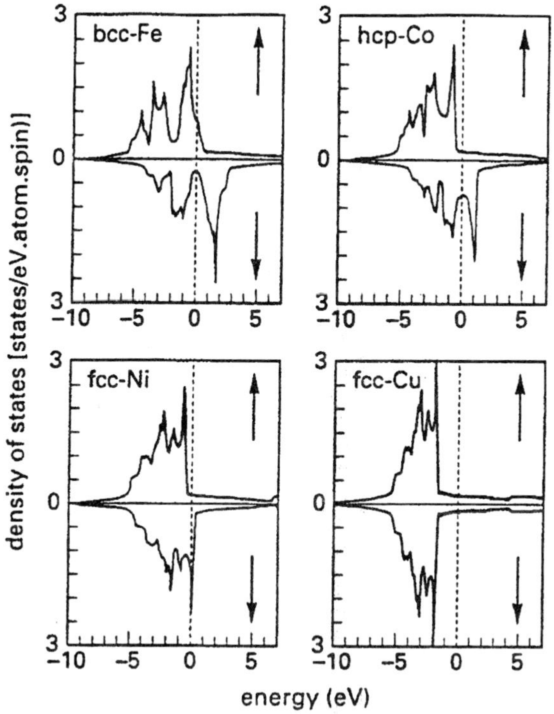

Figure 1. Densities of states of bulk Fe, Co, Ni and Cu.

the electron wave vector parallel to the layers $k_{\|}$ is conserved in tunneling. This restriction will be relaxed later. In contrast to the junction with an amorphous Al_2O_3 barrier used in most experiments, tunneling across a vacuum gap considered here has the great advantage that the real-space Kubo (Keldysh) formula[7–9] can be evaluated without any approximations for a fully realistic band structure of all the components of the junction. The results we obtain for vacuum tunneling are, therefore, exact.

We use a tight-binding parametrization of an *ab initio* band structure of *fcc* Co and Cu (for details, see Ref. 8). The total conductance of the junction Γ^σ in a spin channel σ is expressed[10] in terms of the one-electron Green's functions $G_L^\sigma(E_F, k_{\|})$, $G_R^\sigma(E_F, k_{\|})$ at the surface of the left and right electrodes

$$\Gamma^\sigma = \frac{4e^2}{h} \sum_{k_{\|}} \mathrm{Tr}[T_\sigma \mathrm{Im}\, G_R^\sigma(E_F, k_{\|})][T_\sigma^+ \mathrm{Im}\, G_L^\sigma(E_F, k_{\|})]. \quad (3)$$

The summation in Eq. (3) is over the two-dimensional Brillouin zone and the trace is over the orbital indices corresponding to s, p, d orbitals which are required in a tight-binding parametrization of Co and Cu. Since we use a multi-orbital band structure, G^σ and T_σ are matrices whose size depends on the number of orbitals. The matrix T_σ is given by

$$T_\sigma = t(k_{\|})[I - G_R^\sigma(E_F, k_{\|})t^+(k_{\|})G_L^\sigma(E_F, k_{\|})t(k_{\|})]^{-1}, \quad (4)$$

where I is a unit matrix in the orbital space and $t(k_{\|})$ is the matrix of tight-binding hopping integrals connecting across vacuum gap atomic orbitals in the surface of the right electrode to atomic orbitals in the surface of the left electrode.

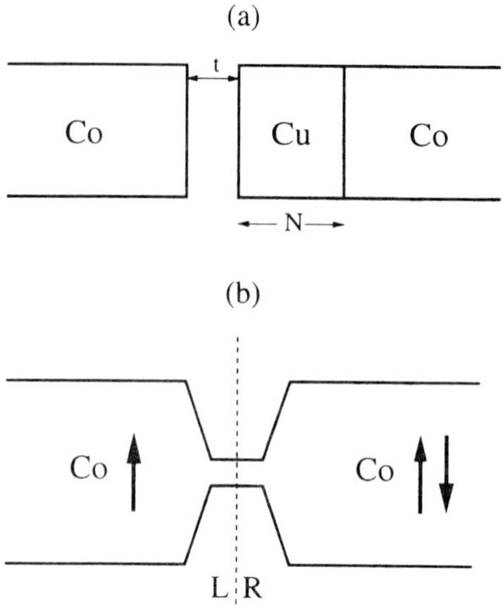

Figure 2. Schematic picture of a Co/vacuum/Cu/Co tunneling junction with hopping matrix t across the vacuum gap (a) and of a metallic contact in the pillar geometry (b). The arrows indicate the orientations of the magnetization.

The Kubo formula (3) has a simple physical interpretation. First of all, since we operate in the linear-response regime, the current is proportional to the conductance and, therefore, Eq. (3) gives effectively the tunneling current. The quantities Im $G^{\sigma}_{L(R)}(E_F, k_{\parallel})$ are (up to a factor$1/\pi$) one-dimensional densities of states of the left (right) electrodes in the channel $(k_{\parallel}, \sigma)$ and the matrix T_{σ} can be regarded as an effective tunneling matrix. Since we assume coherent tunneling (perfect electrodes) the current flows in independent $(k_{\parallel}, \sigma)$ channels which means that all the channels contribute additively to the total current and, hence, the sum over $k_{\parallel}$ in Eq. (3).

With this interpretation, the Kubo formula (3) resembles superficially the Julliere's formula. However, in contrast to the classical theory of tunneling, the Kubo formula (3) does not assume separation of the tunneling junction into two independent left and right electrodes. Although the Green's functions G^{σ}_{L} and G^{σ}_{R} are for disconnected left and right electrodes, the mutual interaction of the two electrodes is described exactly through the matrix T_{σ} defined by Eq. (4). It will be seen that this interaction is essential for correct treatment of the tunneling junction with a nonmagnetic metallic interlayer.[12]

Since the full interaction between the left and right electrodes is contained in Eq. (3), it applies not only to tunneling but also to metallic conduction. In the case of a metallic sample, one has to make sure that the resistance of the electrodes is much lower than the resistance of the sample (which is automatically satisfied for a tunneling junction). This can be achieved experimentally in the 'pillar' geometry shown schematically in Fig. 2(b). The 'sample' in Fig. 2(b) is the narrow region (pillar) made of the same material as the electrodes. It should be noted that calculations made in the slab geometry (Fig. 2a) apply directly to the pillar geometry (Fig. 2b) provided the constriction is adiabatic, *i.e.* the cross section of the structure decreases gradually rather than abruptly.

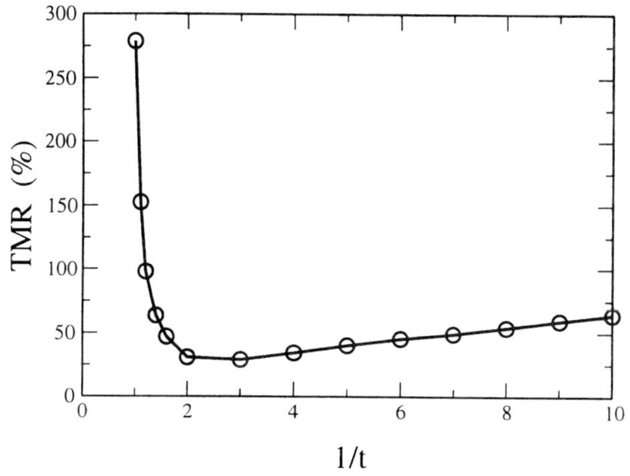

Figure 3. Dependence of the tunneling magnetoresistance of a Co(001) junction on the reciprocal of the reduced s-s electron hopping $1/t$ between the Co electrodes.

We first apply the Kubo formula (3) to calculate the magnetoresistance (MR) ratio for two cobalt electrodes in direct (metallic) contact (Fig. 2b). We then break the contact by introducing a vacuum gap between the left and right electrodes. This allows us to investigate how the magnetoresistance evolves from the ballistic current-perpendicular-to-plane magnetoresistance (CPP GMR) for a metallic system to tunneling magnetoresistance. In calculating the MR we assume that the magnetization in the left electrode, *i.e.* right up to the broken line in Fig. 2(b), points up and that in the right electrode can point either up or down. The MR in the unbroken (metallic) contact is, therefore, due to a completely abrupt domain wall (the orientation of the magnetization changes by 180° from one atomic plane to the next). Following Harrison,[11,10] we model tunneling across vacuum gap (broken contact) by turning off gradually the hopping matrix $t(k_{\|})$ across the gap. As discussed in Ref. 10, hopping between s, p, d orbitals scales differently with the distance between the electrodes (width of the vacuum gap). This has the consequence that tunneling between d orbitals is rapidly suppressed owing to their weak overlap across the gap. In the case of an Al_2O_3 barrier, suppression of d-type tunneling is due to the fact that there are no d orbitals present in the barrier. Given that only the s-s interaction survives in the tunneling regime, it is appropriate to use it as a measure of the width of the vacuum gap between the Co electrodes. It is, therefore, convenient to introduce a dimensionless reduced s-s hopping parameter $0 \leq t \leq 1$ by $t = t^{\text{gap}}_{ss\sigma}/t^{\text{bulk}}_{ss\sigma}$, where $t^{\text{bulk}}_{ss\sigma}$ is the bulk s-s hopping in Co and $t^{\text{gap}}_{ss\sigma}$ is the hopping acros the vacuum gap.

The dependence of the TMR ratio, determined from Eq. (3), on the reciprocal of the reduced hopping parameter $1/t$ is shown in Fig. 3. It can be seen that the TMR ratio drops very rapidly from its metallic value of 280% for the abrupt domain wall ($t = 1$) to about 40% and then remains almost constant in the tunneling regime reaching about 65% for $t = 0.1$. The rapid initial decrease of the MR ratio occurs because, in the metallic limit $t \approx 1$, a significant proportion of the current in Co is carried by d electrons that are highly spin-polarized. This explains a large MR ratio in the metallic regime (abrupt domain wall). In the tunneling regime, the current is carried only by s-p electrons which are weakly spin-polarized and, hence, the TMR ratio is much smaller.

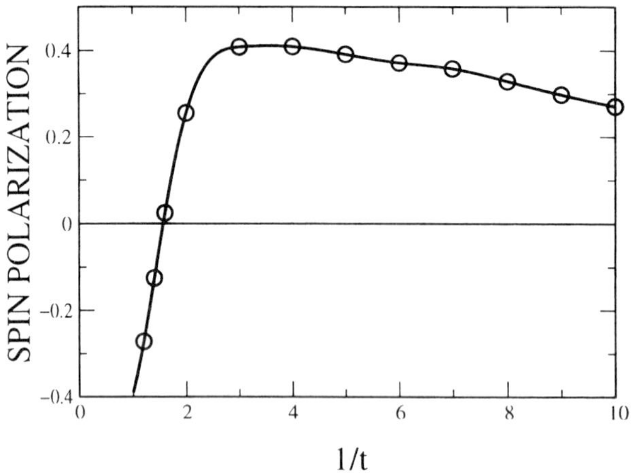

Figure 4. Dependence of the spin polarization of electrons tunneling between two cobalt electrodes across a vacuum gap on the reciprocal of the reduced *s-s* electron hopping $1/t$ between the Co electrodes (width of the gap).

The switching from *d*-type to *s-p*-type conduction, which occurs as one moves from the metallic to the tunneling regime, has also important implications for the sign of the polarization of the tunneling current which can be easily determined from Eq. (3). It should be noted that the correct definition of the spin polarization is not in terms of the densities of states, as assumed in the Julliere's formula, but in terms of the partial tunneling currents carried by $\uparrow$ and $\downarrow$ spin electrons. In the linear-response regime, $P_{L(R)}$ is, therefore, given by

$$P_{L(R)} = \frac{\Gamma^{\uparrow}_{L(R)} - \Gamma^{\downarrow}_{L(R)}}{\Gamma^{\uparrow}_{L(R)} + \Gamma^{\downarrow}_{L(R)}}, \tag{5}$$

where $\Gamma^{\sigma}_{L(R)}$ gives the current of electrons of spin σ tunneling from the left (right) ferromagnet through a barrier (vacuum gap) into a suitable detector of spin-polarized current. In practice, the detector is usually a superconducting aluminum electrode. We stress that the spin polarization of the tunneling current is not just a property of the ferromagnetic electrode (as is assumed in the Julliere's formula) but is instead the joint property of the electrode and the barrier.

The dependence of the spin polarization P of the Co junction on the width of the vacuum gap (reciprocal hopping $1/t$) obtained from Eqs. (3) and (5) is shown in Fig. 4. For a small vacuum gap of the order of the lattice constant ($1/t \approx 1$), the conductance is dominated by d electrons and P has the 'wrong' sign $P < 0$ consistent with the total DOS argument of the classical theory of tunneling.[4] However, there is a rapid crossover to $P > 0$ as the width of the gap increases. It can be seen from Fig. 4 that the calculated P for Co not only has the correct sign in the tunneling regime $1/t \gg 1$ but its magnitude 30–40% is in excellent agreement with the observed[4] $P \approx 35\%$. The crossover from negative to positive P occurs because the overlap of *d*-orbitals decreases with increasing gap much faster than that of *s*-orbitals and it is, therefore, *s* electrons that determine the conductance in the tunneling regime.

We now turn to the bias dependence of the TMR. Under a finite bias V the Kubo formula (3) is no longer applicable. However, in the case of tunneling across a vacuum

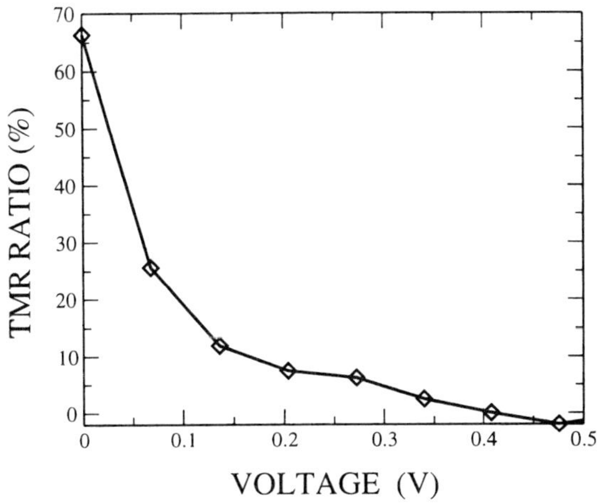

Figure 5. Dependence of the tunneling magnetoresistance of a cobalt junction with a vacuum gap on the applied bias.

gap the Kubo formula can be easily generalized using the nonequilibrium Keldysh formalism.[9] Following Ref. 9, we start with a Co junction in which hopping between the left and right electrodes $t(k_{\|})$ is set equal to zero. The effect of a bias V on the junction with disconnected electrodes is thus calculated in equilibrium. It is clear that it leads only to a rigid shift of the electronic band structure in one of the Co electrodes by eV. We then reconnect adiabatically the two electrodes by switching on the hopping matrix $t(k_{\|})$ and calculate the current between the two electrodes. The Keldysh formalism allows us to perform the calculations in all orders of $t(k_{\|})$. The tunneling current J^{σ} in a spin channel σ is now given by

$$J^{\sigma} = \frac{4e}{h} \sum_{k_{\|}} \mathrm{Tr} \int_{-\infty}^{\infty} [T_{\sigma} \mathrm{Im}\, G_R^{\sigma}(E-eV, k_{\|})][T_{\sigma}^{+} \mathrm{Im}\, G_L^{\sigma}(E, k_{\|})][f(E-eV)-f(E)]\, dE, \quad (6)$$

where f is the Fermi function. Comparing Eqs. (6) and (3), it is easy to see that Eq. (6) reduces in the limit $V \to 0$ to the Kubo formula (3). Equation (6) also has the form one expects from simple physical considerations. Tunneling current is made up from contributions of the states whose energies range from $E_F - eV$ to E_F, *i.e.* those states that are occupied in the right electrodes but empty in the left electrode. The probability of each transition at a fixed energy E is given by the Kubo formula (3).

The dependence of the TMR ratio on the applied bias V determined from Eq. (6) is shown in Fig. 5 for the Co junction with a vacuum gap $t = 0.1$ (hopping between s orbitals is reduced to 10% of the bulk Co value). The calculated TMR ratio decreases very rapidly with increasing bias in good qualitative agreement with experiment (see *e.g.*, Ref. 13). We stress that the rapid decrease of the TMR is due simply to the bias-induced relative shift of the electronic structures of the left and right Co electrodes. This is in contrast to the results obtained using the classical DOS arguments (see, *e.g.*, Ref. 13).

Finally, we shall use the Kubo formula (3) to demonstrate that TMR of a Co junction with a vacuum gap remains nonzero when one of the Co electrodes is covered with a Cu layer (Fig. 2a). The dependence of the TMR ratio obtained by numerical evaluation of the Kubo formula (3) on the thickness of the Cu overlayer is shown in

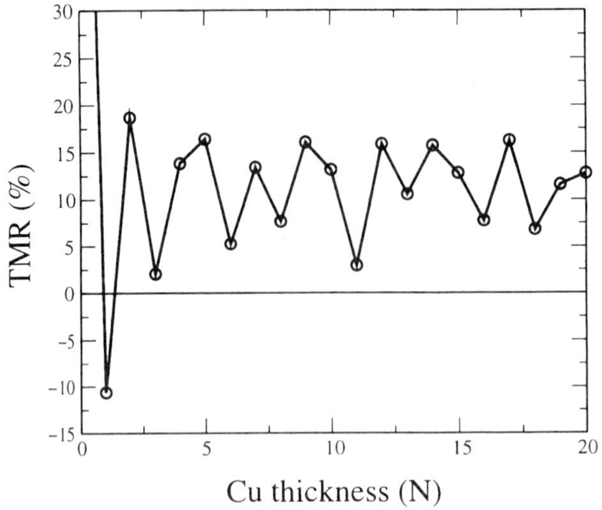

Figure 6. Dependence of the tunneling magnetoresistance of a cobalt junction with a copper interlayer on the thickness of the copper layer.

Fig. 6. The calculation is for (111) orientation of the layers and vacuum gap characterized by reduced hopping $t = 0.1$. In contrast to the Julliere's formula (2), the TMR determined from the Kubo formula (3) is nonzero and oscillates as a function of Cu thickness due to quantum interference of electrons on the Cu interlayer. It is interesting that for a small Cu thickness (two monolayers) the TMR ratio becomes negative. A negative TMR with a very thin gold interlayer has been observed by Moodera.[14]

The physical explanation of a nonzero TMR is that the Cu layer acts as a spin filter. Since the Fermi surfaces of Cu and of the majority-spin electrons in Co are very similar (the Co majority d-band lies below E_F), majority-spin electrons cross easily the Co/Cu interface and participate in tunneling as if there were no intervening Cu layer. On the other hand, there is a poor match between the Cu bands and the minority-spin bands in Co, which results in formation of down-spin quantum well states in the Cu overlayer.[15,16] Since the quantum well states are localized in the Cu layer they do not contribute to transport of charge in the down-spin channel, which gives rise to a spin asymmetry (nonzero polarization P) of the tunneling current and, hence, nonzero TMR.

The apparent paradox that the Julliere's formula predicts zero TMR but the Kubo formula gives a nonzero TMR can now be easily resolved. Since the down-spin quantum well states in the Cu layer contribute to the ordinary density of states (DOS) they are, incorrectly, counted in the Julliere's formula (2) as contributing to the tunneling current. The total DOS of down-spin electrons, which is made up of propagating and quantum well states, is equal to the DOS of up-spin states which are all propagating. There is, therefore, no spin asymmetry in the DOS of the Cu overlayer and, hence, the Julliere's formula gives zero TMR. On the other hand, the Kubo formula excludes automatically all the quantum well states. Since these only occur in the down-spin channel, their loss from transport creates a spin asymmetry of electrons tunneling from a Cu overlayer, *i.e.*, nonzero TMR.

It is clear that for a nonzero TMR effect to occur, one needs a strong scattering at the ferromagnet/nonmagnet interface in one of the spin channels and weak scattering

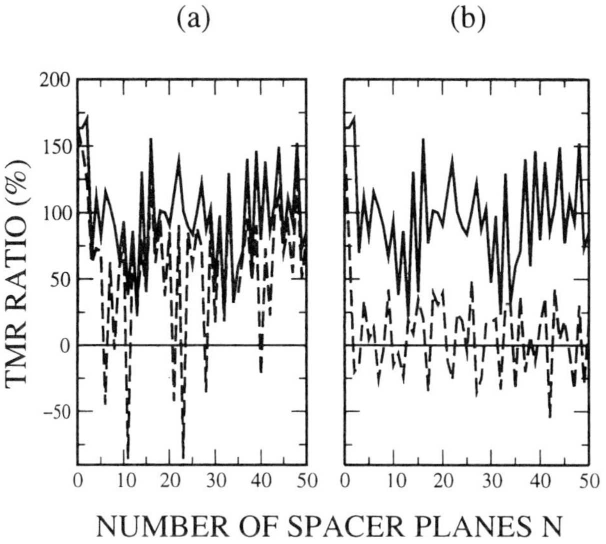

Figure 7. Dependence of the tunneling magnetoresistance on the nonmagnetic interlayer thickness for a junction without (continuous curve) and with interfacial roughness (broken curve): (a) 10% of intermixing at the magnet/nonmagnet interface; (b) 50% of intermixing at the magnet/nonmagnet interface.

in the other spin channel. These are the same conditions as those required for a large GMR in the corresponding ferromagnet/nonmagnet multilayer. It is, therefore, clear that Co/Cu is a particularly good combination but, for example, an Al interlayer should not lead to any sizable TMR since GMR for an Al spacer is very small. This is in agreement with the observation (see, *e.g.*, Ref. 13) that an Al interlayer kills the TMR very effectively.

To observe a nonzero TMR, quantum-well states in one of the spin channels need to be well defined. This is the case when the effect of impurities in the nonmagnetic interlayer is negligible (ballistic transport across the interlayer) and the scattering at the ferromagnet/nonmagnet interface is specular. Scattering from impurities or/and diffuse scattering at the ferromagnet/nonmagnet interface may allow quantum well states to evolve into propagating states, in which case the spin asymmetry of electrons tunneling from the nonmagnetic interlayer may be lost (and with it the TMR effect). The fact that the calculated TMR shown in Fig. 6 is nondecaying as a function of Cu thickness is due to our neglect of impurity/interfacial scattering.

To investigate qualitatively the effect of diffuse scattering at the ferromagnet/nonmagnet interface, we have used a one-band model of the tunneling junction with random intermixing of magnetic and nonmagnetic atoms in one atomic plane of the interface. Since $k_{\|}$ is no longer conserved we 'grew' the whole tunneling junction in real space atom by atom. This method allows us to evaluate the Kubo formula without any approximations for a realistic model of disorder. The price to pay is a simplified band structure—one band model—and a junction with a relatively small cross section (we used 49 atoms in each atomic plane). Figure 7(a) shows the typical TMR ratio as a function of the thickness of a nonmagnetic metallic interlayer for a junction with 10% of intermixing at the magnet/nonmagnet interface. The effect of 50% intermixing is illustrated in Fig. 7(b). In both cases, the calculated TMR was averaged over 50 random configurations of nonmagnetic atoms at the interface. It can be seen that a small amount of intermixing (10%) does not destroy the TMR. When intermixing reaches about 50% all quantum well states are scattered effectively into propagating states and the average TMR vanishes. Rather surprising, large oscillations about zero of the TMR as a function of the nonmagnetic spacer thickness are not washed out by the disorder.

Acknowledgments

It is a pleasure to dedicate this paper to Hans Siegmann on the occasion of his 65th birthday.

References

1. T. Miyazaki and N. Tezuka, *J. Magn. Magn. Mater.* **139**, L231 (1995).
2. J. S. Moodera, L. R. Kinder, T. M. Wong, and R. Mersevey, *Phys. Rev. Lett.* **74**, 3273 (1995).
3. Y. Li, X. W. Li, G. Xiao, R. A. Altman, W. J. Gallagher, A. Marley, K. Roche, and S. S. P. Parkin, *J. Appl. Phys.* **83**, 6515 (1998).
4. R. Mersevey and P. M. Tedrow, *Phys. Rep.* **238**, 173 (1994).
5. M. Julliere, *Phys. Lett. A* **54**, 225 (1975).
6. S. S. P. Parkin, (unpublished).
7. P. A. Lee and D. S. Fisher, *Phys. Rev. Lett.* **47**, 882 (1981).
8. J. Mathon, A. Umerski, and M. A. Villeret, *Phys. Rev. B* **55**, 14378 (1997).
9. C. Caroli, R. Combescot, P. Nozieres, and D. Saint-James, *J. Phys. C* **4**, 916 (1971).
10. J. Mathon, *Phys. Rev. B* **56**, 11810 (1997).
11. W. A. Harrison, *Solid State Theory* (Dover Publications, New York, 1979).
12. J. Mathon and A. Umerski, *Phys. Rev. B* **60**, 2156 (1999).
13. J. S. Moodera, J. Nassar, and J. Mathon, *Ann. Rev. Matter. Sci.* **29**, 381 (1999).
14. J. S. Moodera, J. Nowak, L. R. Kinder, and P. M. Tedrow, *Phys. Rev. Lett.* **83**, 3029 (1999).
15. J. Mathon, M. A. Villeret, A. Umerski, R. B. Muniz, J. d'Albuquerque e Castro, and D. M. Edwards, *Phys. Rev. B* **56**, 11797 (1997).
16. J. E. Ortega, F. J. Himpsel, G. J. Mankey, and R. F. Willis, *Phys. Rev. B* **47**, 1540 (1993); P. Segovia, E. G. Michel, and J. E. Ortega, *Phys. Rev. Lett.* **77**, 3455 (1996).

Spin Polarized Electron Transport and Emission from Strained Semiconductor Heterostructures

Yu. A. Mamaev,[1] A.V. Subashievf,[1] Yu.P. Yashin,[1]
A.N. Ambrazhei,[1] H.-J. Drouhin,[2] G. Lampel,[2] J.E. Clendenin,[3]
T. Maruyama,[3] and G. Mulhollan[3]

[1]*State Technical University*
195251, St.-Petersburg,
RUSSIA
e-mail: mamaev@spes.stu.neva.ru

[2]*Laboratory PMC, UMR 7643-CNRS*
Ecole Polytechnique
91128, Palaiseau
FRANCE

[3]*Stanford Linear Accelerator*
Stanford, 94309 CA
USA

Abstract

High resolution energy distribution curves (EDC) and a polarization *versus* energy distribution curves (PEDC) of the electrons, photoemitted from strained GaAs/GaAsP layers are experimentally studied. In the vicinity of the photoexcitation threshold the polarization does not vary across the energy distribution, which means that no depolarization occurs during energy relaxation in the band bending region (BBR). The electron energy distribution is interpreted in terms of the electron energy relaxation in the band tail states of quantum well formed by the BBR. Polarized electron emission from a series of new strained short-period AlInGaAs/AlGaAs superlattices (SL) is investigated as well. The In layer content was chosen to give minimal conduction-band offset with large strain splitting of the V-band. Simultaneous changing of Al content in both SL layers provides variation of the structure band gap. We demonstrate as well that tuning of the SL to the excitation energy can be achieved without loss of the electron polarization. The polarization of up to 84% was measured at room temperature.

Physics of Low Dimensional Systems
Edited by J. L. Morán-López, Kluwer Academic/Plenum Publishers, New York 2001

I. Introduction

The sources of highly polarized electron beams are actively investigated in a view of their successful and growing applications in high energy physics, atomic physics, studies of thin film, and surface magnetism.[1] In order to produce highly polarized electrons strained GaAs layers on GaAsP pseudo-substrates are most frequently used. The strain-induced valence-band splitting leads to very high (close to 100%) initial electronic optical orientation under excitation by circularly polarized light at the inter-band absorption edge. The emission of polarized electrons in vacuum is provided by activation of the clean GaAs surface by Cs(O) deposition which drastically reduces the electron work function and leads a negative electron affinity (NEA) surface formation.

II. Near-Threshold Emission from Strained GaAs/GaAsP Layer

The polarized electron photoemission from the stressed film is understood as a multi-step process, consisting of: *i*) electron excitation under optical pumping, *ii*) electron energy relaxation to the local equilibrium state, *iii*) capture in the band bending region (BBR), then *iv*) energy relaxation in the potential well formed by the band-bending region and, finally *v*) electron escape into vacuum throughout the weakly transparent surface barrier formed by the activation layer.

The minimum energy of the EDC's are determined by the vacuum level achieved during the activation procedure and the aging processes, whereas the maximum energy corresponds to the ballistic emission of electrons photoexcited from the uppermost valence band level to the conduction band. The precise shapes of the EDC's are determined by the electron kinetics, the details of which are still not completely understood, specially steps (*iii*–*v*). For bulk unstrained GaAs, activated to NEA, the EDC's peak at an energy a few hundreds of millivolt below the bottom of the conduction band is observed revealing the effectiveness of energy relaxation in the BBR.[2–5] For excitation energies *close* to the band gap the polarization of the emitted electrons is essentially constant throughout this distribution and its value decreases with increasing $h\nu$. For *higher* excitation energies the polarization tends to increase at large kinetic energies where electrons are emitted without any energy or spin relaxation.[2]

Additional information is obtained in thin-layer cathodes[6] when the electron emission times are comparable to the energy relaxation times either in the active layer or on the BBR so that a large fraction of electrons are emitted prior to complete thermalisation. The PEDC's reflect the spin relaxation kinetics.

In Ref. 7 the first experimental results for energy and polarization distribution curves of electrons photoemitted from a highly strained GaAs layer were reported. We show that the results for near band-gap excitation are in line with the model of highly localised electron states in the BBR.

II.1. Experimental Results

The experimental set-up was described in Ref. 2. The sample is illuminated by σ^+ or σ^- circularly polarized light from a Ti:Sapphire or He-Ne laser normal to its surface. The photoemitted electrons are energy selected by a cylindrical 90° electrostatic deflector operating in the constant-energy mode. The full width at half maximum of the transmission function is $\Delta E = 20$ meV. The polarization of the energy-selected electrons is measured by Mott scattering on a gold foil at the voltage 30 or 100 kV. NEA

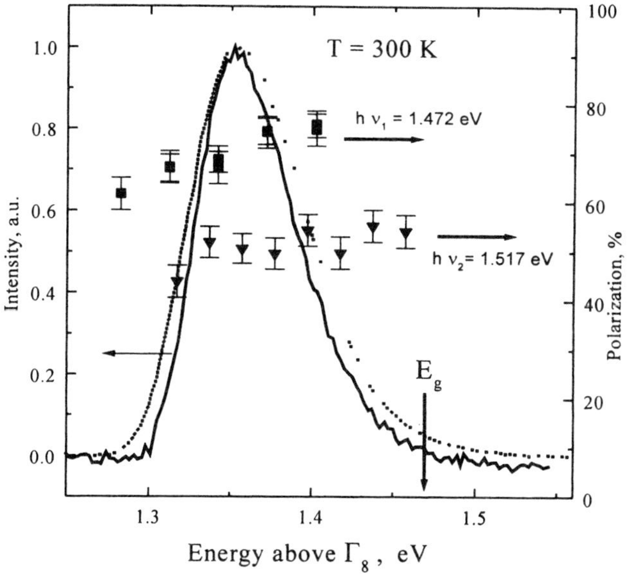

Figure 1. Polarized electron distributions and electron energy distributions of the emitted electrons for GaAs/GaAsP strained cathode. EDCs (solid line, $h\nu = 1.472$ eV; dots, $h\nu = 1.517$ eV) are normalized to the maximum value. $T = 300$ K, energy resolution is 20 meV. The band gap value of the strained overlayer is shown by arrow.

state was achieved by activation of atomically clean surfaces with cesium and oxygen. The thermal cleaning procedure (at residual pressure not exceeding 1×10^{-9} Torr) consisted of gradual heating of the samples up to 400°C and 30 min exposition at this temperature, then heating up to 500°C and 30 min exposition and finally the heating up to 620°C with 1 hour exposition at this temperature. The value of a sample temperature was controlled by a thermocouple. The structure of the sample under investigation was described in Ref. 8. A 140 nm thick GaAs overlayer is MOCVD grown on a $GaAs_{0.72}P_{0.28}$ buffer at the top of commercial GaAs (001) wafer. To reduce the overlayer relaxation effects, a special type of heterostructure with the intermediate $GaAs_{0.55}P_{0.45}/GaAs_{0.85}P_{0.15}$ superlattice has been used. As a result a strain-induced energy splitting of $\delta \approx 40$ meV in the valence band has been achieved so that high values of the photoelectrons spin polarization of $\geq 80\%$ at room temperature were reached even after several re-activation cycles.

The EDC and PEDC data at room temperature for *near band-edge excitation* are presented in Fig. 1. The position of the EDC peak is shifted below the conduction band edge as in the most of unstressed GaAs cathodes, though, EDC are rather narrow (FWHM does not exceed 100 meV) at room temperature and do not change much at 120 K. The shape of the EDC peak remains almost unchanged in the excitation range in the vicinity of the photothreshold $E_g - 0.02\,\text{eV} \leq h\nu \leq E_g + 0.07$ eV, where $E_g = 1.47$ eV. The observed shapes of the EDC are typical for the cathodes with two-stage surface activation procedure.[4,5] It is clearly seen that the polarization remains constant across EDC, so that no depolarization effects for the electrons in BBR region are registered. As a result the integrated values of the electron polarization for the $P(h\nu)$ spectrum and the P values measured at the EDC maximum (both at 20, and 80 meV resolutions) are found to be about equal at given values of $h\nu$.

II.2. Discussion

In the case of near band-edge excitation, $h\nu - E_g \leq kT$, as it is evidenced by Fig. 1, the electrons are captured in BBR prior to the emission. Monte-Carlo modelling of the spatial distribution of the electron potential in BBR[9] reveals large fluctuations of the electronic potential at the surface, originated from randomly distributed ionised acceptors and Cs-originated donor centres. This implies that all the electronic states in the BBR below a certain energy defined as an electron mobility edge (ME) are localised also in the surface plane by the potential fluctuations. The density of the localised states $g(\varepsilon)$ below the ME is a rapidly decreasing function of the localisation energy ε in the bang gap (measured downwards from the ME). The estimated and measured time of the electron emission from the BBR in vacuum $\tau_{\rm emi} \approx 5$ ps is much larger that the time of the delocalised electron energy relaxation due to the emission of the phonons $\tau_{\rm ph} = 1.5 \times 10^{-13}$ s. Therefore the experimental results can be analysed in terms of a model that consider the EDC formation as a result of competition of the processes of the electron emission in vacuum and the electron hopping in tail states with emission of the optical phonons. The model was developed in Ref. 10 for interpretation of the luminescence spectra in mixed crystals and generalized in Ref. 11 for electron emission from BBR.

In this model $g(\varepsilon)$ is approximated by an exponential law with an energy width γ. Therefore, below the ME, the probability for an electron to emit a phonon is proportional to $\exp(-\varepsilon/\gamma)$ and the electron energy relaxation is rapidly suppressed because of the lack of resonant localised final states in the vicinity of the occupied one. To be more specific, we assume that the main energy relaxation process is by phonon emission with an average energy loss δ much smaller than γ, and with a probability $\tau^{-1}(\varepsilon) = \tau^{-1}(0)\exp(-\varepsilon/\gamma)$. It is then possible to write the energy relaxation for the electronic surface density $n(\varepsilon)$ as a Fokker-Planck type equation:

$$\frac{\partial n}{\partial t} = -\delta \frac{\partial}{\partial \varepsilon}\left[\frac{n(\varepsilon)}{\tau(\varepsilon)}\right] - \frac{n(\varepsilon)}{\tau_{\rm emi}}. \qquad (1)$$

In this equation the first term describes the flow of electrons through the states with energy ε: $n(\varepsilon-\delta)/\tau(\varepsilon-\delta) - n(\varepsilon)/\tau(\varepsilon)$ and the second one is just the emission current $J_{\rm emi}(\varepsilon)$. In stationary conditions $\partial n/\partial t = 0$ and one obtains the energy dependence of the emission current in the BBR

$$J_{\rm emi}(\varepsilon) = J_{\rm emi}(0)\exp\{\varepsilon/\gamma - \alpha[\exp(\varepsilon/\gamma) - 1]\}, \qquad (2)$$

where the parameter α is given by $\alpha = (\gamma/\delta)[\tau(0)/\tau_{\rm emi}]$.

The results of the calculations of the EDC together with the experimental curve are shown in Fig. 2. We use fitting of the Eq. (2) dependence to experimental data to estimate the energy relaxation rate in BBR by the phonon emission. First, we note that the value of $\gamma = 28.7$ meV obtained from the fitting is found to be not dependent on the choice of α, whereas the value of α depends on the assumed position of the ME level in BBR below the conduction band. In 2-d system the low-energy shift of ME should be close to γ value. So, we take the position of zero in Eq. (2) at $E_c - \gamma$ (conduction band edge E_c being at 1.47 eV in the doped strained layer) which results in $\alpha = 0.11$.

Then, for the localised states the phonon emission rate at ME $1/\tau(0)$ is equal to the probability of a single jump multiplied by the number of final states, that is, $1/\tau(0) = (1/\tau_{\rm ph})(\delta g_0 \pi a^2)$, where g_0 is the surface electron density of states at ME,

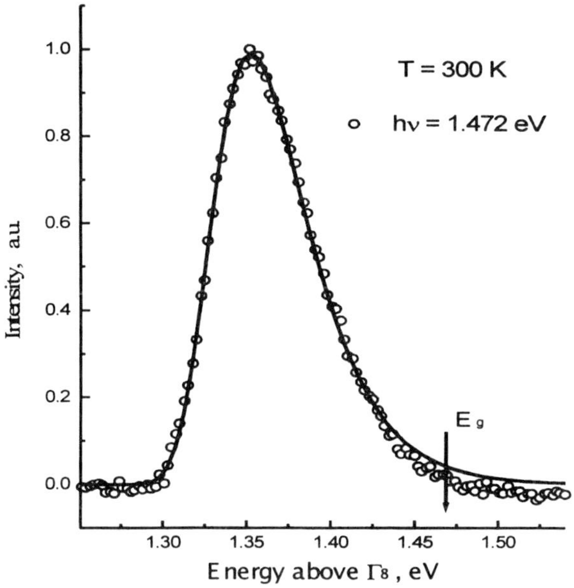

Figure 2. Experimental electron energy distribution curve at $T = 300$ K, together with the results of the calculations (solid line) using Eq. (2) of the localised states model with $\gamma = 28.7$ meV and $\alpha = 0.11$. The band gap value is shown by arrow.

a is an average radius of localised states, so that $g_0 \pi a^2 \gamma \approx 1$. That gives $1/\tau(0) = (1/\tau_{\rm ph})(\delta/\gamma)$ and $\alpha = (\gamma/\delta)^2(\tau_{\rm ph}/\tau_{\rm emi})$. Taking $\delta/\gamma = 0.3$ we get $\tau_{\rm ph} \approx 50$ fs, which is in reasonable agreement with the data for the energy relaxation rate for hot electrons in a narrow p-doped GaAs quantum well.[12]

Finally, note, that the main spin-relaxation mechanism of D'yakonov-Perel (DP) is suppressed for the localised states due to the effective averaging of the odd in k-vector terms in DP Hamiltonian at all directions, so that $\langle k_i k_j k_l \rangle = 0$. Besides, the weak overlap of the electron localised states with hole states out of BBR makes Bir-Aronow-Pikus relaxation due to electron scattering on the holes ineffective. Thus observed depolarization switching off below the conduction band energy is in line with the assumption of localisation of electronic states in this energy region.

III. Strained AlInGaAs/AlGaAs Superlattices with a Minimal Conduction-Band Offset

It has been found that there is a limit to current density that can be extracted from a GaAs film,[13] which results from the photovoltage effects in the band-bending region near the activated surface. A new generation of highly polarized electron sources is associated with semiconductor superlattice structures in which the valence band splitting is achieved as a consequence of hole confinement in the SL quantum wells (QW). The main advantage of SL-based photoemitters is the possibility to vary the properties of the active layer over a wide range by the appropriate choice of layer composition, thickness, and doping profile.[1,14]

Table I

$Al_xIn_yGa_{1-x-y}As/Al_zGa_{1-z}As$ strained superlattice samples.

\Sample	SL1	SL2	SL3	SL4
	(3–336)	(3–657)	(3–895)	(3–896)
Be doping Surface	7×10^{18}	2×10^{19}	5×10^{19}	5×10^{19}
Be doping Inside	4×10^{17}	5×10^{17}	5×10^{17}	5×10^{17}
d, μm	0.136	0.12	0.96	0.96
x, %	20	30	30	32
y, %	18	18	18	18
z, %	0	10	14	17
E_{ex}(pol. max), eV	1.45	1.54	1.58	1.6
max P, %	82.7	83	84	81
Y(pol. max), %	0.094		0.1	0.09

Recently a new strained short-period $Al_xIn_yGa_{1-x-y}As/Al_zGa_{1-z}As$ superlattice with a minimal conduction-band offset was proposed.[1] The main advantage of the SL results from the band line-up between the semiconductor layers of the SL. The Al content determines the formation of a barrier in the conduction band, while adding In leads to conduction band lowering, so the conduction band offset can be completely compensated by appropriate choice of x and y, while barriers for the holes remain uncompensated. As a result a high vertical electron mobility and simultaneously a small spin relaxation rate is achieved while also a large enough valence-band splitting is remained. Additional improvement of the emitter parameters is expected for the SLs with wider band gap. Wider gap ensures higher NEA and therefore higher quantum yield and emitted currents. We show that the performance of this new superlattice exceeds that of the GaAs strained layer cathodes while tuning of the band gap gives additional advantages.

The SL samples were grown by solid-source molecular beam epitaxy on GaAs(100)-oriented substrates. The SL samples consisted of 12–15 pairs of AlGaAs (4 nm) and AlInGaAs (4 nm) doped with Be and were terminated by a 6 nm (or 8 nm) heavily-doped GaAs layer capped with As to ensure stable activation and an NEA surface state. The parameters of the samples are listed in Table I. All the samples were capped by As for the surface protection. The As cap thickness was estimated to be 0.1 mm based on Auger profiling measurements. The characterisation of the samples was done using luminescence and X-ray diffraction techniques. The X-ray diffraction patterns show that in the case of the GaAs substrate and $x \leq 0.2$, the strain relaxation in the thin 4-nm $Al_xIn_yGa_{1-x-y}As$ layers remains negligible for the total SL thickness, d, up to $d \leq 150$ nm.

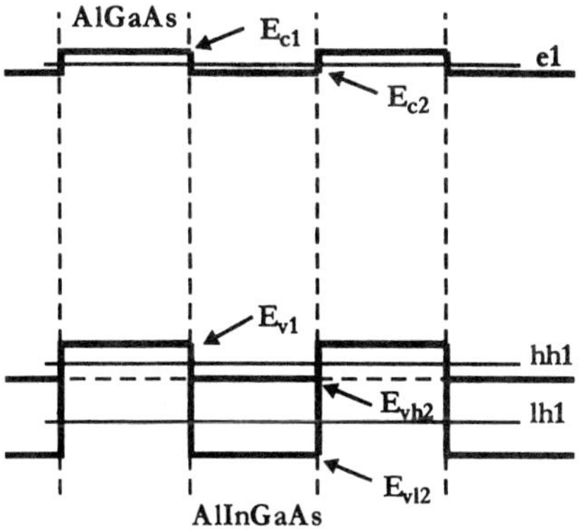

Figure 3. Energy band diagram of $Al_xIn_yGa_{1-x-y}As$/AlGaAs superlattice. The minibands (thin lines) are identified by notation e1 and hh1, lh1 for electrons and holes, respectively.

III.1. Choice of the SL Layer Composition

The miniband spectrum of the SL is determined by the band offsets at the heterointerfaces. In the case of a heterointerface with lattice matched ternary solid solution (*e.g.* $Al_xGa_{1-x}As$-$Al_zGa_{1-z}As$) the conduction-band offset ratio, Q_{c1}, defined as $Q_{c1} = \Delta E_c(x-z)/E_{g1}(x-z)$ [where $E_{g1}(x)$ is the difference in the band gaps of the contacting crystals] is known to remain constant, $Q_{c1} \cong 0.66$. For an $In_yGa_{1-y}As$-GaAs interface the offset is modified by the strain distribution in the contacting layers. For the structure with a thin $In_yGa_{1-y}As$ layer grown on a thick GaAs substrate all the strain is assumed to accumulate in the InGaAs layer. For the case of the $Al_xIn_yGa_{1-x-y}As/Al_zGa_{1-z}As$ SL, a linear interpolation between the values for $Al_xGa_{1-x}As$-GaAs and $In_yGa_{1-y}As$-GaAs interfaces should be valid for small x and y. The schematic of the position of the band edges for $x = 0.30$, $y = 0.18$, $z = 0.1$ is shown in Fig. 3.

We have found that for given band offset parameters the conduction band offset appears to be minimised for $z = 1.1(x - 1.1y)$. For the thermalised electrons at room temperature the influence of the resulting periodical potential should be negligible. Besides, as a result of the conduction-band line up, the 4-nm barriers for the electrons in the SL are transparent. Thus the changes of electron mobility and spin relaxation rate should be small compared to pure GaAs. Using the structures based on quaternary $Al_xIn_yGa_{1-x-y}As$ alloy one can change the band gap by varying the Al content in the layers while In concentration remains unchanged to keep high deformation and strain-induced valence band splitting. It is seen from Fig. 3 that the strain of the $Al_xIn_yGa_{1-x-y}As$ layers produces barriers for both heavy and light holes, the barrier for the light holes being 75 meV higher, which leads to additional hole-miniband splitting favourable for the electron optical orientation. The choice of the layer thickness is dictated by the need to split the hole minibands. The splitting grows when barriers are broad enough and wells are narrow and deep.

III.2. Experimental Results and Discussion

The Mott analysers both at St.-Petersburg Technical University (SPTU) and at the Stanford Linear Accelerator Centre (SLAC) were used to measure the spin polariza-

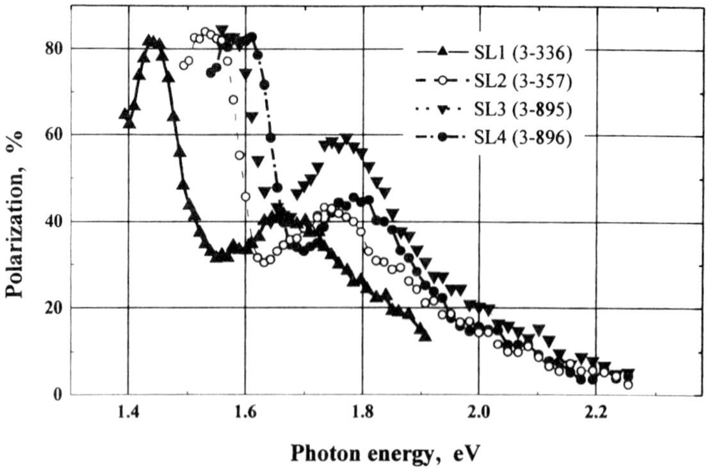

Figure 4. Electron spin polarization spectra for $Al_xIn_yGa_{1-x-y}As/Al_zGa_{1-z}As$ superlattices.

tion of photoelectrons. In Fig. 4 the polarized emission data are shown as a function of the optical excitation energy. The maximum polarization obtained with local optical excitation was 84% and the corresponding quantum yield, Y, was 0.1%. The observed emission spectra can be interpreted in terms of a three-step model linearised for the thin-film emitter case.[15] The polarization dependence on the excitation energy near the excitation edge comes from the initial electron polarization upon excitation by circularly polarized light. The decrease of the polarization from its maximum value with decreasing excitation wavelength starts with electron excitation from the first light-hole miniband. The fall off of polarization as excitation drops below the conduction band minimum can be associated with electron excitation in indirect transitions with electron scattering on the defects or with absorption of the optical phonon in which electrons with both orientations of electron spin are created in the conduction band. A sharp decrease of Y and P below the band edge indicates good quality of the structures.

Thus, the position of the polarization maximum is close to the SL band gap. In the SL the band gap is larger than that in GaAs layers by quantisation energy of the heavy holes and some shift of the conduction band minimum. Calculation of the miniband energies for SL-1 using the model described in Ref. 16 gives absolute values of the hole miniband energies $E_{\mathrm{hh1}} = 13$ meV; $E_{\mathrm{lh1}} = 54$ meV. Therefore the splitting of the valence band is close to 40 meV and it does not change much with simultaneous changes of Al content in both SL layers. The edge of the electronic band in a SL with a small conduction-band offset is close to the average conduction-band energy in the contacting layers.

The calculation of the band gap using the data given in Table I gives for all samples values that exceed the experimentally observed energy of the polarization maximum by $\sim$ 20 meV. This shift of the band gap values is equivalent to a deficit of $\approx$ 3% of Al concentration. The regular difference in the calculated and observed band gap value in the strained quaternary alloy can be attributed to the uncertainties in the conduction band offset calculations and also to some tensile deformation of GaAs layer resulting in less strain in the contacting $Al_xIn_yGa_{1-x-y}As$ layer. This misfit can

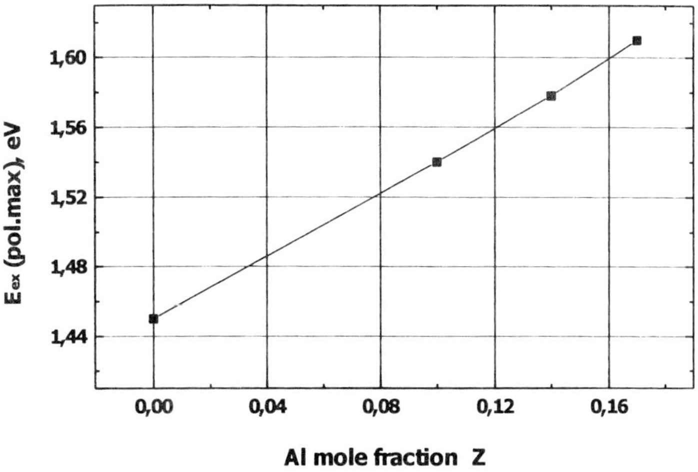

Figure 5. The values of the light energy E_{ex} (pol. max.), corresponding to the polarization maximum as a function of the mole fraction Z of Al in the $Al_xIn_yGa_{1-x-y}As/Al_zGa_{1-z}As$ superlattice.

be rather easily corrected by choosing a SL with larger x when some tuning of the SL band gap to excitation source is needed. Indeed, adding of Al does not influence the deformation so that the band gap variation with x is predictable.

The dependence of the polarization maximum on the excitation energy is shown in Fig 5. Linear dependence is found that makes possible the tuning of the maximum to the excitation wavelength. One can expect larger splitting with thicker barriers and thinner AlGaAs wells. Then, one can expect a smaller spin relaxation rate for optimally chosen doping of the SL, compatible with needed extracted emission current. Thus, the optimisation of the SL structure parameters and doping profile can lead to further improvement of the proposed new SL photoemitter structure. Finally, the maximum current density that can be extracted from these SL samples at high voltage has yet to be determined. Initial measurements using sample 1 gave anomalously low value. Definitive measurements are underway.

IV. Conclusions

The EDC and PEDC measurements for the strained GaAs layer surface activated to NEA, at near band edge excitation demonstrate the electron capture to the band bending region before emission. The shape of the energy distribution peak is in good agreement with the results of the model of the emission from the states localised in the surface plane by the fluctuations of the surface potential. The localisation is also manifested by switching off the spin relaxation across the emission peak.

Electron spin polarization as high as 84% has been reproducibly obtained from strained $Al_xIn_yGa_{1-x-y}As/Al_zGa_{1-z}As$ superlattices with small conduction band offset at the heterointerfaces. The position of the polarization maximum varies linearly with Al concentration and can be easily tuned to an excitation wavelength by an appropiate choice of the SL composition. The conduction band offset at the $Al_zGa_{1-z}As/Al_xIn_yGa_{1-x-y}As$ interface changes its sign at $z \cong 1.1(x - 1.1y)$. The modulation doping of the SL is found to be essential for high polarization and high quantum yield at the polarization maximum. Further improvement of the emitter parameters can be expected with additional optimisation of the SL structure parameters.

References

1. See, *e.g.*, A. V. Subashiev, Yu. A. Mamaev, Yu. P. Yashin, and J. E. Clendenin, *Phys. Low-Dim. Structures* **1/2**, 1 (1999), and references therein.
2. H.-J. Drouhin, C. Hermann, and G. Lampel, *Phys. Rev. B* **31**, 3859 (1985); **31**, 3872 (1985).
3. A. S. Terekhov and D. A. Orlov, *Proc. SPIE* **2550**, 157 (1995).
4. A. W. Baum, W. E. Spicer, R. F. W. Pease, K. A. Costello, and V. W. Aebi, *Proc. SPIE* **2550**, 189 (1995).
5. E. L. Nolle, *Sov. Phys.-Solid State* **31**, 196 (1989).
6. F. Ciccacci, H.-J. Drouhin, C. Hermann, R. Houdre, and G. Lampel, *Appl. Phys. Lett.* **54**, 632 (1989).
7. Yu. A. Mamaev, A. V. Subashiev, Yu. P. Yashin, H.-J. Drouhin, and G. Lampel, *Solid State Commun.*, 2000, accepted for publication.
8. Yu. A. Mamaev, Yu. P. Yashin, A. V. Subashiev, M. S. Galaktionov, B. S. Yavich, O. V. Kovalenkov, D. A. Vinokurov, and N. N. Faleev, *Phys. Low-Dim. Structures* **7**, 27 (1994).
9. B. D. Oskotskij, A. V. Subashie, and G. Lampel, in *AIP Conference Proceedings* **CP421**, Polarized Gas Targets and Polarized Beams, 7-th Intern. Workshop, Urbana, edited by R. Holt (1998), p. 491.
10. E. L. Ivchenko, M. I. Karaman, D. K. Nelson, B. S. Razbirin, A. N. Starukhin, *Phys. Solid State* **36**, 218 (1994).
11. A. V. Subashiev, *Proc. of the Low Energy Polarized Electron Workshop LE-98*, St.-Petersburg, 1998, edited by Yu. A. Mamaev, S. A. Starovoitov, T. V. Vorobyeva, and A.N.Ambrazhei, (SPES-Lab-Pub., 1998), p. 125.
12. D. N. Mirlin, V. I. Perel', and I. I. Reshina, *Semiconductors* **32**, 886 (1998).
13. H. Tang, R. K. Alley, H. Aoyagi, J. E. Clendenin, J. C. Frisch, G. A. Mulhollan, P. J. Saez, D. C. Schultz, and J. L. Turner, *Proc. of the 4th European Particle Accelerator Conf.*, May 17-23, 1994, London, UK, (World Scientific, Singapore, 1994), p. 46.
14. K. Togawa, T. Nakanishi, T. Baba, F. Furita, H. Horinaka, T. Ida, Y. Kurihara, H. Matsumoto, T. Matsuyama, M. Mizuta, S. Okumi, T. Omori, C. Suzuki, Y. Takeushi, K. Wada, M. Yoshioka, *Nucl. Instrum. and Meth.* **A414**, 461 (1998).
15. B. D. Oskotskij, A. V. Subashiev, and Yu. A. Mamaev, *Phys. Low-Dim. Structures* **1/2**, 77 (1997).
16. L. G. Gerchikov, G. V. Rozhnov, and A. V. Subashiev, *Sov. Phys. JETP* **74**, 77 (1992).

Magnetism and Magnetic Anisotropy in Exchange BIAS Systems

A. J. Freeman, K. Nakamura, M. Kim, and W. T. Geng

Department of Physics and Astronomy
Northwestern University
Evanston, IL60208
USA

Abstract

Magnetism in man made ultra-thin materials has become an area of intense activity and excitement, mainly driven by applications in magnetic recording. For the purpose of high-density magnetic recording, the understanding and control of the magnetic properties, such as complex magnetic ordering, enhanced magnetic moments, magneto-crystalline anisotropy and magnetostriction are desired. In this paper, we present results of magnetism and magnetic anisotropy in ferromagnetic NiFe thin film, antiferromagnetic NiMn and their interface determined from first-principles FLAPW (full-potential linearized augmented plane wave) calculations, and demonstrate how their properties behave sensitively on the environment (such as in bulk, at surfaces or interfaces). These results may be important for understanding exchange bias materials.

I. Introduction

Magnetism research has been undergoing a renaissance over the last decade following the discovery of a variety of new scientific phenomena associated with man-made transition metal thin films. Among them are the theoretical prediction of enhanced magnetic moments in ultra-thin films and at surfaces, the discovery of perpendicular magnetic anisotropy in ultra-thin films and layered structures, and the discovery of giant magnetoresistance (GMR) and the accompanying oscillatory exchange coupling in multilayers made by alternating magnetic and nonmagnetic metals. Some of these discoveries are already having a major impact on the magnetic recording industry.

First principles electronic structure studies based on local spin density functional theory, which performs extremely complex simulations of ever increasingly realistic systems, plays a very important role in successfully explaining magnetism in thin films and has led to the facing of even more challenging problems. It is known that the calculations, especially those carried out at Northwestern using the full-potential linearized augmented plane wave (FLAPW) method,[1] played a significant role during the course of developments of the field. The calculations predicted the large enhancement of the magnetic moment for $3d$-transition metal (TM) surfaces or overlayers deposited on inert substrates, and the possible magnetization in some normally non-magnetic materials, for which some results have already been verified experimentally. Complex magnetic structures, especially some antiferromagnetic (AFM) configurations, can now be predicted by comparing total energies with their equilibrium atomic geometries (including multilayer relaxations and reconstructions at surfaces and interfaces) optimized very efficiently using the atomic force approach. Significant progress has been made very recently for the treatment of the weak spin-orbit coupling (SOC) and now we are able to obtain very reliable results of *i*) the magneto-crystalline anisotropy (MCA) energies and magnetostrictive coefficients for transition metal thin films and *ii*) the magneto-optical Kerr effect (MOKE) and soft X-ray magnetic circular dichroism (MCD).

The aim of this paper—dedicated to Hans-Christoph Siegmann—is to present a case study on magnetism and magnetic anisotropy of important materials used in magnetic recording industries, and to demonstrate how their properties depend on the environment in bulk, at surface or interface. We first report results for ferromagnetic (FM) NiFe thin films. The effects of capping a Ta overlayer on NiFe are also discussed, which is commonly used in magnetic recording applications but the effects are not well understood. We next focus on the magnetic anisotropy of antiferromagnetic (AFM) NiMn and the FM/AFM NiFe/NiMn interface, which continues to attract great attention in advanced technological applications—especially the exchange bias which is associated with the magnetic anisotropy created at an interface between ferromagnetic and antiferromagnetic materials.[2,3] NiMn is believed to be a promising candidate as the exchange bias AFM material,[4,5] because it possesses a high crystalline anisotropy, high Neel temperature (T_N) and high corrosion resistance.

The calculations were performed by the FLAPW approach based on the local density approximation (LDA), in which the core states are treated fully relativistically and the valence states are treated semi-relativistically. The MCA energy was calculated by treating the SOC in a second variation way using the state tracking and torque methods.[6–8]

II. Ferromagnetic NiFe Thin Film

In this study, clean surfaces of a 5-layer Ni_xFe_{1-x} slab with $x = 0.5$ and 0.75 and a Ta overlayer of half coverage on each side of the slab were adopted. For the clean surfaces, two different possible geometries for each composition were considered: with either Fe or Ni layers on the surface for $x = 0.5$ in the $L1_0$ structure ($Ni_{0.5}Fe_{0.5}^{Fe}$ and $Ni_{0.5}Fe_{0.5}^{Ni}$) and with both Fe-Ni mixed layers or a Ni layer on the surface for $x = 0.75$ in the $L1_2$ structure ($Ni_{0.75}Fe_{0.25}^{Mix}$ and $Ni_{0.75}Fe_{0.25}^{Ni}$). Experimental bulk lattice constants were employed for the in-plane lattice spacing for both $x = 0.5$ (6.77 a.u.) and 0.75 (6.72 a.u.). The structural optimization was accomplished for each system by performing atomic force calculations[9] to obtain the fully relaxed vertical geometry. For the Ta overlayer case, total energies of several different adsorption sites of the Ni_xFe_{1-x} surface were compared; this revealed that the hollow site is energetically

Table I

Calculated equilibrium vertical positions z (in a.u., measured from the center layer) and spin magnetic moment (M, in μ_B) of each atom for the $Ni_{1-x}Fe_x$ (001) 5-layer thin films with $x = 0.5$ and 0.75. The corresponding values of unrelaxed structures are listed in parentheses for comparison.

	S	S–1	C
$Ni_{0.5}Fe^{Fe}_{0.5}$			
z	6.33	3.21	0.00
M	Fe: 2.79(2.82)	Ni: 0.68(0.65)	Fe: 2.53(2.57)
$Ni_{0.5}Fe^{Ni}_{0.5}$			
z	6.19	3.19	0.00
M	Ni: 0.69(0.69)	Fe: 2.49(2.65)	Ni: 0.68(0.67)
$Ni_{0.75}Fe^{Mix}_{0.25}$			
z	Fe: 6.23	3.14	Fe: 0.00
	Ni: 6.14	–	Fe: 0.00
M	Fe: 2.97(3.03)	Ni: 0.62(0.63)	Fe: 2.80(2.88)
	Ni: 0.65(0.69)	–	Ni: 0.57(0.64)
$Ni_{0.75}Fe^{Ni}_{0.25}$			
z	6.27	Fe: 3.27	0.00
	–	Ni: 3.24	–
M	Ni: 0.70(0.71)	Fe: 2.73(2.84)	Ni: 0.67(0.70)
	–	Ni: 0.63(0.66)	–

the most favorable site for the Ta overlayer rather than the Fe or Ni atomic sites by several eV. Note that there are two different hollow sites for the $Ni_{0.75}Fe^{Ni}_{0.25}$ with either an Fe or a Ni subsurface atom under the Ta adatom.

The optimized positions and the magnetic moments of atoms in the surface (S), the subsurface (S-1) and the center (C) layers are given for the 5-layer clean surface films in Table I. We found that the surface and subsurface layers undergo a large downwards contraction for the clean surfaces (*cf.* Table I), which results in the reduction of the interlayer distance d_{12} (between the surface (S) and subsurface (S-1) layers) and d_{23} (between the subsurface (S-1) and center (C) layers), by 8 to 11% and 3 to 7%, respectively, compared to the bulk alloy. As a result of full relaxation, there appears a slight buckling in the mixed layers of Fe and Ni. The Ni atoms suffer a more downwards relaxation compared to Fe mainly due to the fact that Ni has a smaller atomic volume than Fe. The difference in the vertical positions of Fe and Ni is larger for the mixed layer on the surface (0.09 a.u. for $Ni_{0.75}Fe^{Mix}_{0.25}$) compared to the mixed layer on the subsurface (0.03 a.u. for $Ni_{0.75}Fe^{Ni}_{0.25}$).

Due to surface effects and relaxation, the magnetic moments of these thin films appear to be different from those of the bulk alloys (which were 2.69 (Fe) and 0.68 (Ni) μ_B for $Ni_{0.5}Fe^{Ni}_{0.5}$ in the $L1_0$ structure and 2.94 (Fe) and 0.67 (Ni) μ_B for $Ni_{0.75}Fe^{Ni}_{0.25}$ in the $L1_2$ structure according to our calculations). It was found that the magnetic moments of the Fe atoms are very sensitive to the environment while those of the Ni

Table II

Calculated MCA, ΔE_{sl}, for $Ni_{0.5}Fe_{0.5}$ (001) clean surfaces in meV and their actual atomic compositions in one unit cell in the 5-layer slabs.

	$Ni_{0.5}Fe_{0.5}^{Fe}$	$Ni_{0.5}Fe_{0.5}^{Ni}$	$Ni_{0.75}Fe_{0.25}^{Mix}$	$Ni_{0.75}Ni_{0.25}^{Fe}$
ΔE_{sl}	0.36	−0.27	−0.10	−1.27
Composition	Fe_3Ni_2	Fe_2Ni_3	Fe_3Ni_7	Fe_2Ni_8

atoms are not much influenced. For the $Ni_{0.5}Fe_{0.5}$ film with Fe on the surface, the magnetic moment of the surface Fe is enhanced by 0.13 μ_B and that of the center Fe is reduced by 0.12 μ_B compared to the bulk value with the same lattice spacing while the subsurface Ni magnetic moment is not far from the bulk value (*cf.* the values in parentheses in Table I *versus* bulk values). In the case of the film with the Ni atom on the surface, the magnetic moment of the surface Ni atom does not exhibit a large change compared to the bulk value.

Upon full structural optimization, the magnetic moment of Fe tends to be reduced while that of Ni increases slightly for $x = 0.5$ and decreases for $x = 0.75$. This may be caused by increased interatomic interactions due to the relaxation; the smaller Fe-Fe or Ni-Ni interatomic distance is supposed to reduce the magnetic moments while the increase of interatomic interactions between Ni and Fe is known to enhance the Ni magnetic moments and to reduce the Fe magnetic moments.[10] Therefore, the relaxation acts to reduce the Fe magnetic moments while its influence on the Ni magnetic moments depends on the system; for large Ni composition, the Ni-Ni interatomic interaction dominates over that of Fe-Ni and thus the relaxation reduces the magnetic moment of Ni.

It is clearly seen from Table I that the increase of Ni composition enhances the magnetic moment of Fe as already found from the bulk alloy results. The dependence of magnetic properties on the atomic composition is more clearly demonstrated from magnetocrystalline anisotropy (MCA) calculations. As revealed in previous investigations, MCA is extremely sensitive to the environment. The calculated MCA, defined by the difference in total energies for the magnetic moments oriented perpendicular and in-plane, are given in Table II. These reveal that the MCA changes sensitively depending on the nature of the surface atoms and the atomic composition. For $Ni_{0.5}Fe_{0.5}$, we found that the bulk alloy has the perpendicular (001) direction as the easy axis with MCA = 0.06 meV. For the 5-layer films with Fe on the surface, the perpendicular direction is still the easy axis with an enhanced MCA, while it switches to in-plane with Ni on the surface. The case of $Ni_{0.75}Fe_{0.25}$, which in the bulk case has negligible MCA due to its cubic symmetry, now exhibits a large magnetic anisotropy in keeping with the lowered symmetry, or surface effect, and shows an in-plane preference for the magnetization direction. Consistent with the result of $Ni_{0.5}Fe_{0.5}$ slabs, the $Ni_{0.75}Fe_{0.25}$ thin film with the Fe-Ni mixed surface layer on the surface has a very small MCA, while the one with Ni on the surface shows a large MCA. Also noting that the actual atomic composition in one unit cell is different for each system (as shown in Table II), we can see that the MCA changes with atomic composition; with more Ni included, the in-plane preference becomes stronger (*cf.* $Ni_{0.5}Fe_{0.5}^{Ni}$ and $Ni_{0.75}Fe_{0.25}^{Ni}$).

The optimization results and the magnetic moments of a Ta overlayer on the Ni_xFe_{1-x} substrate show several interesting things. The interlayer distance between

Ta and the surface layer of the substrate (S) shows large relaxation compared to the average interlayer distances of Ta and NiFe (001) bulk (3.26 and 3.24 for $x = 0.5$ and $x = 0.75$, respectively), ranging from 4 to 8%. Due to the Ta overlayer, the relaxation d_{12} of the substrate (between S and S-1) is largely reduced compared to those of the clean surfaces, reaching only up to 6% of the bulk values at most. The magnetism of the Ta overlayer changes dramatically depending on the surface layer; for the systems with an Fe layer and a Fe-Ni mixed layer on the surface, Ta couples antiferromagnetically with the substrate while it orders ferromagnetically for the Ni surfaces. Surprisingly, the induced magnetic moment of Ta is large, ranging from 0.34 to 0.56 μ_B in magnitude. The effect of the Ta overlayer on the magnetic moment of the substrate is also quite interesting: it acts to reduce significantly the magnetic moment of the nearest (*i.e.* surface) atom; the surface Fe and Ni magnetic moments are reduced to about half of their clean surface values. This detrimental effect of the Ta overlayer seems to be confined only to the surface layer and exhibits an efficient metallic screening of the NiFe substrate.

III. Antiferromagnetic NiMn and Ferromagnetic/Antiferromagnetic NiFe/NiMn Interface

Calculations were carried out for bulk NiMn and for five-layer slabs of NiMn, namely Ni_2Mn_3 and Ni_3Mn_2; assumed in the $L1_0$ structure with the experimental lattice parameters, a = 3.70 Å, c = 3.54 Å. For the interface study, NiFe/NiMn and $(NiFe)_2/(NiMn)_2$ superlattice structures were employed. The $L1_0$ atomic ordering in both the NiFe and NiMn layers was assumed with a (001)/(001) crystallographic orientation at the interface, and again the experimental lattice parameters for bulk NiMn were adopted. The lattice parameters in the basal plane of the NiFe were assumed to match those of NiMn, but with the c/a ratio chosen to conserve the experimental atomic volume of the $Ni_{0.5}Fe_{0.5}$ alloy. We confirmed that this structural approximation does not significantly alter the magnetic properties of NiFe. Although very important, the effect of lattice relaxation is outside the scope of the present first step study, and is therefore not discussed.

Table III shows the calculated spin magnetic moments for bulk NiMn, and for five-layer films of NiMn. Note that the Mn atoms are located on two sublattices with moments that have the same magnitude but opposite signs. For bulk NiMn, although the calculated Mn moment (3.13 μ_B) is smaller than experiment (3.8 μ_B),[11,12] the calculations predict the AFM ordering observed in experiments. The magnetic moments at the surface in NiMn are very sensitive to the environment. For the film with Mn on the surface, the magnetic moment is enhanced by 0.28 μ_B. Interestingly, if the Ni atom is at the surface, the magnetic moment does not exhibit a large change, as observed in NiFe case (Table I). The calculated magnetic moments for the NiFe/NiMn and $(NiFe)_2/(NiMn)_2$ superlattice structures are also given in Table III. FM ordering of the Fe magnetic moments in the NiFe layer and AFM ordering of the Mn magnetic moments in the NiMn layer were found. For $(NiFe)_2/(NiMn)_2$, except for the Ni atom at the interface, the magnetic moments recover almost identical values to the bulk. Note that the non-vanishing moment of the Ni atom at the interface, with a ferromagnetic moment alignment, was induced.

Let us now discuss the magnetic anisotropy. Table IV shows the calculated MCA energy, E_{MCA}, defined by the difference in total energies for the magnetic moments oriented perpendicular (c-axis) and in-plane (a-axis). Contrary to bulk NiFe that has

Table III

Calculated magnetic moment (in μ_B) for bulk antiferromagnetic NiMn, for five-layer films of NiMn, and for NiFe/NiMn and $(NiFe)_2/(NiMn)_2$ superlattice structures. S, S$-$1 and C for the five-layer films represent the surface, subsurface and center layers, respectively. For the superlattices, the interface Ni atom between the NiFe and NiMn layers are represented by Ni(I). Because antiferromagnetic structures consist of two sublattices, values in parentheses are the moments in the other sublattice.

bulk	Magnetic moments (μ_B)	
NiMn	Ni: 0.00	Mn: 3.13 (−3.13)

five–layer film	S	S–1	C
Ni_2Mn_3	Mn: 3.41 (−3.41)	Ni: 0.00	Mn: 3.12 (−3.12)
Ni_3Mn_2	Ni: 0.00	Mn: 3.08 (−3.08)	Ni: 0.00

superlattice	Ni	Fe	Ni(I)	Mn	Ni
NiFe/NiMn		2.62 (2.66)	0.43	3.11 (−3.15)	
$(NiFe)_2/(NiMn)_2$	0.78	2.62 (2.66)	0.40	3.14 (−3.13)	0.00

the perpendicular easy magnetic axis with a small magnitude of the MCA energy, bulk NiMn exhibits an in-plane easy axis with a relatively large MCA energy of 0.5 meV. In order to discuss its origin, the MCA energy, $E_{\rm MCA}$ of NiMn, is plotted in Fig. 1 with respect to the change in the number of valence electrons by changing the band filling resulting from varying the highest occupied states around E_F. In this band filling dependence, the change in the number of valence electrons, ΔZ, is given relative to the physical values, *i.e.*, it equals zero when the number of valence electrons is that of NiMn. The figure also shows the ΔE_{dd} and ΔE_{ud} contributions to the MCA energy, which are calculated from the coupling between occupied spin-down and unoccupied spin-down states, and that between occupied spin-up and unoccupied spin-down states, respectively. When $\Delta Z \approx 0$, the negative MCA is dominated by negative ΔE_{ud} contributions. As ΔZ increases, the negative ΔE_{ud} contribution gradually decreases, while the positive ΔE_{dd} contribution is developed because the empty spin-down state is gradually occupied.

For the surface and interface studies, we found that the MCA is sensitive to the environment, as seen in Table IV. The MCA in surface cases exhibits qualitatively the same behavior as bulk that shows in-plane anisotropy, but when the Ni atom is at the surface there is a larger negative MCA energy that creates a strong in-plane anisotropy. The NiFe/NiMn and $(NiFe)_2/(NiMn)_2$ superlattice cases show in-plane

Table IV

Magnetocrystalline anisotropy energy, ΔE_{MCA}, (in meV/cell) for bulk antiferromagnetic NiMn, for five-layer films of NiMn, and for the NiFe/NiMn and $(NiFe)_2/(NiMn)_2$ superlattice structures.

	bulk	five–layer slab		superlattice	
	NiMn	Ni_2Mn_3	Ni_3Mn_{23}	NiFe/NiMn	$(NiFe)_2/(NiMn)_2$
ΔE_{MCA}	−0.51	−1.30	−2.30	−0.14	−0.82

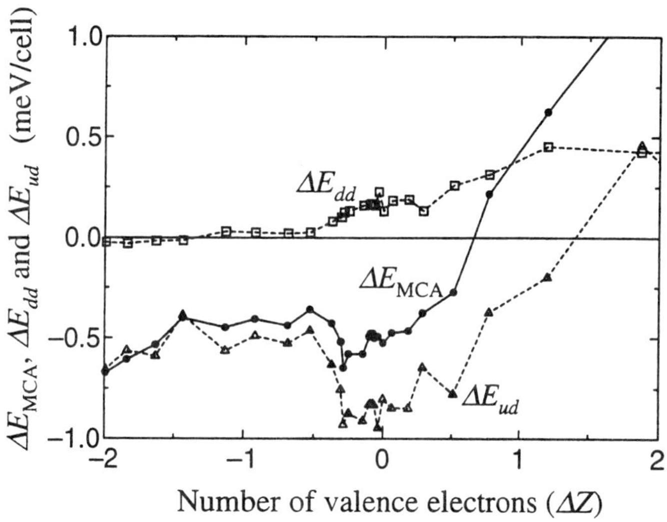

Figure 1. Band filling dependence of the MCA energy (E_{MCA}), and the ΔE_{dd} and ΔE_{ud} contributions to the MCA for NiMn, where ΔE_{dd} and ΔE_{ud} are the contributions from coupling between occupied spin-down and unoccupied spin-down states, and that between occupied spin-up and unoccupied spin-down states, respectively. The number of valence electrons, ΔZ, is given relative to the physical values, *i.e.*, it equals zero when the number of valence electrons is that of NiMn.

magnetic orientation, mainly dominated by the in-plane MCA from the NiMn layer. However, the interface effects on the MCA may not be negligible. In Fig. 2, the band filling dependence of E_{MCA}, ΔE_{dd} and ΔE_{ud} contributions for NiFe/NiMn and $(NiFe)_2/(NiMn)_2$, with respect to the number of their valence electrons, are plotted. Although similar behavior of the band filling dependence in $(NiFe)_2/(NiMn)_2$ with that of bulk NiMn were observed (since the contribution from the NiMn layer tends to dominate the MCA energy), the changes in the band filling dependence from that of bulk are emphasized especially in the NiFe/NiMn case—which might be important for discussing the exchange bias. This is because the magnetic anisotropy at the FM/AFM interface, as well as the interface exchange coupling, may play a crucial role in determining the exchange bias phenomenon.

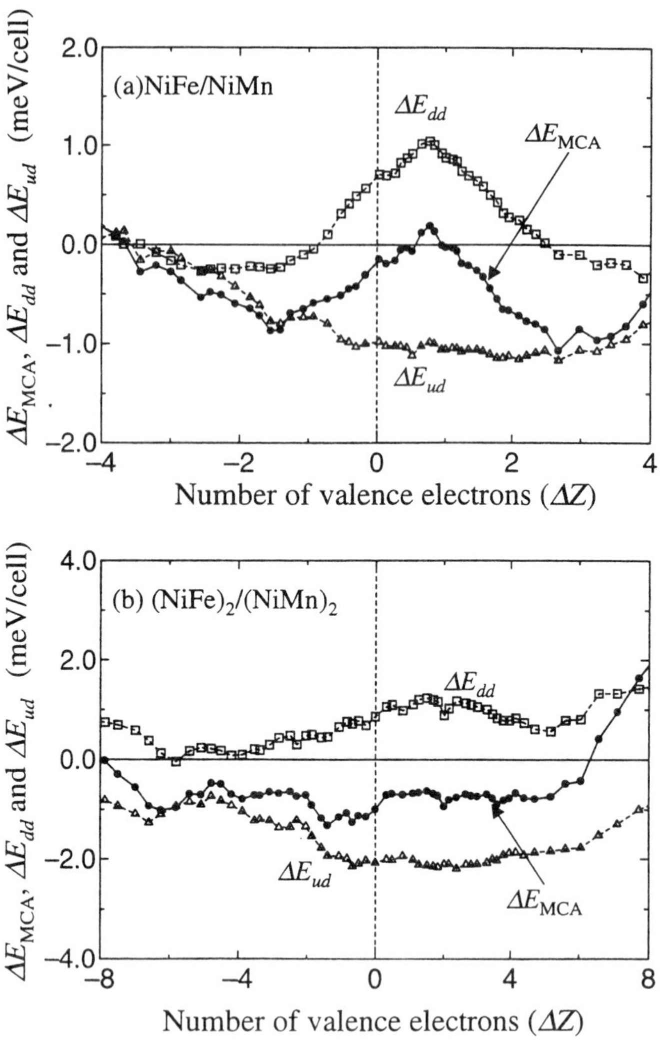

Figure 2. Band filling dependence of the MCA energy (E_{MCA}), and the ΔE_{dd} and ΔE_{ud} contributions to the MCA for (a) NiFe/NiMn and (b) $(NiFe)_2/(NiMn)_2$ superlattice structures. The zero value of ΔZ means the number of valence electrons of their superlattice structures.

IV. Conclusion

We presented results of magnetism and magnetic anisotropy in FM NiMn thin film, AFM NiMn, and their interface obtained from first-principles FLAPW calculations. The magnetic moments of Fe in NiFe and Mn in NiMn were found to depend sensitively on the environment, *i.e.*, surface, relaxation and atomic composition while those of Ni do not change much. Upon contact with the Ta overlayer, the surface atoms of the substrate suffer a significant reduction of their magnetic moments, which is confined to the surface layer due to the efficient metallic screening. The large induced magnetic moment of Ta and the change of its magnetic ordering, due to the surface layer, imply a possible new finding of $5d$ magnetism. Their MCA values are also drastically influenced by their environment. Bulk NiFe and NiMn show slightly perpendicular and in-plane easy magnetic axes, respectively. However, the effects of the surface change their MCA direction and strength. For the interface between NiFe and NiMn, the interface effects on the MCA are not negligible. These researches, coupled with the micro-magnetic modeling at the device level and the microstructure control allowed by advanced process techniques, are expected to offer possible improvement of the magnetic materials desired by high density magnetic recording and sensor materials.

References

1. E. Wimmer, H. Krakauer, M. Weinert, and A. J. Freeman, *Phys. Rev. B* **24**, 864 (1981) and references therein; M. Weinert, *J. Math. Phys.* **22**, 2433 (1981).
2. J. Nogues and I. K. Schuller, *J. Magn. Magn. Mater.* **192**, 203 (1999).
3. A. E. Berkowits and K. Takano, *J. Magn. Magn. Mater.* **200**, 552 (1999).
4. T. Lin, D. Mauri, N. Staud, C. Hawng, J. K. Howard, and G. L. Gorman, *Appl. Phys. Lett.* **65**, 1183 (1994).
5. X. Portier, A. K. Petford-Long, and T. C. Anthony, *IEEE Trans. on Magn.* **33**, 3679 (1997).
6. R. Wu and A. J. Freeman, *J. Magn. Magn. Mater.* **200**, 498 (1999).
7. D. S. Wang, R. Wu, and A. J. Freeman, *Phys. Rev. Lett.* **70**, 869 (1993).
8. X. D. Wang, R. Wu, D. S. Wang, and A. J. Freeman, *Phys. Rev. B* **54**, 61 (1996).
9. R. Yu, D. Singh, and H. Krakauer, *Phys. Rev. B* **43**, 6411 (1992).
10. R. Wu and A. J. Freeman, *Phys. Rev. B* **45**, 7205 (1992).
11. J. S. Kasper and J. S. Kouvel, *J. Phys. Chem. Solids* **11**, 238 (1959).
12. L. Pal, E. Kren, G. Kadar, P. Szabo, and T. Tarnoczi, *J. Appl. Phys.* **59**, 538 (1968).

The Role of Damping in Ultrafast Magnetization Reversal

C. H. Back

Laboratorium für Festkörperphysik,
Eidgenössische Technische Hochschule Zürich,
CH-8093 Zürich
SWITZERLAND

Abstract

Ultrashort magnetic field pulses generated in the final focus test beam facility at the Stanford Linear Accelerator have been used to study fundamental properties of magnetization reversal in thin films with in-plane and perpendicular easy magnetization directions. For perpendicular magnetized samples we observe ring domains with Kerr microscopy, which are reminiscent of the field contour during exposure. Their radii represent switching fields in quantitative agreement with the coherent rotation model. In this case switching is caused by a simple rotation of the magnetization around the effective field while the external field pulse is present and thus damping mechanisms do not play a major role. For films with an uniaxial anisotropy in the plane of the film we observe that smaller fields are sufficient to reverse the magnetization, provided that the field is orthogonal to the magnetization. In this geometry maximum torque is exerted on the spins. Precession of the magnetization around the demagnetizing field completes the reversal after the external field ceases to exist. In this case the remagnetization process takes considerably longer time, so that the effect of damping can no longer be neglected.

I. Introduction

High speed magnetic switching has recently received considerable interest as its understanding is fundamental to high data rate magnetic recording.[1-4] The motion of the magnetization vector $\vec{M}$ can be described by the phenomenological Landau-Lifshitz equation[5] in the modification proposed by Gilbert.[6] This equation describes precessional motion of the magnetization vector around the total magnetic field $\vec{H}_{\rm tot}$. $\vec{H}_{\rm tot}$ is the sum of internal and external field contributions. Thermal fluctuations can be

Physics of Low Dimensional Systems
Edited by J. L. Morán-López, Kluwer Academic/Plenum Publishers, New York 2001

included via a random field term.[7–10] The precessional frequency is simply the Larmor frequency given by the total field. In this approach energy dissipation is taken into account by including a simple damping term in linear approximation which describes a damped relaxation into the magnetic field direction. Recently, Baryakhtar and co-workers[11] have extended this simple approach by introducing a tensor quantity representing dissipative effects. Suhl[12] has described loss mechanisms for large motions of $\vec{M}$ through direct dissipation into the phonon lattice and indirect damping via the production of non-uniform magnetic modes. Hannay *et al.*[13] suggest that intergranular exchange and magnetostatic coupling in longitudinal media effectively increase the damping parameter of the system. The mentioned theoretical papers are based on the Landau-Lifshitz equation. The description of losses or magnetic relaxation through this phenomenological equation does not yield a detailed description of the characteristics of the loss processes, however, even sophisticated micromagnetic theories can only be used to analyze simple excitations like domain walls. The damping term of the Landau-Lifshitz equation, while is not so transparent, can describe a large spectrum of nonlinear excitations.[11]

Only few experiments in strong magnetic fields have been performed at short time scales and thus the role of the damping process has not been studied extensively.[1,14–16] It was shown recently that magnetization reversal below the spin lattice relaxation time τ_{sl} is feasible if the external magnetic field $\vec{H}_{\text{ex}}$ inducing the reversal is applied perpendicular to $\vec{M}$.[17–20] In that case, the torque $\vec{T} = \vec{H}_{\text{ex}} \times \vec{M}$ is maximum and angular momentum is directly transferred from the magnetic sample to the source of the magnetic field. It has also been demonstrated that in-plane magnetized films are favorable for magnetization reversal as the demagnetizing field helps to complete the reversal process.[19,21]

In this paper magnetic switching experiments are revisited. They have been performed at time scales below τ_{sl} with both, perpendicular and in-plane magnetized magnetic films and with a special emphasis on the damping process. For perpendicularly magnetized films switching is caused by a simple rotation of the magnetization around the effective field while the external field pulse is present and thus damping mechanisms do not play a major role.[22] This is no longer the case for in-plane magnetized films where the motion of the magnetization vector takes considerably longer until it is locked into its new equilibrium position. In this case we observe unusual behavior of the effective damping parameter. All calculations have been performed using the single spin Landau-Lifshitz approach as described in Refs. 18 and 19.

II. Experimental

Sample properties and details about the Final Focus Test Beam Facility have been described in detail elsewhere.[18–20] For the purpose of this paper only a short summary of the sample properties is given. The perpendicularly magnetized sample is a Co/Pt multilayer with an effective magnetic anisotropy field $H_{A\text{eff}} = 1320$ kA/m$\pm$130 kA/m. The film for the experiment with in-plane magnetized material consists of Co. It is single crystalline and magnetically uniaxial with the easy direction of $\vec{M}$ in-plane along the x-direction. Its saturation magnetization at room temperature is $\mu_0 M_s = 1.7$ T. The uniaxial anisotropy field lying in the plane of the film is $H_A = 168$ kA/m.

The experiments were performed at the focal point of the final focus test beam (FFTB) facility at the Stanford Linear Accelerator Center (SLAC). The two samples were exposed to shots of the relativistic 46.6 GeV electron bunches with a temporal

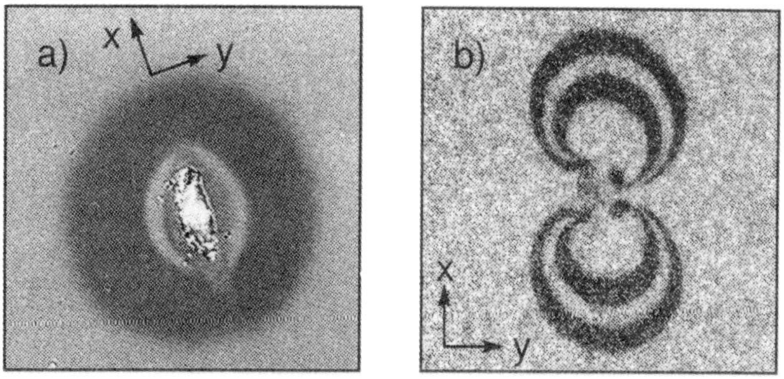

Figure 1. (a) Domain pattern written into a Co/Pt multilayer with an effective perpendicular anisotropy field $H_{A\text{eff}}$ of 1320 kA/m and with $\sigma_t = 2$ ps. Image size 65 μm $\times$ 65 μm. (b) Domain pattern written into a Co film with uniaxial in-plane anisotropy field H_A of 168 kA/m and with $\sigma_t = 2$ ps. Image size 300 μm $\times$ 300 μm.

pulse length σ_t of 2 ps. The Gaussian half widths of the electron beam in the x and y directions were determined to be $\sigma_x = 3.8 \pm 0.4$ μm and $\sigma_y = 0.8 \pm 0.2$ μm, respectively. The number of electrons per bunch was $(9.1 \pm 0.2) \times 10^9$. The magnetic field $H_{\text{ex}}(x, y, t)$ produced by the electron beam moving essentially at the speed of light is computed from the current density $j(x, y, t) = n(x, y, t)ec$ by simply applying Ampere's law. Here $n(x, y, t)$ is the number of electrons at a given position and time, determined by the three dimensional Gaussian beam. The center of the coordinate system is the center of the location of the beam impact.

III. Discussion

Let us first consider the sample with its easy magnetization direction pointing out of the plane of the film. A short but strong magnetic field pulse $\vec{H}_{\text{ex}}$ is applied. $\vec{H}_{\text{ex}}$ lies in the plane of the film and is thus perpendicular to the initial magnetization direction. The total magnetic field $\vec{H}_{\text{tot}}$ comprises of the external field $\vec{H}_{\text{ex}}$ and the effective anisotropy field $\vec{H}_{A\text{eff}}$ of the sample which points out of the plane. The resulting domain configuration after a single shot is displayed in the Kerr micrograph in Fig. 1(a).

Black areas represent areas where the magnetization has switched from $+M$ to $-M$. In this particular geometry reversal occurs by precession of $\vec{M}$ around $\vec{H}_{\text{tot}}$ while $\vec{H}_{\text{ex}}$ acts. The magnetization precesses below the plane defined by the film and then, on a much longer time scale, precesses around $\vec{H}_{A\text{eff}}$ into the opposite easy magnetization direction. The trajectories for this motion at the location $x = 26$ μm and $y = 0$ and for $\alpha = 0.1$ are shown in Figs. 2(a) and 3(a).

For successful reversal the magnetization must precess by an angle $\Phi \geq \pi/2$. The required time for the first stage of the precessional motion is of the order of 5 ps, see Fig. 3(a). This immediately means that damping effects can be neglected.[22] In order for the damping parameter to show any effect it must be increased to large values. In fact a variation of the damping parameter α from 0.08 to 0.3 results in a change of the calculated position of reversal from $x = 27.8$ μm to $x = 23.1$ μm.

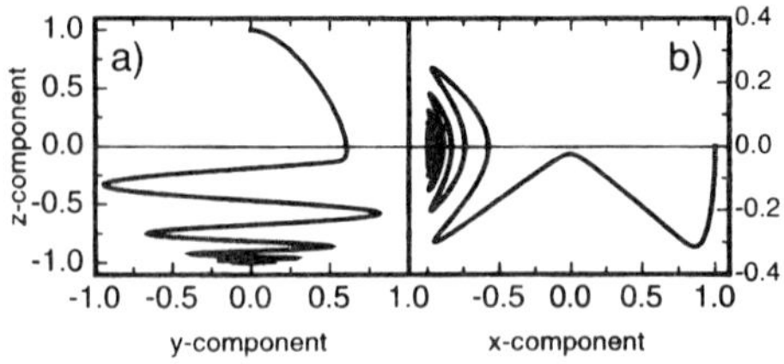

Figure 2. (a) Trajectory for the perpendicularly magnetized film calculated at the position $x = 26$ μm and $y = 0$ and for $\alpha = 0.1$. (b) Trajectory for the in-plane magnetized Co-film calculated at the position $x = 130$ μm and $y = 0$ and for $\alpha = 0.037$.

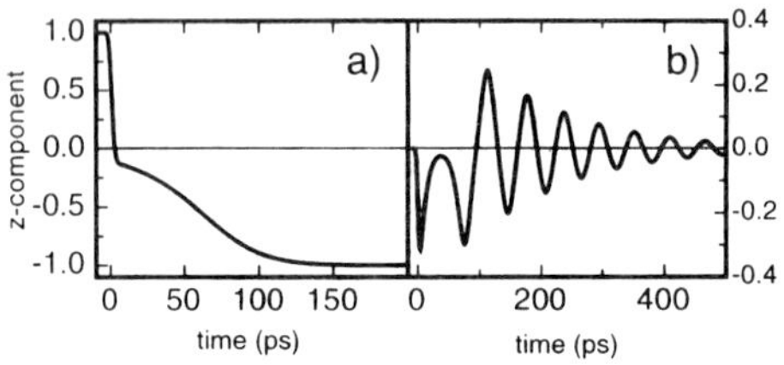

Figure 3. (a) Perpendicular component of the magnetization as a function of time for the switching process of the perpendicularly magnetized film. (b) Perpendicular component of the magnetization as a function of time for the switching process of the in-plane magnetized film.

This change is smaller than the expected error due to the uncertainty in the determination of $\vec{H}_{A\mathrm{eff}}$, see Fig. 2(a) in Ref. 18.

A different scenario is relevant for the switching process in in-plane magnetized samples. In this geometry, magnetization reversal is induced with magnetic field pulses of a few picoseconds duration, but with small field amplitudes of $\cong$ 350 kA/m for a 2 ps field pulse as compared to about 2000 kA/m for the perpendicular film. This is due to the demagnetizing field $\vec{H}_D$ brought about by the precession of $\vec{M}$ during the field pulse out of the plane of the film.[19] When the external magnetic field pulse is terminated, $\vec{H}_D$ persists, and the precession of $\vec{M}$ around $\vec{H}_D$ completes the reversal. Figure 1(b) shows the magnetic pattern generated in the Co film. The initial magnetization direction points along $-x$ (white). Black contrast represents areas that have switched the magnetization direction from $-\vec{M}$ to $+\vec{M}$. The location of impact is at the center of the image, which we also define as the center of the coordinate system. In contrast to conventional reversal and to the experiments described above with perpendicular films, the damping coefficient α describing the relaxation of $\vec{M}$ into the direction of the magnetic field is now the material parameter of crucial importance. The full relaxation process has to be taken into account for the reversal. Figures 2(b) and 3(b) show the trajectories for the motion of $\vec{M}$ recorded at the position $x = 130$ μm and $y = 0$ and for $\alpha = 0.037$. In these figures the motion of the magnetization vector becomes clear: First $\vec{M}$ is kicked out of the plane of the film.

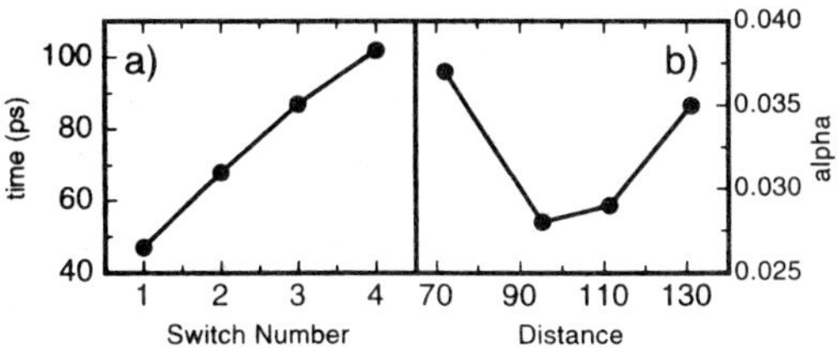

Figure 4. (a) Calculated time until the magnetization vector has precessed into the "locked" magnetization state and (b) calculation of the damping parameter α_{eff} for the four different locations of switching of Fig. 1(b).

The resulting demagnetizing field causes $\vec{M}$ to precess into the $+x$-direction. It locks into this direction (due to the anisotropy field $\vec{H}_A$) after about 50 ps and continues to precess around $\vec{H}_A$. This immediately shows that much longer time is available for damping processes until the "locked" magnetization state is reached. Damping now starts to play a major role. In fact the relaxation process reaches time scales comparable to the spin lattice relaxation time.[23] We have mentioned before[19] that the radii of the experimentally observed inner rings are not exactly reproduced by simple calculations using the Landau-Lifshitz equation. For example the second reversal back to the original direction occurs at smaller field values than expected. The details of the inner structure of the pattern cannot be calculated with the simple Landau-Lifshitz approach unless one assumes that some of the material properties vary with time or magnetic field strength. Supposing that intrinsic properties such as M_s and H_A are constant, one is forced to assume that α is time-dependent or depends on the strength of the magnetic field pulse.

In the following, possible reasons for a non-uniform damping constant α_{eff} are given. At this point it is important to mention that the calculations are performed for decoupled single spins. This means that differences between the oversimplified calculations and the experimental observations can be found in the omission of exchange coupling, magnetostatic coupling, eddy currents, and the generation of non-uniform magnetic modes such as spin waves. However, also more complex processes such as electron-electron scattering can affect the losses. These processes are not easily added to the standard form of the Landau-Lifshitz equation as they do not conserve the length of the vector $\vec{M}$.

The most obvious difference between the switching experiments with in-plane magnetized films and perpendicularly magnetized films is the duration of the precessional motion until a new "locked" state is reached. One can calculate the time needed for the magnetization to lock into its new equilibrium position. The results of these calculations are displayed in Fig. 4(a) for the four different switching locations. In these calculations the damping parameter is varied to fit the experimental location. The resulting effective damping parameter α_{eff} is plotted in Fig. 4(b) for $M_s = 1.7$ T and $H_A = 168$ kA/m as a function of the different switching locations. A linear increase in the switching time can be observed in Fig. 4(a) at time scales much longer than for the perpendicular samples.

Thus, the effect of damping should manifest itself in the experimental data for in-plane magnetized films in contrast to films with perpendicular magnetization. This has been demonstrated in reference[19] where two Co films of different quality have been compared.

Different mechanisms might lead to a time-dependent or spatially dependent effective damping constant α_{eff}. First, it is necessary to mention that the external field excitation is non-uniform in space. This means that the motion of $\vec{M}$ is non-uniform in space as the precessional frequencies vary as a function of external magnetic field. This will lead to a rapid dephasing of $\vec{M}$ and will affect electron-electron scattering.[24] If $\vec{M}$ precesses at a different rate in each location, this scattering will be very strong, leading to a larger effective damping constant. The second mechanism is the excitation of magnons. Field pulses are built with frequencies close to the frequency band of magnons, hence magnon excitation might be enhanced. This can lead to an increase in energy dissipation and thus to an increase in α_{eff}.

Based on experimental advances, magnetization reversal has seen considerable development in recent years. The present experiments would greatly gain in value by ultrafast observation of $\vec{M}$ as it precesses in the various steps of the reversal. Freeman and coworkers[14] have shown that this is indeed possible with the magneto-optic Kerr-effect using picosecond laser pulses. Inductive methods in combination with time resolved second-harmonic magneto-optic Kerr-effect have indeed given hints to the possible existence of two step damping processes.[3,15] Further time-resolved studied are desired and should reveal the underlying principles of magnetic loss processes.

Acknowledgments

Numerous fruitful discussions with H.C. Siegmann are greatfully acknowledged.

References

1. W. D. Doyle, S. Stinnett, C. Dawson, and L. He, *J. Magn. Soc. Jpn.* **22**, 91 (1998).
2. K. B. Klaasen and J. C. L. van Peppen, *IEEE Trans. Magn.* **35**, 625 (1998).
3. N. D.Rizzo, T. J. Silva, and A. B. Kos, *Phys. Rev. Lett.* **83**, 4876 (1999).
4. D. Weller and A. Moser, *IEEE Trans. Magn.* **35**, 2808 (1999).
5. L. D. Landau and E. M. Lifshitz, *Phys. Z. Sowjetunion* **8**, 153 (1935).
6. T. L. Gilbert, *Phys. Rev.* **100**, 1243 (1955).
7. L. Néel, *Ann. Geophys.* **5**, 99 (1949).
8. W. F. Brown, *Phys. Rev.* **130**, 1677 (1963).
9. H. B. Callen and T. A. Welton, *Phys. Rev.* **83**, 34 (1951).
10. J. D. Hannay, N. S. Walmsley, and R. W. Chantrell, *J. Magn. Magn. Mat.* **193**, 245 (1999).
11. V. G. Baryakhtar, B. A. Ivanov, A. L. Sukstanskii, and E. Yu. Melikhov, *Phys. Rev. B* **56**, 619 (1997).
12. H. Suhl, *IEEE Trans. Magn.* **34**, 1834 (1998).
13. J. D. Hannay, R. W. Chantrell, H. J. Richter, *J. Appl. Phys.* **85**, 5012 (1999).
14. W. K. Hiebert, A. Stankiewicz, and M. R. Freeman, *Phys. Rev. Lett.* **79**, 1134 (1997).
15. T. M. Crawford, T. J. Silva, C. W. Teplin, and C. T. Rogers, *Appl. Phys. Lett.* **74**, 3386 (1999).
16. Complete switching experiment differ from FMR measurements where a small (as compared to the intrinsic fields) RF excitation causes precession of the magnetic moments around a constant effective field composed of internal and external field

contributions [G. T. Rado, *J. Appl. Phys.* **32**, 129 (1960)]. In FMR the magnetization vector decribes a coherent precessional motion around the saturation magnetization direction. Thus we might expect different damping behavior as compared to complete switching experiments where non uniform local magnetization structures develop.[13]
17. H. C. Siegmann, E. L. Garwin, C. Y. Prescott, J. Heidmann, D. Mauri, D. Weller, R. Allenspach, and W. Weber, *J. Magn. Magn. Mat.,* **151**, L8 (1995).
18. C. H. Back, D. Weller, J. Heidmann, D. Mauri, D. Guarisco, E. L. Garwin, and H. C. Siegmann, *Phys. Rev. Lett.* **81**, 3251 (1998).
19. C. H. Back, R. Allenspach, W. Weber, S. S. P. Parkin, D. Weller, E. L. Garwin, and H. C. Siegmann, *Science* **285**, 864 (1999).
20. C. H. Back, and H. C. Siegmann, *J. Magn. Magn. Mat.* **200**, 774 (1999).
21. H. B. Callen, International Symposium on the Theory of Switching, Harvard University, April 1957, V1 and V2, 179 (1957).
22. L. He and W. D. Doyle, *J. Appl. Phys.* **79**, 6489 (1996).
23. Values for τ_{sl} are found to lie in the 100 ps range. A. Vaterlaus, T. Beutler, D. Guarisco, M. Lutz, and F. Meier, *Phys. Rev. B* **46**, 5280, (1992). A. Scholl, L. Baumgarten, R. Jacquemin, and W. Eberhardt, *Phys. Rev. Lett.* **79**, 5146 (1997) and references therein.
24. D. Oberli, R. Burgermeister, S. Riesen, W. Weber, and H. C. Siegmann, *Phys. Rev. Lett.* **81**, 4228 (1998).

Magnetised Foil as a Spin Filter

P. S. Farago[1] and K. Blum[2]

[1] *Department of Physics and Astronomy*
University of Edinburgh
Edinburgh
SCOTLAND

[2] *Universität Münster*
Institut für Theoretische Physik I
Münster
GERMANY

Abstract

The elastic transmission of low-energy spin-polarised electrons through thin ferromagnetic films is treated in terms of elastic forward scattering of electrons. General symmetry arguments are used to derive the interaction operator which governs the process. Experimental results reported to date rule out spin-exchange scattering of the projectile electrons as a mechanism for the spin filtering action. The analogy of the observed rotation of the spin component normal to the magnetisation and Faraday effect in magneto-optics is discussed.

I. Introduction

It has been reported[1,2] that a thin ferromagnetic foil permanently magnetised in its plain transmits low-energy electrons as a spin filter. If the electrons are incident at right angles to the foil and are initially polarised parallel or antiparallel to the magnetisation, the elastic transmission takes place with different probability, $\omega(+)$ or $\omega(-)$, respectively. Moreover, if the electrons are initially polarised in the plane normal to the magnetisation, the polarisation of the transmitted electrons is rotated about the direction of the magnetisation in a manner reminiscent of Faraday effect in magneto-optics.

The purpose of this paper is to describe these phenomena in terms of elastic electron scattering without attempting a dynamic description of the underlying interactions. In Sec. II the basic properties of the spin filter is introduced on a grossly simplified model. In Sec. III. account is taken of the finite thickness of the target yielding a qualitative agreement between theoretical and experimental results. Some remarks are made on the possible role of spin exchange collision in Sec. IV. In Sec. V the rotation of the electron polarisation is discussed in some detail in comparison with the classical magneto-optical effect of Faraday rotation.

II. The Spin Filter

We shall consider a scattering target which is characterised by its magnetisation $\vec{\mu}$, and is situated at the origin of a right handed Cartesian reference frame fixed in the laboratory. The axes are defined by the unit vectors

$$\hat{e}_1,\, \hat{e}_2,\, \hat{e}_3 \;=\; \hat{e}_1 \times \hat{e}_2,$$

and will be referred to as the 1, 2 and 3 axis. The magnetisation is an axial vector taken as right handed and parallel to the 1-axis

$$\vec{\mu} \;=\; \mu_0\, \hat{e}_1\,.$$

The electrons are incident with a momentum $\vec{p}$ of arbitrary direction. As we are interested in elastic transmission, elastic forward scattering is considered only, *i.e.* the momentum of the transmitted electrons is the same.

If we consider a single incident electron in the spin state $|\chi\rangle$, its spin state after transmission will be

$$|\chi\rangle \;=\; \hat{M}\,|\chi\rangle.$$

As the electron is a spin π-$\frac{1}{2}$ particle, $|\chi\rangle$ is a two element column vector, the operator $\hat{M}$ is a 2×2-matrix, and can be expanded in terms of the 2×2-unit matrix σ_0 and the Pauli matrixs. We take quantisation direction along the 1-axis, so that

$$\sigma_0 \;=\; \begin{vmatrix} 1 & 0 \\ 0 & 1 \end{vmatrix}\,, \quad \sigma_1 \;=\; \begin{vmatrix} 1 & 0 \\ 0 & -1 \end{vmatrix}\,, \quad \sigma_2 \;=\; \begin{vmatrix} 0 & 1 \\ 1 & 0 \end{vmatrix}\,, \quad \sigma_3 \;=\; \begin{vmatrix} 0 & -i \\ i & 0 \end{vmatrix}\,,$$

i.e. the operator takes the form:

$$\hat{M} \;=\; g_0\,\sigma_0 \;+\; a_1\,\sigma_1 \;+\; a_2\,\sigma_2 \;+\; a_3\,\sigma_3\;.$$

The task now is to determine the components of the vector $\vec{a}$ in such a manner that the operator $\hat{M}$ will be a scalar operator, *i.e.* invariant under spatial inversion *and* time reserval. The arguments used in the construction of the operator are similar to those applied in the treatment of electron scattering from chiral and oriented molecules.[3] For our purpose there are three vectors available. A polar vector: $\vec{p}$, which changes sign on spatial inversion and changes sign under time reversal, and two axial vectors: $\vec{\mu}$ and $\vec{\sigma}$ both of which are invariant under spatial inversion (like magnetic field generated by a current loop, and the angular momentum of a spinning top), but change sign under time reversal. The following scalar combinations of these vectors have the required invariance:

$$\vec{\mu}\cdot\vec{\sigma},\; (\hat{p}\cdot\vec{\mu})\,(\hat{p}\cdot\vec{\sigma}),$$

where $\hat{p} \;=\; \vec{p}/|p|$ is the unit vector parallel to the momentum. Hence the required operator is

$$\hat{M} \;=\; \tilde{g}_0\,\sigma_0 \;+\; \tilde{g}_1\,(\vec{\mu}\cdot\vec{\sigma}) \;+\; \tilde{g}_2\,(\hat{p}\cdot\vec{\mu})\,(\hat{p}\cdot\vec{\sigma})\,, \tag{1}$$

where the three quantities, $\tilde{g}_j$ $(j \;=\; 0,\, 1,\, 2)$, are in general complex numbers determined by the dynamics of the scattering process. It should be noted that $\hat{M}$ is invariant under the reversal of the direction of $\hat{p}$.

It is seen

$$\vec{\mu} \cdot \vec{\sigma} = \mu_0 \sigma_1 , \quad \hat{p} \cdot \vec{\mu} = \mu_0 \cos \theta,$$

$$\hat{p} \cdot \vec{\sigma} = \sigma_1 \cos \theta + \sigma_2 \sin \theta \cos \phi + \sigma_3 \sin \theta \sin \phi,$$

θ being the polar angle and ϕ the azimuth of the momentum vector. Substituting these expressions, the interaction operator takes the form:

$$\hat{M} = \tilde{g}_0 \sigma_0 + (\tilde{g}_1 + \tilde{g}_2 \cos^2 \theta) \mu_0 \sigma_1 + \frac{1}{2} \tilde{g}_2 \mu_0 \sin 2\theta (\sigma_2 \cos \phi + \sigma_3 \sin \phi) \ldots \quad (1a)$$

We shall investigate more closely two cases of special interest:

- Case 1: $\vartheta = 0$, the momentum is parallel to the magnetisation.
- Case 2: $\vartheta = (\frac{\pi}{2})$, the momentum is at right angles to the magnetisation.

The third term in Eq. (1a) vanishes in both cases, so we find

$$\hat{M}_1 = \hat{M}(\theta = 0) = \tilde{g}_0 \sigma_0 + \tilde{h}_1 \mu_0 \sigma_1 , \quad \tilde{h}_1 = (\tilde{g}_1 + \tilde{g}_2);$$

$$\hat{M}_2 = \hat{M}(\theta = \pi / 2) = \tilde{g}_0 \sigma_0 + \tilde{h}_2 \mu_0 \sigma_1 , \quad \tilde{h}_2 = \tilde{g}_1 .$$

Note that in matrix form, both operators are diagonal:

$$\hat{M}_j = \begin{pmatrix} \tilde{g}_0 + \tilde{h}_j \mu_0 & 0 \\ 0 & \tilde{g}_0 - \tilde{h}_j \mu_0 \end{pmatrix}, \quad j = 1, 2. \quad (2)$$

These diagonal operators describe spin filters, which transmit electrons in the eigenstates of the diagonal spin operator σ_1, usually referred to as electrons with "spin up" or "spin down" with respect to the 1-axis:

$$\hat{M}_j \begin{pmatrix} 1 \\ 0 \end{pmatrix} = (\tilde{g}_0 + \tilde{h}_j \mu_0) \begin{pmatrix} 1 \\ 0 \end{pmatrix}, \quad \hat{M}_j \begin{pmatrix} 0 \\ 1 \end{pmatrix} = (\tilde{g}_0 - \tilde{h}_j \mu_0) \begin{pmatrix} 0 \\ 1 \end{pmatrix}.$$

The corresponding transmission probabilities are:

$$\omega_j (+) = |\tilde{g}_0 + \tilde{h}_j \mu_0|^2 , \quad \omega_j (-) = |\tilde{g}_0 - \tilde{h}_j \mu_0|^2 .$$

Note that the transmission probabilities arise from the interference of transmission via different channels characterised by different "scattering amplitudes". This becomes more explicit if the moduli and phase angles ε of the two scattering amplitudes are introduced:

$$\tilde{g}_0 = g_0 \exp(-i \varepsilon / 2) , \quad \tilde{h}_j = h_j \exp(i \varepsilon / 2);$$

and hence

$$\omega_j (+) = {g_0}^2 + (h_j \mu_0)^2 + 2 g_0 h_j \mu_0 \cos(\varepsilon),$$

$$\omega_j (-) = {g_0}^2 + (h_j \mu_0)^2 - 2 g_0 h_j \mu_0 \cos(\varepsilon). \quad (3)$$

The quality or efficiency of the filter can be described by the fractional difference of the two transmission probabilities

$$q_j = \frac{\omega_j (+) - \omega_j (-)}{\omega_j (+) + \omega_j (-)} = \frac{2 g_0 h_j \mu_0 \cos \varepsilon}{{g_0}^2 + (h_j \mu_0)^2}. \quad (4)$$

There is no filter action if the two amplitudes are at quadrature ($\varepsilon = \pi / 2$).

In order to describe the polarisation of the transmitted beam even if the incident beam is only partially polarised the use of the density matrices is an appropriate device[4] or the closely related Stokes vector formalism.[5,6]

Let us denote the beam intensity I and the components of the electron polarisation, $\vec{P}\,(|P| \leq 1)$, which is a vector in real space, by

$$I = S_0, \qquad P_\alpha = \frac{S_\alpha}{S_0};$$

$$I' = S_0', \qquad P_\alpha' = \frac{S_\alpha'}{S_0'} \quad (\alpha = 1, 2, 3),$$

for the incident beam and the transmitted beam, respectively. Then it can be shown that S and S' form "four-vectors", linked to one another by a linear transformation:

$$S_n' = \sum_{m=0}^{3} s_{nm}\, S_m,$$

where

$$s_{nm} = \mathrm{tr}\,[\sigma_m\, \hat{M}_j\, \sigma_n\, \hat{M}_j^{\,*}]\,, \qquad (m,\, n = 0 \ldots 3)\,, \qquad (j = 1,\, 2).$$

Here (*) denots the Hermitian conjugate.

The elements of the 4×4 "transfer-matrix" s_{nm} are real numbers, formed by bilinear combinations of the scattering amplitudes. As a consequence of the diagonal structure of the interaction operator $\hat{M}_j$ the transfer matrix turns out to consist of two non-zero 2×2 submatrices:

$$\begin{pmatrix} I' \\ I'P_1' \\ I'P_2' \\ I'P_3' \end{pmatrix} = \begin{pmatrix} g_0^2 + (h_j\,\mu_0)^2 & 2\,g_0 h_j \mu_0 \cos\varepsilon & 0 & 0 \\ 2\,g_0 h_j \mu_0 \cos\varepsilon & g_0^2 + (h_j \mu_0)^2 & 0 & 0 \\ 0 & 0 & g_0^2 - (h_j \mu_0)^2 & -2\,g_0 h_j \mu_0 \sin\varepsilon \\ 0 & 0 & 2\,g\,h_j \mu_0 \sin\varepsilon & g_0^2 - (h_j \mu_0)^2 \end{pmatrix} \begin{pmatrix} I \\ IP_1 \\ IP_2 \\ IP_3 \end{pmatrix} \tag{5}$$

where beam intensities and polarisation components have been re-introduced. Subscripts $j = 1, 2$ refer to case 1 or case 2, respectively.

The diagonal structure of the matrix (5) follows from the symmetries of the process. In particular, it follows from the axial vector nature of the magnetisation and the spin polarisation, and from the independence of the $\hat{M}$-matrix (1) on the sign of $\hat{p}$ (the sign of $\hat{p}$ is therefore physically irrelevant).

One must bear in mind the geometrical difference between cases 1 and 2 in the laboratory frame.

- Case 1: Beam axis is parallel to the magnetisation, *i.e.* the 1-axis. Hence the longitudinal component of the polarisation is aligned with the magnetisation; while any transverse polarisation component is at right angles to the magnetisation.
- Case 2: Beam axis is at right angles to the magnetisation, *i.e.* lies in the (2,3) plane. Hence the polarisation component P_1 is now transverse component but is aligned with the magnetisation, while polarisation components in the (2,3) plane are normal to the magnetisation.

Yet the structure of the transfer matrix is the same in both cases, which means that with respect to the magnetisation direction the polarisation vector behaves in the same way. This behaviour is readily read off (5) and is summarized in the subsequent paragraphs (A–D). It should be stressed again that the numerical values with subscripts $j = 1$ and $j = 2$ are in general different, so that while the qualitative behaviour is the same, the quantitative values can be different in the two cases.

(A) If the incident beam is unpolarised, the transmitted beam will be attenuated and polarised parallel to the direction of magnetisation

$$I_j' = g_0^2 + (h_j m_0)^2 I \,, \qquad \vec{P}_j' = q_j \hat{e}_1 \,, \qquad (j = 1, 2). \qquad (6a)$$

(B) If the incident beam has a polarisation component parallel to the magnetisation, the reversal of the polarisation direction leads to a transmission asymmetry

$$A_j = \frac{I'(\vec{P}) - I'(-\vec{P})}{I'(\vec{P}) + I'(-\vec{P})} = q_j P_1, \qquad (6b)$$

as follows from Eqs. (4) and (5). Furthermore, the component P_1 will in general change. We obtain from Eqs. (4) and (5)

$$P_1' = \frac{q_j + P_1}{1 + q_j P_1} .$$

For $P_1 = 0$ the polarisation component

$$P_1' = q_j, \qquad (6c)$$

will be produced.

(C) From Eqs. (6b) and (6c) follows

$$A_j (P_1 = 1) = P_1' (P_1 = 0) = q_j . \qquad (6d)$$

This is characteristic of the performance of spin filters and is usually stated as the "analysing power" of the filter equal to its "polarising power".

(D) Let us now consider the case where the incident beam is polarised entirely in the plane normal to the magnetisation ($P_1 = 0$). Two events will then take place simultaneously. A polarisation component P_1' will be produced, and the polarisation in the plane normal to $\vec{\mu}$ will be rotated by an angle δ_j around the 1-axis.

Because of the diagonal structure of the matrix (5) P_1' is given by Eq. (6c). Let us denote the angle between the total final polarisation P' and the 1-axis by ζ, that is

$$\tan \zeta = \frac{P_1'}{\sqrt{P_1'^2 + P_2'^2 + P_3'^2}}.$$

If the incident electrons are completely polarised ($P_2^2 + P_3^2 = 1$), then the final electrons are necessarily completely polarised too ($P' = 1$) (see *e.g.* Kesler[4]). Hence, we obtain

$$\sin \zeta = P_1' = q_j. \qquad (7)$$

Let us now consider the change of the transverse components. From the relation between the initial and final polarisation components it follows that if

$$\frac{P_3}{P_2} - \tan \phi \quad \text{then} \quad \frac{P_3'}{P_2'} = \tan (\phi + \delta_j),$$

$$\tan \delta_j = \left(\frac{(P_3'/P_2') - (P_3/P_2)}{1 + (P_3'/P_2')(P_3/P_2)} \right)_j = \frac{2 g_0 h_j \mu_0}{g_0^2 - h_j \mu_0^2} \sin \varepsilon . \qquad (8)$$

Note that the precession angle δ_j is equal to that maintained between the resultant amplitudes

$$\tilde{a}_j = \tilde{g}_0 + \tilde{h}_j \, \mu_0 \, , \qquad \tilde{b}_j = \tilde{g}_0 - \tilde{h}_j \, \mu_0$$

in the complex plane.

Suppose the direction of the electron beam, after its first passage through the foil were reversed for another passage without changing the final polarisation. Neither the quantisation axis nor μ would be affected. Thus if the polarisation vector in the plane normal to the magnetisation was rotated in the first passage by an angle δ_j, the second passage would *increase* the rotation by the same amount. This behaviour is characteristic of the rotation of the plane of polarisation in the Faraday effect of classical magneto-optics.

III. Assembly of Scatterers

In order to obtain a more realistic view of the polarisation effects in the elastic transmission through a magnetised foil, one has to take into account the finite thickness of the target.

Let use assume that the target consists of an assembly of elements of magnetisation μ and they are aligned with the z-axis. We shall continue to assume that the electrons observed after transmission had undergone a single elastic interaction, but in the target some attenuation takes place due to scattering off the forward direction. The transmission will be described in terms reminiscent of an optical treatment, namely as the propagation of a plane wave through a refractive optical medium. The essence of the argument is to associate a *difference* of the wave vector (inverse of the electron wavelength) within the medium and in free space with the elastic forward scattering amplitude. This amounts to saying that one associates the forward scattering amplitude with a "refractive index". But as it was seen, there are two coherent elastic forward scattering channels, and correspondingly there are two different refractive indices which govern the propagation of the "electron waves".

Consider the passage of polarised electrons as the propagation of a plane wave of number $k = 1 / \lambda$ through a thin magnetised ferromagnetic film ($\lambda = h / p$ is the electron wavelength). It is assumed that the wave is incident normal to the film and in it only single scattering takes place. It can be shown[7] that the vectors within the medium, k' and in free space, k are related by

$$k'^2 + k^2 = 4 \pi \rho F(0),$$

where ρ is the number density of scattering atoms and $F(0)$ is the elastic forward scattering amplitude determined by the dynamics of the interaction involved. As $F(0)$ is in general complex, k' is complex, too. Assuming the $\mathrm{Im}(k') \ll \mathrm{Re}(k') \approx k$, one obtains to the first approximation:

$$k' = \left(k + \frac{2 \pi \rho}{k}\right) F(0) \, .$$

If the incident wave has an amplitude $|\psi\rangle$ then, from sample of thickness d, it emerges with the amplitude $|\psi'\rangle = \exp(i k' d) |\psi\rangle$.

This relation can be applied to link the initial and final spin states in the case of spin dependent transmission by replacing the forward scattering amplitude with the interaction operator $\hat{M}$ for forward scattering. Thus one obtains a new operator to determine the final spin state $|\psi'\rangle$ from the initial state $|\psi\rangle$:

$$|\psi'\rangle = \hat{N}\,|\chi\rangle\,.$$

Omitting a common phase factor, the new operator will be

$$\hat{N} = \exp\left[i\,\alpha\,d\,(g_0\,\sigma_0 + \mu_0\,h_j\,\sigma_1)\right] \quad \text{with} \quad \alpha = \frac{2\,\pi\,\rho}{k}. \tag{9}$$

By expanding the exponential function in a power series, one readily finds that this operator has the same structure as its "parent" (2) with diagonal elements

$$\exp\left[i\,\alpha\,d\,(g_0 + \mu_0\,h_j)\right] \quad \text{and} \quad \exp\left[i\,\alpha\,d\,(g_0 - \mu_0\,h_j)\right].$$

Following the same procedure as before, one finds for the polarisation independent attenuation

$$\left(\frac{I'}{I}\right)_{j,\,(P_1=0)} = \exp\left[-2\,\alpha\,d\,\mathrm{Im}\,(g_0)\right]\,\cosh\left[2\,\alpha\,d\,\mu_0\,\mathrm{Im}\,(h_j)\right]. \tag{10a}$$

Note that if the target is demagnetised $\mu_0 = 0$ gives the exponential attenuation of the beam expected from the total cross section

$$Q = \left(\frac{4\,\pi}{k}\right)\,\mathrm{Im}\,(g_0),$$

according to the "optical theorem".

For the transmission asymmetry A_j and the angle of rotation Θ_j, respectively, one finds:

$$A_j = -\tanh\left[2\,\alpha\,d\,\mu_0\,\mathrm{Im}\,(h_j)\right], \tag{10b}$$

$$\theta_j = -2\,\alpha\,d\,\mu_0\,\mathrm{Re}\,(h_j). \tag{10c}$$

For the polarisation rotation into the direction of the magnetisation Eq. (7) remains valid with (10b) for A_j.

Equations (10a)–(10c) with (7) qualitatively reproduce the experimental results in their energy and target thickness dependence. Moreover given the target parameters which define d and α together with three measured quantities, (I'/I), A, and Θ, these equations allow the determination of the moduli of the forward scattering amplitudes and their phase shifts.

IV. Spin Exchange Interaction

In the context of elastic electron-atom scattering it has been pointed out, that the rotation of electron polarisation in Mott scattering is analogous to the rotation of the plane of polarisation of light by optically active substances, while the spin rotation in exchange scattering is analogous to the Faraday rotation.[8]

Although there is a world of difference between an assembly of polarised one-electron atoms aligned in a magnetic field and a ferromagnetic medium, there is a similarity between them. From the point of view of elastic forward scattering both are targets with non uniform population of the target spin states. In contrast to an assembly of spin polarised atoms created by some external process of polarisation (Stern-Gerlach-effect, optical pumping, etc.), in a ferromagnetic film nature produces an unbalanced spin polarisation in the split d-shell by exchange interaction.

In exploring the possible role of spin exchange scattering in the elastic transmission of polarised electrons through ferromagnetic films one would start by amending the mathematical formalism for the treatment of spin exchange scattering from atoms[4] by replacing the "target polarisation vector" by the magnetisation vector. The outcome of such considerations is that, as in the case of electron-atom scattering, the "analysing power" and the "polarising power" of the spin exchange interaction are different. This is in contrast to currently available experimental evidence.

V. The Rotation of Electron Polarisation and Faraday Effect

It is interesting to compare the general results derived in Sec. II for the behaviour of the electron polarisation with the Faraday effect in optics which can be summarised as follows.

If a static magnetic field is applied to any transparent material, the plane of polarisation of a linearly polarised beam of light gets rotated if the light is propagating parallel to the magnetic field. If the light is reflected back through the medium, the rotation is multiplied in each transit. (This is in contrast to that observed in natural optical rotation where the rotation is cancelled of reflecting the light back through the medium.) If the angle between the direction of the light beam and of the magnetic field, say κ, varies the angle of the polarisation rotation varies with $\cos\kappa$, *i.e.* it vanishes if the light beam is at right angles to the magnetic field.[9] What is similar between the behaviour of electron polarisation and Faraday effect and what is different between them calls for some comments.

It was seen that an initial spin polarisation perpendicular $\vec{\mu}$ will be rotated about $\vec{\mu}$ by an angle δ. This effect, which can be observed in both cases 1 and 2, may well be called the electron-optic analogue of Faraday rotation. Yet there is a subtle difference between the two. In case 1 the momentum vector $\vec{p}$ and the magnetisation $\vec{\mu}$ are parallel (or antiparallel) to one another. Their product $\vec{p} \cdot \vec{\mu}$ (a pseudo-scalar) changes its sign under space inversion and defines a "screw sense", or handedness. Hence the inverted system (like a left handed glove) cannot be transformed back to its original form (into a right handed glove) by a rotation.

In case 2 the momentum vector $\vec{p}$ and the magnetisation $\vec{\mu}$ are at right angles to one another, hence $\vec{p} \cdot \vec{\mu} = 0$, no handed system is defined. But look separately at the two factors: under space inversion the polar vector $\vec{p}$ transforms into $-\vec{p}$, but it leaves the axial vector $\vec{\mu}$ invariant. Therefore the inverted system now can be transformed back to its original form by a rotation about the vector $\vec{\mu}$, in contrast to what applied in case 1.

Nevertheless, a rotation of the polarisation about the magnetisation can be observed in both cases. This shows that in experiments under consideration it is not necessary to define a screw-sense by the geometry of the experiment. It is sufficient to introduce a "sense of rotation". This is provided here by the magnetisation vector $\vec{\mu}$, an axial vector, primarily defined by a sense of rotation (*e.g.*, its current loop). The direction of $\vec{p}$ is without relevance.

These considerations are reflected by the structure of the interaction matrix (1). The first two terms would suffice in principle to produce "Faraday rotation", and the matrix $\hat{M}$ is invariant under the reversal of $\vec{p}$.

There are however other experiments in which the effect of choosing different geometries, as in cases 1 and 2, show up explicitly. For example if the incident electron beam is initially unpolarised, in case 1 the transmitted beam would be longitudinally polarised; in case 2 it would emerge transversely polarised.

Why does the polarisation rotation behave differently in the cases of electrons and light in the "transverse" geometry of case 2? The answer lies in the fundamental difference between electrons and photons and their polarisation states. Suppose linearly polarised light is incident at right angles to $\vec{\mu}$. A rotation about $\vec{\mu}$ would violate Maxwells transversality condition (which is related to the fact that photons have zero rest mass). Hence in an optical experiment in the "transverse" set-up as in case 2 no rotation of the plane polarisation can be expected.

References

1. D. Oberli, R. Burgermeister, S. Riesen, and H. C. Siegmann, *Phys. Rev. Lett.* **81**, 4228 (1988)
2. W. Weber, D. Oberli, and H.C. Siegmann, *New J. Phys.* **1**, 9.1-9.6 (1999), (http://www.njp.org).
3. K. Blum and D. G. Thompson, *Adv. At. Mol. Opt. Phys.* **38**, 39 (1997)
4. J. Kessler, *Polarized Electrons*, 2nd edition (Springer, Berlin, 1985)
5. W. H. McMaster, *Am. J. Phys.* **22**, 351 (1954)
6. P. S. Farago, *Comments Atom. Mol. Phys.* **6**, 99 (1977)
7. M. F. Mott and H. S. W. Massey, *The Theory of Atomic Collisions*, 2nd edition, (Clarendon Press, Oxford, 1965).
8. J. Byrne, P. S. Farago, *J. Phys. B* **4**, 954 (1971).
9. M. Born, *Optiks* (Springer, Berlin, 1933), Chap. 7, p. 363.

Influence of an Atomic Grating on a Magnetic Fermi Surface

T. Greber, W. Auwärter, and J. Osterwalder

Physik-Institut
Universität Zürich
Winterthurerstrasse 190
CH-8057 Zürich
SWITZERLAND

Abstract

The spin-split electronic band structure of a ferromagnet-insulator interface is investigated. For the model system of a monolayer of hexagonal boron nitride on Ni(111) it is found that the insulator strongly influences the Fermi surface of the underlying ferromagnet. Angular resolved photoemission around the Fermi level provides direct insight into the modification of the electronic structure in the whole surface Brillouin zone. The commensurate *h*-BN overlayer acts as an atomic grating that produces strong surface umklapp processes. The intensity ratio between the spin-split *s*-*p* bands at the Fermi level is discussed in view of spin scattering asymmetry.

I. Introduction

The coupling between two magnets may be mediated by magnetic stray fields, electron tunneling or scattering. The electron mediated coupling mechanisms give rise to phenomena like the giant magneto resistance (GMR) *i.e.* an electrical resistance that strongly depends on the relative orientation of the magnetizations in two separated layers.[1,2] The physics of this coupling mechanism involves a spin dependent transmission coefficient for electrons moving parallel to or across the junction and is still under debate. In this context the knowledge of the Fermi surface and its evolution across the interface is of key importance.

The present paper shall contribute to a better understanding of spin dependent coupling in non-magnetic tunneling junctions. It foots on previous angular resolved photoemission experiments on clean nickel[3,4] and investigates the influence of a non-magnetic insulator on the Fermi surface. Cuts across the Fermi surfaces—as measured

Physics of Low Dimensional Systems
Edited by J. L. Morán-López, Kluwer Academic/Plenum Publishers, New York 2001

with photoemission[5]—of the bare Ni(111) and a monoatomic layer of hexagonal boron nitride on Ni(111) are compared. Perfect single layer h-BN films can be grown on Ni(111) and they are expected to behave like an insulator since the h-BN conduction band lies above the Fermi level.[6] Here, it is shown that the non-magnetic insulator acts as a two-dimensional grating that multiplies states at the Fermi level and modifies the Ni bulk Fermi surface near the interface. The direct photoemission transitions from the exchange split s-p bands are exploited as spin polarized electron sources. A strong spin dependent emission asymmetry is found in comparing the down to up intensity ratio in Ni(111) and h-BN/Ni(111). It has to be emphasized that in this kind of experiments the spin has not to be measured—this would require a magnetized sample (and a spin detector)—but is resolved in k-space *i.e.* from photoemission angle and energy. The assignment of majority or minority spin is deduced from band structure calculations.

II. Experimental

The experiments were performed in a modified VG ESCALAB 220 photoelectron spectrometer with He Iα ($\hbar\omega = 21.2$ eV) and He Iβ ($\hbar\omega = 23.1$ eV) radiation.[7,8] These two photon energies are used in order to probe the same k-space sections of the two samples which have different work functions. In order to get the work function, spectra are taken at normal emission and a bias voltage of -9 V is applied to the sample in order to resolve the secondary electron emission cut off. From the width ΔE of the spectra the work function $\Phi = \hbar\omega - \Delta E$ is determined. For Ni(111) a work function of 5.1 eV and for h-BN/Ni(111) of 3.3 eV is found. The overall energy/momentum resolution was better than 50 meV/ 0.02 Å^{-1} FWHM and all presented data were taken at room temperature. The vector $k_{\|}$ is determined from the electron kinetic energy in the vacuum E_{kin} and the polar emission angle θ with respect to the optical surface: $k_{\|} = \sin(\theta)\sqrt{2m_e E_{\text{kin}}}/\hbar$. The sample preparation and characterization is described elsewhere.[6,9] It is found that h-BN forms perfect (1×1) commensurate monolayers on Ni(111).

III. Results and Discussion

In Fig. 1 Fermi surface maps of bare Ni(111) and h-BN/Ni(111) are shown. Essentially the photoemission intensity from the Fermi level is displayed on a linear gray scale for different emission angles in parallel projection. The maps correspond to a two dimensional cut across the k-space. In the free electron final state picture this two dimensional cut is the surface of a sphere with radius $k_f = \sqrt{2m_e(\hbar\omega - E_B - \Phi + U}/\hbar$, where m_e is the free electron mass, $\hbar\omega$ the photon energy, E_B the binding energy ($\equiv 0$ at the Fermi level), Φ the work function and U the inner potential (10.7 eV).[5] The momentum of the photon is neglected and direct transitions obey the momentum coservation

$$\mathbf{k}_f = \mathbf{k}_i + \mathbf{G}, \tag{1}$$

where $\mathbf{k}_f$ is the final state wave vector inside the solid, $\mathbf{k}_i$ the initial state wave vector and $\mathbf{G}$ is a reciprocal lattice vector of the three dimensional bulk lattice *and/or* two dimensional surface lattice. Clearly the Fermi surface map of Ni(111) has three fold rotational symmetry [Fig. 1(a)]. This makes sure that at least two atomic layers of nickel are probed. The Fermi surface map of h-BN/Ni(111) is as well three fold symmetric as it has to be expected from the atomic structure of the h-BN overlayer.[9,10]

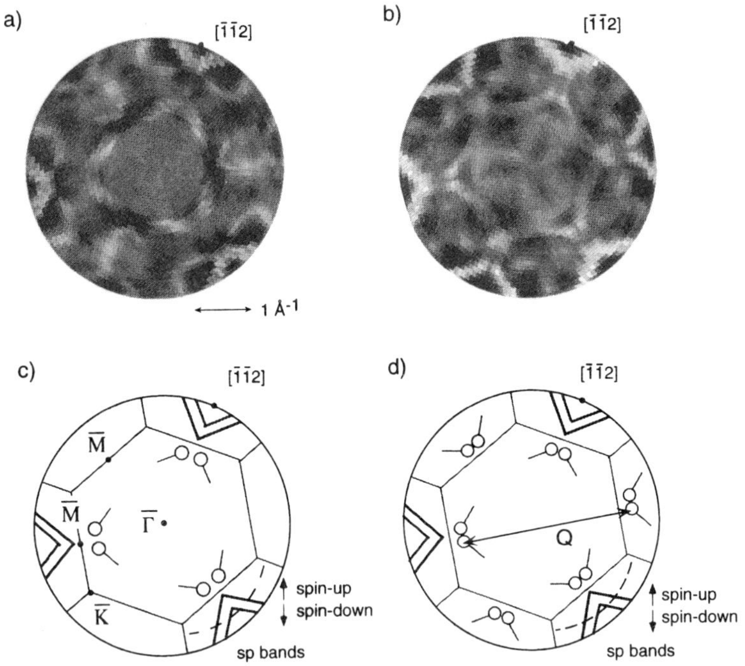

Figure 1. Fermi surface maps as measured by angular resolved photoemission. The intensity on a linear gray scale (white means high intensity) is displayed as a function of $\sin\theta$ and φ, θ being the polar and φ the azimuthal emission angle. The high symmetry direction $[\bar{1}\bar{1}2]$ is indicated. (a) Ni(111) measured with He Iβ radiation ($k_{\|} \leq 2.15$ Å^{-1}). (b) h-BN/Ni(111)) measured with He Iα radiation ($k_{\|} \leq 2.15$ Å^{-1}). (c) Schematic diagram of the sampled k-space in (a). The high symmetry points of the surface Brillouin zone are $\bar{\Gamma}$, $\bar{M}$, $\bar{K}$. The $\wedge$ wedge like features and the pairs of "signal disks" indicate s-p derived, spin-split bands. The dashed line across the wedges indicates the azimuthal cut in Figs. 2 and 3. (d) Schematic diagram of the sampled k-space in (b). The three signal disk pairs appear shifted by a primitive reciprocal surface lattice vectors $|\mathbf{Q}| = (2\pi/a)(2/\sqrt{3})$, where a is the surface lattice constant.

However, in comparing Figs. 1(a) and 1(b) it can be seen that this insulating overlayer with no electronic states on the Fermi surface, strongly influences the shape of the Fermi surface and therefore the electronic coupling across the interface. The Fermi surface gets distorted and new features emerge. In the following we would like to highlight a particular spin-split feature from the s-p band and a surface umklapp. In Figs. 1(c) and 1(d) the surface Brillouin zones of the corresponding maps in Figs. 1(a) and 1(b) are shown. The two features that are further discussed are schematically sketched. There are the two slightly shifted $\wedge$ wedge like features in the second surface Brillouin zones along the $[\bar{1}\bar{1}2]$ direction and the features in the first surface Brillouin zone that are mimicked as a pair of "signal disks". Both features are mainly s-p derived bands.[11] They are cuts across the Fermi surface in the Γ_{002} (wedges) and the Γ_{111} (signal disks) bulk Brillouin zones close to the $L_{\frac{1}{2}\frac{1}{2}\frac{3}{2}}$ point. The signal-disk like features change their shape in going from Ni(111) to h-BN/Ni(111). More importantly, in the h-BN/Ni(111) case [Figs. 1(b) and 1(d)] it is seen that the three signal disk pairs are replicated three more times. They are shifted by a primitive reciprocal surface lattice vector $|\mathbf{Q}| = (2\pi/a)(2/\sqrt{3}) = 2.9$ Å^{-1}, where $a = 2.49$ Å

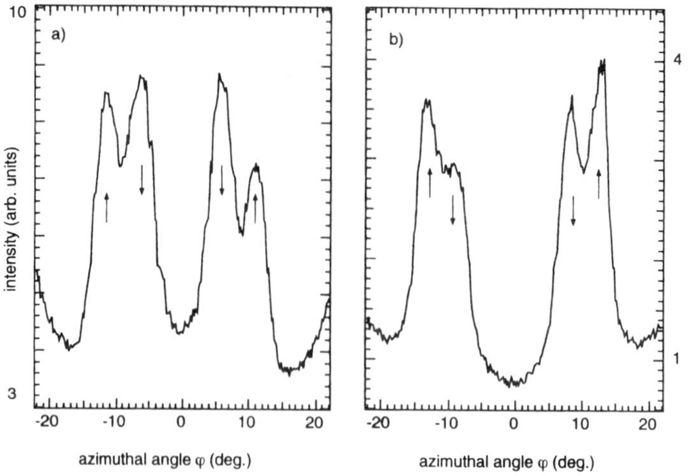

Figure 2. The photoemission intensity at the Fermi level is plotted *versus* azimuthal angle. The polar emission angle ($\theta = 78°$) is kept constant. $\varphi = 0$ is defined in the $(\bar{1}10)$ plane. The arrows indicate the minority $\downarrow$ and majority $\uparrow$ bands, respectively. (a) Ni(111) measured with He Iβ radiation. (b) h-BN/Ni(111) measured with He Iα radiation. Note the strong change in the intensity ratio between minority and majority spin transitions.

is the surface lattice constant. This is a surface umklapp process where the lattice vector **G** in the photoemission [see Eq. (1)] contains as well an element of the two dimensional reciprocal surface lattice. Note that the h-BN/Ni(111) system has still the same (1×1) symmetry as the clean Ni(111) surface. We take the occurence of such surface umklapps as an indication that the h-BN layer acts as an efficient grating for any electrons that cross this interface and that therefore such umklapps influence the tunneling characteristics of such junctions. Since the s-p bands are exchange splitted (see below) this will affect as well the spin asymmetry in the tunneling current.

In Fig. 2 two cuts across the s-p wedges in the second surface Brillouin zone of Ni and h-BN/Ni(111) are displayed. For Ni (h-BN/Ni(111)) the same polar emission angle $\theta = 78°$ and the photon energies of 23.1 (21.2) eV were chosen in order to sample the same initial state locations.

The azimuthal emission angle was scanned $\pm 22°$ around the $(\bar{1}10)$ plane. The four peaks represent cuts across the Fermi-surface. They are characterized by their positions (δk_{ex} and $\delta k_{\uparrow\uparrow}$), their width γ and their area A. In Table I the left/right averaged quantities are summarized. The large group velocity (see Fig. 3) identifies the bands as the s-p bands. The polarization can be assigned from comparison with band structure calculations where the inner peaks reflect the minority s-p band while the outer ones those with majority spin.[11] The exchange splitting δk_{ex} between the majority and the minority s-p band is 0.19 (0.16) Å^{-1}. From the change in the angular splitting between the majority s-p bands ($\delta k_{\uparrow\uparrow}$) it can be seen that the Fermi surface gets slightly distorted by the h-BN interface. It has to be noted that the low work function of h-BN/Ni(111) with respect to the instrument work function may cause a slight violation of the nominal $k_{\|}$ conservation. This is due to the electrostatic "lensing effect" since the different work functions induce electric fields in the vacuum. At these large polar emission angles we guess that in the measurement $k_{\|}$ is overestimated by about 4%. Therefore the $\delta k_{\uparrow\uparrow}$ value of h-BN/Ni(111) has to be considered as slightly overestimated, while control experiments with He Iα radiation on Ni(111) indicate that δk_{ex} is not affected by this shift in $k_{\|}$. The minority *s-p* band peak width $\gamma_{\downarrow}$ is about 30% larger than that of the majority bands ($\gamma_{\uparrow}$). This is in line with a shorter lifetime of minority excitations.[12] In three dimensional systems, however, the connection between the angular broadening and the initial-and final state liftimes is quite involved.[13] The intensity variation of the spin up/spin down doublets left and right from the high symmetry plane are caused by the loss of mirror symmetry due to the oblique incidence of the photons in our experimental set up. It provides a rough estimate for the change of the photoemission matrix element with respect to the orientation of the incoming light. In the following the area ratio $\Upsilon = A_{\downarrow} : A_{\uparrow}$ shall be discussed. For Ni(111) ($\Upsilon_{Ni(111)}$=1.7) it does not correspond to that on Ni(100) ($\hbar\omega$ = 44 eV) found by Petrovykh *et al.*[14,16] (0.56) nor to that on Ni(110) (0.8).[15,16] Therefore Υ may bank on the experimental parameters and/or the crystal face.[17] As it is shown below even the change of Υ in going from Ni(111) to *h*-BN/Ni(111) is influenced by more than one physical mechanism. The change in the area ratio $\Upsilon = A_{\downarrow} : A_{\uparrow}$ of 1.7 for Ni(111) to 1.1 for *h*-BN/Ni(111) bears information on the spin dependent electron transmission coefficient. The *h*-BN overlayer clearly alters the intensities of the spin polarized direct photoemission transitions at this particular place in *k*-space. This behaviour of decreasing Υ upon adsorption of a non-magnetic layer can be related to spin-dependent scattering of the electrons during the propagation to the detector. The data shown in Fig. 2 indicate that minority photoelectrons get much more efficiently scattered. This is in line with Siegmann's rule[18] stating that at low kinetic energies the electron scattering cross section is essentially proportional to the number of valence band *d*-holes of a material. Thus for ferromagnets where the *d*-holes are polarized a spin filter effect is expected and minority spins are scattered more efficiently. In the case of *h*-BN Siegmann's rule predicts no spin filtering since *h*-BN has no polarized *d*-shell. Therefore our finding of a strong asymmetry of the spin-transition intensities calls for an extension of Siegmann's rule: it signals that spin filtering may also occur in non-magnetic overlayers that are coupled to a magnetic substrate. This coupling may be mediated by hybridisation of delocalized *s-p* states extending into the interface and/or by the surface umklapps of the *h*-BN grating on Ni(111) that increase the available phase space for electron hole pair excitations. Though the umklapps may play a crucial role for the understanding of the magnetic coupling across this interface it has to be emphasized that these very same umklapps may differently influence the matrix elements of the two spin channels shown in Fig. 2. Therefore the (essential) assumption of a constant matrix element for a quantitative determination of the spin-scattering asymmetry may be hampered by surface umklapp processes.

In Fig. 3 two dispersion plots of the *s-p* bands of Ni and *h*-BN/Ni(111) in the second surface Brillouin zone are shown. In order to make transitions from thermally populated states above the Fermi level visible, the data have been normalized with an experimentally determined Fermi function. The contrast is the best near the Fermi level since there the broadening due to the hole lifetime is minimal. The steep slope of the bands indicates a large group velocity and identifies the bands to be *s-p* like. In Fig. 3(b) new features at azimuthal angles of $\pm 10°$ are resolved. These features have a maximum contrast at about 100 meV above the Fermi level. With the present data set the origin of these extra features can not be unambiguosly determined. It is likely that their origin is a surface umklapp. They may thus influence the matrix element of the photoemission process and alter the intensities of the *s-p* transitions at the Fermi level as shown in Fig. 2. Before any quantitative statement on the spin-scattering asymmetry such features have to be understood.

Table I

Parameters from the fits of four Gaussians to *s*-*p* bands at the Fermi level in the second surface Brillouin zone (see Fig. 2). The k-values are calculated from $k = \sin(\varphi) \cdot k_{\|}$, where φ is the angle measured from the $(\bar{1}10)$ plane and $k_{\|} = \sin(\theta)\sqrt{2m_e E_{\rm kin}}/\hbar$, ($\theta = 78°$).

Sample	$\delta k_{\rm ex}$	$\delta k_{\uparrow\uparrow}(\text{Å}^{-1})$	$\gamma_{\uparrow}(\text{Å}^{-1})$	$\gamma_{\downarrow}(\text{Å}^{-1})$	$A_{\downarrow} : A_{\uparrow}$
$Ni(111)$	0.19(2)	0.83(2)	0.13(2)	0.17(2)	1.71(8)
h–BN/Ni(111)	0.16(2)	0.94(8)	0.12(2)	0.15(2)	1.08(5)

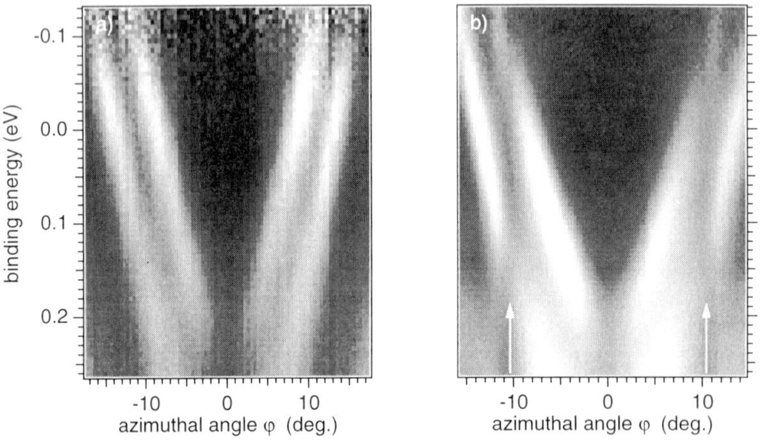

Figure 3. Photoemission around the Fermi level. Dispersion plots along the azimuthal angles of Fig. 2. The bright lines correspond to the spin-split *s*-*p*-bands. In order to visualize the bands above the Fermi level the data are normalized with an experimentally determined Fermi function. (a) Ni(111) measured with He Iβ radiation. (b) h-BN/Ni(111) measured with He Iα radiation. The arrows at $\varphi \sim \pm 10°$ indicate the positions where new spectral features (best seen 100 meV above the Fermi level) start to interfere with the 'native' *s*-*p* bands.

Conclusions

In conclusion, it has been shown how an advanced photoemission experiment may contibute to the better understanding of magnetic tunneling. It is of key importance to know the electronic states in k-space in order to comprehend electronic coupling across interfaces. For the model system of a layer of hexagonal boron nitiride on nickel it is demonstrated that a non-magnetic atomic grating may produce additional features on the Fermi surface. These surface umklapps are proposed to play a non-negligible role for the detailed understanding of tunneling in general and spin-polarized tunneling *i.e.* magnetic coupling in particular. For the understanding of intensity ratios and their exploitation for the determination of relevant physical quantities as *e.g.* spin scattering anisotropies, k-space has to be explored in detail. Support from high quality *ab initio* calculations is required.

Acknowledgments

Technical assistance from B. Schmid, P. Treier and W. Deichmann and financial support from the Schweizerischen Nationalfonds are gratefully acknowledged.

References

1. *Physics Today* **45**, April 1995.
2. M. N. Baibich, J. M. Broto, A. Fert, F. Nguyen Van Dau, F. Petroff, P. Etienne, G. Creuzet, A. Friedrich and J. Chazelas, *Phys. Rev. Lett.* **61**, 2472 (1988).
3. P. Aebi, T. J. Kreutz, J. Osterwalder, R. Fasel, P. Schwaller, and L. Schlapbach, *Phys. Rev. Lett.* **76**, 1150 (1996).
4. T. Greber, T. J. Kreutz, and J. Osterwalder, *Phys. Rev. Lett.* **79**, 4465 (1997).
5. P. Aebi, J. Osterwalder, R. Fasel, D. Naumović, and L. Schlapbach, *Surf. Sci.* **307**, 917 (1993).
6. A. Nagashima, N. Tejima, Y. Gamou, T. Kawai, and C. Oshima, *Phys. Rev. B* **51**, 4606 (1995).
7. T. Greber, O. Raetzo, T. J. Kreutz, P. Schwaller, W. Deichmann, E. Wetli, and J. Osterwalder, *Rev. Sci. Instrum.* **68**, 4549 (1997).
8. J. Osterwalder, T. Greber, J. Kröger, J. Wider, H.-J. Neff, F. Baumberger, M. Hoesch, W. Auwärter, R. Fasel, and P. Aebi, these proceedings, p. 245.
9. W. Auwärter, T. J. Kreutz, T. Greber and J. Osterwalder, *Surf. Sci.* **429**, 229 (1999).
10. Y. Gamou, M. Terai, A. Nagashima, and C. Oshima, *Sci. Rep. RITU* **A44**, 211 (1997).
11. T. J. Kreutz, T. Greber, P. Aebi, and J. Osterwalder, *Phys. Rev. B* **58**, 1300 (1998).
12. M. Aeschlimann, M. Bauer, S. Pawlik, W. Weber, R. Burgermeister, D. Oberli, and H. C. Siegmann, *Phys. Rev. Lett.* **79**, 5158 (1997).
13. J. K. Grepstadt, B.J. Slagsvold, and I. Bartos, *J. Phys. F* **12**, 1679 (1982).
14. D. Y. Petrovykh, K. N. Altmann, H. Höchst, M. Laubscher, S. Maat, G. J. Mankey, and F. Himpsel, *Appl. Phys. Lett.* **73**, 3459 (1998).
15. T. J. Kreutz, P. Aebi, and J. Osterwalder, *Sol. State Commun.* **96**, 339 (1995).
16. F. J. Himpsel, K. N. Altmann, G. J. Mankey, J. E. Ortega, and D. Y. Petrovykh, *J. Magn. Magn. Mater.* **200**, 456 (1999).
17. "In electron spectroscopies the intensity is, in contrast to energy, spin and momentum, a "bad quantum number" *i.e.* the matrix element and the transmission may vary from experiment to experiment. H. C. Siegmann private Communication, Berlin 1994.
18. H. C. Siegmann, *J. El. Spectr. and Rel. Phen.* **68**, 505 (1994).

Non-Equilibrium Physics in Solids: Hot-Electron Relaxation

K. H. Bennemann

Institut für Theoretische Physik
Freie Universität Berlin Arnimallee 14
D-14 195 Berlin
GERMANY
e-mail: khb@physik.fu-berlin.de

Abstract

We discuss the time scales for relaxation of excited electrons due to laser irradiation in transition-metals, ferromagnets like Ni, Co, high T_c-superconductors, and semiconductors like diamond and graphite. Ultrafast relaxation faster than ps may occur as a result of strong electron-electron interactions, while magnetoelastic forces controlling magnetic reorientation and domain dynamics, for example, involve relaxation times of the order of 100 ps or more.

I. Introduction

Due to recent advances in time resolved optical studies, laser physics, one may observe the non-equilibrium behaviour of solids, in particular metals and semiconductors, in which many electrons have been excited by strong laser irradiation. Then, using pump-probe spectroscopy one observes the time-resolved response. Thus, two-photon photoemission (2PPE) studied the lifetimes of excited electrons in Cu,[1] in the ferromagnets Ni, Co,[2] the relaxation of the magnetization in Ni with many hot electrons caused by laser irradiation,[3] the relaxation of excitations in high T_c-superconductors,[4] and laser induced phase transitions in carbon, graphitization of diamond by electron-hole pairs.[5] Thus, ultrafast fs-responses were observed, while earlier experiments studying magnetic reorientation dynamics and spin lattice relaxation at surfaces and in thin films observed a slower response of the order of 100 ps.[6] In the following we discuss the general situation and the time scales for the various relaxation phenomena. For simplicity, we use "golden-rule" type arguments.

The physical situation is illustrated in Fig. 1. Upon laser irradiation many excited electrons are present and put the solid in a non-equilibrium state. After a short

Physics of Low Dimensional Systems
Edited by J. L. Morán-López, Kluwer Academic/Plenum Publishers, New York 2001

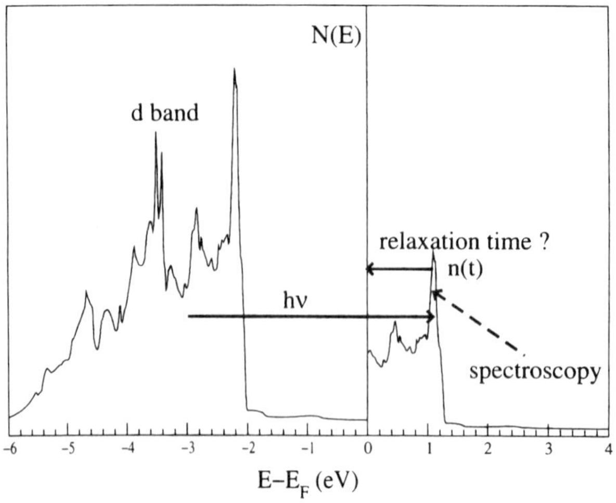

Figure 1. Illustration of the scenario for the non-equilibrium state produced by exciting optically electrons out of the ground state (the Fermi sea). The hot electrons relax first due to electron-electron interactions. The electron thermalization is characterized by the electron temperature $T_{\mathrm{el}}(t)$. Energy transfer from the electrons to the lattice occurs via the electron-lattice coupling and which controls $T_{\mathrm{el}}(t)$ for longer times.

time the hot electrons thermalize again due to electron-electron interactions. This thermalization may involve up to 100 fs or more. *d*-electrons will thermalize faster than *s*-electrons, for example. As a result the electrons will aquire an electronic temperature $T_{\mathrm{el}}(t)$ which will change with time and become different from the lattice temperature $T_{\mathrm{latt}}(t)$, $T_{\mathrm{el}} > T_{\mathrm{latt}}$. As time progresses energy is transferred from the hot electron system to the lattice via the electron lattice coupling. This occurs during a time of $t \geq$ 200–300 fs, roughly, and continues for several ps and more, and finally $T_{\mathrm{el}}(t) \to T_{\mathrm{latt}}(t)$ again after many ps. Hence, after laser irradiation T_{el} increases first, reaches a maximum at a time which is set by the intensity of the laser light and the duration of the laser pulse, the electron-electron interaction strength and the one of the electron-lattice coupling.[3,7]

Due to the strong electron-electron interaction in transition-metals and noble metals one expects for the excited electrons relaxation times $\tau(E) \sim (E - E_F)^{-2} \sim N^{-1}(0)$, hence of the order of 100–10 fs depending on $(E - E_F)$ of course, the matrix-elements and the density of states at the Fermi-energy E_F, $N(0)$. Thus, τ in Cu and Ag is expected to be larger than in Ni, Co, and Fe.[8]

Due to the hot electrons the magnetization in itinerant ferromagnets will change and becomes time dependent, $M(t)$. On general grounds, one expects

$$M(t) \to M[T_{\mathrm{el}}(t)],$$

as soon as the excited electrons have thermalized again. Consequently response times of the electron relaxation and of the (itinerant) magnetization should be of the same order, possibly faster than 100 fs in view of the strength of the electron-electron interaction and the exchange interaction. Note, changes of the magnetization must obey angular momentum conservation and thus possibly may involve atomic spin-orbit coupling (which is of the order of 50–70 meV in Ni and Fe, for example). We expect the magnetization to decrease upon laser irradiation up to a minimum, when

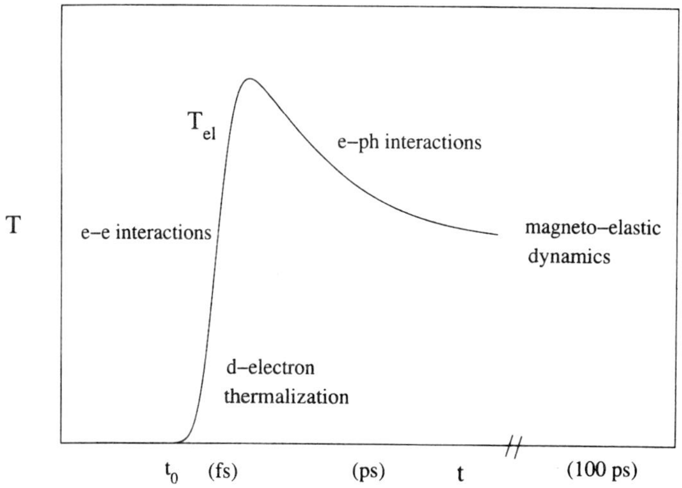

Figure 2. Time scales for the dynamics of the non-equilibrium state. Electron thermalization occurs during 10–100 fs. Thus, $T_{\text{el}}(T) \to T_{\text{latt}}$ after some ps. After a longer time, $t \geq 100$ ps, magnetoelastic dynamics involving domains, magnetic reorientation and spin-relaxation in a warm lattice occurs. Note, for itinerant metals changes of the magnetization happen on the same time scale as those of the electronic distribution.

$T_{\text{el}}(t) \to$ max. and then to relax again when $T_{\text{el}}(t) \to T_{\text{latt}}(t)$. Note, hot electrons affect also the exchange coupling J and then the coupling between magnetic films, for example.

In contrast, magnetoelastic responses controlled by spin-lattice coupling and magnetic anisotropy energies (of the order of 10^{-5} eV in transition-metals) will occur during times of the order of 100 ps or longer, for example. This will be the case for magnetic reorientation at surfaces and in thin films spin-relaxations in hot lattices, and magnetic domain dynamics.

In semiconductors like graphite, diamond electron-hole excitations cause a change of bonding, for example of $sp^3 \to sp^2$ bonding in graphitization of diamond upon laser irradiation.[9] Again, such structural changes involve relatively strong electron-electron interactions and, in terms of phonons, Brillouin-zone boundary phonons. Thus response times of the order of several 100 fs and faster than ps-times are expected. The optical control of $sp^3 \rightleftharpoons sp^2$ bonding is of course a fascinating problem and may lead to optical switching of amorphous $\rightleftharpoons$ crystalline transitions and diamond $\rightleftharpoons$ graphite under special conditions.[10]

For high T_c-superconductors one expects an interesting dynamical behaviour, since Cooper-pairing and antiferromagnetic interactions are involved.[11] Thus, upon laser irradiation destroying superfluid density n_s, Cooper-pair density, and also causing antiferromagnetic excitations one expects corresponding relaxations as where recently observed by Kaindl *et al.*[4] On general grounds one expects $\tau_1 \sim n_s^{-1} \sim T_c^{-1} \sim$ several ps for the superfluid relaxation, and $\tau_3 \sim \Delta E_{\text{mag}}^{-1} \sim (T^*)^{-1}$ for the antiferromagnetic relaxation time, $\tau_3 \simeq (T_c/T^*)\tau_1$. Here, ΔE_{mag} is the characteristic energy for a magnetic excitation also characterized by the temperature T^*. Since $(T_c/T^*) \ll 1$ one expects $\tau_3 < \tau_1$, as is observed.[4]

In summary, we have outlined the time-scales for important relaxation processes in solids occuring upon laser irradiation. This is illustrated in Fig. 2.

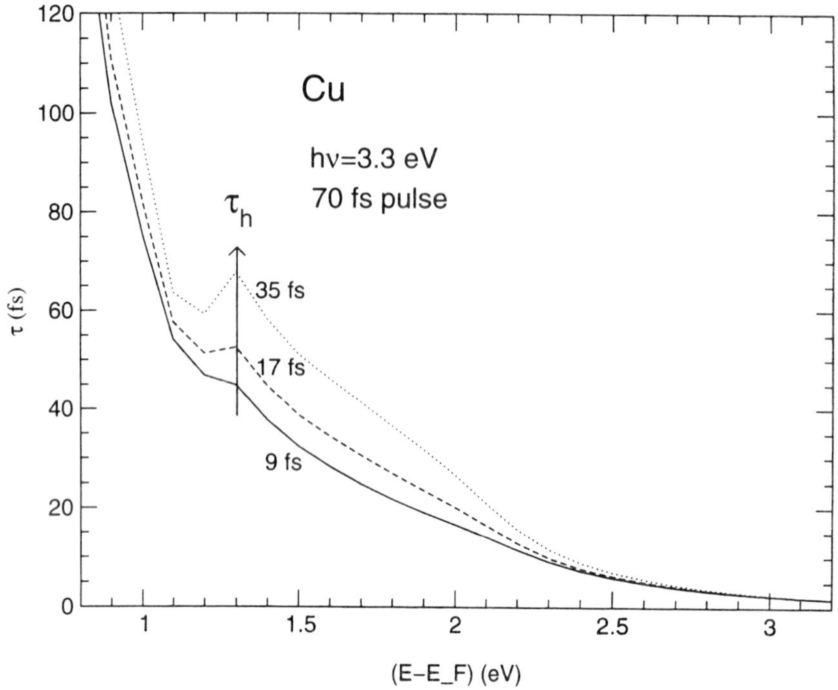

Figure 3. 2PPE spectra for Cu and dependence of the lifetime τ referring to the level occupation dynamics on the lifetime τ_h of holes. The latter controls the contribution of Auger electrons to τ. The structure may not appear for Au, for example, where τ_h is smaller.

In the following we present in an illustrative way results for the dynamics in non-equilibrium solids, supposedly illustrating the general situation. For calculating the non-equilibrium distribution of electrons we use the Boltzman-equation.[2] Similarly, $T_{\rm el}(t)$ is calculated using two coupled equations for $\dot{T}_{\rm el}(t)$ and $\dot{T}_{\rm latt}(t)$.[3]

First, in Fig. 3 we show for Cu the relaxation time τ referring to the dynamics of the electron occupation $n(E,t)$ as deduced from 2PPE.[2] Note, τ includes contributions from secondary electrons, in particular Auger-electrons. Of course, one expects τ to depend on the lifetime τ_h of the holes near E_F. From experiment one estimates $\tau_h \simeq$ 15 fs. The structure on $\tau(E)$ results from the contribution of the Cu-d band which begins approximately 2 eV below E_F. The exciting photons have the energy $\hbar\omega \simeq$ 3.6 eV. The position of the side peak is hence as expected. Height and position of the structure is in fair agreement with experiment. The results indicate the role played by Auger electrons.[12]

In Fig. 4 we show how 2PPE results depend on the duration T of the exciting laser pulse. Experiments should verify this. We conclude that Auger electrons are not negligible for the 2PPE spectrum of Cu, smaller fingerprint of this is also expected for Au, but not for Ni, Co, etc.[12]

In Fig. 5 we show results for the lifetime τ of hot electrons in Cu and Co, for example.[2] The smaller values of τ in Co result from the larger density of states $N(E)$ near E_F. However, note, agreement with experimental results by Aeschlimann *et al.*[2] is only obtained if the characteristic matrix-element M for electronic collisions is ~ 0.5 in Co rather than 0.9 as in Cu. This may result from the fact that the d-electrons

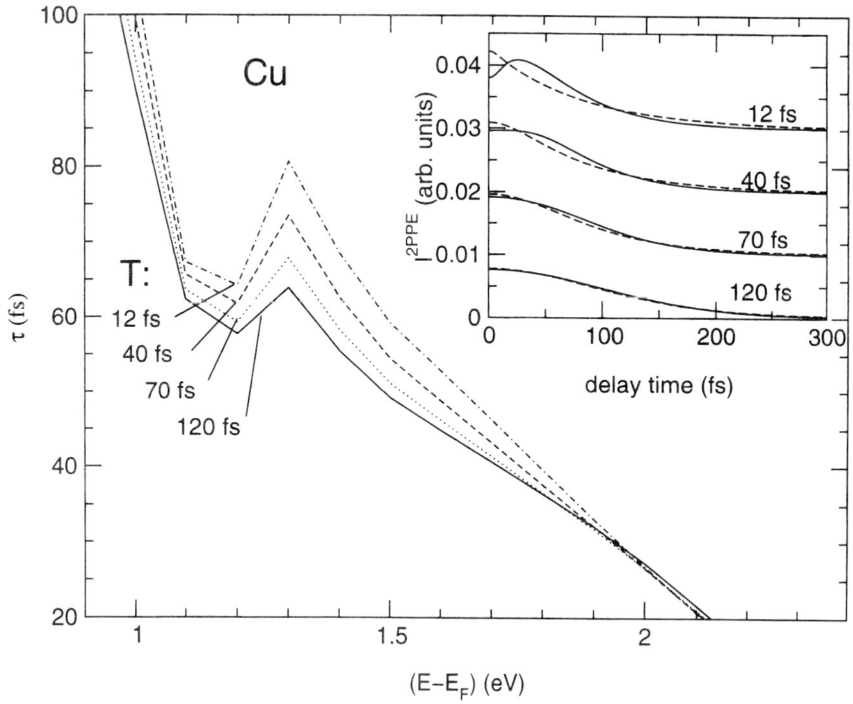

Figure 4. Dependence of relaxation of hot electrons in Cu on duration T of the exciting laser pulse.

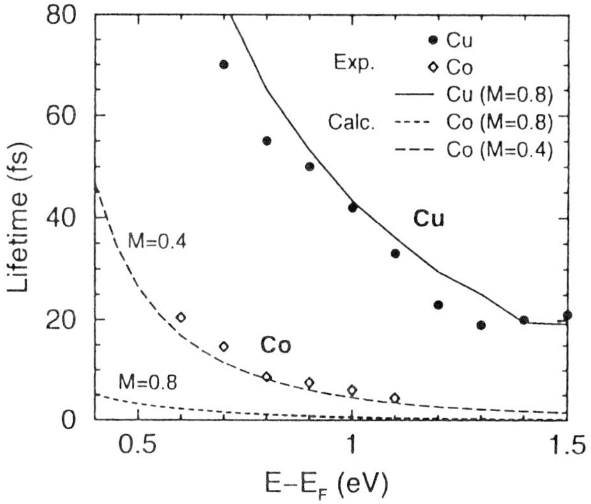

Figure 5. Lifetime τ of excited electrons in Cu and Co. Note, the electron-electron interaction matrix-element M is taken to be smaller for Co then for Cu due to d-electron screening.

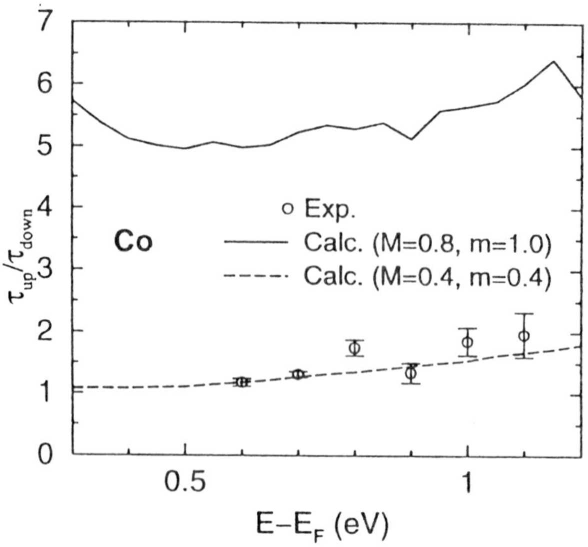

Figure 6. Spin-dependent lifetime $\tau_\sigma(E)$ for Co. Note the dependence on M.

in Co contribute strongly to the screening of the Coulomb interactions between the electrons.[2] Note, for $(E - E_F) \sim 1$ eV one finds already fast relaxations of the order of 40 fs or smaller. This sheds also light on the time required for thermalization of hot electrons.[2]

In Fig. 6 results are presented for the spin-dependent lifetime $\tau_\sigma(E)$. These seem to suggest that spin-flips are not playing a dominating role for the hot electrons.

The magnetic response upon laser irradiation is illustrated in the following. Generally we expect in itinerant ferromagnets

$$M(T_{\text{latt}}) \to M(t) \to M[T_{\text{el}}(t)] \to M(T_{\text{latt}}), \tag{1}$$

for non-equilibrium. Here, the electronic temperature is determined from the coupled master-type equations $\dot{T}_{\text{el}}(t)$ and $\dot{T}_{\text{latt}}(t)$.

To compare with experiments we analyze the dynamics expected for nonlinear magneto-optics, magnetic dichroism, Kerr-rotation, etc.[13] The second harmonic signal (SHG) $I(2\omega)$ is analyzed in the form[3,13]

$$\delta^- = I(M) - I(-M), \tag{2}$$

where $I \propto |E_i(2\omega)|^2$ and $E_1(2\omega) \propto F_{ijl}\chi_{ijl}E_jE_l$. Here, F_{ijl} refer to Fresnel factors and χ_{ijl} is the nonlinear susceptibility. Note, in general different sensor elements χ_{ijl} having generally different phases may contribute to $E_i(2\omega)$ and to the non-linear light intensity $I(2\omega)$. This involves phases of χ_{ijl} which may depend on the excitations and on time.[3,14] Also, in principle χ_{ijl} should be calculated for the non-equilibrium distribution of electrons. In order to simplify, quasi as a test of the situation, we assume[13,7]

$$\chi_{ijl}(M) = \chi^1_{ijl} + \chi^2_{ijl}M + \cdots, \tag{3}$$

splitting χ into an even and odd contribution in the magnetization M. Then, approximately,

$$\delta^-(t) \approx \frac{D(t)}{D(t_0)} - 1 \underset{t}{\longrightarrow} \frac{M(t)}{M(t_0)} - 1, \tag{4}$$

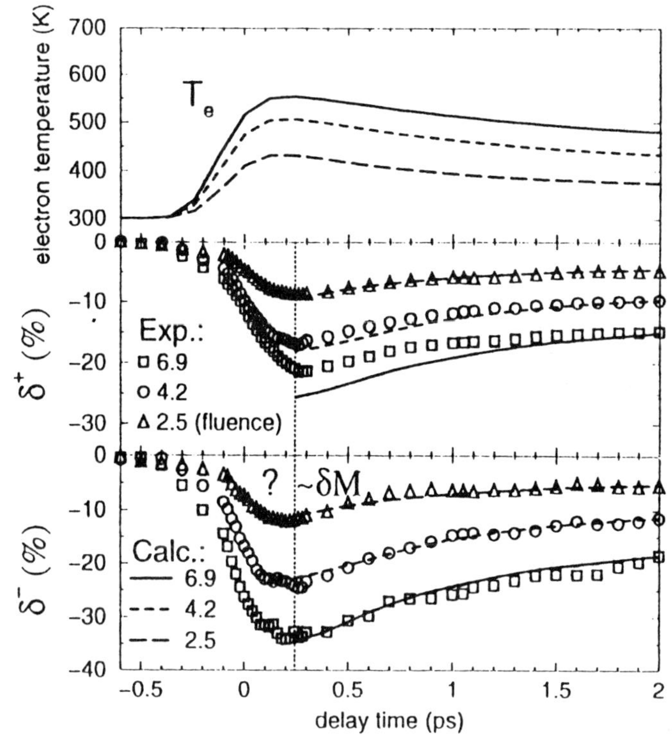

Figure 7. Results for $\delta^-(t) \sim I(\vec{M}) - I(-M)$ using SHG. $T_e(t)$ is the electronic temperature, $\delta^+(t) \sim I(\vec{M}) + I(-\vec{M})$. Note, T_e is max. when $\delta^-(t)$ is minimal. Theory and experiment compare well for $t \geq 200$ fs, but not for small t.

where the probe laser heating up the electrons starts at time t_0 and where $D(t) = I(M) - I(-M)$. The analysis assumes no time dependent phase effects, and essentially a few dominating χ_{ijl} determining $E_i(2\omega)$.[3,7]

In Fig. 7 we show results for $T_{el}(t)$ and the SHG-signal $\delta^-(t)$.[7] One concludes that for somewhat longer delay times in the pump probe analysis, $\Delta(t) \geq$ 100–200 fs, $\delta^-(t) \sim M(T_{el})/M(t_0) - 1$ holds approximately.

The magnetic response is very fast. Note, $M(t) \to 0$ upon laser irradiation for times $t \geq 50$ fs and $M(t) \to$ min. when $T_{el} \to$ max.[3,7]

In Fig. 8 we show that $M(t) \to 0$ for $T_{el} \geq T_c$, T_c is the Curie-temperature. This is in agreement with recent experiments.[15]

Further experiments are expected to support this general behaviour, namely, that electronic and magnetic relaxations have a similar time scale in itinerant ferromagnetic metals. Of course, this can also be deduced by using the Landau-Lifshitz Bloch equation.[7]

High-T_c-superconductors like $La_{1-x}Sr_xCuO_4$, YBCO, etc. seem to achieve Cooper pairing via exchange of an antiferromagnetic excitation between holes. Figure 9 characterizes the phase diagram with T_c being the superconducting transition temperature, $T_c \propto n_s$ for underdoped superconductors, and T^* being a critical temperature characteristic for a.f. excitations.[16] Hence, in such systems one expects dynamics upon laser irradiation due to breaking up Cooper-pairs and due to antiferromagnetic excitations.

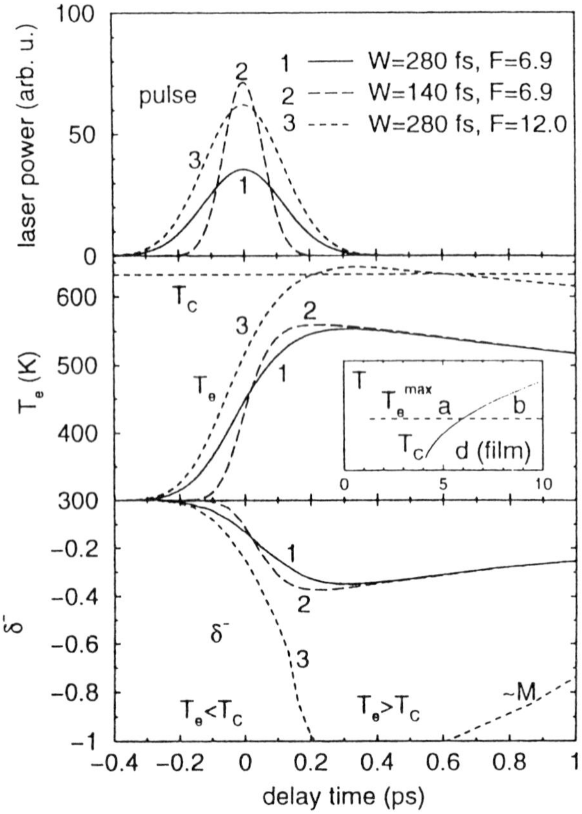

Figure 8. SHG results showing the dependence of the magnetic response $\delta^-(t)$ on the duration T of the exciting laser pulse and its intensity. Behaviour of the magnetic response $\delta^-(t)$ for $T > T_c$ being the Curie-temperature, is also shown.

In Fig. 9 we illustrate the expected dynamics. Regarding relaxation towards superfluidity, for the underdoped superconductors with doping $x < 0.15$, we have

$$\tau_1 \sim (\Delta\overline{e^{i\varphi}})^{-1} \sim T_c^{-1}, \tag{5}$$

where φ refers to the phase of the Cooper pairs and approximately $\overline{e^{i\varphi}} = 1$ for $T < T_c$ and $\overline{e^{i\varphi}} = 0$ for $T > T_c$. In accordance with experiments we estimate $\tau_1 \geq$ 1–3 ps. Furthermore, one expects for the magnetic excitations (ΔE_{af})

$$\tau_3 \sim \frac{1}{\Delta E_{\mathrm{af}}} \sim \frac{1}{T^*}, \tag{6}$$

which may be rewritten as

$$\tau_3 \simeq \frac{T_c}{T^*}\tau_1. \tag{7}$$

Hence, $\tau_3 < \tau_1$ and one gets a fast ($\tau_3 \leq 1$ ps) and a slower relaxation τ_1 as is observed.[4] If phase coherent Cooper pairs continue to exist above T_c, one would expect further relaxations characterized by τ_2. We note again, that the dynamic sheds light on the important physics behind high T_c-superconductivity.

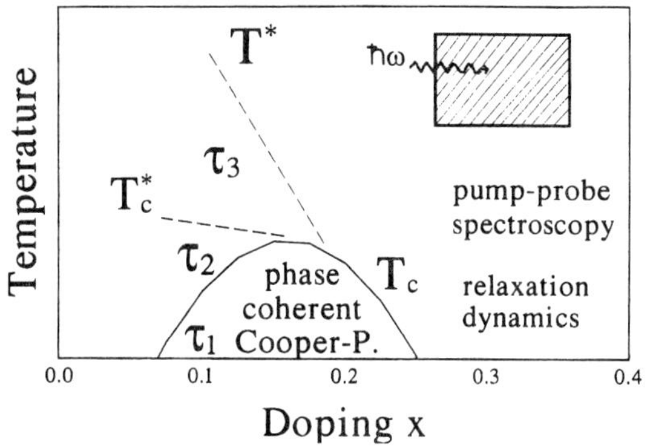

Figure 9. Dynamics in underdoped high-T_c-superconductors with transition temperature $T_c \sim n_s$ and temperature T^* characterizing a.f. excitations. τ_1 refers to relaxation of phase-coherent Cooper pairs and τ_3 to relaxation of a.f. excitations.

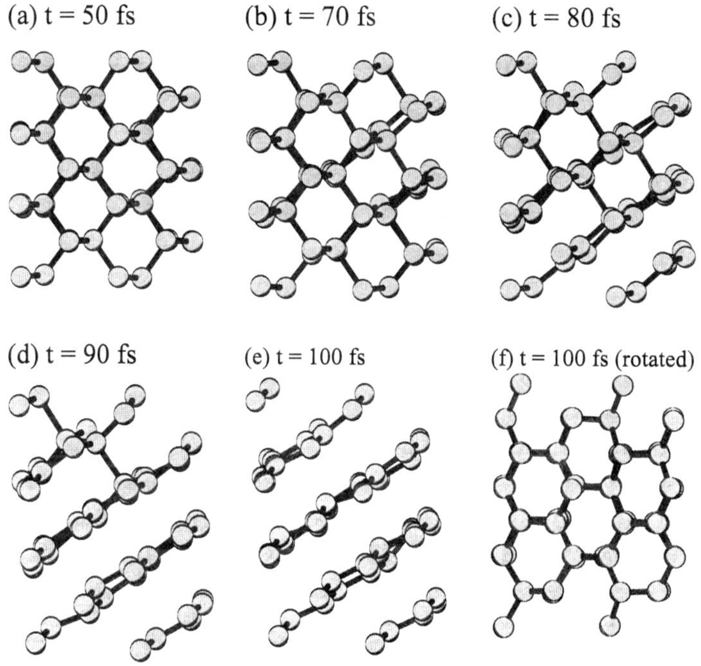

Figure 10. Graphitization of diamond upon laser irradiation producing electron-hole excitations. Note, the diamond structure has changed significantly for $t \geq 80$ fs.

Finally, we discuss bond-breaking dynamics, in particular the one related to $sp^3 \rightleftharpoons sp^2$ bond formation and controlled optically. Obviously, $sp^3 \rightarrow sp^2$ involves metallization and structural change like graphitization of diamond.[5,9]

In Fig. 10 we show by using a tight-binding approximation for the s, p electrons and molecular-dynamics for the atomic structure, the ultrafast structural changes in diamond upon laser irradiation creating many electron-hole pairs and thus $sp^3 \rightarrow sp^2$ bond changes. After times of the order of 80 fs or more the structure changes.

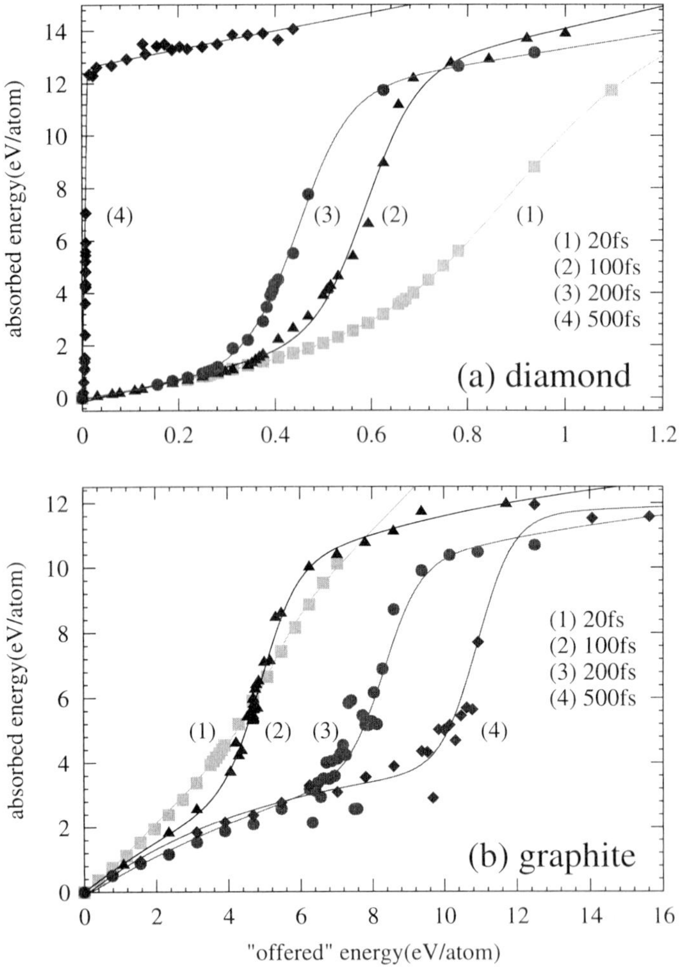

Figure 11. Absorbed energy from a laser pulse in diamond (a) and in graphite (b).

In Fig. 11 we show how the light absorption in diamond and graphite depends on the duration T of the exciting laser pulse.[17] This sheds light on the role played by T for the non-equilibrium behaviour.

Structural changes which take some time explain the opposite behaviour observed for diamond and graphite. In the case of graphite more energy is absorbed from short pulses than from long ones, for diamond the opposite is true. Atomic disorder becoming effective for longer pulses causes in graphite less absorption due to bond changes, $sp^2 \rightarrow sp^2+sp^3$, while in diamond disorder consisting of partial graphitization and creation of states in the diamond gap between valence and conduction band causes more absorption. This shows again how time-dependent behaviour on a fs-time scale may reveal interesting physics.

In summary, the discussion of the ultrafast response of metals and semiconductors shows that pump and probe spectroscopy and relaxation of non-equilibrium states produced by laser irradiation may help to understand better important physical processes. For magnetic recording in particular magnetic reorientation transitions due to $T_{\rm el}(t)$ and domain dynamics is important.[18]

Acknowledgments

I thank first of all Prof. H. C. Siegmann for many helpful discussions during many years. Furthermore, I have to thank many people for help, in particular R. Knorren, H. Jeschke, Prof. E. Matthias, Dr. Manske, M. Wöste and E. v. Sulzburg.

References

1. M. Aeschlimann, M. Bauer, S. Pawlick, W. Weber, R. Burgemeister, D. Oberli, and H.C. Siegmann, *Phys. Rev. Lett.* **79**, 5158 (1997); E. Knoesel, A. Hotzel and M. Wolf, *Phys. Rev. B* **57**, 12812 (1998).
2. R. Knorren, K.H. Bennemann, R. Burgemeister, and M. Aeschlimann, *Phys. Rev. B*, April 2000.
3. J. Hohlfeld, E. Matthias, R. Knorren, and K. H. Bennemann, *Phys. Rev. Lett.* **78**, 4861 (1997); R. Knorren, and K. H. Bennemann, *Appl. Phys. B* **68**, 501 (1999).
4. R. A. Kaindl, M. Woerner, T. Elsaesser, D. C. Smith, J. F. Ryan, G. A. Farnan, M. P. McCurry, and D. G. Walmsley, *Science* **287**, 470 (2000).
5. H. Jeschke, FU thesis, Berlin (2000); J. Jeschke, M. E. Garcia, and K. H. Bennemann, *Appl. Phys. A* **69**, 49 (1999).
6. W. Hübner and K. H. Bennemann, *Phys. Rev. B* **53**, 1 (1996); H. C. Siegmann, private communication; D. R. Penn, S. P. Apell, and S. M. Girvin, *Phys. Rev. B* **32**, 7753 (1985); M. Aeschlimann, M. Bauer, S. Pawlik, W. Weber, R. Burgermeister, D. Oberli, and H. C. Siegmann, *Phys. Rev. Lett.* **79**, 5158 (1997).
7. R. Knorren and K. H. Bennemann, *Appl. Phys. B* **68** 401 (1999); *Proc. Stockholm, Conf. Magnetism*, August (1999), to be publ. (ed. K. V. Rao); R. Knorren, FU thesis, Berlin (2000).
8. H. C. Siegmann and M. Aeschlimann, private communication (these authors are amongst the first to use $\tau^{-1} \propto N(0)$ for the analysis).
9. H. Jeschke, M. Garcia, and K. H. Bennemann, to be published.
10. During crystal growth electron-hole excitations may play an important role; See also H. Jeschke, FU thesis, Berlin (2000).
11. D. Manske and K. H. Bennemann, *Low. Temp. Conference*, February (2000), Houston.
12. R. Knorren, G. Bouzerar, and K. H. Bennemann, *Phys. Rev. B*, to be publ. (2000).
13. T. A. Luce and K. H. Bennemann, in *Nonlinear Optics in Metals* (Clarendon Press, Oxford, 1998) p. 437.
14. J. Hohlfeld, R. Knorren, J. Güdde, and E. Matthias, private communication (1999).
15. U. Konrad, J. Güdde, V. Jähnke, and E. Matthias, *Appl. Phys. B* **68**, 51 (1999).
16. D. Manske, T. Dahm, and K. H. Bennemann, cond-mat/9912062.
17. H. Jeschke, M. Garcia, and K. H. Bennemann, to be publ. (2000).
18. Clearly, SHG and pump probe spectroscopy can be used to study such dynamics, and from $\tau \sim \Delta E_{\mathrm{anis}}^{-1}$, where ΔE_{anis} is the anisotropy energy, one estimates times of the order of 100 ps.

Density Functional Theory of the Lattice Fermion Model

R. Lopez-Sandoval and G. M. Pastor

Laboratoire de Physique Quantique,
Unité Mixte de Recherche 5626 du CNRS
Université Paul Sabatier
118 route de Narbonne
F-31062 Toulouse Cedex
FRANCE

Abstract

Recent developments on the density functional theory (DFT) of lattice fermion models are reviewed. Formally exact, self-consistent equations are derived for the single-particle density matrix γ_{ij} that involve derivatives of the interaction-energy functional $W[\gamma]$ and fractional occupations of natural orbitals $\eta_{k\sigma}$. The dependence of the correlation energy functional $E_C(\gamma_{12}) = W - E_{HF}$ on the nearest-neighbors (NN) density matrix γ_{12} is analyzed. A pseudo-universal scaling behavior of $\varepsilon_C = E_C/E_{HF}$ as a function of $g_{12} = (\gamma_{12} - \gamma_{12}^{\infty}) / (\gamma_{12}^{0} - \gamma_{12}^{\infty})$ is revealed, where γ_{12}^{0} (γ_{12}^{∞}) stands for the uncorrelated (strongly correlated) value of γ_{12}. Based on exact dimer results and on scaling properties of $E_C(\gamma_{12})$, a simple, explicit approximation to $W(\gamma_{12})$ is proposed for the Hubbard model. Ground-state energies and charge-excitation gaps of one- and two-dimensional lattices are obtained, in remarkable agreement with available exact results or accurate numerical simulations. The scope of DFT is thereby extended to the limit of strong electron correlations.

I. Introduction

Density functional theory (DFT) has been the subject of remarkable developments since its original formulation by Hohenberg and Kohn (HK).[1,2] After formal improvements, extensions, and an uncountable number of applications to a wide variety of physical problems, this theoretical approach has become the most successful method of determining the electronic properties of matter from first principles.[3,4] The main innovation of DFT, actually the origin of its breakthrough, is to replace the wave

Physics of Low Dimensional Systems
Edited by J. L. Morán-López, Kluwer Academic/Plenum Publishers, New York 2001

function by the electronic density $\rho(\vec{r})$ as the fundamental variable of the many-body problem. Most density functional (DF) calculations are based on the Kohn-Sham (KS) scheme which allows to determine several ground state properties from the solution of a set of self-consistent single-particle equations.[2] Although this transformation is formally exact, the practical implementations always require approximations, since the KS equations involve functional derivatives of the unknown interaction energy $W[\rho(\vec{r})]$ or, equivalently, of the exchange and correlation (XC) energy $E_{XC}[\rho(\vec{r})]$. Therefore, the functional dependence of $W[\rho(\vec{r})]$ or $E_{XC}[\rho(\vec{r})]$ are central to the development of DF methods. The currently most widespread approximations—the local density approximation (LDA)[2] with spin polarized[5] and gradient corrected extensions[6]—were originally derived from exact results for the homogeneous electron gas. It is one of the goals of the present work to investigate the properties of the interaction-energy functional from an intrinsically inhomogeneous point of view by considering many-body lattice models.

Despite the remarkable success of the local spin density approximation, present DFT fails systematically in accounting for phenomena where strong electron correlations are important, for example, in heavy-fermion materials or high-T_c superconductors.[7] These systems are usually described by simplifying the low-energy electron dynamics using parameterized lattice models such as Pariser-Parr-Pople, Anderson, or Hubbard Hamiltonians.[8–11] Being in principle an exact theory, the limitations of the DF approach have to be ascribed to the approximations used for exchange and correlation, but not to the underlying Hohenberg-Kohn-Sham formalism. Consequently, it is a very interesting challenge to extend the range of applicability of DFT to strongly correlated systems, and to characterize the properties E_{XC} in the limit of strong correlations. Studies of the XC functional on simple models should provide useful insights for future extensions to realistic Hamiltonians. Moreover, taking into account the demonstrated power of the DF approach in *ab initio* calculations, one may also expect that DFT, with an appropriate E_{XC}, could become an efficient tool for studying the physics of many-body models, a subject of theoretical interest on its own.

The concepts of DFT have been applied to lattice models already in previous works.[12–14] Gunnarsson and Schönhammer were, to our knowledge, the first to propose a DF approach on a semiconductor model in order to study the band-gap problem.[12] In this case, the local site occupancies γ_{ii} were taken as the basic variables of the DF formalism. Some years later Schindlmayr and Godby[13] provided a different formulation of DFT on a lattice by considering as basic variables both diagonal elements γ_{ii} and off-diagonal elements γ_{ij} of the single-particle density matrix (see also Refs. 15–17). Schönhammer *et al.* then derived a more general framework that unifies the two previous approaches.[14] Using Levy's constrained search method,[18] they showed that different basic variables and different W functionals can be considered depending on the type of model or perturbation under study. Site occupations alone may be used as basic variables, if only the orbital energies are varied (*i.e.*, if all hopping integrals t_{ij} are kept constant for $i \neq j$). However, off-diagonal elements γ_{ij} must be included explicitly if W is intended to be applied to more general situations where the values of t_{ij} are varied. As an example, consider the Hubbard model for different lattice structures or interaction regimes (*i.e.*, different U/t).

The present paper reviews some recent developments on the density-matrix functional theory (DMFT) of lattice models.[19] The basic formalism is recalled in Sec. II. Exact self-consistent equations for determining the ground-state γ_{ij} are derived that involve derivatives of the interaction-energy functional $W[\gamma]$, and fractional occupations of natural orbitals $\eta_{k\sigma}$. In Sec. III, an explicit form of Levy's constrained search of $W[\gamma]$ is given in terms of many-body eigenvalue equations. Exact results for the dependence of W on the nearest-neighbors (NN) density matrix γ_{12} are given for the Hubbard model, and a pseudo-universal scaling behavior of $\varepsilon_C = E_C/E_{HF}$ is identified. Based on exact dimer results and on the scaling properties of $E_C(\gamma_{12})$, a simple explicit approximation to $W(\gamma_{12})$ is proposed for the Hubbard Hamiltonian. Sec. IV discusses representative applications of this *Ansatz*. The performance of the present DF approach is analyzed in particular by comparison with exact results and accurate numerical simulations. Finally, Sec. V summarizes our conclusions.

II. Density Functional Theory on a Lattice

We consider a general many-body lattice Hamiltonian given by

$$H = \sum_{ij\sigma} t_{ij}\, \hat{c}_{i\sigma}^{\dagger}\hat{c}_{j\sigma} + \frac{1}{2}\sum_{\substack{ijkl\\ \sigma\sigma'}} V_{ijkl}\, \hat{c}_{i\sigma}^{\dagger}\hat{c}_{k\sigma'}^{\dagger}\hat{c}_{l\sigma'}\hat{c}_{j\sigma}\,, \tag{1}$$

where $\hat{c}_{i\sigma}^{\dagger}$ ($\hat{c}_{i\sigma}$) is the usual creation (annihilation) operator for an electron with spin σ at site (or orbital) i. H can be regarded as the second quantization of Schrödinger's equation on a basis.[20] However, in the present paper, the hopping integrals t_{ij} and the interaction matrix elements V_{ijkl} are taken as parameters to be varied independently. V_{ijkl} defines the type of many-body interactions which may be repulsive (Coulomb like) or attractive (in order to simulate pairing), and which are usually approximated as short ranged (*e.g.*, intra-atomic in the Hubbard model). The hopping matrix t_{ij} defines the lattice (*e.g.*, one dimensional chains, square or triangular two-dimensional lattices) and the range of the single-particle interactions (*e.g.*, up to first or second neighbors). Once the basis set has been chosen, t_{ij} is determined by the external potential.[20] Consequently, t_{ij} plays the role given in conventional DFT to the external potential $V_{\text{ext}}(\vec{r})$. Thus, the single-particle density-matrix γ_{ij} replaces the electronic density $\rho(\vec{r})$ as the basic variable of the DF formalism in a lattice. Note that γ_{ij} measures both the distribution of the electronic density among the sites, for $i = j$, and the degree of electron delocalization, for $i \neq j$.

The energy E_{gs} and density-matrix γ_{ij}^{gs} of the ground-state are obtained by minimization of the energy functional

$$E[\gamma] = E_K[\gamma] + W[\gamma], \tag{2}$$

with respect to γ.[13] The first term

$$E_K[\gamma] = \sum_{ij} t_{ij}\gamma_{ij}, \tag{3}$$

is the kinetic energy associated with the electronic motion in the lattice. The second term is the interaction-energy functional given by

$$W[\gamma] = \min_{\Psi \to \gamma} \left[\sum_{\substack{ijkl\\ \sigma\sigma'}} V_{ijkl} \langle \Psi[\gamma] | \hat{c}_{i\sigma}^{\dagger}\hat{c}_{k\sigma'}^{\dagger}\hat{c}_{l\sigma'}\hat{c}_{j\sigma} | \Psi[\gamma] \rangle \right]. \tag{4}$$

The minimization in Eq. (4) implies a search over all N-particle states $|\Psi[\gamma]\rangle$ satisfying $\langle \Psi[\gamma]| \sum_\sigma \hat{c}_{i\sigma}^{\dagger}\hat{c}_{j\sigma} |\Psi[\gamma]\rangle = \gamma_{ij}$ for all i and j.[18] Therefore, $W[\gamma]$ represents the minimal interaction energy compatible with a given density distribution and degree of delocalization as quantified by the density-matrix γ. Note that this involves both diagonal γ_{ii} and off-diagonal γ_{ij}.[13,15,16] $W[\gamma]$ is universal in the sense that it is independent of t_{ij}, *i.e.*, of the system under study. It depends on the considered model interaction V_{ijkl}, on the number of electrons N_e or band-filling $n = N_e/N_a$, and on the number of sites or orbitals N_a.

In order to derive the equations for determining E_{gs} and γ_{ij}^{gs}, the density matrix $\gamma_{ij} = \gamma_{ij\uparrow} + \gamma_{ij\downarrow}$ may be written, without loss of generality, in terms of its eigenvalues $\eta_{k\sigma}$ (occupation numbers) and eigenvectors $u_{ik\sigma}$ (natural orbitals) as

$$\gamma_{ij\sigma} = \sum_k u_{ik\sigma}\, \eta_{k\sigma}\, u_{jk\sigma}^* \,. \tag{5}$$

For the minimization of $E[\gamma]$, we introduce Lagrange multipliers μ and $\lambda_{k\sigma}$ to impose the constraints $\sum_{k\sigma} \eta_{k\sigma} = N_e$ and $\sum_i |u_{ik\sigma}|^2 = 1$. Derivation of

$$F = E[\gamma] + \sum_{k\sigma} \left[\lambda_{k\sigma}(1 - \sum_i |u_{ik\sigma}|^2)\right] + \mu\left(N_e - \sum_{k\sigma} \eta_{k\sigma}\right), \tag{6}$$

with respect to $u_{ik\sigma}^*$ and $\eta_{k\sigma}$ yields the eigenvalue equations

$$\sum_j \left(t_{ij} + \frac{\partial W}{\partial \gamma_{ij\sigma}}\right) u_{jk\sigma} = \varepsilon_{k\sigma} u_{ik\sigma} \,, \tag{7}$$

with $\varepsilon_{k\sigma} = \lambda_{k\sigma}/\eta_{k\sigma}$, and

$$\sum_{ij} \left(\frac{t_{ij} + \partial W}{\partial \gamma_{ij\sigma}}\right) u_{ik\sigma}^* u_{kj\sigma} = \mu \,. \tag{8}$$

Since $\partial F/\partial \eta_{k\sigma} = \varepsilon_{k\sigma} - \mu$, one of the following three cases applies at the minimum: *i)* $\eta_{k\sigma} = 1$ if $\varepsilon_{k\sigma} < \mu$, *ii)* $\eta_{k\sigma} = 0$ if $\varepsilon_{k\sigma} > \mu$, or *iii)* $0 < \eta_{k\sigma} < 1$ if $\varepsilon_{k\sigma} = \mu$. Putting aside peculiar situations such as vanishing V_{ijkl}, one observes that the case $0 < \eta_{k\sigma} < 1$ is actually the only relevant one. Therefore, the ground state corresponds to a scalar effective single-particle operator

$$\left(t_{ij} + \frac{\partial W}{\partial \gamma_{ij\sigma}}\right) = \mu\, \delta_{ij} \,. \tag{9}$$

Notice the difference with the KS equations which involve integer occupations of the KS orbitals.[2,15]

III. Interaction Energy Functional for the Hubbard Model

For the applications we consider the single-band Hubbard model with nearest-neighbor (NN) hoppings, which in the usual notation is given by[10]

$$H = -t \sum_{\langle i,j\rangle\sigma} \hat{c}_{i\sigma}^\dagger \hat{c}_{j\sigma} + U \sum_i \hat{n}_{i\downarrow}\hat{n}_{i\uparrow} \,. \tag{10}$$

In this case the interaction energy functional reads

$$W[\gamma_{ij}] = \min\left(U \sum_l \langle \Psi[\gamma_{ij}]|\; \hat{n}_{l\uparrow}\hat{n}_{l\downarrow}\; |\Psi[\gamma_{ij}]\rangle\right), \tag{11}$$

where the minimization is performed with respect to all N-particle $|\Psi[\gamma_{ij}]\rangle$ satisfying $\langle\Psi[\gamma_{ij}]|\sum_\sigma \hat{c}^\dagger_{i\sigma}\hat{c}_{j\sigma}|\Psi[\gamma_{ij}]\rangle = \gamma_{ij}$ for all i and j. The functional dependence of $W[\gamma]$ can be simplified at the expense of its universality if the hopping integrals t_{ij} are short ranged. For example, if only NN hoppings are considered as in Eq. (10), the kinetic energy is independent of the density-matrix elements γ_{ij} between sites i and j that are not NN's. Therefore, the constrained search in Eq. (4) may restrict $|\Psi[\gamma_{ij}]\rangle$ to satisfy $\langle\Psi[\gamma_{ij}]|\ \sum_\sigma \hat{c}^\dagger_{i\sigma}\hat{c}_{j\sigma}\ |\Psi[\gamma_{ij}]\rangle = \gamma_{ij}$ only for $i = j$ and for NN ij. While this is a considerable simplification, it also implies that W loses its universal character since the dependence on the NN γ_{ij} is now different for different lattices. Nevertheless, as discussed in the following, a nearly universal behavior of W can be recovered by an appropriate scaling of γ_{ij} within the domain of representability corresponding to each lattice.[19]

In order to determine $W[\gamma_{ij}]$ we look for the extremes of

$$F = U\sum_i [\langle\Psi|\hat{n}_{i\uparrow}\hat{n}_{i\downarrow}|\Psi\rangle] + \varepsilon(1 - \langle\Psi|\Psi\rangle) + \sum_{i,j}\tau_{ij}\Big(\langle\Psi|\sum_\sigma \hat{c}^\dagger_{i\sigma}\hat{c}_{j\sigma}|\Psi\rangle - \gamma_{ij}\Big), \quad (12)$$

with respect to $|\Psi\rangle$. Lagrange multipliers ε and τ_{ij} have been introduced to enforce the normalization of $|\Psi\rangle$ and the representability of γ_{ij}. Derivation with respect to $\langle\Psi|$, ε, and τ_{ij} yields the eigenvalue equations

$$\sum_{ij\sigma}\tau_{ij}\hat{c}^\dagger_{i\sigma}\hat{c}_{j\sigma}|\Psi\rangle + U\sum_i \hat{n}_{i\uparrow}\hat{n}_{i\downarrow}|\Psi\rangle = \varepsilon|\Psi\rangle, \quad (13)$$

and the auxiliary conditions $\langle\Psi|\Psi\rangle = 1$ and $\gamma_{ij} = \langle\Psi|\sum_\sigma \hat{c}^\dagger_{i\sigma}\hat{c}_{j\sigma}|\Psi\rangle$. The Lagrange multipliers τ_{ij} play the role of hopping integrals to be chosen in order that $|\Psi\rangle$ yields a given γ_{ij}. The pure-state representability of γ_{ij} ensures that there is always a solution.[21] In practice, however, one usually varies τ_{ij} in order to scan the domain of representability of γ_{ij}. The associated eigenvalue problem can be solved numerically for not too large clusters with different lattice structures and periodic boundary conditions by using Lanczos iterative method.[22]

For a one-dimensional (1D) chain with NN hoppings $t_{ij} = t$, translational symmetry implies that the density-matrix elements γ_{ij} between NN's are all identical. Therefore, one may set $\tau_{ij} = \tau$ for all NN ij. Eq. (13) has then the same form as the 1D Hubbard model which was solved exactly by Lieb and Wu.[23] In this case the lowest eigenvalue ε is determined following the work by Shiba.[24] The coupled Bethe-Ansatz equations are solved as a function of τ by means of a simple iterative procedure.

III.1 Exact Results for $W(\gamma_{12})$

In Fig. 1 the correlation energy E_C is given as a function of the NN density-matrix element γ_{12} for a ring having N_a=12 sites and different band-fillings n=N_e/N_a. Results are shown only for $N_e \leq N_a$, since for $N_e \geq N_a$, $E_C(\gamma_{ij}, N_e)$=$E_C(-\gamma_{ij}, 2N_a - N_e)$ due to electron-hole symmetry. Although $E_C(\gamma_{12})$ depends strongly on N_e, several qualitative properties are shared by all band fillings: *i)* The domain of representability of γ_{12} is bound by the bond orders in the uncorrelated limits. In fact, $\gamma_{12}^{0-} \leq \gamma_{12} \leq \gamma_{12}^{0+}$, where γ_{12}^{0+} (γ_{12}^{0-}) corresponds to the ground state of the $U = 0$ tight-binding model for $t > 0$ ($t < 0$). On bipartite lattices $\gamma_{12}^{0+} = -\gamma_{12}^{0-} = \gamma_{12}^{0}$. Notice that γ_{12}^{0} increases monotonously with N_e as the single-particle band is filled up. This is an important contribution to the band-filling dependence of E_C (see Fig. 1). *ii)* In the delocalized limit, $E_C(\gamma_{12}^0) = 0$ for all the N_e for which $W(\gamma_{12}^0)$ derives from a single Slater

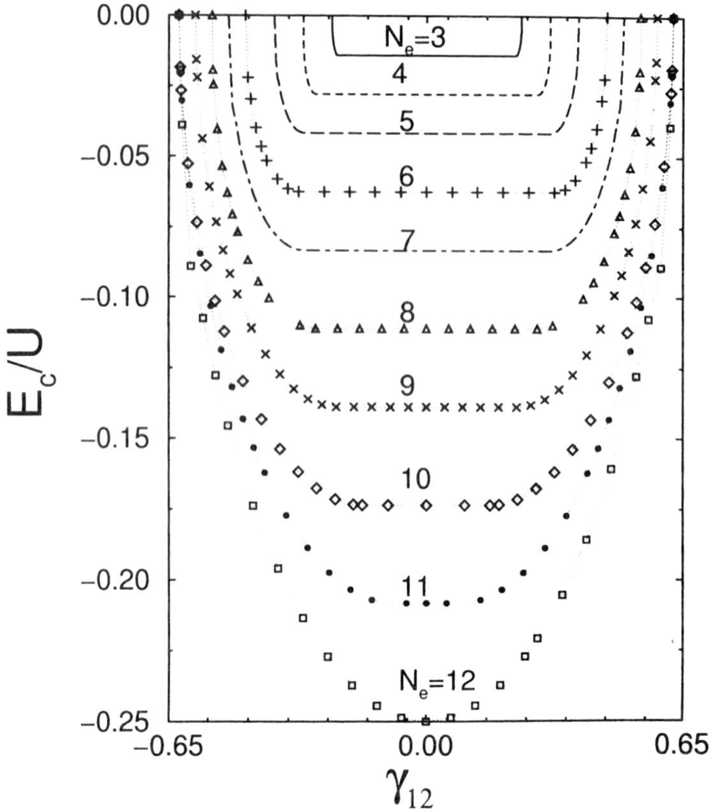

Figure 1. Correlation energy $E_{\rm C}(\gamma_{12}) = W - E_{\rm HF}$ of the Hubbard model ($U > 0$) on a 12-site ring as a function of the density matrix element γ_{ij} between NN's. $\gamma_{ij} = \gamma_{12}$ for all NN pairs ij. The corresponding numbers of electrons N_e are indicated. Note that on bipartite lattices $E_{\rm C}(\gamma_{12}, N_e) = E_{\rm C}(-\gamma_{12}, N_e) = E_{\rm C}(\gamma_{12}, 2N_a - N_e)$.

determinant.[25] Moreover, the divergence of $\partial E_{\rm C}/\partial\gamma_{12}$ for $\gamma_{12} = \gamma_{12}^0$ indicates that $\gamma_{12}^{gs} < \gamma_{12}^0$ for arbitrary small $U > 0$, as expected from perturbation theory. *iii*) Starting from $\gamma_{12} = \gamma_{12}^0$, $E_{\rm C}(\gamma_{12})$ decreases with decreasing γ_{12} reaching its lowest possible value $E_{\rm C}^\infty = -E_{\rm HF}$ for $\gamma_{12} = \gamma_{12}^{\infty +}$ ($N_e \leq N_a$). The same behavior is of course observed for $\gamma_{12} < 0$. In particular, $E_{\rm C} = -E_{\rm HF}$ also for $\gamma_{12} = \gamma_{12}^{\infty -}$. As shown in Fig. 1, $E_{\rm C}^\infty$ decreases rapidly with increasing N_e, since $E_{\rm HF}$ increases quadratically with the electron density.[26] *iv*) On bipartite lattices $\gamma_{12}^{\infty +} = -\gamma_{12}^{\infty -} = \gamma_{12}^\infty$, while on non-bipartite structures one generally has $|\gamma_{12}^{\infty +}| \neq |\gamma_{12}^{\infty -}|$, since the single-particle spectrum is different for positive and negative energies. The decrease of $E_{\rm C}$ with decreasing $|\gamma_{12}|$ shows that the reduction of the Coulomb energy due to correlations is done at the expense of kinetic energy or electron delocalization. *v*) $\gamma_{12}^\infty > 0$ for all $N_e \neq N_a$, and $\gamma_{12}^\infty = 0$ for $N_e = N_a$. γ_{12}^∞ represents the largest NN bond order that can be constructed under the constraint of vanishing Coulomb repulsion energy. A lower bound for γ_{12}^∞ is given by the bond order $\gamma_{12}^{\rm FM}$ in the fully-polarized ferromagnetic state ($\gamma_{12}^\infty \geq \gamma_{12}^{\rm FM}$). This state is obtained by occupying the lowest single-particle states with electrons of the same spin ($N_e \leq N_a$). Therefore, $\gamma_{12}^{\rm FM}$ increases with N_e for $N_e \leq N_a/2$, and then decreases for $N_a/2 < N_e \leq N_a$ reaching $\gamma_{12}^{\rm FM} = 0$ at half-band filling ($\gamma_{12}^{\rm FM} > 0$ for $N_e < N_a$). In this way the non-monotonous dependence of γ_{12}^{∞} on N_e can be explained (see Fig. 1). *vi*) The correlation energy is constant and equal to $-E_{\rm HF}$ for $\gamma_{12}^{\infty -} \leq \gamma_{12} \leq \gamma_{12}^{\infty +}$. These values of γ_{12} can never correspond to the ground-state energy of the Hubbard model, since in this range increasing γ_{12} always lowers the kinetic energy ($t > 0$) without increasing the Coulomb repulsion ($\gamma_{12}^{\infty} \leq \gamma_{12}^{gs} \leq \gamma_{12}^{0}$). For $\gamma_{12}^{\infty -} < \gamma_{12} < \gamma_{12}^{\infty +}$, γ_{12} cannot be represented by a ground state of Eq. (13). In this range γ_{12} can be derived from a linear combination of states having minimal Coulomb repulsion.[19]

In order to compare the functional dependences of the correlation energy for different band fillings, it is useful to scale $E_{\rm C}$ in units of the Hartree-Fock energy and to bring the relevant domains $\gamma_{12}^{\infty} \leq \gamma_{12} \leq \gamma_{12}^{0}$ of different N_e to a common range. In Fig. 2, $\varepsilon_{\rm C} = E_{\rm C}/E_{\rm HF}$ is shown as a function of $g_{12} = (\gamma_{12} - \gamma_{12}^{\infty})/(\gamma_{12}^{0} - \gamma_{12}^{\infty})$. We observe that the results for $\varepsilon_{\rm C}(g_{12})$ are remarkably similar for all band-fillings. The largest deviations from the common trend are found for $N_e = 4$. This anomalous behavior is related to the degeneracy of the single-particle spectrum and to the finite size of system. Figure 2 reveals that for the Hubbard model the largest part of the dependence of $E_{\rm C}(\gamma_{12})$ on band filling comes from $E_{\rm HF}$, γ_{12}^{0} and γ_{12}^{∞}. Similar conclusions are derived from the results for the infinite 1D chain.[19] One concludes that $\varepsilon_{\rm C}(g_{12})$ is a useful basis for introducing practical approximations applicable to more complex systems.

Equation (11) may be solved analitically for very small clusters by performing the constrained minimization explicitly. For a Hubbard dimer with $N_e = N_a = 2$ one obtains

$$\frac{W(\gamma_{12})}{N_a} = \frac{U}{4}\left(1 - \sqrt{1 - \gamma_{12}^2}\right). \tag{14}$$

This is also shown in Fig. 2 for the sake of comparison. The strong similarity between $\varepsilon_{\rm C}(g_{12})$ for $N_a = 2$ and $N_a = 12$ is a remarkable result. Systematic studies including the infinite 1D chain confirm that $\varepsilon_{\rm C}$ depends weakly on band-filling n and chain length N_a.[19] Therefore, the exact functional for finite clusters could be used to derive accurate approximations to $E_{\rm C}(\gamma_{12})$ for extended systems.

III.2 General Ansatz for $W(\gamma_{12})$

The interaction energy of the dimer (Eq. 14) already describes the fundamental interplay between electron delocalization and charge fluctuations, and satisfies several important properties of $W(\gamma_{12})$ that are valid for arbitrary lattices: *i*) The domain of definition of $W(\gamma_{12})$ is limited by the representability of γ_{12}. In fact, $\gamma_{12} \leq \gamma_{12}^{0} = 1$, where γ_{12}^{0} corresponds to the extreme of the kinetic energy (maximum degree of delocalization), and thus to the $U = 0$ ground-state for a given lattice and a given n. *ii*) For $\gamma_{12} = \gamma_{12}^{0}$, the underlying electronic state $\Psi[\gamma_{12}^{0}]$ is a single Slater determinant, and therefore $W(\gamma_{12}^{0}) = E_{\rm HF} = U/4$, where $E_{\rm HF}$ is the Hartree-Fock energy. Moreover, $\partial W/\partial\gamma_{12} = \infty$ for $\gamma_{12} = \gamma_{12}^{0}$, since $\gamma_{12}^{gs} < \gamma_{12}^{0}$ already for arbitrary small U/t, as expected from perturbation theory. *iii*) Starting from $\gamma_{12} = \gamma_{12}^{0}$, $W(\gamma_{12})$ decreases with decreasing γ_{12} reaching its lowest possible value, $W = 0$, for $\gamma_{12} = \gamma_{12}^{\infty}$ [$\gamma_{12}^{\infty} = 0$ in Eq. (14), since $n = 1$]. The decrease of W with decreasing $|\gamma_{12}|$ shows that the correlation-induced reduction of the Coulomb energy is done at the expense of kinetic energy or electron delocalization. *iv*) γ_{12}^{∞} represents the largest NN bond order that can be obtained under the constraint of vanishing Coulomb energy. Although it is possible, away from half-band filling, to represent density matrices with $|\gamma_{12}| < |\gamma_{12}^{\infty}|$, γ_{12}^{gs} can never be found in this range.[19]

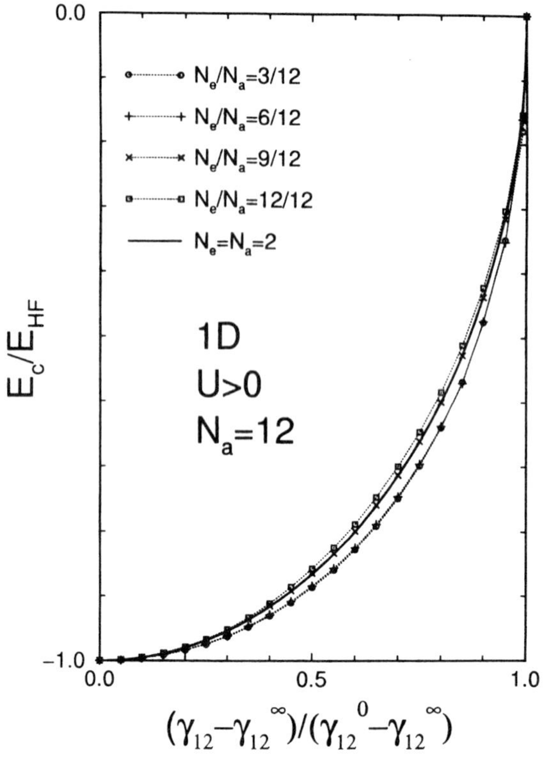

Figure 2. Correlation energy $E_C = W - E_{HF}$ in units of the Hartree-Fock energy E_{HF} for the Hubbard model on a 12-site ring. Results are given as a function of the degree of delocalization $g_{12} = (\gamma_{12} - \gamma_{12}^{\infty})/(\gamma_{12}^{0} - \gamma_{12}^{\infty})$, where γ_{12}^{0} (γ_{12}^{∞}) refers to the density-matrix element between NN's in the uncorrelated (strongly correlated) ground state corresponding to the $U = 0$ ($U = +\infty$) limit. The dimer result is also shown ($N_a = N_e = 2$).

Taking these properties into account and the pseudo-universal behavior $\varepsilon_C(g_{12})$, we propose the following *Ansatz* for the Hubbard model on arbitrary periodic lattices:

$$W(\gamma_{12}) = E_{HF}\left(1 - \sqrt{1 - \frac{(\gamma_{12} - \gamma_{12}^{\infty})^2}{(\gamma_{12}^{0} - \gamma_{12}^{\infty})^2}}\right). \tag{15}$$

This is equivalent to scaling the physically relevant domain, $\gamma_{12}^{\infty} \leq \gamma_{12} \leq \gamma_{12}^{0}$, where γ_{12} may be represented by a ground state of Eq. (10). Notice that γ_{12}^{0} and γ_{12}^{∞} depend on the considered lattice structure and on band filling n. In practice, γ_{12}^{∞} may be approximated by the ferromagnetic fully-polarized γ_{12}^{FM} ($|\gamma_{12}^{\infty}| \geq |\gamma_{12}^{FM}|$) which is obtained, as γ_{12}^{0}, by integration of the single-particle spectrum. In the following section several applications of DMFT are discussed using Eq. (15) as approximation to the interaction energy functional.

IV. Applications to One- and Two-Dimensional Lattices

As a first application of the theory we consider the 1D Hubbard model. In Fig. 3 the ground-state energy is given as a function of U/t for different band fillings n.

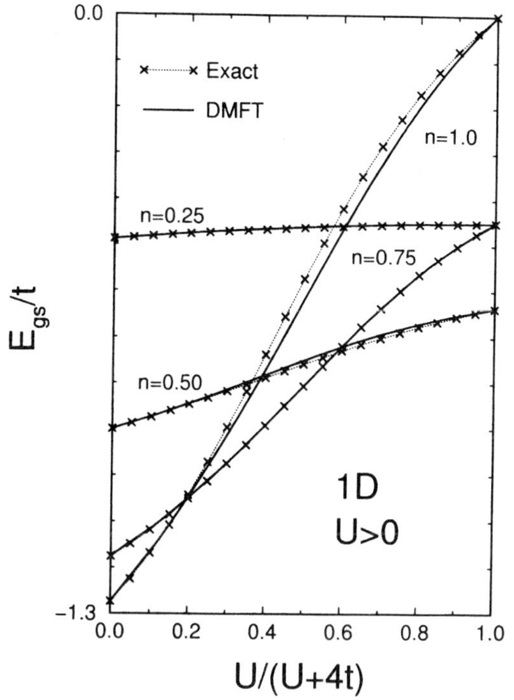

Figure 3. Ground-state energy $E_{\rm gs}$ of the one-dimensional (1D) Hubbard model as a function of Coulomb-repulsion strength U/t for different band-fillings n. The solid curves correspond to present density-matrix functional theory (DMFT)[19] and the crosses to Bethe-Ansatz exact results.[23]

Comparison with exact results shows a very good agreement for all values of U/t. A similar level of accuracy is achieved for the kinetic energy and Coulomb energy separately. Thus, local moments and kinetic energy renormalizations are also very well described. Notice that no artificial symmetry breaking is invoked in order to reproduce the effects of correlation and localization. This contrasts with other approaches, like unrestricted Hartree Fock, that yield an antiferromagnetic spin-density-wave state for $n = 1$. For $n \leq 0.8$, our results are almost indistinguishable from the exact ones. Appreciable quantitative discrepancies are found for half band-filling and intermediate values of U/t. Moreover, in the Heisenberg limit ($n = 1$), we obtain $E_{\rm gs} \simeq -\alpha t^2/U$ with $\alpha \simeq 3.24$ while in the exact solution $\alpha = 4 \ln 2 \simeq 2.77$. These discrepancies can be corrected by including in Eq. (15) a 4th-order term in g_{12}.[19]

In Fig. 4 results are given for the 2D square lattice. As in the 1D case, our simple analytical Ansatz for W compares very well with far more demanding quantum Monte Carlo (QMC) calculations.[27] One concludes that the theory provides a correct description of the changes in the correlation energy for different dimensions.

A more rigorous test for a DF method is the determination of band gaps such as the charge-excitation energy

$$\Delta E_{\rm c} = E_{\rm gs}(N_e + 1) + E_{\rm gs}(N_e - 1) - 2E_{\rm gs}(N_e). \quad (16)$$

This quantity is particularly interesting in strongly correlated systems, since it measures the insulating character of the electronic spectrum as a function of U/t. More-

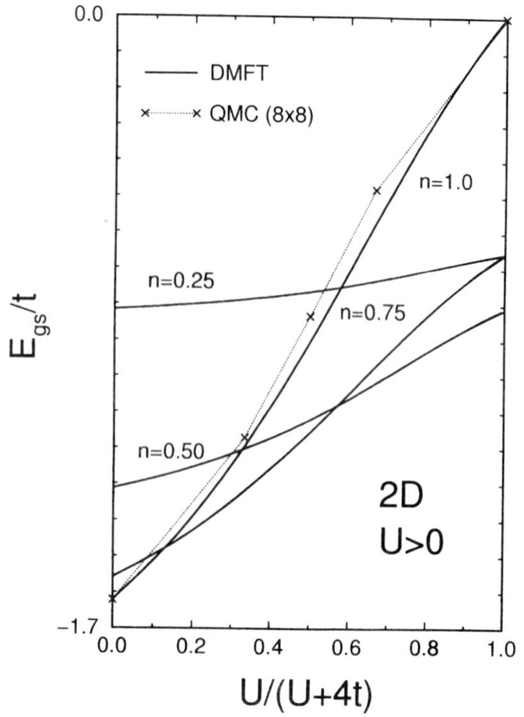

Figure 4. Ground-state energy E_{gs} of the two-dimensional Hubbard model (square lattice) as a function of Coulomb-repulsion strength U/t for different band-fillings n. The curves correspond to the present DMFT[19] and the crosses to quantum Monte Carlo calculations ($\beta = 4.5/t$).[27]

over, ΔE_c can be related to the discontinuity in the XC energy at half-band filling. In Fig. 5 the charge gap of the 1D chain is shown as a function of U/t. Although the overall agreement with with the Bethe-Ansatz solution is good, there is a significant overestimation of the gap for $U/t \ll 1$. Here we obtain $\Delta E_c \propto (U/t)^2$, while in the exact result ΔE_c increases exponentially with U/t. This discrepancy reflects the difficulty in describing long range effecs with XC functional derived from the dimer. For larger U/t the accuracy improves as localization starts to set in, and the relative error in ΔE_c vanishes. DMFT describes quite accurately the development of a Mott insulator in the strongly correlated limit.

Finally, we would like to comment briefly on a few other applications. Dimerized chains with hoppings $t \pm \delta t$ have been considered by extending Eq. (15) to allow for alternations of γ_{12}. Comparison with exact results for finite rings shows that accuracy improves with increasing dimerization (*e.g.*, for $N_a = N_e = 12$ and $U/t = 4$, $|E_{gs} - E_{gs}^{ex}|/|E_{gs}^{ex}| = 3.3 \times 10^{-2}$, 1.4×10^{-2}, and 2.6×10^{-3} for $\delta t/t = 0$, 1/2, and 3/4, respectively). In fact, the non-dimerized case, shown in detail in Fig. 3, is the most difficult one, since for a collection of dimers ($\delta t = t$) the exact W is recovered. Eq. (4) has also been applied successfully to non-bipartite structures and compact clusters. For example, for the triangular lattice with $n = 1$ and $U/t = 1$ one obtains $|E_{gs} - E_{gs}^{ex}|/|E_{gs}^{ex}| = 1.7 \times 10^{-4}$. As in the 1D case, accuracy tends to improve away from half-band filling. Thus, the changes in correlations associated to changes in the lattice

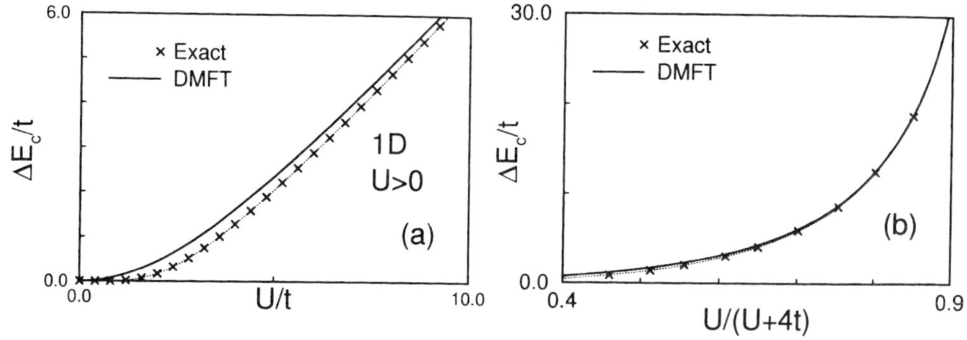

Figure 5. Charge excitation gap ΔE_c of the 1D Hubbard model at half-band filling as a function of (a) U/t and (b) $U/(U+4t)$. DMFT (solid curve)[19] is compared with the Bethe-Ansatz solution (crosses).[23]

structure are correctly described.[19] The attractive (negative U) Hubbard model has been investigated in 1D and 2D lattices. This is of considerable interest, since the nature of electronic correlations is very different from that of the repulsive case. An accurate approximation to $W(\gamma_{12})$ can be derived from the exact solution of the dimer problem, in a similar way as for $U > 0$. Levy's functional corresponds here to the maximum average number of double occupations for a given γ_{12} since pairing is favored. The calculations show that the accuracy of DMFT is in general better for $U < 0$ than in the repulsive case. For instance, in a 1D ring with $N_a = N_e = 12$ we find $|E_{gs} - E_{gs}^{ex}|/|E_{gs}^{ex}| = 1.2 \times 10^{-3}$, 7.5×10^{-3} and 1.5×10^{-4} for $|U|/t = 1$, 4, and 64, respectively.[28]

V. Conclusions

A new DF approach to fermion correlations has been developed, which is by all means independent of the homogeneous electron gas, and which provides with a unified description of weak and strong coupling regimes. In this framework the basic variable is the single-particle density matrix γ_{ij} and the key functional is the correlation energy $E_C[\gamma_{ij}]$. In this brief review we have presented representative rigorous results for $E_C(\gamma_{12})$ on periodic Hubbard models, where γ_{12} is the density-matrix element between NN's ($\gamma_{ij} = \gamma_{12}$ for all NN ij). These have been derived from finite-cluster exact diagonalizations and from the Bethe-Ansatz solution of the 1D chain. A more detailed account of our calculations of $E_C(\gamma_{12})$ as a function of the number of sites N_a, band filling N_e/N_a, and lattice structure may be found elsewhere.[19] The observed pseudo-universal behavior of $\varepsilon_C(g_{12}) = E_C/E_{HF}$ as a function of $g_{12} = (\gamma_{12} - \gamma_{12}^{\infty})/(\gamma_{12}^{0} - \gamma_{12}^{\infty})$ encourages transferring $\varepsilon_C(g_{12})$ from finite-size systems to infinite lattices or even to different lattice geometries. A basis for applications of density functional theory to many-body lattice models is thereby provided.

The theory has been used to study the properties of lattice Hamiltonians taking the Hubbard model as a particularly relevant example. Based on exact dimer results and on the scaling properties of $E_C(\gamma_{12})$, a simple, explicit approximation to $W(\gamma_{12})$ has been proposed. Quantitative results for the ground-state energies and charge-excitation gaps of 1D and 2D Hubbard models demonstrate the ability of the method to incorporate the subtle competition between kinetic energy and correlation-induced localization. The scope of DFT is thereby extended to include the limit of strong electron correlations.

This work opens the possibility of extensions in several interesting directions. For example, one may explore more general *Ansätze* for $W[\gamma]$ by including next NN contributions explicitly, or one may apply the present formulation to richer physical situations like dimerized chains, disorder systems, magnetic impurities, or multiband Hamiltonians. Such developments are expected to be relevant, both for the study lattice fermion models, and in view of a DF description of strong correlations from first-principles.

Acknowledgments

The authors acknowledge support from CONACyT-Mexico (R.L.S.) and from the Alexander von Humboldt Foundation (G.M.P.). Computer resources were provided by IDRIS (CNRS).

References

1. P. Hohenberg and W. Kohn, *Phys. Rev. B* **136**, 864 (1964).
2. W. Kohn and L. J. Sham, *Phys. Rev. A* **140**, 1133 (1965).
3. R. G. Parr and W. Yang, *Density-Functional Theory of Atoms and Molecules* (Oxford University, New York, 1989).
4. R. M. Dreizler and E. K. U. Gross, *Density Functional Theory* (Springer, Berlin, 1990).
5. U. von Barth and L. Hedin, *J. Phys. C: Solid State Phys.* **5**, 1629 (1972).
6. D. C. Langreth and M. J. Mehl, *Phys. Rev. B* **28**, 1809 (1983); A. D. Becke, *Phys. Rev. A* **38**, 3098 (1988); J. P. Perdew, K. Burke, and M. Ernzerhof, *Phys. Rev. Lett.* **77**, 3865 (1996) and references therein.
7. W. E. Picket, *Rev. Mod. Phys.* **61**, 433 (1989); O. K. Andersen, H. L. Skriver, H. Nohl, and B. Johansson, *Pure Appl. Chem.* **52**, 93 (1979); K. Tekakura, A. R. Williams, T. Oguchi, and J. Kübler, *Phys. Rev. Lett.* **52**, 1830 (1984).
8. R. Pariser and R. G. Parr, *J. Chem. Phys.* **21**, 466 (1953); **21** 767 (1953); J. A. Pople, *Trans. Faraday Soc.* **49**, 1375 (1953).
9. P. W. Anderson, *Phys. Rev.* **124**, 41 (1961).
10. J. Hubbard, *Proc. R. Soc. London, Ser. A* **276**, 238 (1963); **281**, 401 (1964); J. Kanamori, *Prog. Theo. Phys.* **30**, 275 (1963); M. C. Gutzwiller, *Phys. Rev. Lett.* **10**, 159 (1963).
11. See, for instance, P. Fulde, *Electron Correlations in Molecules and Solids* (Springer, Berlin, 1991).
12. O. Gunnarsson and K. Schönhammer, *Phys. Rev. Lett.* **56**, 1968 (1986); K. Schönhammer and O. Gunnarsson, *Phys. Rev. B* **37**, 3128 (1988)
13. A. Schindlmayr and R. W. Godby, *Phys. Rev. B* **51**, 10427 (1995).
14. K. Schönhammer, O. Gunnarsson, and R. M. Noack, *Phys. Rev. B* **52**, 2504 (1995).
15. T. L. Gilbert, *Phys. Rev. B* **12**, 2111 (1975).
16. R. A. Donnelly and R. G. Parr, *J. Chem. Phys.* **69**, 4431 (1978); R. A. Donnell, *J. Chem. Phys.* **71**, 28744 (1979). See also, Ref. 3, p. 213, and references therein.
17. S. M. Valone, *J. Chem. Phys.* **73**, 1344; *J. Chem. Phys.* 4653 (1980).

18. M. Levy, *Proc. Natl. Acad. Sci. U.S.A.* **76**, 6062 (1979).
19. R. López-Sandoval and G. M. Pastor, *Phys. Rev. B* **61**, 1764 (2000); and to be published.
20. Given a basis set $\{\phi_i\}$, the hopping integrals are expressed as

$$t_{ij} = \int d^3r \phi_i^*(\vec{r}) \Big[-\frac{\hbar^2}{2m}\nabla^2 + V_{\text{ext}}(\vec{r}) \Big] \phi_j(\vec{r}), \tag{17}$$

and the matrix elements of the Coulomb interaction as

$$V_{ijkl} = e^2 \int d^3r\, d^3r' \phi_i^*(\vec{r}) \phi_k^*(\vec{r}\,') \frac{1}{|\vec{r} - \vec{r}\,'|} \phi_l(\vec{r}\,') \phi_j(\vec{r}). \tag{18}$$

21. A single-particle density matrix γ_{ij} is said to be pure-state N-representable if an N-particle state $|\Psi\rangle$ exists, such that $\gamma_{ij} = \langle\Psi| \sum_\sigma \hat{c}_{i\sigma}^\dagger \hat{c}_{j\sigma} |\Psi\rangle$ for all i and j.
22. C. Lanczos, *J. Res. Nat. Bur. Stand.* **45**, 255 (1950); B. N. Parlett, *The Symmetric Eigenvalue Problem* (Prentice-Hall, Engelwood Cliffs, NJ, 1980); J. K. Collum and R. A. Willoughby, *Lanczos Algorithms for Large Symmetric Eigenvalue Computations* (Birkhauser, Boston, 1985), Vol. I.
23. L. H. Lieb and F. Y. Wu, *Phys. Rev. Lett.* **20**, 1445 (1968).
24. H. Shiba, *Phys. Rev. B* **6**, 930 (1972).
25. In the presence of degeneracies in the single-particle spectrum of finite systems one may find that the minimum of $W[\gamma_{ij}]$ does not correspond to a single Slater determinant, and that $W(\gamma_{12} \to \gamma_{12}^0) < E_{\text{HF}}$. Such a behavior is observed, for example, in rings with $N_a = N_e = 4m$. This is a finite-size effect which decreases with increasing N_a.
26. In the nonmagnetic case the Hartree-Fock energy of the Hubbard model is $E_{\text{HF}}/N_a = (U/4)(N_e/N_a)^2$ for N_e even and $E_{\text{HF}}/N_a = (U/4)(N_e/N_a)^2[1 - (1/N_e)^2]$ for N_e odd. Notice that in this model the difference between E_{HF} and the Hartree energy $E_{\text{H}}/N_a = (U/2)(N_e/N_a)^2$ is only the self interaction.
27. J. E. Hirsch, *Phys. Rev. B* **31**, 4403 (1985).
28. R. López-Sandoval and G. M. Pastor, to be published.

Action Density Functional Theory

Jaime Keller

División de Ciencias Básicas, Facultad de Química
and
Facultad de Estudios Superiores-Cuautitlán
Universidad Nacional Autónoma de México
Apartado Postal 70-528, 05410 México, D.F.
MEXICO
e-mail: keller@servidor.unam.mx
keller@cms.tuwien.ac.at

Abstract

From the definition of a density of action over spacetime we derive a theory to obtain the energy as a function of particle density, which contains the standard Density Functional Theory (DFT) and enlarges it by considering in one functional both the density of particles and the energy density per particle. The formulation presents a mathematical structure where the particle density is a factor of the energy per particle. The starting point is the definition of a global action density $\mathbf{K}(\mathbf{x})$ from which, by successive projections, the density $\mathbf{K}_j$ for a given type j of particles is projected. This density contains two factors, which by construction correspond to the $[\mathbf{K}(\mathbf{x})_j]_R = n_j(\mathbf{x})$ particle density and to the average energy per particle. The minimization of the energy density functional, with respect to changes in the density, contains as a consequence two terms: the first corresponds to the standard density functional theory for non-interacting particles, and the second to the optimization of the kinetic and the interparticle interaction energy, terms that go beyond the standard DFT.

I. Introduction

In the calculation of the properties of many particles systems the methods based on the use of the particle density as the basic variable have been specially successful. The method started with the suggestions of Thomas (1927) and of Fermi (1927, 1928) to use the density as the variable to compute the total energy as a functional of this density. The original idea was considered to be an approximated method where the guidance could come from wave functions quantum mechanics. These ideas have also been extended to the calculation of transition probabilities and response functions. (see for example Parr and Yang[7] and Ref. 4). In a series of papers[3] we have stated that the theory required a third theorem which introduces the correct boundary conditions and the poles of the Green's function of the system.

In practice the use of DFT has been very successful because, besides the use of the density functional equations, an auxiliary set of equations were introduced by considering that the density could have been constructed from a set of auxiliary functions and occupation numbers. The analysis of this procedure, now known as the Kohn-Sham equations, revealed that the eigenfunctions and eigenvalues of this procedure were related to the elementary excitation of the system where one particle was removed at a time.[2] This is quite similar to the case of standard quantum mechanics with wave functions, where in the Hartree approximation the different states also are in one to one correspondence to the same type of excitations of the system or in the Hartree-Fock method where, according to the Koopmanns theorem, the eigenfunctions and the eigenvalues correspond to this, removal of one particle, elementary excitations of the system. It has been clear from the analysis of these cases that the inclusion of these excitations correspond to the more significant contributions to the density. Other elementary excitations contribute by amounts that are one or two orders of magnitude smaller, for example in the case of resonance energies and densities in polycyclic hydrocarbons.[8]

In our present paper we will take a different starting point. The equations will be a systematic derivation from the consideration of the existence for any physical system of a density of action in spacetime and a theory relating that density of action (to the desired degree of complexity) to a representation of the physical system to which that action corresponds.

In Sec. II we present the concept of a spacetime-action geometry in a form suitable for our purposes. In Sec. III we define the action distribution function and its main properties. In Sec. IV we derive the equations for the new description of matter and interactions fields. We particularize to a case similar to standard DFT theory and present one example of the practical use of the present theory as applied to the study of the electronic structure of atoms.

II. Spacetime Action Geometry

In mathematical physics it is usual to consider spacetime as a frame of reference for the description of matter and the interaction fields, its use corresponds to postulating a specific approach to Geometry and in particular to Geometrical Analysis.[10] In fact spacetime, having a multivector structure and containing a spinor (and dual spinor) space, not only describes our perception of the physical nature but is also a powerful mathematical tool. Here, we present both, a motivation for the use of a more general geometry corresponding to space-time-action (STA).

To visualize the geometrical nature of action, consider that there exists a numerical and a dimensional relationship between the fundamental constants:

$$g_D e = \left(\frac{\hbar}{2}\right) c = \frac{hc}{4\pi}, \tag{1}$$

relating the Dirac monopole charge g_D, the electrical charge e, the unit of action h and the numerical value of the velocity of light c.

In spacetime geometry the constants e and c are invariant scalar quantities. That is, they are invariant under proper or improper Lorentz transformations. Then, in the basic relations of electromagnetism a moving charge would appear as a four current $\mathbf{j} = j^{\mu}e_{\mu}$, where $j^{\mu} = ev^{\mu}$ with v^{μ} the four velocity. The Maxwell equations would then relate $\mathbf{j}$ to a vector four potential $\partial^{\mu}\partial_{\mu}\mathbf{j} = \mathbf{A}$, and to the (bivector) electromagnetic strengths $\mathbf{F} = F^{\mu\nu}e_{\mu}e_{\nu} = \frac{1}{2}(\partial^{\mu}A^{\nu})e_{\mu}e_{\nu}$. On the other hand, if a magnetic monopole g_{D} exists, it will, when moving, originate a **trivector** current (usually called axial vector four current) $\mathbf{k}$

$$\mathbf{k} = \mathbf{g}_{D}\mathbf{v} = \mathbf{g}_{D}v^{\mu}e_{\mu} = k^{\lambda\nu\rho}e_{\lambda}e_{\nu}e_{\rho}, \tag{2}$$

which will, through the natural extensions of the Maxwell equations to magnetic currents, originate an axial vector potential $\partial^{\mu}\partial_{\mu}\mathbf{k} = \mathbf{B} = b^{\nu\lambda\rho}e_{\nu}e_{\lambda}e_{\rho}$, where all indexes μ, ν, λ, ρ are different among themselves. Then, geometrically, the "constant" $\mathbf{g}_{D}$ must have a pseudoscalar property, which results in $\mathbf{k}$ and $\mathbf{B}$ being axial vectors and bivector electromagnetic fields strengths. Here we have used the multivector scalar product property $\frac{1}{2}(e_{\mu}e_{\nu} + e_{\nu}e_{\mu}) = g_{\mu\nu}\mathbf{1}$, with $\mathbf{1}$ the scalar unit of the geometric algebra. This is the reason to have used $\mathbf{g}_{D} = g_{D}e_{\mathrm{ps}}$ in Eq. (2) above, that is the magnetic monopole strength multiplied by the unit pseudoscalar $e_{\mathrm{ps}} = e_0e_1e_2e_3$.

Then, in (1) there are two possibilities, either there is some factor missing or action has also Spacetime Geometry pseudoscalar properties. Our analysis[9] have shown that, if considered jointly with spacetime, this second option is the more useful identification of the geometric properties of the action. Otherwise, when action is considered separately, it is always properly represented as a scalar quantity. The action constant is then, geometrically, both the constant relating energy-momentum to spacetime and the key to the construction of an unified geometry of space,time and action. The new geometry is derived from the introduction of an action coordinate $x^5 = a(\mathbf{x})d_0/h$ where a is the density of action at a given point $\mathbf{x}$ of spacetime, h is the Planck's constant and d_0 is an invariant basic length to be determined below, basically $x^5 = \kappa a(\mathbf{x})$. We use the traditional indexes 0,1,2,3 for time and space and, also, the isomorphism between the Dirac gamma symbols with the vectors in the geometry of spacetime.

In fact, the special property of the pseudoscalar in spacetime (in the notation above $e_5 = ie_0e_1e_2e_3$) is that $e_5e_{\mu} = -e_{\mu}e_5$ (from $e_{\mu}e_{\nu} = -e_{\nu}e_{\mu}$, $\mu \neq \nu$) and then it has the same commuting properties with the generating vectors than the generating vectors among themselves. The linearly independent multivector $i\gamma_5$ is then the immediate candidate to introduce an additional basis vector, adding one more dimension and, through its use, obtain the five dimensional carrier space spanned by the basic vectors e_v, $v = 0, 1, 2, 3, 5$ with metric $g_{uv} = \mathrm{diag}\,(1, -1, -1, -1, 1)$. Its use allows the construction of a geometrical frame of work for the description of physical processes: a unified space-time-action geometry G_{STA}.

The product of $e_5 = ie_{\mathrm{ps}}$ with any element of the original Spacetime Geometry G_{ST} is a purely imaginary quantity with the result that the elements of the new (space-time-action) geometry are equivalent to $G_{\mathrm{STA}} = G_{\mathrm{ST}} \otimes C$.

The value of d_0 we use (taken from the theory of the electron) is $d_0 = r_0/2\alpha$, the Compton radius, where $r_0 = e^2/m_0c^2$, the radius that relates the mass of the electron to an electromagnetic equivalent energy and also $r_0 = \mu/g_{D}c$, the ratio of the electron magnetic moment to the Dirac monopole magnetic charge. With this choice the presentation of the theory will immediately be suitable for the study of elementary particles. Nevertheless, we have shown that the same units are practical in the study of gravitational interactions.

There is a simple picture of STA: it consists of the spacetime continuum with an action density at every point of spacetime. This should be in fact the description of the space of physical phenomena. In spacetime, even if a geometrical unity has been achieved, space and time coordinates are not really equivalent for physical phenomena. While we can move in space in any direction and we can place systems in different points of space which can be changed according to the physical circumstances, time is always flowing, for all observers and systems, without any possibility of a back in time displacement. Even for an event as a light ray, where a congruence of points is defined from the property $ds^2 = g_{\mu\nu}dx^\mu dy^\nu = (ct)^2 - x^2 = 0$, that is points where the spacetime distance vanishes, there is an evolution in time which can not be either put to zero or, even less, made negative. Then, to have an action defined at every point of spacetime corresponds in practice to an energy ε defined at each point of space for a given observer, because the observer has to consider the local and relative rate of change of action A: $\partial A/\partial t = \varepsilon$, as far as for it time is always flowing. A density of action in spacetime corresponds to a density of energy in space for a given observer. Of course density of action in spacetime is a relativistic covariant expression. In fact for another observer in relative motion with respect to the first the energy density will appear as a fourvector energy-momentum density. The theory is called START.

Below we will call this density $\mathbf{K} = \mathbf{K}(\mathbf{X}, \mu, \nu)$, where the arguments of the function are the spacetime coordinates $\mathbf{X}$, and two collective indexes for each one of the factors of action (energy-momentum) and (time-space).

III. Action Density Functional Theory and Wave Function Theory

In the space defined above we now introduce a dimensionless function $\mathbf{K} = \mathbf{K}(\mathbf{X}, \mu, \nu)$ defined at each point $\mathbf{X}$ of spacetime.

Here we have to make two crucial considerations about what we know about action and about quantum mechanics that will be basic for the systematic procedure presented in this section.

As we mentioned in the previous section for each observer the concept of action is locally the study of the energy attributed to physical phenomena. Relativity theory showed that an energy content in a given volume presents an inertia mass M, through the basic relation $M = E/c^2$. This total energy content is independent of the form we have chosen to describe that matter. Let us for example start by considering that we have a particle or a molecule, which means that we have chosen a macroscopic point of view and a separation into the particle itself and some external forces representing the mutual interaction between the material and the rest of the physical system. For some practical purposes this could be a sufficient description where only shape and density will be required. We also know that if we consider that this piece of material is in movement relative to some measuring device, quantum mechanics can be applied to the material as a unit. Otherwise we may consider that the particle consists of atoms or molecules in interaction, and that there is an effective interaction potential between the molecules (this is a very common case in the study, for example, of rare gas solids). Quantum mechanics should be applied again for this system of interacting molecules. If our decision is to describe the material as electrons and nuclei we will again apply quantum mechanics at this level of detail in the description. The next step, the study of the nucleus or the study of the nucleons are again admissible. In every case we will have a total energy which should be equal to the total energy of the previous steps and we will have the choice of separation into: constitutional energy or mass, kinetic energy, and interaction energy.

That is: quantum mechanics is a universal description of the phenomena, valid for any degree of detail we might have chosen for the description and can not be a property of the components but a basic property for the description of nature. Action and spacetime are fundamental concepts in the description of nature and not concepts dependent on the system we are describing. Our separation into kinetic, potential and constitutive energy is correct but arbitrary.

Particle density and density of action are gauge invariant physical quantities. We then need to develop a procedure which can allow gauge freedom; that is, allow for arbitrary but correct and useful descriptions. This is possible with the introduction of the probability amplitude known as wave function ψ, required to contain the necessary information in a form compatible to the basic concept that the energy-momentum is obtained from the operator $i\hbar\partial_\mu$ applied to the function which describes the splitting of the action density into a particle density ρ and the action per particle. The definition $\rho = \psi^2$ allows gauge independence. This procedure can be carried at any level of description, hence the universality of the possibilities to use wave quantum mechanic: it corresponds to

$$E(\mathbf{x}) = i\hbar\partial_t\mathbf{K}(\mathbf{X}) = i\hbar\frac{1}{2}\left\{(\partial_t\psi^\dagger)\psi - \psi^\dagger i\hbar\partial_t\psi\right\} = \psi^\dagger\psi k(\mathbf{x}) = \rho(\mathbf{x})k(\mathbf{x}). \quad (3)$$

Once we have established that:

1.- We are defining a density of action in spacetime which corresponds to an energy density in space $E(\mathbf{x}) = i\hbar\partial\mathbf{K}(\mathbf{X})/\partial t$ for a fixed observer.

2.- The universality of the description, which allows a choice of the level of detail (for example: molecules$\rightarrow$ atoms$\rightarrow$ electron and nuclei $\rightarrow$ nucleons $\rightarrow$ quarks, provided that at each step the decision is formally made by selecting of the type of "particle", and both the type of interaction between particle and the internal energy of the particles) we can now proceed to the steps creating a practical density functional theory.

The energy density should be written as a product of a particle density and a (global) energy per particle

$$E = \int E(\mathbf{x})d\mathbf{v} = \int \rho(\mathbf{x})\varepsilon\, d\mathbf{v} = \int \rho(\mathbf{x})[\mathbf{k}(\mathbf{x}) + \mathbf{V}(\mathbf{x}) + \mathbf{V}_{xc}(\mathbf{x}) + \varepsilon_0(\mathbf{x})]\, d\mathbf{v}, \quad (4)$$

where we have defined the energy density $E(\mathbf{x})$, the particle density $\rho(\mathbf{x})$, the actual kinetic energy per particle $\mathbf{k}(\mathbf{x})$, the external and average internal potential energy per particle $\mathbf{V}(\mathbf{x})$, the correction to the average kinetic and potential energy per particle arising from the statistics of the type of particles under consideration $\mathbf{V}_{xc}(\mathbf{x})$, and a local energy $\varepsilon_0(\mathbf{x})$, a basic term required to compensate for any difference in the sum of the previous terms with respect to the average energy per particle ε. Density functional theory describes then the self organization of the system with density $\rho(\mathbf{x})$.

The fact that we are arbitrarily defining the terms above requires the possibility of changing the description of the energy partitioning without changing the description of the density. That is that the density $\rho(\mathbf{x})$ is required to be gauge invariant whereas the description of the energy (action) is gauge dependant. This is achieved by constructing the energy density as the product of two conjugated quantities $\Psi(\mathbf{x})$ and $\Psi^\dagger(\mathbf{x})$ such that $\rho(\mathbf{x}) = \Psi^\dagger(\mathbf{x})\Psi(\mathbf{x})$ is gauge invariant. Here we have defined an auxiliary quantity which can be essentially written in terms of the basic action $a(\mathbf{x})$ and the action introduced by the gauge freedom $\phi(\mathbf{x})$ in units of $\hbar$, as

$$\Psi(\mathbf{x}) = \sqrt{\rho(\mathbf{x})} f(\mathbf{x}) e^{-ia(\mathbf{x})+i\phi(\mathbf{x})}, \tag{5}$$

where $f(\mathbf{x})$ introduces the symmetry constrains, showing the gauge freedom of the description of the energy associated with the particle.

The definition of a system is equivalent to the requirement that at all positions $\mathbf{x}$ we must have the same average energy per particle ε, which only in the simplest cases would be the sum of a kinetic and a potential energy part in the traditional sense. A well known example is the case of electron density functional theory where the kinetic energy is assumed to correspond to the kinetic energy of the free electron gas and then our term $\varepsilon_0(\mathbf{x})$ will contain, among other terms, the difference between the actual kinetic energy and the free electron gas term. The term $\varepsilon_0(\mathbf{x})$ exists either from the incomplete description of the other terms, the usual case, or from inaccuracies in the computational procedure. It acts locally to distribute the density in the form which minimizes the total energy and corresponds then to a variational procedure in the formulation of the theory. In the definition above $\int \rho(\mathbf{x})\varepsilon_0(\mathbf{x}) d\mathbf{v} = 0$. The use of the Hohenberg-Kohn Theorems[1] and the Kohn-Sham minimization procedure[5] for the definition of this two terms

$$\delta\left(E[\rho] - \varepsilon\left\{ \int \rho(\mathbf{x})\, d\mathbf{v} - 1 \right\} \right) = 0, \tag{6}$$

allows the direct self-consistent determination of $\rho(\mathbf{x})$ and ε (see Ref. 5).

In the formalism we have a very useful self-consistency relation

$$\Psi(\mathbf{x}) = \int w(\mathbf{x}) G(\mathbf{x}, \mathbf{x}') \Psi(\mathbf{x}') + \Psi_0(\mathbf{x}) \quad \text{where} \quad \mathrm{w}(\mathbf{x}) = \Delta\rho(\mathbf{x})\varepsilon + \rho(\mathbf{x}) \boldsymbol{\Delta}\varepsilon(\mathbf{x}), \tag{7}$$

where we have introduced both, the response function of the system $G(\mathbf{x}, \mathbf{x}')$, and the effective potential which would be caused either by fluctuations of the density or by differences in the local definition of energy per particle. This reaction would propagate to all points of the distribution to achieve stability. We have used elsewhere an approximation for the response function in terms of the lowest elementary excitations of an electron gas.

IV. Restriction to Standard DFT

The auxiliary function Ψ in (3) is identical to the standard quantum mechanical ψ in the case of a "one" particle system, otherwise we should consider it as the function which represents the gauge dependent square root of the density of action, obeying the equations

$$h_{\text{eff}} \Psi(\mathbf{r}) = \mu \Psi(\mathbf{r}). \tag{8}$$

If, instead of considering a system of independent particles a system of interacting particles is chosen for description, the Ψ will be non-local and non-diagonal density matrix will be obtained, as a usual in wave function quantum mechanichs. Otherwise we have (8), where we have defined an effective operator

$$h_{\text{eff}} = -\nabla^2 + v(\mathbf{r}) + v_{\text{eff}}(\mathbf{r}), \tag{9}$$

such that the energy of the system

$$T_s[\rho] = \int \Psi^\dagger(\mathbf{r})(-\nabla^2)\Psi(\mathbf{r}) + T_\theta[\rho], \tag{10}$$

defining, variationally, the effective potential

$$\frac{\delta T_s[\rho]}{\delta\rho(\mathbf{r})} = -\frac{\Psi^\dagger\nabla^2\Psi}{\Psi^\dagger\Psi} + v_\theta([\rho];\mathbf{r}) + V_{ss}([\rho];\mathbf{r}), \tag{11}$$

$$v_\theta([\rho];\mathbf{r}) + V_{ss}([\rho];\mathbf{r}) = \frac{\delta T_\theta[\rho]}{\delta\rho(\mathbf{r})} \quad , \quad \frac{\delta T_s[\rho]}{\delta\rho(\mathbf{r})} + v_{KS}([\rho];\mathbf{r}) = \varepsilon_M. \tag{12}$$

Then (8) reads

$$-\nabla^2\Psi + v_{KS}([\rho];\mathbf{r})\Psi + \{v_\theta([\rho];\mathbf{r}) + V_{ss}([\rho];\mathbf{r})\}\Psi, \tag{13}$$

$$\Delta\bar{\varepsilon}(\mathbf{r})\rho(\mathbf{r}) - \sum_i \varepsilon_i |\psi_i(\mathbf{r})|^2 - \bar{\varepsilon}\rho(\mathbf{r}). \tag{14}$$

This is equivalent to the optimization of

$$\Omega = E[\rho] - \mu\left[\int \rho(\mathbf{r})d\mathbf{r} - \mathbf{N}\right] - \lambda\left[\int \sum \varepsilon_i |\psi_i(\mathbf{r})|^2 d\mathbf{r} - \int \bar{\varepsilon}\rho(\mathbf{r})d\mathbf{r}\right], \tag{15}$$

using a set of auxiliary functions ϕ_i, with the index i running through all possible forms of extracting one particle from the system

$$-\nabla^2\phi_i + [v_{xc}(\mathbf{r}) + v_{Coul}(\mathbf{r}) + v_{ext}(\mathbf{r})]\phi_i = \varepsilon_i\phi_i, \tag{16}$$

which will define the elementary excitations of the system corresponding to the removal of one particle with rate of change of the energy ε_i

$$\sum_i h_i(\mathbf{r})^{KS} = \sum_i \varepsilon_i |\phi_i|^2. \tag{17}$$

This is equivalent to the Kohn–Sham procedure and the use of the Kohn-Sham effective hamiltonian $h_i(\mathbf{r})^{\mathrm{KS}}$

$$h_i(\mathbf{r})^{\mathrm{KS}} = -\phi_i^*\nabla^2\phi_i + \phi_i^*[v_{xc}(\mathbf{r}) + v_{\mathrm{Coul}}(\mathbf{r}) + v_{\mathrm{ext}}(\mathbf{r})]\phi_i. \tag{18}$$

and then for the gauge dependent square root of the density auxiliary function (see Refs. 10 and 11)

$$-\nabla^2\Psi + v^{\mathrm{FK}}(\mathbf{r})\Psi = \bar{\varepsilon}\Psi, \tag{19}$$

$$v^{\mathrm{FK}}(\mathbf{r}) = v_{xc}(\mathbf{r}) + v_{\mathrm{Coul}}(\mathbf{r}) + v_{\mathrm{ext}}(\mathbf{r}) + \Psi^\dagger\nabla^2\Psi + k(\mathbf{r}) - \Delta\bar{\varepsilon}(\mathbf{r}). \tag{20}$$

The last three terms correspond, the first two to the correct kinetic energy density and the last one the symmetry constraint potential arising from the actual values of the energy necessary to remove one electron from the system and the average energy per electron:

$$\Delta\bar{\varepsilon}(\mathbf{r}) = \sum \frac{n_i\varepsilon_i |\phi_i(\mathbf{r})|^2}{\rho(\mathbf{r}) - \bar{\varepsilon}}. \tag{21}$$

V. Conclusions

We have presented an action-density functional formalism, and shown that not only the standard density functional theory is recovered and that in a sense it is more fundamental than wave function wave mechanics, but also that the analysis[11] of the mapping of the density matrix into a density allowable for density functional theory requires the introduction of auxiliary terms which represent the internal symmetries of the system.

Acknowledgment

The ideas of this paper were developed from a sugestion of Prof. Hans Christoph Siegmann to present the basis of quantum mechanics and DFT in a direct form; this paper is respectfully dedicated to him. JK is a member of the Sistema Nacional de Investigadores, Conacyt, México.

References

1. R. G. Parr and W. Yang, *Density-Functional Theory of Atoms and Molecules* (New York, Oxford University Press, 1989).
2. J. Keller and J. L. Gazquez, (Eds.). *Density Functional Theory* (Springer-Verlag, Berlin, 1983).
3. J. Keller, *Int. J. Quantum Chem. Symp.* **20**, 767 (1986); J. Keller and E. Ludeña, *Int. J. Quantum Chem. Symp.* **21**, 171 (1987).
4. J. F. Janak, *Phys. Rev. B* **18**, 7165 (1978).
5. A. Theophilou, *J. Phys. C* **12**, 5419 (1979).
6. J. Keller and P. Weinberger, (2000), *Phil. Mag.* To be published.
7. J. Keller, (2000). *Proc. Int. Conf. Geometrization of Physics.* Kazan State University, Kazan, October 4-8 (1999).
8. P. Hohenberg and W. Kohn, *Phys. Rev. B* **136**, 864 (1964).
9. J. Keller, *Physics. Atomic Nuclei* **62**, (1999).
10. J. A. Flores and J. Keller, *Phys. Rev. A.* **45**, 6259 (1992).
11. J. Keller, J. A. Flores, and A. Keller, *Folia Chemica Theorica Latina (Madrid)* **XVIII** (24), 175 (1990).

A GaAs Type Source of Polarized Electrons at the Mainz Race Track Microtron MAMI

K. Aulenbacher,[1] P. Drescher,[2] H. Euteneuer,[1] D. V. Harrach,[1]
P. Hartmann,[1] J. Hoffmann,[1] K.-H. Kaiser,[1] H. J. Kreidel,[1]
M. Leberig,[1] Ch. Nachtigall,[2] E. Reichert,[2] M. Schemies,[2]
J. Schuler,[1] M. Steigerwal,[2] and Ch. Zalto[1]

[1] *Institut für Kernphysik*
Joh. Gutenberg Universität
D–55099 Mainz
GERMANY

[2] *Institut für Physik*
Joh. Gutenberg Universität
D–55099 Mainz
GERMANY

Abstract

The source of polarized electrons attached to the 855 MeV race track microtron MAMI in Mainz is based on photoemission from III-V-semiconductors. Strained layers of GaAsP are used at present. Typical degree of spin-polarization of emitted electrons is 75% with 830 nm light irradiating the cathode of the source gun. The light is produced by a gain switched laser diode followed by a laser diode amplifier. The diode is driven by the 2.45 GHz radiofrequency of the accelerator. So a 2.45 GHz pulsed beam in synchronism with the accelerator-rf is injected into MAMI. Thereby a capture efficiency of injection better than 93% is achieved. Beam currents up to 20 μA that are spin-polarized to a degree of 75% are available at MAMI for long term experiments. Total beam time scheduled for experiments is above 2000 h/a.

Physics of Low Dimensional Systems
Edited by J. L. Morán-López, Kluwer Academic/Plenum Publishers, New York 2001

I. Introduction

In 1974 the history[1] of the photoemission of spin-polarized electrons from III-V-semiconductor surfaces started at ETH in Zürich. GaAs, prepared for negative electron affinity (NEA), or one of its derivatives emits highly spin-polarized electrons if irradiated with circularly polarized light of a photon energy just above the gap energy of the material.[2,3] Perhaps there is no other discovery from the laboratory of H. C. Siegmann that found so wide applications in so many different fields of fundamental physics research. Early examples are in solid state physics the investigation of magnetic structure of surfaces,[4] in atomic physics the study of spin-dependent electron atom scattering,[5] in nuclear and elementary particle physics parity violation experiments[6] and the investigation of nucleon electromagnetic structure.[7,8]

Particularly at electron accelerator centers like Bates,[9] Jefferson Lab,[10] KEK,[11] MAMI,[12] NIKHEF,[13] and SLAC[14] the interest in experiments that exploit the electron spin degree of freedom in scattering experiments is steeply growing in recent years. Investigations that ask for polarized electron beams take more than 90% of available beam time in some of these laboratories.

The Mainz accelerator consists of three cascaded so-called race track microtrons that accelerate electrons to energies up to 855 MeV.[15] Beams of longitudinally polarized electrons are currently applied in a series of medium energy experiments. Topics are the measurement of the nucleon electric form factors via quasielastic reactions $^1\mathrm{H}(\vec{e},e'\vec{p})$, $^2\mathrm{D}(\vec{e},e'\vec{p})$, $^2\mathrm{D}(\vec{e},e'\vec{n})$, and $^3\vec{\mathrm{He}}(\vec{e},e'n)$,[16–18] of strange contributions to the proton form factor in a parity experiment,[19] or the test of the so-called Gerasimov-Drell-Hearn sum rule in the spin-dependence of the photon-proton cross section.[20]

II. Strained $GaAs_{0.95}P_{0.05}$ Cathodes

The standard cathode installed in the gun of the MAMI source of polarized electrons consists of strained layer NEA $GaAs_{0.95}P_{0.05}$, whose epilayer has a thickness of 150 nm.[21] The cathode samples are produced by MOCVD at the Ioffe Institut in St. Petersburg.[22] Figure 1 shows the quantum efficiency QE and spin-polarization P of emitted electrons as a function of photon energy of light irradiating a typical sample of this material (curves from Ref. 23) The maximum degree of spin-polarization of emitted electrons is 76% at a wavelength of 835 nm. Maximum values of polarization scatter from sample to sample in the range 70% to 84% with a mean at 75%. The quantum efficiency at 835 nm amounts to 1×10^{-3} in Fig. 1 which means that one gets 0.67 μA highly polarized emission current per mW light power from this cathode. Quantum efficiencies up to 4×10^{-3} at the wavelength of maximum polarization have been observed with these St. Petersburg cathodes.

III. Source Set Up

Figure 2 shows the set up of the vacuum components and the electron optical elements of the beam line of the source.[24] The upper part consisting of the gun, the preparation chamber, and the load lock is a copy of the system described in Ref. 12. In Fig. 3 the triode configuration of the gun is sketched. An exhausted cathode may be replaced by a fresh sample stored in the preparation chamber. This is done by lowering the *elevator* indicated in Fig. 3 and passing the cathode crystals through a side opening of the anode with help of a vacuum linear movement manipulator (not shown in Fig. 3).

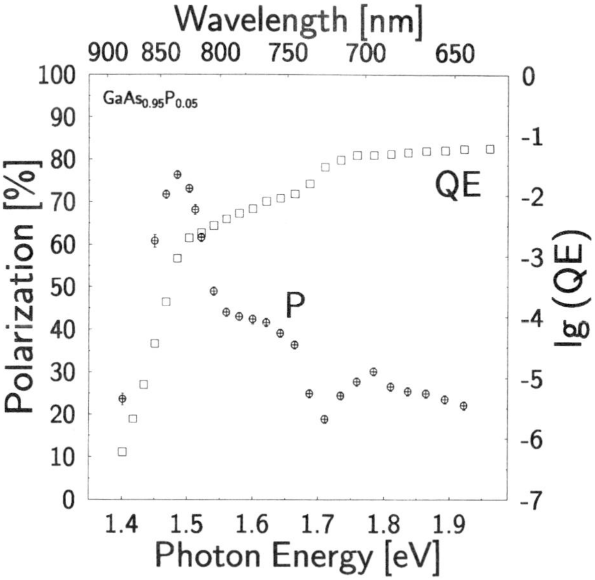

Figure 1. Photo electron emission from $GaAs_{0.95}P_{0.05}$. Quantum efficiency **QE** and polarization **P** of emitted electrons as a function of photon energy.

The light source described below consists of a gain switched laser diode followed by a laser diode amplifier. It produces linearly polarized radiation that is transformed to circular polarization by a Pockels cell. The light passes through a window in the vacuum chamber of the α-magnet *alpha 1* in Fig. 2 to the gun and is finally focused to a spot of 0.3 mm diameter at the cathode. A short beamline consisting of several magnetic quadrupoles brings the 100 keV-beam produced by the gun to the α-magnet *alpha 1*. The beam is deflected here through 270° and directed to the entrance of the injector linac of the accelerator. An unpolarized beam from a thermionic gun may also be injected wich is indicated at the right side of Fig. 2. In this case *alpha 1* is shut off. The transmission of the polarized beam from cathode to the entrance of the injection linac is better than 93%. The transverse emittance of the beam at injection was measured with help of a wire scanner to be 0.3 π·mm·mrad at 100 keV.

Vacuum is produced by combinations of ion getter and NEG pumps. The partial pressure of hydrogen is 4×10^{-11} mbar in the gun chamber. Partial pressures of all other contributions to the residual gas are in the low 10^{-12} mbar range. With these vacuum conditions the operational life time of a cathode run at 10 μA is about one week.

IV. The 2.45 GHz pulsed light source

The accelerating elements in MAMI are linear accelerators that are driven by 2.45 GHz microwave power. The linacs accept electrons for acceleration in a narrow range of their microwave period only. So the capture efficiency is very low if one injects dc-beams. The efficiency can be improved by a radio frequency buncher system that transforms the primary dc-beam into a 2.45 GHz train of electron bunches before injection into the accelerator. In case of MAMI a sophisticated buncher[25] increases

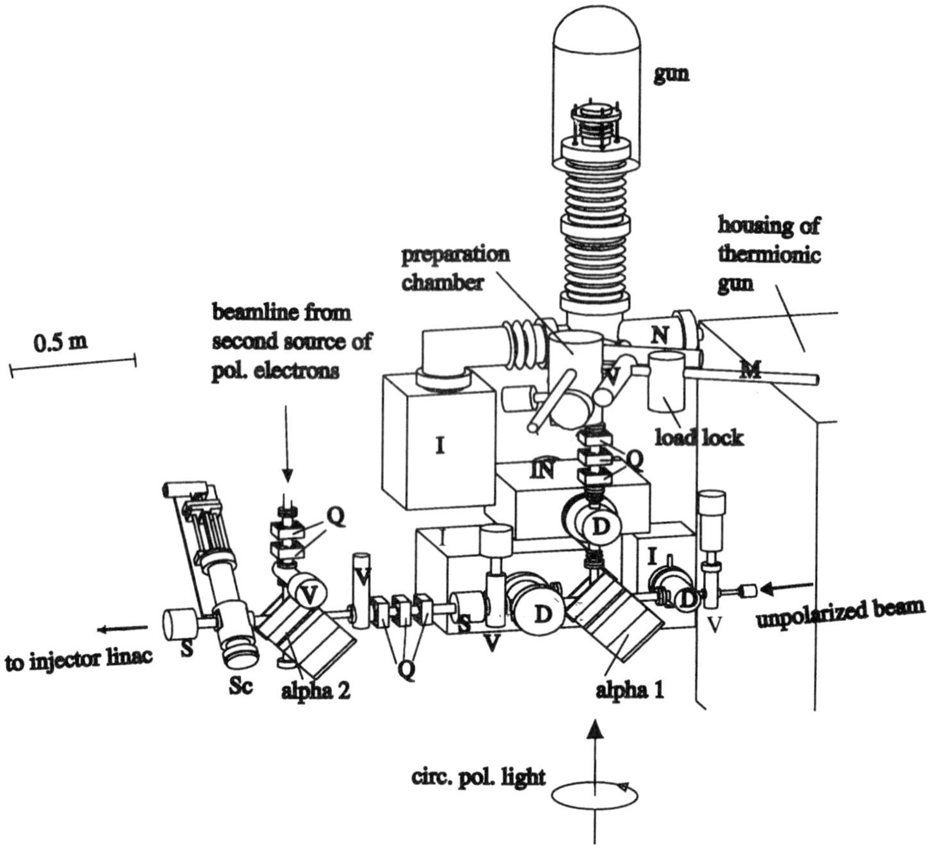

Figure 2. MAMI source of polarized electrons. alpha $=\alpha$-magnet, D=differential pumping stage, I=ion getter pump, N=NEG pump, M=linear movement manipulator, Q=quadrupol, S=solenoid. Sc=wire scanner, V=all metal valve.

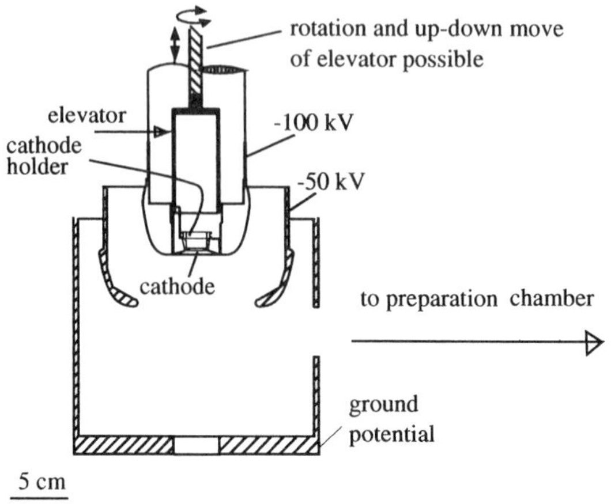

Figure 3. Section through the 100 keV gun.

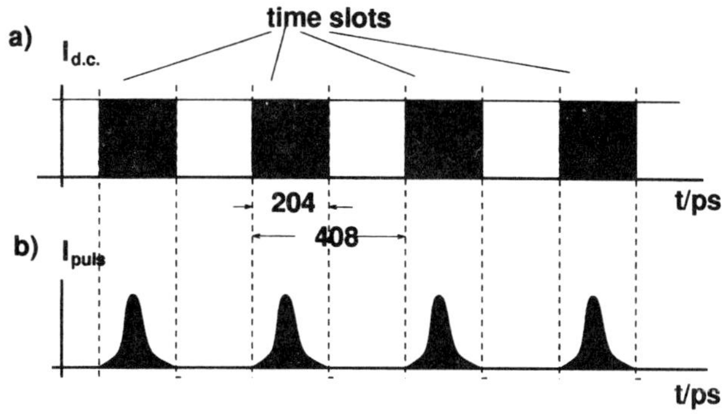

Figure 4. Time dependence of 100 keV beam current before injection into accelerator. (a) DC mode. (b) Pulsed mode. The charge in the hatched regions is captured by the accelerator.

the capture efficiency up to 50%. But still half of the precious polarized electrons are lost. Further improvement is obtained by driving the polarized source in 2.45 GHz pulsed mode. This is indicated in Fig. 4. Electrons are captured for acceleration during half of a rf-period only, that is in time slots of 204 ps per period. This leads to a capture efficiency of 50% for a dc-beam as indicated in the upper part of Fig. 4. An efficiency near 100% is obtainable if one succeeds in producing a 2.45 GHz pulsed beam with pulses that are not wider than 204 ps at the base as shown in the lower part of Fig. 4.

Peter Hartmann of our group showed that thin layer photocathodes of GaAs or GaAsP have response times that are fast enough for producing electron bunches with durations of a few picoseconds only by irradiating the cathodes with picosecond light pulses.[26,27] Therefore the situation indicated in the lower part of Fig. 4 can be obtained by shining pulsed laser light onto the cathode of the source gun that is synchronised with the radio frequency of the accelerator. A very elegant solution for an adequate laser source is the gain switched diode laser system published by Poelker.[28,29] A 2.45 GHz version built at our laboratory[30] is sketched in Fig. 5. A low power laser diode (seed laser in Fig. 5) is directly gain switched by the 2.45 GHz radio frequency of the MAMI master oscillator and produces a 2.45 GHz train of light pulses that is in exact synchronism with the accelerator-rf. The pulses are directed to a laser diode amplifier purchased from SDL (model 8630). The system was successfully operated up to an average output power of 150 mW until now. Figure 6 shows the shape of the electron bunches produced. The shape is Gaussian, FWHM is 53 ps. This is by far short enough for producing a pulsed beam like that in Fig. 4. The capture efficiency for the polarized 100 keV beam injected into MAMI was increased to values better than 93% by this measure.

V. Spin Tuning

Most experiments at MAMI energies need a longitudinally polarized electron beam at the interaction target. Indeed, the beam delivered by the source is longitudinally polarized, but the orientation of the polarization vector with respect to momentum is changed while the electrons are travelling through the dipol magnets of the race track microtrons of MAMI. The electron g-factor is slightly different from two, therefore the polarization vector precesses faster than the momentum vector in transverse magnetic

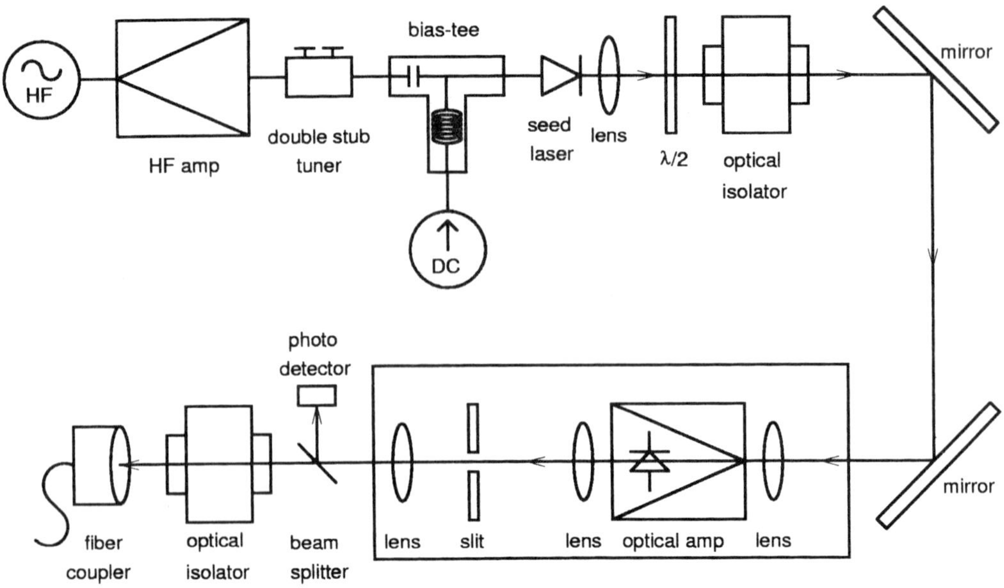

Figure 5. Set up of 2.45 GHz gain-switched laser diode followed by laser diode amplifier.

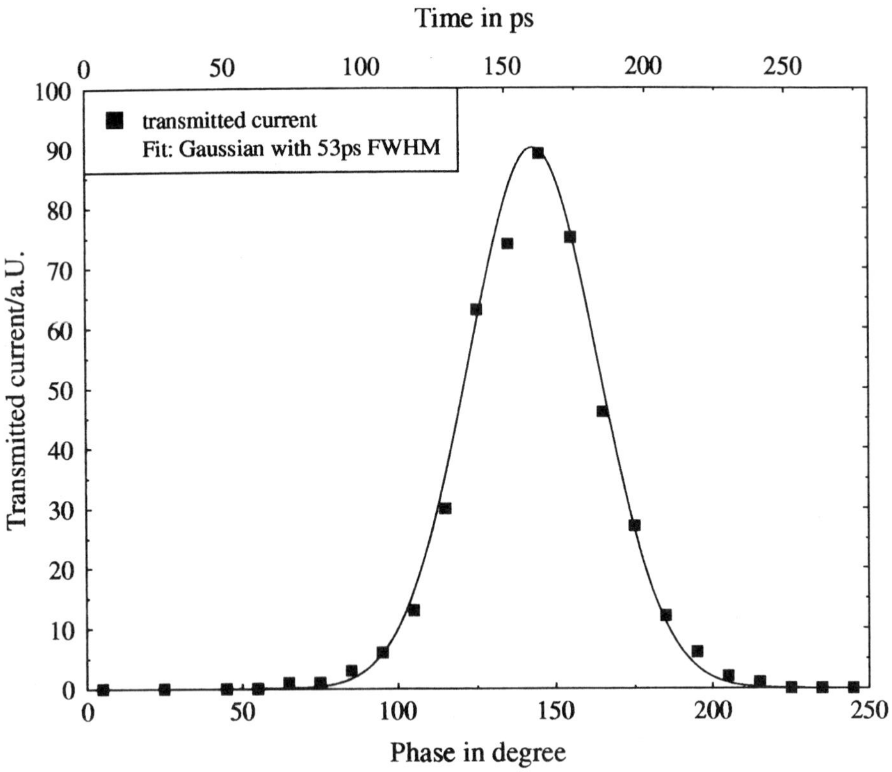

Figure 6. Time dependence of an electron bunch produced with the laser system of Fig. 5.

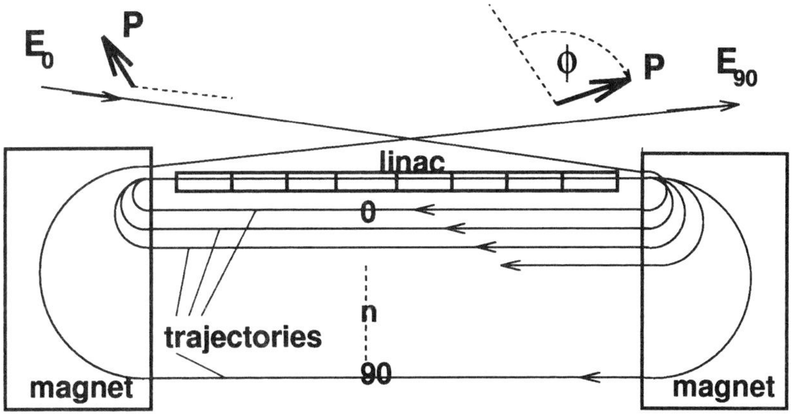

Figure 7. Schematic sketch of electron trajectories in the final race track stage of MAMI. Input energy $E_0 = 180$ MeV, energy gain per turn $\Delta = 7.5$ MeV, number of turns $= n$, final energy $E_{90} = 855$ MeV. Polarization vector **P** of the entrance and exit beam is indicated.

fields.[31] A rough picture of the electron trajectories in the third stage of MAMI is shown in Fig. 7. The stage consists of a linear accelerator that accelerates electrons by $\Delta = 7.5$ MeV at every passage. The electron trajectories are bent back to the entrance of the linac by two dipol magnets so that 90 passages of electrons are possible in this stage. Magnetic field strengths and distances are chosen in such a way that the electrons always enter the linac in the right phase of the accelerating microwave field.[15] The injection energy is $E_0 = 180$ MeV for stage three giving a maximum end energy of 855 MeV for 90 passages. The orientation of the polarization vector is changed during acceleration as mentioned above. Using the expressions published in Ref. 31 one gets for the change Φ of angle between momentum and polarization **P**

$$\Phi = 2\pi a(n+1)\left(\gamma_0 + \frac{n}{2}\delta\right), \tag{1}$$

where

$$\gamma_0 = \frac{E_0}{m_0c^2}; \quad \delta = \frac{\Delta}{m_0c^2}; \quad a = \frac{g-2}{2};$$

$g =$ g-factor of the electron; m_0 =rest mass of the electron; $n =$ number of turns.

So one has to provide means that reduce the angle between momentum and polarization **P** of the accelerated beam down to zero. A spin manipulator that adjusts the spin orientation at low energies before injection into the accelerator has been used at our laboratory for that purpose.[32] But it turned out that it is simpler to tune the orientation of **P** by tuning the end energy of the accelerator. E_0 as well as Δ may be variied slightly by a common factor. Using equation (1) one gets for the change in Φ

$$\Delta\Phi = 2\pi a(n+1)\left(\gamma_0 + \frac{n}{2}\delta\right)\frac{\Delta E_n}{E_n}. \tag{2}$$

With n = 90 and $E_{90} = 855$ MeV follows

$$\frac{\Delta\Phi}{\Delta E} = 45\frac{\text{deg}}{\text{MeV}}. \tag{3}$$

Relation (3) has been successfully applied in several recent experiments at MAMI to adjust **P** in longitudinal direction at the interaction point of beam and target.

Conclusion

The MAMI source of polarized electrons based on photo electron emission from strained $GaAs_{0.95}P_{0.05}$ has demonstrated excellent performance in a series of experiments lasting several hundred hours of beam time each. Beam currents up to 20 μA spin-polarized to a degree of 75% are available to experimenters.

Higher currents are feasible but limit the life time of the cathodes by ion sputtering. Ions are produced by ionization of the residual gas in the gun. It is hoped that implementation of NEG elements in the cathode region[33] of the gun chamber will reduce the residual gas pressure far enough to push the current limit to 100 μA, the maximum current that MAMI can stand.

Acknowledgments

This work is supported by *Deutsche Forschungsgemeinschaft, Sonderforschungsbereiche* 201/443.9.

References

1. E. Garwin, D. T. Pierce, and H. C. Siegmann, *Helv. Phys. Acta* **47**, 393 (1974).
2. D. T. Pierce and F. Meier, *Phys. Rev. B* **13**, 5484 (1976).
3. D. T. Pierce, in *Experimental Methods in the Physical Sciences*, Vol. 29A, edited by R. Celotta and T. Lucatorto (Academic Press, New York, London, Toronto, Sydney, San Francisco, 1995), p. 1.
4. J. Kirschner, *Polarized Electrons at Surfaces*, (Springer, Berlin, 1985).
5. J. Kessler, *Polarized Electrons* (Springer, Berlin, 1985).
6. C. Prescott, W. Atwood, R. Cottrell, H. DeStaebler, E. Garwin, A. Gomidec, R. Miller, l. S. Rochester, T. Sato, D. Sherden, C. Sinclair, S. Stein, R. E. Taylor, J. Clendenin, V. Hughes, N. Sasao, K. Schüler, M. Borghini, K. Lübbelsmeyer, and W. Jentschke, *Phys. Lett. B* **77**, 347 (1978).
7. A. K. Thompson, A. Berstein, T. Chupp, D. Angelis, G. Dodge, G. Dodson, K. Dow, M. Farkondeh, W. Fong, J. Kim, R. Loveman, J. Richardson, H. Schmieden, D. Tieger, T. Yates, M. Wagshul, and J. Zumbro, *Phys. Rev. Lett.* **68**, 2901 (1992).
8. M. Meyerhoff, D. Eyl, A. Frey, H. G. Andresen, J. R. M. Annand, K. Aulenbacher, J. Becker, J. Blume-Werry, T. Dombo, P. Drescher, J. E. Ducret, H. Fischer, P. Grabmayr, S. Hall, P. Hartmann, T. Hehl, W. Heil, J. Hoffmann, J. D. Kellie, F. Klein, M. Leduc, H. Möller, C. Nachtigall, M. Ostrick, E. W. Otten, R. O. Owens, S. Plützer, E. Reichert, D. Rohe, M. Schäfer, L. D. Schearer, H. Schmieden, K.-H. Steffens, R. Surkau, and T. Walcher, *Phys. Lett. B* **327**, 201 (1994).
9. G. D. Cates, V. W. Hughes, R. Michaels, H. R. Schäfer, T. J. Gay, M. S. Lubell, R. Wilson, G. W. Dodson, K. A. Dow, S. B. Kowalski, K. Isakovich, K. S. Kumar, M. E. Schulze, P. A. Souder, and D. H. Kim, *Nucl. Instr. Meth. A* **278**, 293 (1989).
10. D. A. Engwall, B. M. Dunham, L. S. Cardman, D. P. Heddle, and C. K. Sinclair, *Nucl. Instr. Meth. A* **324**, 409 (1993).
11. T. Nakanishi, S. Okumi, K. Togawa, C. Takahashi, C. Suzuki, F. Furuta, T. Ida, K. Wada, T. Omori, Y. Kurihara, M. Tawada, M. Yoshioka, H. Horinaka, K. Wada, T. Matsuyama, T. Baba, and M. Mizuta, in *7th International Workshop*

on Polarized Gas Targets and Polarized Beams, Urbana 1997, AIP Conf. Proc. 421, edited by R. Holt and M. Miller, (AIP, Woodbury, New York, 1997), p. 300.

12. K. Aulenbacher, C. Nachtigall, H. G. Andresen, J. Bermuth, T. Dombo, P. Drescher, H. Euteneuer, H. Fischer, D. v. Harrach, P. Hartmann, J. Hoffmann, P. Jennewein, K.-H. Kaiser, S. Köbis, H. J. Kreidel, J. Langbein, M. Petri, S. Plützer, E. Reichert, M. Schemies, H.-J. Schöpe, K.-H. Steffens, M. Steigerwald, H. Trautner, and T. Weis, *Nucl. Instr. Meth. A* **391**, 498 (1997).
13. Y. Bolkhovityanov, A. Gilinsky, C. deJager, E. Konstantinov, S. Konstantinov, I. Koop, V. Korchagin, V. Kozak, F. Kroes, B. Lazarenko, A. Mamutkin, B. Militsyn, A. Nikiforov, V. Osipov, N. Papadakis, S. Popov, E. Pozdeev, M. vanden Putte, G. Serdobintsev, Y. Shatunov, D. Shevelev, T. Sluijk, A. Terekhov, and N. Vodinas, in *Proceedings of SPIN96*, 14–16 September 1996, Amsterdam, edited by C. deJager, T. Ketel, P. Mulders, J. Oberski, and M. Oskam-Tamboezer (World Scietific, Singapore, New Jersey, London, Hong Kong, 1996), p. 730.
14. R. Alley, H. Aoyagi, J. Clendenin, J. Frisch, C. Garden, E. Hoyt, R. Kirby, L. Klaisner, A. Kulikov, R. Miller, G. Mulhollan, C. Prescott, P. Sáez, D. Schultz, H. Tang, J. Turner, K. Witte, M. Woods, A. D. Yeremian, and M. Zolotorev, *Nucl. Instr. Meth. A* **365**, 1 (1995).
15. H. Herminghaus, A. Feder, K.-H. Kaiser, W. Manz, and H. v. d. Schmitt, *Nucl. Instr. Meth.* **138**, 1 (1976).
16. M. Ostrick, C. Herberg, H. Andresen, J. Annand, K. Aulenbacher, J. Becker, P. Drescher, D. Eyl, A. Frey, P. Grabmayr, P. Hartmann, T. Hehl, W. Heil, J. Hoffmann, D. Ireland, J. Kellie, F. Klein, K. Livingston, C. Nachtigall, A. Natter, E. Otten, R. Owens, E. Reichert, D. Rohe, H. Schmieden, R. Sprengard, M. Steigerwald, K.-H. Steffens, T. Walcher, and R. Watson, *Phys. Rev. Lett.* **83**, 276 (1999).
17. J. Becker, H. Andresen, J. Annand, K. Aulenbacher, K. Beuchel, J. Blume-Werry, T. Dombo, P. Drescher, M. Ebert, D. Eyl, A. Frey, P. Grabmayr, T. Grossmann, P. Hartmann, T. Hehl, W. Heil, C. Herberg, J. Hoffmann, J. Kellie, F. Klein, K. Livingston, M. Leduc, M. Meyerhoff, H. Möller, C. Nachtigall, A. Natter, M. Ostrick, E. Otten, R. Owens, S. Plützer, E. Reichert, D. Rohe, M. Schäfer, H. Schmieden, R. Sprengard, M. Steigerwald, K.-H. Steffens, R. Surkau, T. Walcher, R. Watson, and E. Wilms, *Eur. Phys. J. A* **6**, 329 (1999).
18. C. Herberg, M. Ostrick, H. Andresen, J. Annand, K. Aulenbacher, J. Becker, P. Drescher, D. Eyl, A. Frey, P. Grabmayr, S. Hall, P. Hartmann, T. Hehl, W. Heil, J. Hoffmann, D. Ireland, J. Kellie, F. Klein, K. Livingston, C. Nachtigall, A. Natter, E. Otten, R. Owens, E. Reichert, D. Rohe, H. Schmieden, R. Sprengard, M. Steigerwald, K.-H. Steffens, T. Walcher, and R. Watson, *Eur. Phys. J. A* **5**, 131 (1999).
19. *Jahresbericht 1996–1997 des Inst. f. Kernphysik*, Joh. Gutenberg Univ. Mainz (1998).
20. J. Ahrens, S. Altieri, J. Annand, G. Anton, H.-J. Arends, K. Aulenbacher, R. Beck, C. Bradtke, A. Braghieri, N. Degrande, N. d'Hose, H. Dutz, S. Goertz, P. Grabmayr, K. Hansen, J. Harmsen, D. von Harrach, S. Hasegawa, T. Hasegawa, E. Heid, K. Helbing, H. Holvoet, L.V. Hoorebeke, N. Horikawa, T. Ivata, P. Jennewein, T. Kageya, B. Kiel, F. Klein, R. Kondratiev, K. Kossert, J. Krimmer, M. Lang, B. Lannoy, R. Leukel, V. Lisin, T. Matsuda, J. McGeorge, A. Meier, D. Menze, W. Meyer, T. Michel, J. Naumann, R. Owens, A. Panzeri, P. Pedroni, T. Pinelli, I. Preobrajenski, E. Radtke, E. Reichert, G. Reicherz, C. Rohlof, D. Ryckbosch, F. Sadiq, M. Sauer, B. Schoch, M. Schumacher, B. Seitz, T. Speckner, M.

Steigerwald, N. Takabayashi, G. Tamas, A. Thomas, R. vande Vyver, A. Wakai, W. Weihofen, F. Wissmann, F. Zapadtka, and G. Zeitler, *Phys. Rev. Lett.*, to be submitted, (2000).

21. P. Drescher, H. G. Andresen, K. Aulenbacher, J. Bermuth, T. Dombo, H. Euteneuer, N.N. Faleev, H. Fischer, M.S. Galaktionov, D. v.Harrach, P. Hartmann, J. Hoffmann, P. Jennewein, K.-H. Kaiser, S. Köbis, O. V. Kovalenkov, H. J. Kreidel, J. Langbein, Y. A. Mamaev, C. Nachtigall, M. Petri, S. Plützer, E. Reichert, M. Schemies, K.-H. Steffens, M. Steigerwald, A. V. Subashiev, H. Trautner, D. A. Vinokurov, Y. P. Yashin, and B. S. Yavich, *Applied Physics A* **63**, 203 (1996).
22. B. Yavich, I. Kochnev, V. Komin, O. Kovalenkov, D. Vinokurov, Y. Mamaev, and Y. Yashin, in *Proceedings of MOVPE VI*, Gent, Belgium, 1995, p. E14.
23. M. Schemies, PhD-thesis, Inst. f. Physik, Joh. Gutenberg Universität Mainz, 1998.
24. M. Steigerwald, PhD-thesis, Inst. f. Physik, Joh. Gutenberg Universität Mainz, 1998.
25. V. T. Shvedunov, M. O. Ihm, H. Euteneuer, K.-H. Kaiser, and T. Weis, in *EPAC96*, Vol. 2, edited by S. Myers, A. Pacheco, R. Pascual, C. Petit-Jean-Genaz, and J. Poole, (Institute of Physics Publishing, Bristol and Philadelphia, 1996), p. 1556.
26. P. Hartmann, J. Bermuth, J. Hoffmann, S. Köbis, E. Reichert, H.G. Andresen, K. Aulenbacher, P. Drescher, H. Euteneuer, H. Fischer, P. Jennewein, K. H. Kaiser, H. J. Kreidel, C. Nachtigall, S. Plützer, M. Schemies, K.-H. Steffens, M. Steigerwald, H. Trautner, D. v.Harrach, I. Altarev, R. Geiges, K. Grimm, T. Hammel, E. Heinen-Konschak, H. Hofmann, E.-M. Kabuss, A. Lopes-Ginja, F. Maas, and E. Schilling, *Nucl. Instr. Meth. A* **379**, 15 (1996).
27. P. Hartmann, PhD-thesis, Inst. f. Kernphysik, Joh. Gutenberg Universität Mainz, 1998.
28. M. Poelker, *Appl. Phys. Lett.* **67**, 2762 (1995).
29. M. Poelker and J. Hansknecht, in *7th International Workshop on Polarized Gas Targets and Polarized Beams*, Urbana 1997, AIP Conf. Proc. 421, edited by R. Holt and M. Miller, (AIP, Woodbury, New York, 1997)
30. M. Zalto, Diplom-thesis, Inst. f. Kernphysik, Joh. Gutenberg Universität Mainz, 1998.
31. V. Bargman, L. Michel, and V. L. Telegdi, *Phys. Rev. Lett.* **2**, 435 (1959).
32. K. H. Steffens, H. G. Andresen, J. Blume-Werry, F. Klein, K. Aulenbacher, and E. Reichert, *Nucl. Instr. Meth. A* **325**, 378 (1993).
33. Ch. K. Sinclair in *Proceedings of the International Workshop on Polarized Sources and Targets*, Erlangen 1999.

The Physics of Low Dimensions: Summary

S. F. Alvarado

IBM Research
Zürich Research Laboratory
8803 Rüschlikon
SWITZERLAND

More than 100 scientists gathered in the city of Oaxaca to discuss the present status of low-dimensional physics. Many came from foggy and influenza-stricken Europe, coughing and sneezing, but the blue skies and mild breeze in Oaxaca pesuaded all these ailments to disappear very quickly. The flowering trees, the beautiful hotel Victoria, the Mexican hospitality, and the meeting with old friends soon returned everybody to a good humor and created a unique atmosphere at the conference.

About half of the contributions dealt with magnetism and spin-polarized electrons. The increased attention of the research community to these topics is due to the new applications of magnetism in "spin electronics" in which electric currents are governed by the spin state of electrons. Ferromagnetic metals act as the sources and as detectors of the spin polarization. But one has to consider that there are two types of ferromagnetic metals, those with positive polarization at the Fermi level and with negative spin polarization. The field is based on deepening our understanding of magnetism in metals concerning the spin-polarized electronic structure, the mechanism, and the correct interpretation of electron tunneling through oxide barriers, and, most of all, the spin dependence of the scattering and motion of the spin-polarization vector in the transport of electrons through solids.

In the opening talk by H. C. Siegmann, the tunneling of spin-polarized electrons and the scattering of electrons in ferromagnetic metals were focused upon. Spin-polarized spectroscopies such as threshold photoemission of spin-polarized electrons from ferromagnets, pioneered in the laboratory of H. C. Siegmann 31 years ago, allows one to study the spin dependence of electron-electron scattering and to determine the sign of the spin-polarization at the Fermi energy. More recent photoemission experiments use pump-probe excitation of the electrons with femto-second laser pulses. This can be used to determine the lifetimes of hot electrons and their dependence on the spin state. A most complete picture of the transport of polarized electrons can be gained from a new experiment in which a low-energy, spin-polarized electron beam is prepared by means of a GaAs-type photocathode and transmitted through a very thin ferromagnetic film. This experiment demonstrates spin-selective absorption of electrons as well as, for the first time, the precession of the spin in the exchange field of the ferromagnet. K. Bennemann presented the theory of ultrafast dynamics in ferromagnetic transition metals and explained time-resolved magneto-optics. A. J. Freeman showed the progress made in calculating magnetic properties at surfaces and interfaces, predicting even such complex properties such as magnetostriction.

Stuart Parkin, winner of the AIP-prize 1999-2000 for industrial applications of giant magnetoresistance, gave a compeling talk on magnetic tunnel junctions and their possible use in magnetic random-access memories. However, quite a number of questions remain open concerning the tunneling of electrons between ferromagnetic metals through oxide layers. The imaging of magnetic structures in STM with atomic resolution still remains a dream. Since the first reports of spin-polarized STM by R. Wiesendanger in 1990 and by myself in 1992, little progress has been made. B. Heinrich described an impressive application of an atomic force microscope equipped with a conducting tip for the characterization of magnetic tunnel junctions. With this, the local magnetic tunneling characteristics have been probed with a lateral resolution of a few nanometers. G. Mathon showed convincingly that there is a way to understand many basic features of spin-polarized electron tunneling, such as the dependence of the tunneling current on the parentage of the electron states and on interference in nonmagnetic spacer layers. This agrees nicely with the findings of my clean experiments in 1995 using the fluorescence in GaAs as an analyzer for the spin polarization of the tunneling current and showing that the highly polarized $3d$-electrons can contribute significantly to tunneling.

Progress made in imaging magnetization patterns with unprecedented spatial and temporal resolution has contributed significantly to the flourishing of low-dimensional magnetism. U. Gradmann had prepared monolayer nano-strips of Fe, and A. Vaterlaus discussed domain formation in two-dimensional magnetic particles. The phase transitions were imaged impressively using scanning electron microscopy with spin-polarized analysis, SEMPA. Similarly beautiful magnetization patterns were shown by D. Pierce proving that the Néel temperature of Cr is raised significantly between two Fe layers. G. Schönhense presented the photoelectron microscope, PEEM, and showed that it delivers element-specific imaging of the magnetization. B. Sincoviç presented the first pictures of domain walls in antiferromagnets using PEEM at the Advanced Light Source (ALS) in Berkeley. Finally, C. Back showed that one can also achieve high temporal resolution in magnetic imaging using the magneto-optic Kerr effect in conjunction with femtosecond laser pulses. By disturbing the steady-state magnetization with an external magnetic field pulse one can see that the magnetization performs oscillations, which he demonstrated using the example of a permalloy film. This opens the way to increase the data rate in magnetic recording. In fact with the help of the Stanford Linear Accelerator it has been demonstrated that magnetic switching can be induced in the time scale of a few picoseconds by very short magnetic pulses of moderate strength if applied correctly. In this experiment the magnetic field pulses of a few picoseconds are produced by highly energetic electron bunches. D. Weller presented the current status of high-speed switching phenomena in magnetic recording and its future prospects. The density of the magnetic bits, the readout of the magnetization direction, and the data rate for writing are all intimately connected with and profit directly from the basic knowledge gained.

It is not only magnetism that has advanced due to the use of spin-polarized electric currents: E. Reichert showed that experiments in high-energy physics that exploit the spin degree of freedom in the scattering of electrons have produced significant results to explore parity violation and to establish the nucleon electromagnetic structure. In the neutron, for instance, scattering from the magnetic moment can be distinguished from scattering on the charge distribution with the help of the GaAs-type, spin-modulated electron source developed in H. C. Siegmann's laboratory in 1974. The amazing perfection that these sources have reached lately, delivering almost completely polarized electron beams of very high luminosity, was demonstrated in the contribution of Y. Mamaev.

The other main topic at this meeting, besides magnetism, was small particles. Their magnetic and superconducting properties were discussed by B. Mühlschlegel. M. J. Yacaman gave a superb talk on superlattices formed by nanoparticles, whereas K. Sattler presented nanodots and nanowires of Si. Several talks dealt with nanoparticles in gas suspension, their characterization, formation in combustion, and effects on human health. This represented another much respected research field pursued at H. C. Siegmann's laboratory.

Further excellent contributions were devoted to surface science, in particular to surface and interface spectroscopy. P. S. Bagus explained how one must take a complete configuration interaction wave function to understand the general features of the x-ray photoelectron spectra of transition metal oxides. This topic is of growing concern for many applications. C. S. Fadley presented a recently discovered phenomenon called multi-atom resonant photoemission, which provides a unique tool to study near-neighbor identities, bonding, and magnetism. One-dimensional systems were also discussed in impressive presentations, for instance the one given by D. Menzel.

Space does not allow me to describe all the excellent presentations given in this fine and well organized workshop, and I apologize for the omissions. Nevertheless I want to mention that after each talk, the speaker was presented with an *alebrije*, one of the colorful hand carvings fantasy creatures and animals for which Oaxaca is famous. It would not be fitting to conclude this report on a such remarkable event without mentioning the lovely open-air conference banquet in the moonlight under flowering tropical trees and featuring the typical dance of the state of Oaxaca, the *Guelaguetza*, performed in traditional costumes. Later, the banquet participants themselves danced to the happy and infectious music of a mariachi band. Not to forget the soothing daily influence of an hour's rest at noon at the *zocalo* of Oaxaca or buying a beautiful souvenir at one of the numerous markets. Every participant felt gratitude towards those who organized the meeting and brought us all together.

Index